Lecture Notes in Computer Science   8614

Commenced Publication in 1973
Founding and Former Series Editors:
Gerhard Goos, Juris Hartmanis, and Jan van Leeuwen

Helmut Jürgensen   Juhani Karhumäki
Alexander Okhotin (Eds.)

Descriptional Complexity of Formal Systems

16th International Workshop, DCFS 2014
Turku, Finland, August 5-8, 2014
Proceedings

 Springer

Volume Editors

Helmut Jürgensen
Western University
Department of Computer Science
London, ON, Canada
E-mail: hjj@csd.uwo.ca

Juhani Karhumäki
University of Turku
Department of Mathematics and Statistics
Turku, Finland
E-mail: karhumak@utu.fi

Alexander Okhotin
University of Turku
Department of Mathematics and Statistics
Turku, Finland
E-mail: alexander.okhotin@utu.fi

ISSN 0302-9743   e-ISSN 1611-3349
ISBN 978-3-319-09703-9   e-ISBN 978-3-319-09704-6
DOI 10.1007/978-3-319-09704-6
Springer Cham Heidelberg New York Dordrecht London

Library of Congress Control Number: 2014944667

LNCS Sublibrary: SL 1 – Theoretical Computer Science and General Issues

Typesetting: Camera-ready by author, data conversion by Scientific Publishing Services, Chennai, India

Printed on acid-free paper

Springer is part of Springer Science+Business Media (www.springer.com)

Preface

The 16th International Workshop on Descriptional Complexity of Formal Systems (DCFS 2014) was organized by the research center on Fundamentals of Computing and Discrete Mathematics (FUNDIM) at the Department of Mathematics and Statistics of the University of Turku (Turku, Finland). The subject of the workshop is the size of mathematical models of computation, which serves as a theoretical representation of such things as the engineering complexity of computer software and hardware. It also models similar complexity phenomena in other areas of computer science, including unconventional computing and bioinformatics.

The DCFS workshop series is a result of merging together two workshop series: Descriptional Complexity of Automata, Grammars and Related Structures (DCAGRS) and Formal Descriptions and Software Reliability (FDSR). These precursor workshops were DCAGRS 1999 in Magdeburg, Germany, DCAGRS 2000 in London, Ontario, Canada, and DCAGRS 2001 in Vienna, Austria, as well as FSDR 1998 in Paderborn, Germany, FSDR 1999 in Boca Raton, Florida, USA, and FSDR 2000 in San Jose, California, USA. These workshops were merged in DCFS 2002 in London, Ontario, Canada, which is regarded as the 4th DCFS. Since then, DCFS workshops were held in Budapest, Hungary (2003), in London, Ontario, Canada (2004), in Como, Italy (2005), in Las Cruces, New Mexico, USA (2006), in Nový Smokovec, Slovakia (2007), in Charlottetown, Canada (2008), in Magdeburg, Germany (2009), in Saskatoon, Canada (2010), in Limburg, Germany (2011), in Braga, Portugal (2012), and again in London, Ontario, Canada (2013).

In 2014, the DCFS workshop was for the first time held in Finland, in the city of Turku. The new location attracted a record number of submissions, more than any DCFS in the past. This volume contains extended abstracts of four invited talks and 27 contributed talks presented at the workshop. The latter were selected from the submissions by the Program Committee on the basis of at least three reviews per submission.

This workshop is a result of combined efforts of many people, to whom we wish to express our gratitude. In particular, we are indebted to our invited speakers—Andris Ambainis, Oscar Ibarra, Manfred Kufleitner, and Nikolay Vereshchagin—and to all the speakers and participants of the workshop. We are grateful to all our Program Committee members and to all reviewers for their work on selecting the workshop program, which was carried out using the Easy-Chair conference management system. Thanks are due to the members of our Organizing Committee, Mikhail Barash and Markus Whiteland, for taking care of all local matters, and to the staff of the University of Turku for administrative assistance. We gratefully acknowledge the financial support from our sponsors:

- The Federation of Finnish Learned Societies
- The Finnish Academy of Science and Letters (mathematics foundation)
- The Turku University Foundation
- The City of Turku

Finally, we wish to thank the editorial team at Springer, and personally Alfred Hofmann, Anna Kramer and Ingrid Beyer, for efficient production of this volume.

June 2014   Alexander Okhotin
                                                         Helmut Jürgensen
                                                         Juhani Karhumäki

Organization

Program Committee

Cristian Calude	University of Auckland, New Zealand
Frank Drewes	Umeå University, Sweden
Viliam Geffert	Šafárik University, Košice, Slovakia
Helmut Jurgensen	Western University, Canada
Christos Kapoutsis	Carnegie Mellon University, Qatar
Juhani Karhumaki	University of Turku, Finland
Michal Kunc	Masaryk University, Brno, Czech Republic
Martin Kutrib	University of Giessen, Germany
Christof Loeding	RWTH Aachen, Germany
Pierre McKenzie	University of Montréal, Canada; ENS Cachan, France
Alexander Okhotin	University of Turku, Finland
Giovanni Pighizzini	Università degli Studi di Milano, Italy
Kai Salomaa	Queen's University, Canada
Klaus Sutner	Carnegie Mellon University, USA
Damien Woods	California Institute of Technology, USA

Additional Reviewers

Barash, Mikhail
Berglund, Martin
Bianchi, Maria Paola
Bille, Philip
Blondin, Michael
Campeanu, Cezar
Domaratzki, Mike
Fefferman, Bill
Finkel, Alain
Freivalds, Rusins
Goc, Daniel
Goldwurm, Massimiliano
Holzer, Markus
Jakobi, Sebastian
Jeż, Artur
Jiraskova, Galina
Klauck, Hartmut
Klima, Ondrej
Ko, Sang-Ki
Konstantinidis, Stavros

Krebs, Andreas
Lang, Martin
Meckel, Katja
Mereghetti, Carlo
Mráz, František
Murphy, Niall
Ng, Timothy
Otto, Friedrich
Palano, Beatrice
Palioudakis, Alexandros
Prusa, Daniel
Sebej, Juraj
Seki, Shinnosuke
Siromoney, Rani
Vaszil, György
Watson, Bruce
Wendlandt, Matthias
Winter, Sarah
Yakaryilmaz, Abuzer

Table of Contents

Recent Developments in Quantum Algorithms and Complexity

Andris Ambainis*

Faculty of Computing, University of Latvia, Raina bulv. 19, Riga, LV-1586, Latvia
`andris.ambainis@lu.lv`

Abstract. We survey several recent developments in quantum algorithms and complexity:

- Reichardt's characterization of quantum query algorithms via span programs [15];
- New bounds on the number of queries that are necessary for simulating a quantum algorithm that makes a very small number of queries [2];
- Exact quantum algorithms with superlinear advantage over the best classical algorithm [4].

1 Introduction

Quantum computing (and, more broadly, quantum information science) is a new area at the boundary of computer science and physics. Quantum computers compute by encoding information into a quantum state and performing quantum transformations on it. For certain problems, quantum computers can be exponentially faster than conventional (classical) computers, due to a larger space of states and transformations that they can use.

The field of quantum computing was shaped by two major quantum algorithms discovered in mid-1990s: Shor's algorithm and Grover's algorithm.

Shor's algorithm [19] solves two very hard number theoretic problems: factoring and discrete logarithm. Both of those problems are thought to require exponential time classically. In contrast, Shor's algorithm solves them in polynomial time quantumly. This indicates that quantum computers may be exponentially faster than classical. (Another indication of that is the oracle result of Simon [18] who constructed a problem that can be solved in polynomial time given an oracle A but provably requires exponential time classically, given the same oracle A.)

Grover's quantum algorithm solves a generic exhaustive search problem with N possible solutions in time $O(\sqrt{N})$. This provides a quadratic quantum speedup for a range of search problems, from ones that are solvable in polynomial time classically to NP-complete ones.

---

* Supported by FP7 FET projects QALGO and RAQUEL and ERC Advanced Grant MQC.

H. Jürgensen et al. (Eds.): DCFS 2014, LNCS 8614, pp. 1–4, 2014.

2 Recent Results

A number of developments (both new quantum algorithms and complexity-theoretic results about the power of quantum algorithms) have followed after those two discoveries. In this talk, I will survey 3 recent developments in quantum complexity theory that have happened over the last 6 years:

1. **Characterization of Quantum Query Algorithms**
   Most of quantum algorithms (including Grover's algorithm and quantum part of Shor's algorithm) can be analyzed in a query model where the input data $x_1, \ldots, x_N$ are given by a black box that answers queries about the values of input bits x_i and the complexity of the algorithm is measured by the number of queries to the black box that it makes. This model is a quantum generalization of the classical *decision tree* model of computation and has been studied in detail [10].

   Let $Q_2(f)$ be the quantum query complexity of $f(x_1, \ldots, x_N)$ (the smallest number of queries for a quantum algorithm that computes f). Recently, Reichardt [15,16] discovered that $Q_2(f)$ can be characterized (up to a constant factor) by a semidefinite program (SDP). This is a very interesting result because no similar characterization is known for its classical counterpart, the decision tree complexity $D(f)$. As a result, there are many f for which we know $Q_2(f)$ up to a constant factor but determining $D(f)$ is still open. (For example, this is true for the iterated majority function studied in [14].)

   Reichardt's work was a culmination of two lines of work: quantum algorithms for formula evaluation and span programs [11,7,17] and "quantum adversary" lower bound method [3,13]. It turns out that Reichardt's is equal (again, up to a constant factor) to both the complexity of the best span program based quantum algorithm and the best lower bound provable by the most general form of "quantum adversary" method. Thus, we have both a universal algorithmic method for quantum query algorithms (span programs) and a universal lower bound method ("quantum adversary"). Again, no similar characterizations are known classically.

   Currently, one of challenges in quantum algorithm is using Reichardt's characterization to develop new quantum algorithms. Here, most interesting results have been obtained using learning graphs, an easy to use framework for quantum algorithms developed by Belovs [8,9] which is a special case of Reichardt's SDP.

2. **Power of Quantum Algorithms with a Very Small Number of Queries**
   Some quantum query algorithms achieve very big advantages over the classical algorithms. For example, Shor's algorithm for factoring is based on a query problem called *period-finding* in which we have to find a period of a sequence $x_1, \ldots, x_N$, given a promise that the period is of the order $O(\sqrt{N})$. Many instances of period finding can be solved with $O(1)$ quantum queries. At the same time, period-finding can be shown to require

$\Omega(N^{1/4})$ queries classically. This leads to a question: if a computational problem $f(x_1, \ldots, x_N)$ can be solved with a constant number of queries k quantumly, how many queries can it require classically?

We have shown [2] that the "Forrelation" problem of Aaronson [1] (in which $x_1, \ldots, x_N$ describe two vectors and one has to check whether the Fourier transform of the second vector is similar to the first vector) is solvable with 1 query quantumly but requires $\Omega(\sqrt{N})$ queries classically. This is tight: by a random sampling argument, we can also show that $O(\sqrt{N})$ classical queries are enough to simulate any 1-query quantum algorithm on input data of size N. More generally, $O(N^{1-1/2k})$ classical queries suffice to simulate any k-query quantum algorithm on input data of size N.

3. **Exact Quantum Algorithms**

A quantum algorithm is exact if, on any input data, it outputs the correct answer with certainty (probability 1). While many quantum algorithms are known for the bounded-error model (where the algorithm is allowed to output an incorrect answer with a small probability), only a small number of exact quantum algorithms are known. Until recently, in the query model, for computing total functions $f(x_1, \ldots, x_N)$, the biggest advantage for exact quantum algorithms was just a factor of 2: N queries classically vs. $N/2$ for an exact quantum algorithm.

In [4], we presented the first example of a Boolean function $f(x_1, ..., x_N)$ for which exact quantum algorithms have superlinear advantage over deterministic algorithms. Any deterministic algorithm that computes our function must use N queries but an exact quantum algorithm can compute it with $O(N^{0.8675\ldots})$ queries. We have also discovered a number of other exact algorithms that are better than classical algorithms [5,6]. This shows that the advantages for exact algorithms are more common than previously thought.

References

1. Aaronson, S.: BQP and the Polynomial Hierarchy. In: Proceedings of the ACM Symposium on the Theory of Computing (STOC 2010), pp. 141–150 (2010); Also arXiv:0910.4698
2. Aaronson, S., Ambainis, A.: Forrelation: a problem capturing the power of quantum algorithms. Manuscript in preparation
3. Ambainis, A.: Quantum Lower Bounds by Quantum Arguments. Journal of Computer and System Sciences 64(4), 750–767 (2002) Also quant-ph/0002066
4. Ambainis, A.: Superlinear advantage for exact quantum algorithms. In: Proceedings of the ACM Symposium on the Theory of Computing (STOC 2013), pp. 891–890 (2013); Also arXiv:1211.0721
5. Ambainis, A., Gruska, J., Zheng, S.: Exact query complexity of some special classes of Boolean functions. arXiv:1404.1684
6. Ambainis, A., Iraids, J., Smotrovs, J.: Exact Quantum Query Complexity of EXACT and THRESHOLD. In: Proceedings of the 8th Conference on the Theory of Quantum Computation, Communication and Cryptography (TQC 2013), pp. 263–269 (2013); Also arXiv:1302.1235

7. Ambainis, A., Childs, A., Reichardt, B., Spalek, R., Zhang, S.: Any AND-OR formula of size N can be evaluated in time $N^{1/2+o(1)}$ on a quantum computer. SIAM Journal on Computing 39(6), 2513–2530 (2010); Also FOCS 2007
8. Belovs, A.: Span programs for functions with constant-sized 1-certificates: extended abstract. In: Proceedings of the ACM Symposium on the Theory of Computing (STOC 2012), pp. 77–84 (2012); Also arXiv:1103.0842
9. Belovs, A.: Learning-Graph-Based Quantum Algorithm for k-Distinctness. In: Proceedings of the IEEE Conference on Foundations of Computer Science (FOCS 2012), pp. 207–216 (2012); Also arXiv:1205.1534
10. Buhrman, H., de Wolf, R.: Complexity measures and decision tree complexity: a survey. Theoretical Computer Science 288(1), 21–43 (2002)
11. Farhi, E., Goldstone, J., Gutman, S.: A Quantum Algorithm for the Hamiltonian NAND Tree. Theory of Computing 4(1), 169–190 (2008); Also quant-ph/0702144
12. Grover, L.: A fast quantum mechanical algorithm for database search. In: Proceedings of the ACM Symposium on the Theory of Computing (STOC 1996), pp. 212–219 (1996)
13. Høyer, P., Lee, T., Špalek, R.: Negative weights make adversaries stronger. In: Proceedings of the ACM Symposium on the Theory of Computing (STOC 2007), pp. 526–535 (2007); Also quant-ph/0611054
14. Magniez, F., Nayak, A., Santha, M., Sherman, J., Tardos, G., Xiao, D.: Improved bounds for the randomized decision tree complexity of recursive majority. arXiv:1309.7565. Earlier version: ICALP 2011 (2011)
15. Reichardt, B.: Span programs and quantum query complexity: The general adversary bound is nearly tight for every boolean function. In: Proceedings of the IEEE Conference on Foundations of Computer Science (FOCS 2009), pp. 544–551 (2009); Also arXiv:0904.2759
16. Reichardt, B.: Reflections for quantum query algorithms. In: Proceedings of SIAM-ACM Symposium on Discrete Algorithms (SODA 2011), pp. 560–569 (2011); Also arXiv:1005.1601
17. Reichardt, B., Špalek, R.: Span-Program-Based Quantum Algorithm for Evaluating Formulas. Theory of Computing 8(1), 291–319 (2012); Also arXiv:0710.2630
18. Simon, D.: On the Power of Quantum Computation. SIAM Journal on Computing 26(5), 1474–1483 (1997)
19. Shor, P.: Algorithms for Quantum Computation: Discrete Logarithms and Factoring. SIAM Journal on Computing 26(5), 1484–1509 (1997)

Automata with Reversal-Bounded Counters: A Survey

Oscar H. Ibarra

Department of Computer Science,
University of California, Santa Barbara, CA 93106, USA
ibarra@cs.ucsb.edu

Abstract. We survey the properties of automata augmented with reversal-bounded counters. In particular, we discuss the closure/non-closure properties of the languages accepted by these machines as well as the decidability/undecidability of decision problems concerning these devices. We also give applications to several problems in automata theory and formal languages.

Keywords: finite automaton, pushdown automaton, visibly pushdown automaton, transducer, context-free grammar, reversal-bounded counters, semilinear set, finitely-ambiguous, finite-valued, decidable, undecidable.

1 Introduction

A counter is an integer variable that can be incremented by 1, decremented by 1, and tested for zero. It starts at zero and can only store nonnegative integer values. Thus, one can think of a counter as a pushdown stack with a unary alphabet, in addition to the bottom of the stack symbol which is never altered.

An automaton (DFA, NFA, DPDA, NPDA, etc.) can be augmented with multiple counters, where the "move" of the machine also now depends on the status (zero or non-zero) of the counters, and the move can update the counters. See [27] for formal definitions.

It is well known that a DFA augmented with two counters is equivalent to a Turing machine (TM) [39]. However, when we restrict the operation of the counters so that during the computation, the number of times each counter alternates between nondecreasing mode and nonincreasing mode is at most some fixed number r (such a counter is called *reversal-bounded*), the computational power of a DFA (and even an NPDA) augmented with reversal-bounded counters is significantly weaker than a TM. When $r = 1$, we refer to the counters as 1-reversal (thus, once a counter decrements, it can no longer increment). Note that a counter that makes r reversals can be simulated by $\lceil \frac{r+1}{2} \rceil$ 1-reversal counters.

A k-NCM (resp., k-DCM, k-NPCM, k-DPCM) is an NFA (resp., DFA, NPDA, DPDA) augmented with k reversal-bounded counters. For machines with two-way input head operating on an input with left and right end markers, we use the notation k-2NCM, k-2NPCM, etc. When the number of counters is not

H. Jürgensen et al. (Eds.): DCFS 2014, LNCS 8614, pp. 5–22, 2014.

specified, we just write NCM, NPCM, etc. So, e.g., a 2NPCM is a two-way NPDA augmented with some number of reversal-bounded counters.

Automata with reversal-bounded counters were first studied in [4,27], where closure and decision properties were investigated. For example, in [27], it was shown that the class of languages accepted by DCMs is effectively closed under union, intersection, and complementation (closure under the first two operations also hold for NCMs), and emptiness, finiteness, disjointness, containment, and equivalence problems are decidable.

NCMs and NPCMs and their two-way versions have been extensively studied. Many generalizations have been introduced, see, e.g., [15,5,7,8]. They have found applications in areas like timed-automata [12,10,11,6], transducers [21,23,2,3,16,38,29], membrane computing [33], DNA computing [30], verification [12,37,31,11,13,46,5,24], and Diophantine equations [22,14,32,45].

Here, we give a survey of important results concerning machines augmented with reversal-bounded counters and summarize some recent developments involving the use of these machines for solving problems in automata theory and formal languages.

2   Decidability of the Emptiness and Infiniteness Problems

Let $\mathbb{N}$ be the set of nonnegative integers and m be a positive integer. A subset Q of $\mathbb{N}^m$ is a *linear set* if there exist vectors $v_0, v_1, \ldots, v_n \in \mathbb{N}^m$ such that $Q = \{v_0 + i_1 v_1 + \cdots i_n v_n \mid i_1, \ldots, i_n \in \mathbb{N}\}$. The vectors v_0 (referred to as the *constant*) and $v_1, \ldots, v_n$ (referred to as *periods*) are called the *generators* of the linear set Q. A finite union of linear sets is called a *semilinear set*. Every finite subset of $\mathbb{N}^m$ (including the empty set $\varnothing$) is semilinear – it is just a finite union of linear sets with no periods.

Let $\Sigma = \{a_1, \ldots, a_m\}$. For a word w over Σ and a letter $a \in \Sigma$, we denote by $|w|_a$ the number of occurrences of a's in w. The *Parikh map* of w is the m-dimensional vector $\phi(w) = (|w|_{a_1}, \ldots, |w|_{a_m})$. The *Parikh map* of a language $L \subseteq \Sigma^*$ is defined as $\phi(L) = \{\phi(w) \mid w \in L\}$. (The reader is assumed to be familiar with basic notions in formal languages and automata theory; see, e.g., [26].)

The following fundamental result was first shown in 1978 [27]:

Theorem 1. *The emptiness and infiniteness problems for NPCMs are decidable.*

Proof. We briefly sketch the proof given in [27]. Let M be a k-NPCM with input alphabet $\Sigma = \{a_1, \ldots, a_m\}$. We may assume that each counter is 1-reversal and acceptance is when the machine enters an accepting state when all the counters are zero. We construct an NPDA (without counters) M' over alphabet $\Sigma \cup \Delta$, where $\Delta = \{(i, +1), (i, -1) \mid 1 \le i \le k\}$. The symbol $(i, +1)$ (resp. $(i, -1)$) encodes the possible action of counter i in one step of the computation, i.e., encountering $(i, +1)$ (resp., $(i, -1)$) on the input represents incrementing (resp., decrementing) counter i. If w is an input to M, then an input to M' would be a string w' which would consist of w shuffled with some string z in Δ^*, where for each i, all the $(i, +1)$'s appear before all the $(i, -1)$'s. M' simulates M and whenever counter i increments (resp., decrements), M' checks that its input head is on symbol $(i, +1)$ (resp., $(i, -1)$) before moving right. M' accepts w' if M accepts w. Since M is an NPDA, the Parikh map of $L(M')$ is an effectively computable semilinear set $Q_1 \subseteq \mathbb{N}^{m+2k}$, where the first m coordinates correspond to the symbols in Σ and the last $2k$ coordinates correspond to the symbols in Δ (in the order: all the $(i, +1)$'s followed by all the $(i, -1)$'s). Let $Q_2 = \{(i_1, \ldots, i_m, j_1, \ldots, j_k, j_1, \ldots, j_k) \mid i_r \geq 0 \text{ for } 1 \leq r \leq m, j_s \geq 0 \text{ for } 1 \leq s \leq k\}$. Clearly, Q_2 is semilinear. Hence, $Q_3 = Q_1 \cap Q_2$ is semilinear, since semilinear sets are closed under intersection. Semilinear sets are also closed under projection; hence, we can obtain the semilinear set $Q \subseteq \mathbb{N}^m$ corresponding to the Parikh map of $L(M)$ by deleting the last $2k$ coordinates in Q_3. The result follows, since the emptiness and infiniteness problems for semilinear sets are decidable [18].   □

When the number k of counters and the bound r on the counter reversals are *fixed*, an upper bound of NPTIME for the nonemptiness decision procedure was shown by Filiot et al at MFCS 2010 [16]. Their proof consisted of a careful analysis of the procedure in [27] (which is briefly described in the proof above) for showing that the Parikh map of the language accepted by an NPCM M whose counters are 1-reversal counters is semilinear and then appealing to two results in [41]: constructing an existential Presburger formula for representing the Parikh map of the language accepted by the NPCM M can be accomplished in $O(|M|)$ time, and satisfiability of existential Presburger formula is in NPTIME.

In the above improvement, k and r are assumed to be fixed; otherwise, the procedure would be in NEXP. However, recently, the complexity of the emptiness problem has been improved. In their CAV 2012 paper, Hague et al [24] showed that the emptiness problem for NPCMs (thus the number of counters and the bound r on reversals are not fixed, but r is assumed to be written in unary) is NP-complete, even when the counters can be compared against and incremented/decremented by constants that are given in binary. The construction in [24] used a direct polynomial-time reduction to satisfaction over existential Presburger formulas.

For the case of NCMs, the following was shown in [20]:

Theorem 2. *For fixed k and r, the emptiness problem for k-NCMs whose counters are r-reversal is in NLOGSPACE (hence, also in PTIME).*

For the machine models studied in the rest of the paper, we will just show the decidability of emptiness and infiniteness, and not discuss their complexity. However, one should be able to obtain lower and upper bounds on the complexity of the emptiness problems using the results and techniques in [20,24].

A 2NCM is finite-crossing if there is an integer c such that the input head crosses the boundary between any two adjacent input symbols at most c times. The following was shown in [20]:

Theorem 3. *Every finite-crossing 2NCM can effectively be converted to an equivalent NCM. Hence, the emptiness and infiniteness problems for finite-crossing 2NCMs are also decidable.*

Finite-crossing 2NCMs are quite powerful. For example, they can accept languages that are not context-free. However, from Theorem 3, every NCM can be converted to an NCM (i.e., one-way). By using the result of Baker and Book [4] that every NCM can converted to one that runs in linear time, it is easy to show that the context-free language $L = \{x\#x^r \mid x \in (a + b)^+\}$ cannot be accepted by an NCM and, hence, cannot be accepted by a finite-crossing 2NCM.

Theorems 1 and 3 can be generalized. Define a 3-phase finite-crossing 2NPCM M which operates in three phases: In the first phase, M operates as a finite-crossing 2NCM without using the stack. In the second phase, with the configuration (state, input head position, and counter values) the first phase left off, M operates as an NPCM where the head can only move right on the input. Finally, in the third phase with the configuration (state, head position, counter values but not the stack) the second phase left off, M operates again as a finite-crossing 2NCM without using the stack. It is possible that the machine has only one or two phases. So, e.g., M can accept with only Phase 1, or with only Phases 1 and 2. A 3-phase finite-crossing 2DPCM is one in which all phases are deterministic. The following was shown in [38]:

Theorem 4. *The emptiness and infiniteness problems for 3-phase finite-crossing 2NPCMs are decidable.*

Theorem 4 seems to be the strongest result that we can prove in the sense that we cannot generalize the NPCM in the second phase to be a finite-crossing 2NPCM. In fact, it can be shown that a 2DPDA which makes only 3 reversals on the stack and 2-turns on the input has an undecidable emptiness problem.

An automaton M is over a letter-bounded language (resp., word-bounded language) if $L(M) \subseteq a_1^* \cdots a_k^*$ for some $k \geq 1$ and distinct letters $a_1, \ldots, a_k$ (resp., $L(M) \subseteq w_1^* \cdots w_k^*$ for some $k \geq 1$ and not-necessarily distinct words $w_1, \ldots, w_k$).

The following result appeared in [38]:

Theorem 5. *The emptiness and infiniteness problems for finite-crossing 2DPCMs over word-bounded languages are decidable.*

When we remove the "finite-crossing" condition in Theorem 3, we get the following:

Theorem 6. *1. The emptiness problem for 2DFAs (even over letter-bounded languages) with two reversal-bounded counters is undecidable. [27].*
2. The emptiness problem for 2DFAs with one reversal-bounded counter is decidable [34].
3. The emptiness problem for 2NFAs with one reversal-bounded counter over word-bounded languages is decidable [14].

2DFAs and 2NFAs augmented with one reversal-bounded counter are rather powerful. For example the non-semilinear language $L = \{a^{m_i} b^i \mid i, m \geq 1\}$ can be accepted by a 2DFA with one counter that only reverses once. Applications of the decidability of the emptiness problem for these machine (over word-bounded languages for the nondeterministic case) to decision problems in verification and Diophantine equations, etc. can be found in [22,32,45].

Open: Is emptiness decidable for 2NFAs with one reversal-bounded counters (i.e., the inputs are unrestricted)?

Theorem 3 does not hold for 2DPDAs, even when they have no reversal-bounded counters and the input head only makes 1 turn on the input: a left-to-right sweep of the input followed by a right-to-left sweep. In fact, we can even restrict the stack to make only one reversal. Consider, e.g., the language $L = \{a^1 \# a^2 \# a^3 \# \cdots \# a^{n-1} \# a^n \# \mid n \geq 1\}$. As shown in [36], one can construct a 2DPDA M which makes only 1-turn on the input and one reversal on the stack to accept L. This language can be accepted also by a 2DCA (two-way deterministic one-counter automaton) which makes only 1-turn on the input.

3   VPDAs with Reversal-Bounded Counters

The model of a visibly pushdown automaton (VPDA) was first introduced and studied in [1]. It is an NPDA where the input symbol determines the (push/stack) operation of the stack. The input alphabet Σ is partitioned into three disjoint alphabets: $\Sigma_c, \Sigma_r, \Sigma_{int}$. The machine pushes a specified symbol on the stack if it reads a call symbol in Σ_c on the input; it pops a specified symbol (if the specified symbol is at top of the stack) if it reads a return symbol in Σ_r on the input; it does not use the (top symbol of) the stack and can only change state if it reads an internal symbol in Σ_{int} on the input. The machine has no ε-moves, i.e, it reads an input at every step. An input $x \in \Sigma^*$ is accepted if the machine, starting from one of a designated set of initial states with a distinguished symbol $\perp$ at bottom of the stack (which is never altered), eventually enters an accepting sate after processing all symbols in x. For details, see [1]. In this paper, we assume without loss of generality, that the VPDA has only one initial state.

Recall that a 1-ambiguous NPDA is one where every input is accepted in at most one accepting computation. A 1-ambiguous NPDA is more powerful than a visibly pushdown automaton (VPDA) since the former can accept languages, like $L_1 = \{x \# x^R \mid x \in (a+b)^+\}$ and $L_2 = \{a^k b a^k \mid k \geq 1\}$, that cannot be accepted by the latter. There are languages accepted by 1-ambiguous NPDAs that are not deterministic context-free languages (DCFLs), whereas languages accepted by VPDAs are DCFLs [1]. Note also that a 1-ambiguous NPDA can have ε-moves.

Now consider a VPDA augmented with k reversal-bounded counters. We allow the machine to have ε-moves, but in such moves, the stack is not used, only the state and counters are used and updated. Acceptance of an input string is when machine eventually falls off the right end of the input in an accepting state. Thus, a VPDA M with k reversal-bounded counters operates like a VPDA but can now use reversal-bounded counters as auxiliary memory.

Consider the language $L_1 = \{w \mid w = xy$ for some $x \in (a+b)^+, y \in (0+1)^+, x$ when a, b are mapped to $0, 1$, respectively, is the reverse of $y\}$. Clearly, L_1 can be accepted by a VPDA M_1. However, the language $L_2 = \{w \mid w \in L_1$, the number of a's + number of 0's in $w =$ the number of b's + number of 1's in $w\}$ cannot be accepted by a VPDA. But L_2 can be accepted by a VPDA M_2 with two 1-reversal counters C_1 and C_2 as follows: On a given input w, M_2 simulates M_1, and stores the number of a's and 0's (resp., the number of b's and 1's) it sees on the input in counters C_1 and C_2, respectively. Then when M_1 accepts, M_2 (on ε-moves) decreases the counters simultaneously and accepts if the counters become zero at the same time.

Denote a VPDA with k reversal-bounded counters by k-VPCM, and by VPCM if the number of reversal-bounded counters is not specified. Clearly, VPCMs are a special case of NPCMs. Hence, from Theorem 1, we have:

Theorem 7. *The emptiness and infiniteness problems for VPCMs are decidable.*

Theorem 8. *The class of languages accepted by VPCMs is closed under union, intersection, renaming, concatenation, Kleene-*.*

Proof. The constructions are similar to the those for VPDAs without reversal-bounded counters in [1]. For example, for intersection, suppose M_1 and M_2 are two VPCMs with k_1 and k_2 reversal-bounded counters. We construct a VPCM with $k_1 + k_2$ reversal-bounded counters to simulate M_1 and M_2 in parallel. However, since the machines can have ε-transitions (which do not involve the stack), whenever one machine, say M_1, wants to make an ε-move but the other machine M_2 wants to make a non-ε-move, the simulation of M_2 is suspended temporarily until M_1 decides to make a non-ε-move, at which time the parallel simulation of both machine can be resumed.   □

With respect to complementation, unlike for VPDAs, we have:

Theorem 9. *The class of languages accepted by VPCMs is not closed under complementation.*

Proof. Consider the language $L = \{x \# y \mid x, y \in (a + b)^+, x \neq y\}$. Clearly, L can be accepted by a VPCM (where $a, b, \#$ are internal inputs) that does not use the stack, but uses a reversal-bounded counter to verify if an input string is in L. Suppose the complement, L^c, of L, can be accepted by a VPCM. Let $L_1 = (a+b)^+ \# (a+b)^+$. Since L_1 is regular it can be accepted by a VPDA. Then, by Theorem 8, $L_2 = L^c \cap L_1 = \{x \# x^R \mid x \in (a + b)^+\}$ can be accepted by a VPCM M_2. Since the symbols $a, b, \#$ are internal, the stack is not used by M_2 in its computation, and it only uses reversal-bounded counters. Thus we can remove the stack from M_2, and it becomes an NCM. From [4], M_2 can be converted to an equivalent NCM M_3 that runs in linear time. We get a contradiction, since it is easy to show (by a simple counting argument) that L_2 cannot be accepted by M_3.   □

Every VPDA can be converted to an equivalent deterministic VPDA [1]. In contrast, from Theorem 9, we have:

Corollary 1. *There are VPCMs that cannot be converted to equivalent deterministic VPCMs.*

We can define a deterministic VPCM (DVPCM) in the obvious way: The machine has at most one choice of move at each step. In particular, if there is a transition on ε, there is no transition on any $a \in \Sigma$. We also assume that the machine always halts. Thus a DVPCM is a deterministic VPDA with reversal-bounded counters. (Note that a VPDA has no ε-transitions.) Since a VPDA can always be made deterministic, it follows that DVPCMs are strictly more powerful than deterministic VPDAs.

The first part of the next result is obvious. The second part follows from the first using Theorem 7.

Corollary 2

1. *The class of languages accepted by DVPCMs is closed under Boolean operations.*
2. *The containment and equivalence problems for DVPCMs are decidable.*

The second part of the above corollary does not hold for VPCMs, since the universe problem (does a given transducer accept all strings?) is already undecidable even for an NFA augmented with just one 1-reversal counter [4].

4   Applications

In this section we discuss some applications of the decidability of the emptiness and infinite problems for machines augmented with reversal-bounded counters.

4.1   Multiple Morphism Equivalence on Context-Free Languages

A morphism g is a mapping from $\Sigma^* \to \Delta^*$ such that $g(\varepsilon) = \varepsilon$, and $g(a_1 \cdots a_n) = g(a_1) \cdots g(a_n)$ for $n \geq 1, a_1, \ldots, a_n \in \Sigma$. The multiple morphism problem on CFLs is defined as follows: Given k pairs of morphisms $(g_1, h_1), \ldots, (g_k, h_k)$ and a CFL $L \subseteq \Sigma^*$, is it the case that for every $w \in L$, there exists an i such that $g_i(w) = h_i(w)$?

The following theorem was shown in [25]. We include a proof using the decidability of the emptiness problem for finite-crossing 3-phase 2NPCM.

Theorem 10. *The multiple morphism equivalence problem on CFLs is decidable.*

Proof. Let L be a language accepted by an NPDA M and $(g_1, h_1), \ldots, (g_k, h_k)$ be k pairs of morphisms. We construct a finite-crossing 3-phase 2NPCM M' with $2k$ counters, $C_1, \ldots, C_k, D_1, \ldots D_k$. M', when given input $w = a_1 \cdots a_n$, first makes $2k$ sweeps of the input without using the stack. On sweep $1 \le i \le k$, M' applies morphism g_i on w without recording the $g_i(w)$ but nondeterministically guessing some position p_i on the input and storing this position in counter C_i and the symbol a_{p_i} in the state. Then on the next k sweeps, M' applies morphism h_i on w and again nondeterministically guessing some position q_i on the input and storing this position in counter D_i and the symbol a_{q_i} in the state. Next, M checks that for each $1 \le i \le k$, the values of C_i and D_i are the same (this can be done by simultaneously decrementing the counters and confirm that they become zero at the same time), but that $a_{p_i} \ne a_{q_i}$. Finally, M' makes one last sweep of the input simulating the NPDA M on w (now using the stack) and accepts if M accepts. Clearly, $L(M') \ne \varnothing$ if and only if there is a $w \in L$ such that $g_i(w) \ne h_i(w)$ for all $1 \le i \le k$. The result follows, since the emptiness problem for finite-crossing 3-phase 2NPCMs is decidable (Theorem 4).   □

Clearly, the above theorem still holds when the NPDA M is replaced by an NPCM (i.e., NPDA with reversal-bounded counters). Variations of the problem above can also be shown decidable. For example, it is decidable if $L' = \{w \mid w \in L,$ there is no i such that $g_i(w) = h_i(w)\}$ is infinite, since the infiniteness problem for 3-phase 2NPCMs is decidable (Theorem 4).

4.2   Finite-Valuedness in Transducers

A transducer T is an acceptor with outputs. For example, an NPDT is a nondeterministic pushdown automaton with outputs. So the transitions are rules of the form:

$(q, a, Z) \to (p, x, y)$

where q, p are states, a is an input symbol or ε, Z is the top of the stack symbol, x is a (possibly null) string of stack symbols, and y is an output string (possibly ε). In this transition, T in state q, reads a, 'pops' Z and writes x on the stack, outputs string y, and enters state p.

We say that (u, v) is a transduction accepted by T if, when started in the initial state q_0, with input u, and the stack contains only the initial stack symbol Z_0, T enters an accepting state after reading u and producing v. The set of transductions accepted by T is denoted by $R(T)$. An NPCMT is an NPCM with outputs. Similarly, An NCMT, NFT, etc., is an NCM, NFA, etc. with outputs.

A transducer T is k-ambiguous $(k \ge 1)$ if T with outputs ignored is a k-ambiguous acceptor. (Note that 1-ambiguous is the same as unambiguous). T is k-valued $(k \ge 1)$ if for every u, there are at most k distinct strings v such that (u, v) is in $R(T)$. T is finite-valued if T is k-valued for some k.

The following result was recently shown in [29]. We include the proof in [29] to illustrate the reduction of the problem to the emptiness problem for NPCMs (which is decidable by Theorem 1).

Theorem 11. *The k-valuedness problem for 1-ambiguous NPCMT T is decidable.*

Proof. Consider the case $k = 1$, Given T with m reversal-bounded counters, we construct an NPCM M with two additional 1-reversal counters C_{12} and C_{21}.

Hence, M will have $m + 2$ reversal-bounded counters. M on input x, simulates T suppressing the outputs and accepts x if it finds two outputs y_1 and y_2 such that y_1 and y_2 disagree in some position p and x is accepted by T. In order to do this, during the simulation, M uses C_{12} and C_{21} to record the positions i and j (chosen nondeterministically) in y_1 and y_2, respectively, and the symbols a and b in these positions, such that $i = j$ and $a \neq b$. Clearly, a and b can be remembered in the state. Storing i and j need only increment C_{12} and C_{21} during the simulation. To check that $i = j$, after the simulation, M decrements C_{12} and C_{21} simultaneously and verifies that they reach zero at the same time. Note that since T is 1-ambiguous, M's accepting computation on x (except for the outputs) is unique and therefore the procedure just described can be accomplished by M on a single accepting run on the input. Clearly, T is 1-valued if and only if $L(M) = \varnothing$, which is decidable, since emptiness of NPCMs is decidable (by Theorem 1).

The above construction generalizes for any $k \geq 1$. Now M on input x, checks that there are at least $k + 1$ distinct outputs $y_1, \ldots, y_{k+1}$. M uses $k \times (k + 1)$ additional 1-reversal counters (so now M will have $m + k \times (k + 1)$ reversal-bounded counters.) In the simulation, for $1 \leq i \leq k + 1$, M nondeterministically selects k positions $p_{i1}, \ldots, p_{i(i-1)}, p_{i(i+1)}, \ldots, p_{i(k+1)}$ in output y_i and records these positions in counters $C_{i1}, \ldots, C_{i(i-1)}, C_{i(i+1)}, \ldots, C_{i(k+1)}$ and the symbols at these positions in the state. At the end of the simulation, M accepts x if for all $1 \leq i, j \leq k + 1$ such that $i \neq j$, the symbol in position p_{ij} is different from the symbol in position p_{ji} and the value of counter C_{ij} is the same as the value of C_{ji}. $\qquad\qquad\square$

The construction in the proof above does not work when T is k-ambiguous for any $k \geq 2$. This is because the computation of T on an input x may not be unique, so it is possible, e.g., that one accepting run on x produces output y_1 and a different accepting run on x produces output y_2. So to determine if $y_1 \neq y_2$, we need to simulate two runs on input x, i.e., M will no longer be one-way. In fact, the following was shown in [29]:

Theorem 12. *For any $k \geq 1$, it is undecidable, given a $(k + 1)$-ambiguous 1-reversal NPDT T (i.e., its stack makes only one reversal), whether T is k-valued.*

4.3   VPDTs with Reversal-Bounded Counters

Visibly pushdown transducers (VPDT) were introduced in [40], where ε-transitions that can produce outputs were allowed. Allowing such transitions makes some decision problems (e.g., single-valuedness) undecidable. Later, in [16], VPDTs that do not allow ε-transitions were investigated, where it was shown that the k-valuedness problem for VPDTs is decidable.

We can generalize the result above for VPDTs with reversal-bounded counters. A VPCMT T is a VPCM with outputs. Since a VPCM is allowed ε-transitions (where the stack is not used), we assume that on an ε-transition, the VPCMT can only output ε.

Theorem 13. *The k-valuedness problem for VPCMTs is decidable.*

4.4 Context-Free Transducers

A context-free transducer (CFT) T is a CFG with outputs, i.e., the rules are of the form $A \to (\alpha, y)$, where α is a string of terminals and nonterminals, and y is an output string (possibly ε). We assume that the underlying CFG G of T, i.e., the grammar obtained by deleting the outputs, has no ε-rules (i.e., rules of the form $A \to \varepsilon$) and unit-rules (i.e., rules of the form $A \to B$), where A, B are nonterminals). We also assume that all nonterminals are useful, i.e., reachable from the start nonterminal S and can reach a terminal string.

We consider only *leftmost* derivations in T, i.e., at each step, the leftmost nonterminal is expanded). Thus T generates transductions (u, v) (where u is a terminal string and v is an output string) derived in a sequence of rule applications in a *leftmost* derivation: $(S, \varepsilon) \Rightarrow^+ (u, v)$

A nonterminal A in the underlying CFG of a CFT is self-embedding if there is some leftmost derivation $A \Rightarrow^+ \alpha A \beta$ where α, β are strings of terminals and nonterminals. (Note that $|\alpha \beta| > 0$, since there are no ε-rules and unit-rules.)

The notions of ambiguous and finite-valuedness can also be defined for context-free grammars (CFGs) with outputs.

A CFT T is k-ambiguous for a given k (resp., finitely-ambiguous) if its underlying CFG is k-ambiguous (resp., finitely-ambiguous).

The next three lemmas were recently shown in [29]. We sketch the proofs given in [29] for completeness.

Lemma 1. *Let T be a finitely-ambiguous CFT with terminal and nonterminal alphabets Σ and N, respectively. Let G be its underlying finitely-ambiguous CFG. Let A be a nonterminal such that $A \Rightarrow^+ \alpha A \beta$, where $\alpha, \beta \in (\Sigma \cup N)^*$. Then this derivation (of $\alpha A \beta$) is unique.*

Proof. It is straightforward to show that if two distinct derivations $A \Rightarrow^+ \alpha A \beta$ exist, then A would not be finitely-ambiguous.   □

Lemma 2. *It is decidable, given a finitely-ambiguous CFT T, whether there exist a nonterminal A and a leftmost derivation $A \Rightarrow^+ \alpha A \beta$ for some $\alpha, \beta \in (\Sigma \cup N)^*$ (note that $\alpha \beta| > 0$), such that there are at least two distinct outputs generated in the derivation.*

Proof. Let A be a nonterminal and $L = \{w \mid w = \alpha A \beta$, for some $\alpha, \beta \in (\Sigma \cup N)^*$ such that $|\alpha \beta| > 0$, and $A \Rightarrow^+ \alpha A \beta$ produces at least two distinct outputs$\}$. (Thus $L \subseteq (\Sigma \cup N)^*$.)

We construct an NPCM M to accept L. M, when given input w, tries to simulate a leftmost derivation $A \Rightarrow^+ \alpha A \beta$ (which, if it exists, is unique by Lemma 1) and checks that there are at least two distinct outputs generated in the derivation. Initially, A is the only symbol on the stack. Each derivation step is of the form $B \to x\varphi$, where x is in Σ^* and φ is in $N(\Sigma \cup N)^* \cup \{\varepsilon\}$. If B is the top of the stack, M simulates the step by popping B, checking that the remaining input segment to be read has prefix x (if $x \neq \varepsilon$), and pushing φ on the pushdown stack (if $\varphi \neq \varepsilon$). It uses two 1-reversal counters C_1 and C_2 to check that there is a discrepancy in the outputs corresponding to the derivation $A \Rightarrow^+ \alpha A \beta$. Since the derivation $A \Rightarrow^+ \alpha A \beta$ is unique, this can be done in the same manner as described in the proof of Theorem 11.

At some point in the derivation, M guesses that the stack contains a string of the form $z = \gamma_1 A \gamma_2$, where $\gamma_1, \gamma_2 \in (\Sigma \cup N)^*$. M then pops the stack and checks that the remaining input yet to be read is $\gamma_1 A \gamma_2$ and accepts if there were two distinct outputs generated in the derivation.

It is easily verified that $L(M) = L$. The result follows since the emptiness problem for NPCMs is decidable by Theorem 1. □

Lemma 3. *Let T be a finitely-ambiguous CFT and G be its underlying CFG. Suppose for some nonterminal A, there is a leftmost derivation $A \Rightarrow^+ \alpha A \beta$ for some $\alpha, \beta \in (\Sigma \cup N)^*$, with $|\alpha\beta| > 0$ such that there are at least two distinct outputs generated in the derivation. Then T is not finite-valued.*

Proof. Suppose there is a self-embedding nonterminal A and a derivation $A \Rightarrow^+ \alpha A \beta$, where $|\alpha\beta| > 0$ and there are at least two distinct strings y_1 and y_2 that are outputted in the derivation. Since all nonterminals are useful, we have in G, $S \Rightarrow^* wAx \Rightarrow^+ w\alpha A \beta x \Rightarrow^* w\alpha^k A \beta^k x \Rightarrow^+ wu^k z v^k x$ for some terminal strings w, u, z, v, x, for all $k \geq 1$. Let y_0, y_3, y_4, y_5 be the outputs in the derivations $S \Rightarrow^* wAx$, $\alpha \Rightarrow^* u$, $A \Rightarrow^+ z$, and $\beta \Rightarrow^* v$, respectively. There are two cases: *Case 1.* $|y_1| = |y_2|$ but $y_1 \neq y_2$, and *Case 2.* $|y_1| \neq |y_2|$. In both cases, the CFT T can be shown to be finite-valued. □

4.4.1 Linear Context-Free Transducers

A linear context-free transducer (LCFT) is a CFT whose underlying grammar is a linear CFG (LCFG). Thus, the rules are of the form $A \rightarrow (uBv, y)$ or $A \rightarrow (u, y)$, where A, B are nonterminals, u, v are terminal strings with $|uv| > 0$, and y is an output string. The following converse of Lemma 3 was shown in [29]:

Lemma 4. *Let T be a finitely-ambiguous LCFT and G be its underlying LCFG. Suppose there is no nonterminal A for which there is a leftmost derivation $A \Rightarrow^+ uAv$ for some $u, v \in \Sigma^*$, with $|uv| > 0$ such that there are at least two distinct outputs generated in the derivation. Then T is finite-valued.*

From Lemmas 2, 3, and 4:

Theorem 14. *It is decidable, given a finitely-ambiguous LCFT T, whether it is finite-valued.*

When the LCFT is not finitely-ambiguous, Theorem 14 does not hold:

Theorem 15. *It is undecidable, given a LCFT T, whether it is finite-valued.*

Proof. In [44], it was shown that there is a class of LCFGs for which every grammar in the class is either unambiguous or unboundedly ambiguous, but determining which is the case is undecidable. Let G be a LCFG in this class. Number the rules in G and construct a LCFT T which outputs the rule number corresponding to each rule. The result then follows. □

Lemmas 1, 2, and 3 were proved for finitely-ambiguous CFTs. However, Lemma 4 does not hold for finitely-ambiguous CFTs. For consider the 1-ambiguous CFT: $S \to (SA, 0) \mid (a, 0)$, $A \to (a, \{0, 1\})$, where S, A are nonterminals, a is a terminal symbol, and 0, 1 are output symbols. This CFT satisfies the hypothesis of Lemma 4, but it is not finite-valued.

4.4.2 Left-Derivation-Bounded Context-Free Transducers
A CFG is left-derivation-bounded (LDBCFT) if there is an $s \geq 1$ such that for any nonterminal A, every sentential form derivable from A when restricted to *leftmost* derivation has at most s nonterminals [42]. If the derivation is *not restricted* to leftmost derivation, the grammar is called derivation-bounded [17]. If the upper bound s must hold for *all* derivations of the sentential form, the CFG is called non-terminal bounded. Clearly, every nonterminal-bounded CFG is left-derivation-bounded, which, in turn, is derivation-bounded. It was shown in [42] that the language classes defined by these grammars form a strict hierarchy.

The following was proved in [29]:

Theorem 16. *It is decidable, given a finitely-ambiguous LDBCFT T, whether it is finite-valued.*

4.4.3 Context-Free Transducers
Let T be a CFT and G be its underlying CFG. For $A \in N$ and $\alpha \in (\Sigma \cup N)^*$, let $NT(A, \alpha)$ be the number of A's in α. Let $\#$ be a symbol not in the terminal alphabet of G. Let $L_A = \{\alpha \#^d \mid S \Rightarrow^+ \alpha, NT(A, \alpha) \geq d\}$. Clearly, G is left-derivation-bounded if and only if there is a nonterminal A such that L_A is infinite. Call such an A an abundant nonterminal.

The proofs of Lemma 5 and Corollary 3 below use the decidability of the infiniteness problem for NPCMs (Theorem 1).

Lemma 5. *It is decidable, given a CFG G and a nonterminal A, whether A is abundant. Hence, it is decidable whether G is left-derivation-bounded.*

The above lemma generalizes to:

Corollary 3. *It is decidable, given a CFG G and distinct nonterminals $A_1, \ldots, A_k$, whether $L_{(A_1, \ldots, A_k)} = \{\alpha \#^d \mid S \Rightarrow^+ \alpha, NT(A_1, \alpha) \geq d, \ldots, NT(A_k, \alpha) \geq d\}$ is infinite.*

We now assume that the CFT is 1-ambiguous.

Theorem 17. *It is decidable, given a 1-ambiguous CFT T, whether it is finite-valued.*

Proof. (Sketch) Let G be the underlying (1-ambiguous) CFG of T. If G is left-derivation-bounded (which is decidable by Lemma 5), then we can decide if T is finite-valued (by Theorem 16).

If G is not left-derivation-bounded, let A be an abundant nonterminal. Determine if there is a string of terminals and nonterminals β such that $A \Rightarrow^+ \beta$ and in this derivation, there are at least two distinct output strings. Call such a nonterminal multi-valued. Since G is 1-ambiguous, this property can be decided by a construction similar to those in the proofs of Theorem 11 and Lemma 2. [NOTE: This is the part where we need T (hence G) to be 1-ambiguous, since this is unlike a self-embedding derivation for which Lemma 2 applies, where the derivation is unique if T is finitely ambiguous.]

Clearly, if G has an abundant multi-valued nonterminal, then T is not finite-valued. Now suppose G does not have any abundant multi-valued nonterminal. We consider two cases:

Case 1. G has a self-embedding nonterminal B such that $B \Rightarrow^+ \gamma_1 B \gamma_2$ for some γ_1, γ_2 and in this derivation, there are at least two distinct outputs. Then T is not finite-valued (by Lemma 3).

Case 2. Assume G does not have a self-embedding nonterminal as described in case 1. We can show that T is finite-valued. We omit the proof.

From (1) and 2(a) and 2(b), it follows that T is finite-valued.   □

4.5   Containment and Equivalence Problems for Transducers

It is known that the equivalence problem for two-way deterministic finite transducers (2DFTs) is decidable [21]. However, if nondeterminism is allowed, the problem becomes undecidable even for one-way nondeterministic finite transducers (NFTs) [19]. The undecidability holds even for NFTs operating on a unary input (or output) alphabet [28]. For single-valued (i.e., 1-valued) NFTs, the problem becomes decidable [23], and the decidability result was later extended to finite-valued NFTs in [9]. The complexity of the problem was subsequently derived in [43].

In [38], we generalized the result of [21] for some models of two-way transducers (with input end markers # and $) augmented with reversal-bounded counters. Call the nondeterministic (resp., deterministic) version 2NCMT (resp., 2DCMT). The relation defined by such a transducer T is $R(T) = \{(x, y) \mid T,$ when started in its initial state on the left end marker of $\#x\$$, outputs y and falls off the right end marker in an accepting state$\}$. The transducer is finite-crossing if there is some fixed k such that in every accepting computation on any input $\#x\$$, the number of times the input head crosses the boundary between any two adjacent symbols of $\#x\$$ is at most k. Note that the number of turns (i.e., changes in direction from left-to-right and right-to-left and vice-versa) the input head makes on the input may be unbounded. Also note that the requirement is only for accepting computations. So if $R(T) = \varnothing$, then T is finite-crossing and, in fact, k-crossing for any k. We assume that when we are given a finite-crossing machine, the integer k for which the machine is k-crossing is also specified.

Clearly, every 2DFT is finite-crossing because, by definition, a valid computation must always fall off the right end marker in an accepting state. So no input cell can be visited twice in the same state; otherwise, the machine will never fall off the right end marker. Hence, a 2DFT is a special case of finite-crossing 2DCMT. But the latter is more powerful. Consider the relation $R = \{(xd^k, y) \mid x \in \Sigma^+, k > 0, |x| \geq 2k, x = x_1x_2x_3, |x_1| = |x_3| = k, y = x_3x_2x_1\}$, where d is a symbol not in alphabet Σ. R can be implemented on a finite-crossing 2DCMT using two reversal-bounded counters (in fact, the head need only make finite-turn on the input tape), but cannot be implemented on a 2DFT. On the other hand, for 2NFT, we can no longer say it is finite-crossing since the machine can always decide to fall off the right end marker, even though a cell has been visited more than once in the same state; e.g., $\{(x, x^n) \mid x \in \Sigma^+, n > 0\}$ can be implemented on a 2NFT. However, a finite-crossing 2NCMT is more powerful than a finite-crossing 2NFT, as R cannot also be implemented on a finite-crossing 2NFT.

It was shown in [38] that the following problems are decidable:

1. Given a finite-crossing 2NCMT T_1 and a finite-crossing 2DCMT T_2, is $R(T_1) \subseteq R(T_2)$? Hence, equivalence of finite-crossing 2DCMTs is decidable.
2. Given a one-way nondeterministic pushdown transducer with reversal-bounded counters (NPCMT) T_1 and a finite-crossing 2DCMT T_2, is $R(T_1) \subseteq R(T_2)$?
3. Given a finite-crossing 2NCMT T_1 and a DPCMT T_2 (the deterministic version of NPCMT), is $R(T_1) \subseteq R(T_2)$?
4. Given a finite-crossing 2DCMT T_1 and a DPCMT T_2, is $R(T_1) = R(T_2)$?

In the above results, the "finite-crossing" assumption is necessary, since it can be shown that when the two-way input is unrestricted, the equivalence problem becomes undecidable. Also, the NPCMT and DPCMT in (2), (3) and (4) above cannot be generalized to be two-way, since it can be shown that it is undecidable to determine, given a 2DPCMT T, whether $R(T) = \varnothing$, even when T makes only two turns on the input. However, the following was proved in [38]:

5. It is decidable to determine, given two finite-crossing 2DPCMTs whose inputs come from a bounded language (i.e., from $w_1^* \cdots w_k^*$ for some non-null strings $w_1, \ldots, w_k$) T_1 and T_2, whether $R(T_1) \subseteq R(T_2)$. (Hence, equivalence is also decidable.)

The proofs for the results above use the decidability of the emptiness problems in Section 2, in particular, the decidability of emptiness for 3-phase finite-crossing 2NPCMs.

The containment and equivalence between single-valued finite-crossing two-way nondeterministic finite transducers (2NFTs) and various finite-crossing two-way transducers with reversal-bounded counters were investigated in [38], where the following problems were shown to be decidable:

6. Given a finite-crossing 2NCMT (or an NPCMT) T_1 and a single-valued finite-crossing 2NFT T_2, is $R(T_1) \subseteq R(T_2)$?

7. Given a single-valued finite-crossing 2NFT T_1 and a finite-crossing 2DCMT (or a DPCMT) T_2, is $R(T_1) \subseteq R(T_2)$?
8. Given a single-valued finite-crossing 2NFT T_1 and a finite-crossing 2DCMT (or a DPCMT) T_2, is $R(T_1) = R(T_2)$?

In [2,3], deterministic and nondeterministic versions of streaming string transducers (SSTs, for short) were investigated. An SST uses a finite set of variables ranging over strings from the output alphabet in the process of making a single pass through the input string to produce the output string. It was shown in [2] that deterministic SSTs and two-way deterministic finite transducers (2DFTs) are equally expressive. For nondeterministic SSTs, they are incomparable with 2-way nondeterministic finite transducers (2NFTs) in terms of the expressiveness, although they are more expressive than 1-way nondeterministic finite transducers (NFTs).

As "variables" used in SSTs can be regarded as a form of auxiliary memory, it would be of interest to further investigate the relationship between two-way (finite-crossing) transducers augmented with auxiliary memory and SSTs with respect to the expressiveness as well as various decision problems. As was pointed out earlier, the relation $T = \{(xd^k, y) \mid x \in \Sigma^+, k > 0, |x| \geq 2k, x = x_1x_2x_3, |x_1| = |x_3| = k, y = x_3x_2x_1\}$, where d is a symbol not in alphabet Σ can be implemented on a finite-crossing 2DCMT, but cannot be implemented on a 2DFT. Hence, finite-crossing 2DCMTs are more expressive than deterministic SSTs, as the latter are equivalent to 2DFTs. The relationship between nondeterministic SSTs and various 2-way (finite-crossing) transducers with auxiliary memory remains unknown.

Finally, we consider the containment and equivalence problems for VPCMTs and DVPCMTs. As we mentioned earlier, the containment and equivalence problems for NFTs is undecidable. Hence these problems are also undecidable for VPCMTs. However, we can prove:

Theorem 18. *The following problems are decidable:*

1. *Given a VPCMT T_1 and a DVPCMT T_2, is $R(T_1) \subseteq R(T_2)$?*
2. *Given two DVPCMTs T_1 and T_2, is $R(T_1) = R(T_2)$?*

Proof. Clearly, we only need to prove (1). Let M_1 be the underlying VPCM of T_1 and M_2 be the underlying DVPCM of T_2. Thus, $L(M_1) = domain(R(T_1))$ and $L(M_2) = domain(R(T_2))$.

First we determine if $L(M_1) \subseteq L(M_2)$. This is decidable, since from Theorem 2, we can construct a DVPCM M_3 to accept the complement of $L(M_2)$. Then from Theorem 8 we can construct a VPCM M_4 to accept $L(M_1) \cap L(M_3)$. Clearly $L(M_1) \subseteq L(M_2)$ if and only if $L(M_4) = \varnothing$, which is decidable by Theorem 7.

Obviously, if $L(M_1) \not\subseteq L(M_2)$, then $R(T_1) \not\subseteq R(T_2)$. Otherwise, $R(T_1) \not\subseteq R(T_2)$ if and only if there exists an x such that the following two conditions are satisfied:

(a) For some y, (x, y) is in $R(T_1)$, and

(b) x is in $domain(R(T_2))$ and the only z such that (x, z) is in $R(T_2)$ is different from y.

Given T_1, M_1, T_2, M_2, we construct a VPCM M such that $L(M) \neq \varnothing$ if and only if the conditions above are satisfied (we omit the construction here). The result follows since we can decide if $L(M) = \varnothing$ by Theorem 7.   □

References

1. Alur, R., Alur, R., Madhusudan, P.: Visibly pushdown languages. In: Proc. of STOC 2004, pp. 202–211 (2004)
2. Alur, R., Cerny, P.: Expressiveness of streaming string transducers. In: Proc. 30th Annual Conf. on Foundations of Software Technology and Theoretical Computer Science, pp. 1–12 (2010)
3. Alur, R., Deshmukh, J.: Nondeterministic streaming string transducers. In: Aceto, L., Henzinger, M., Sgall, J. (eds.) ICALP 2011, Part II. LNCS, vol. 6756, pp. 1–20. Springer, Heidelberg (2011)
4. Baker, B., Book, R.: Reversal-bounded multipushdown machines. J. Comput. System Sci. 8, 315–332 (1974)
5. Bouchy, F., Finkel, A., San Pietro, P.: Dense-choice counter machines revisited. In: Proc. of INFINITY 2009, pp. 3–22 (2009)
6. Bouchy, F., Finkel, A., Sangnier, A.: Reachability in timed counter systems. Electr. Notes Theor. Comput. Sci. 239, 167–178 (2009)
7. Cadilhac, M., Finkel, A., McKenzie, P.: Affine Parikh automata. RAIRO - Theor. Inf. and Applic. 46(4), 511–545 (2012)
8. Cadilhac, M., Finkel, A., McKenzie, P.: Unambiguous constrained Automata. Int. J. Found. Comput. Sci. 24(7), 1099–1116 (2013)
9. Culik, K., Karhumaki, J.: The equivalence of finite valued transducers (on HDTOL languages) is decidable. Theoret. Comput. Sci. 47, 71–84 (1986)
10. Dang, Z.: Binary reachability analysis of pushdown timed automata with dense clocks. In: Berry, G., Comon, H., Finkel, A. (eds.) CAV 2001. LNCS, vol. 2102, pp. 506–517. Springer, Heidelberg (2001)
11. Dang, Z., Bultan, T., Ibarra, O.H., Kemmerer, R.A.: Past pushdown timed automata. In: Watson, B.W., Wood, D. (eds.) CIAA 2001. LNCS, vol. 2494, pp. 74–86. Springer, Heidelberg (2003)
12. Dang, Z., Ibarra, O.H., Bultan, T., Kemmerer, R.A., Su, J.: Binary reachability analysis of discrete pushdown timed automata. In: Emerson, E.A., Sistla, A.P. (eds.) CAV 2000. LNCS, vol. 1855, pp. 69–84. Springer, Heidelberg (2000)
13. Dang, Z., Ibarra, O.H., Miltersen, P.B.: Liveness verification of reversal-bounded multicounter machines with a free counter. In: Hariharan, R., Mukund, M., Vinay, V. (eds.) FSTTCS 2001. LNCS, vol. 2245, pp. 132–143. Springer, Heidelberg (2001)
14. Dang, Z., Ibarra, O.H., Sun, Z.-W.: On two-way nondeterministic finite automata with one reversal-bounded counter. Theor. Comput. Sci. 330(1), 59–79 (2005)
15. Finkel, A., Sangnier, A.: Reversal-bounded counter machines revisited. In: Ochmański, E., Tyszkiewicz, J. (eds.) MFCS 2008. LNCS, vol. 5162, pp. 323–334. Springer, Heidelberg (2008)
16. Filiot, E., Raskin, J.-F., Reynier, P.-A., Servais, F., Talbot, J.-M.: Properties of visibly pushdown transducers. In: Hliněný, P., Kučera, A. (eds.) MFCS 2010. LNCS, vol. 6281, pp. 355–367. Springer, Heidelberg (2010)

17. Ginsburg, S., Spanier, E.H.: Derivation-bounded languages. J. Comput. System Sci. 2(3), 228–250 (1968)
18. Ginsburg, S., Spanier, E.H.: Bounded ALGOL-like languages. T. Am. Math. Soc. 113, 333–368 (1964)
19. Griffiths, T.: The unsolvability of the equivalence problem for Λ-free nondeterministic generalized sequential machines. J. Assoc. Comput. Mach. 15, 409–413 (1968)
20. Gurari, E., Ibarra, O.H.: The complexity of decision problems for finite-turn multicounter machines. J. Comput. Syst. Sci. 22, 220–229 (1981)
21. Gurari, E.: The equivalence problem for deterministic two-way sequential transducers is decidable. SIAM J. Comput. 11, 448–452 (1982)
22. Gurari, E., Ibarra, O.H.: Two-way counter machines and Diophantine equations. J. ACM 29(3), 863–873 (1982)
23. Gurari, E., Ibarra, O.H.: A note on finite-valued and finitely ambiguous transducers. Math. Systems Theory 16, 61–66 (1983)
24. Hague, M., Lin, A.W.: Model checking recursive programs with numeric data types. In: Gopalakrishnan, G., Qadeer, S. (eds.) CAV 2011. LNCS, vol. 6806, pp. 743–759. Springer, Heidelberg (2011)
25. Harju, T., Ibarra, O.H., Karhumaski, J., Salomaa, A.: Some decision problems concerning semilinearity and commutation. J. Comput. Syst. Sci. 65(2), 278–294 (2002)
26. Hopcroft, J.E., Ullman, J.D.: Introduction to Automata, Languages and Computation. Addison-Wesley (1978)
27. Ibarra, O.H.: Reversal-bounded multicounter machines and their decision problems. J. of the ACM 25(1), 116–133 (1978)
28. Ibarra, O.H.: The unsolvability of the equivalence problem for ε-free NGSM's with unary input (output) alphabet and applications. SIAM J. Computing 7, 524–532 (1978)
29. Ibarra, O.H.: On the ambiguity, finite-valuedness, and lossiness problems in acceptors and transducers. In: Proc. of CIAA 2014 (to appear, 2014)
30. Ibarra, O.H.: On decidability and closure properties of language classes with respect to bio-operations. In: Proc. of DNA 20 (to appear)
31. Ibarra, O.H., Bultan, T., Su, J.: Reachability analysis for some models of infinite-state transition systems. In: Palamidessi, C. (ed.) CONCUR 2000. LNCS, vol. 1877, pp. 183–198. Springer, Heidelberg (2000)
32. Ibarra, O.H., Dang, Z.: On the solvability of a class of diophantine equations and applications. Theor. Comput. Sci. 352(1-3), 342–346 (2006)
33. Ibarra, O.H., Dang, Z., Egecioglu, O., Saxena, G.: Characterizations of Catalytic Membrane Computing Systems. In: Rovan, B., Vojtáš, P. (eds.) MFCS 2003. LNCS, vol. 2747, pp. 480–489. Springer, Heidelberg (2003)
34. Ibarra, O.H., Jiang, T., Tran, N., Wang, H.: New decidability results concerning two-way counter machines. SIAM J. Computing (24), 123–137 (1995)
35. Ibarra, O.H., Seki, S.: Characterizations of bounded semilinear languages by one-way and two-way deterministic machines. In: Proc. 13th Int. Conf. on Automata and Formal Languages, pp. 211–224 (2011)
36. Ibarra, O.H., Seki, S.: Semilinear sets and counter machines: a brief survey (submitted)
37. Ibarra, O.H., Su, J., Dang, Z., Bultan, T., Kemmerer, R.A.: Counter machines: decidable properties and applications to verification problems. In: Nielsen, M., Rovan, B. (eds.) MFCS 2000. LNCS, vol. 1893, pp. 426–435. Springer, Heidelberg (2000)

38. Ibarra, O.H., Yen, H.-C.: On the containment and equivalence problems for two-way transducers. Theor. Comput. Sci. 429, 155–163 (2012)
39. Minsky, M.: Recursive unsolvability of Post's problem of Tag and other topics in the theory of Turing machines. Ann. of Math. (74), 437–455 (1961)
40. Raskin, J.-F., Servais, F.: Visibly pushdown transducers. In: Aceto, L., Damgård, I., Goldberg, L.A., Halldórsson, M.M., Ingólfsdóttir, A., Walukiewicz, I. (eds.) ICALP 2008, Part II. LNCS, vol. 5126, pp. 386–397. Springer, Heidelberg (2008)
41. Verma, K.N., Seidl, H., Schwentick, T.: On the complexity of equational horn clauses. In: Nieuwenhuis, R. (ed.) CADE 2005. LNCS (LNAI), vol. 3632, pp. 337–352. Springer, Heidelberg (2005)
42. Walljasper, S.J.: Left-derivation bounded languages. J. Comput. and System Sci. 8(1), 1–7 (1974)
43. Weber, A.: Decomposing finite-valued transducers and deciding their equivalence. SIAM. J. on Computing 22, 175–202 (1993)
44. Wich, K.: Exponential ambiguity of context-free grammars. In: Proc. of 4th Int. Conf. on Developments in Language Theory, pp. 125–138. World Scientific (1999)
45. Xie, G., Dang, Z., Ibarra, O.H.: A Solvable class of quadratic diophantine equations with applications to verification of infinite-state systems. In: Baeten, J.C.M., Lenstra, J.K., Parrow, J., Woeginger, G.J. (eds.) ICALP 2003. LNCS, vol. 2719, pp. 668–680. Springer, Heidelberg (2003)
46. Xie, G., Dang, Z., Ibarra, O.H., Miltersen, P.B.: Dense counter machines and verification problems. In: Hunt Jr., W.A., Somenzi, F. (eds.) CAV 2003. LNCS, vol. 2725, pp. 93–105. Springer, Heidelberg (2003)

Star-Free Languages and Local Divisors

Manfred Kufleitner*

FMI, University of Stuttgart, Germany
kufleitner@fmi.uni-stuttgart.de

Abstract. A celebrated result of Schützenberger says that a language is star-free if and only if it is is recognized by a finite aperiodic monoid. We give a new proof for this theorem using local divisors.

1 Introduction

The class of regular languages is built from the finite languages using union, concatenation, and Kleene star. Kleene showed that a language over finite words is definable by a regular expression if and only if it is accepted by some finite automaton [3]. In particular, regular languages are closed under complementation. It is easy to see that a language is accepted by a finite automaton if and only if it is recognized by a finite monoid. As an algebraic counterpart for the minimal automaton of a language, Myhill introduced the *syntactic monoid, cf.* [6].

An extended regular expression is a term over finite languages using the operations union, concatenation, complementation, and Kleene star. By Kleene's Theorem, a language is regular if and only if it is definable using an extended regular expression. It is natural to ask whether some given regular language can be defined by an extended regular expression with at most n nested iterations of the Kleene star operation — in which case one says that the language has generalized star height n. The resulting decision problem is called the *generalized star height problem*. Generalized star height zero means that no Kleene star operations are allowed. Consequently, languages with generalized star height zero are called *star-free*. Schützenberger showed that a language is star-free if and only if its syntactic monoid is aperiodic [7]. Since aperiodicity of finite monoids is decidable, this yields a decision procedure for generalized star height zero. To date, it is unknown whether or not all regular languages have generalized star height one.

In this paper, we give a proof of Schützenberger's result based on *local divisors*. In commutative algebra, local divisors have been introduced by Meyberg in 1972, see [2,4]. In finite semigroup theory and formal languages, local divisors were first used by Diekert and Gastin for showing that pure future local temporal logic is expressively complete for free partially commutative monoids [1].

---

* The author gratefully acknowledges the support by the German Research Foundation (DFG) under grant DI 435/5-1 and the support by ANR 2010 BLAN 0202 FREC.

H. Jürgensen et al. (Eds.): DCFS 2014, LNCS 8614, pp. 23–28, 2014.
© Springer International Publishing Switzerland 2014

2 Preliminaries

The set of finite words over an alphabet A is A^*. It is the free monoid generated by A. The empty word is denoted by ε. The *length* $|u|$ of a word $u = a_1 \cdots a_n$ with $a_i \in A$ is n, and the *alphabet* $\mathrm{alph}(u)$ of u is $\{a_1, \ldots, a_n\} \subseteq A$. A language is a subset of A^*. The concatenation of two languages $K, K' \subseteq A^*$ is $K \cdot K' = \{uv \mid u \in K, v \in K'\}$, and the set difference of K by K' is written as $K \setminus K'$. Let A be a finite alphabet. The class of *star-free languages* $\mathrm{SF}(A^*)$ over the alphabet A is defined as follows:

- $A^* \in \mathrm{SF}(A^*)$ and $\{a\} \in \mathrm{SF}(A^*)$ for every $a \in A$.
- If $K, K' \in \mathrm{SF}(A^*)$, then each of $K \cup K'$, $K \setminus K'$, and $K \cdot K'$ is in $\mathrm{SF}(A^*)$.

By Kleene's Theorem, a language is regular if and only if it can be recognized by a deterministic finite automaton [3]. In particular, regular languages are closed under complementation and thus, every star-free language is regular.

Lemma 1. *If $B \subseteq A$, then $\mathrm{SF}(B^*) \subseteq \mathrm{SF}(A^*)$.*

Proof. It suffices to show $B^* \in \mathrm{SF}(A^*)$. We have $B^* = A^* \setminus \bigcup_{b \notin B} A^* b A^*$. □

A monoid M is *aperiodic* if for every $x \in M$ there exists a number $n \in \mathbb{N}$ such that $x^n = x^{n+1}$.

Lemma 2. *Let M be aperiodic. Then $x_1 \cdots x_k = 1$ in M if and only if $x_i = 1$ for all i.*

Proof. If $xy = 1$, then $1 = xy = x^n y^n = x^{n+1} y^n = x \cdot 1 = x$. □

A monoid M *recognizes* a language $L \subseteq A^*$ if there exists a homomorphism $\varphi : A^* \to M$ with $\varphi^{-1}(\varphi(L)) = L$. A consequence of Kleene's Theorem is that a language is regular if and only if it is recognizable by a finite monoid, see *e.g.* [5]. The class of *aperiodic languages* $\mathrm{AP}(A^*)$ contains all languages $L \subseteq A^*$ which are recognized by some finite aperiodic monoid.

The *syntactic congruence* $\equiv_L$ of a language $L \subseteq A^*$ is defined as follows. For $u, v \in A^*$ we set $u \equiv_L v$ if for all $p, q \in A^*$ we have $puq \in L \Leftrightarrow pvq \in L$. The *syntactic monoid* $\mathrm{Synt}(L)$ of a language $L \subseteq A^*$ is the quotient $A^* / \equiv_L$ consisting of the equivalence classes modulo $\equiv_L$. The *syntactic homomorphism* $\mu_L : A^* \to \mathrm{Synt}(L)$ with $\mu_L(u) = \{v \mid u \equiv_L v\}$ satisfies $\mu_L^{-1}(\mu_L(L)) = L$. In particular, $\mathrm{Synt}(L)$ recognizes L and it is the unique minimal monoid with this property, see *e.g.* [5].

Let M be a monoid and $c \in M$. We introduce a new multiplication $\circ$ on $cM \cap Mc$. For $xc, cy \in cM \cap Mc$ we let

$$xc \circ cy = xcy.$$

This operation is well-defined since $x'c = xc$ and $cy' = cy$ implies $x'cy' = xcy' = xcy$. For $cx, cy \in Mc$ we have $cx \circ cy = cxy \in Mc$. Thus, $\circ$ is associative and c is the neutral element of the monoid $M_c = (cM \cap Mc, \circ, c)$. Moreover,

$M' = \{x \in M \mid cx \in Mc\}$ is a submonoid of M such that $M' \to cM \cap Mc$ with $x \mapsto cx$ becomes a homomorphism. It is surjective and hence, M_c is a divisor of $(M, \cdot, 1)$ called the *local divisor of M at c*. Note that if $c^2 = c$, then M_c is just the local monoid $(cMc, \cdot, c)$ at the idempotent c.

Lemma 3. *If M is a finite aperiodic monoid and $1 \neq c \in M$, then M_c is aperiodic and $|M_c| < |M|$.*

Proof. If $x^n = x^{n+1}$ in M for $cx \in Mc$, then $(cx)^n = cx^n = cx^{n+1} = (cx)^{n+1}$ where the first and the last power is in M_c. This shows that M_c is aperiodic. By Lemma 2 we have $1 \notin cM \cap Mc$ and thus $|M_c| < |M|$. $\qquad\square$

3   Schützenberger's Theorem on Star-Free Languages

The following proposition establishes the more difficult inclusion of Schützenberger's result $\mathrm{SF}(A^*) = \mathrm{AP}(A^*)$. Its proof relies on local divisors.

Proposition 1. *Let $\varphi : A^* \to M$ be a homomorphism to a finite aperiodic monoid M. Then for all $p \in M$ we have $\varphi^{-1}(p) \in \mathrm{SF}(A^*)$.*

Proof. We proceed by induction on $(|M|, |A|)$ with lexicographic order. The claim is obvious for $A = \emptyset$. For $p = 1$ we have $\varphi^{-1}(1) = \{a \in A \mid \varphi(a) = 1\}^*$. Here, the inclusion from left to right follows from Lemma 2 and the other inclusion is trivial. By Lemma 1, we conclude $\varphi^{-1}(1) \in \mathrm{SF}(A^*)$. This also covers both the case $|M| = 1$ and the situation where $\varphi(a) = 1$ for all $a \in A$.

Let now $p \neq 1$ and let $c \in A$ with $\varphi(c) \neq 1$. We set $B = A \setminus \{c\}$ and we let $\varphi_c : B^* \to M$ be the restriction of φ to B^*. We have

$$\varphi^{-1}(p) = \varphi_c^{-1}(p) \cup \bigcup_{p = p_1 p_2 p_3} \varphi_c^{-1}(p_1) \cdot \left[\varphi^{-1}(p_2) \cap cA^* \cap A^*c\right] \cdot \varphi_c^{-1}(p_3). \quad (1)$$

The inclusion from right to left is trivial. The other inclusion can be seen as follows: Every word w with $\varphi(w) = p$ either does not contain the letter c or we can factorize $w = w_1 w_2 w_3$ with $c \notin \mathrm{alph}(w_1 w_3)$ and $w_2 \in cA^* \cap A^*c$, i.e., we factorize w at the first and the last occurrence of c. Equation (1) is established by setting $p_i = \varphi(w_i)$. By induction on the size of the alphabet, we have $\varphi_c^{-1}(p_i) \in \mathrm{SF}(B^*)$, and thus $\varphi_c^{-1}(p_i) \in \mathrm{SF}(A^*)$ by Lemma 1.

Since $\mathrm{SF}(A^*)$ is closed under union and concatenation, it remains to show $\varphi^{-1}(p) \cap cA^* \cap A^*c \in \mathrm{SF}(A^*)$ for $p \in \varphi(c)M \cap M\varphi(c)$. Let

$$T = \varphi_c(B^*).$$

The set T is a submonoid of M. In the remainder of this proof, we will use T as a finite alphabet. We define a substitution

$$\sigma : \quad (B^* c)^* \to T^*$$
$$v_1 c \cdots v_k c \mapsto \varphi_c(v_1) \cdots \varphi_c(v_k)$$

for $v_i \in B^*$. In addition, we define a homomorphism $\psi : T^* \to M_c$ with $M_c = (\varphi(c)M \cap M\varphi(c), \circ, \varphi(c))$ by

$$\psi : \quad T^* \to M_c$$
$$\varphi_c(v) \mapsto \varphi(cvc)$$

for $\varphi_c(v) \in T$. Consider a word $w = v_1 c \cdots v_k c$ with $k \geq 0$ and $v_i \in B^*$. Then

$$\psi\big(\sigma(w)\big) = \psi\big(\varphi_c(v_1)\varphi_c(v_2) \cdots \varphi_c(v_k)\big)$$
$$= \varphi(cv_1 c) \circ \varphi(cv_2 c) \circ \cdots \circ \varphi(cv_k c)$$
$$= \varphi(cv_1 cv_2 \cdots cv_k c) = \varphi(cw). \tag{2}$$

Thus, we have $cw \in \varphi^{-1}(p)$ if and only if $w \in \sigma^{-1}\big(\psi^{-1}(p)\big)$. This shows $\varphi^{-1}(p) \cap cA^* \cap A^*c = c \cdot \sigma^{-1}\big(\psi^{-1}(p)\big)$ for every $p \in \varphi(c)M \cap M\varphi(c)$. In particular, it remains to show $\sigma^{-1}\big(\psi^{-1}(p)\big) \in \mathrm{SF}(A^*)$. By Lemma 3, the monoid M_c is aperiodic and $|M_c| < |M|$. Thus, by induction on the size of the monoid we have $\psi^{-1}(p) \in \mathrm{SF}(T^*)$, and by induction on the size of the alphabet we have $\varphi_c^{-1}(t) \in \mathrm{SF}(B^*) \subseteq \mathrm{SF}(A^*)$ for every $t \in T$. For $t \in T$ and $K, K' \in \mathrm{SF}(T^*)$ we have

$$\sigma^{-1}(T^*) = A^*c \cup \{1\}$$
$$\sigma^{-1}(t) = \varphi_c^{-1}(t) \cdot c$$
$$\sigma^{-1}(K \cup K') = \sigma^{-1}(K) \cup \sigma^{-1}(K')$$
$$\sigma^{-1}(K \setminus K') = \sigma^{-1}(K) \setminus \sigma^{-1}(K')$$
$$\sigma^{-1}(K \cdot K') = \sigma^{-1}(K) \cdot \sigma^{-1}(K').$$

Only the last equality requires justification. The inclusion from right to left is trivial. For the other inclusion, suppose $w = v_1 c \cdots v_k c \in \sigma^{-1}(K \cdot K')$ for $k \geq 0$ and $v_i \in B^*$. Then $\varphi_c(v_1) \cdots \varphi_c(v_k) \in K \cdot K'$, and thus $\varphi_c(v_1) \cdots \varphi_c(v_i) \in K$ and $\varphi_c(v_{i+1}) \cdots \varphi_c(v_k) \in K'$ for some $i \geq 0$. It follows $v_1 c \cdots v_i c \in \sigma^{-1}(K)$ and $v_{i+1} c \cdots v_k c \in K'$. This shows $w \in \sigma^{-1}(K) \cdot \sigma^{-1}(K')$.

We conclude that $\sigma^{-1}(K) \in \mathrm{SF}(A^*)$ for every $K \in \mathrm{SF}(T^*)$. In particular, $\sigma^{-1}\big(\psi^{-1}(p)\big) \in \mathrm{SF}(A^*)$. $\qquad\square$

Remark 1. A more algebraic viewpoint of the proof of Proposition 1 is the following. The mapping σ can be seen as a length-preserving homomorphism from a submonoid of A^*—freely generated by the infinite set B^*c—onto T^*; and this homomorphism is defined by $\sigma(vc) = \varphi_c(v)$ for $vc \in B^*c$. The mapping $\tau : M\varphi(c) \cup \{1\} \to M_c$ with $\tau(x) = \varphi(c) \cdot x$ defines a homomorphism. Now, by Equation (2) the following diagram commutes:

$$
\begin{array}{ccc}
(B^*c)^* & \xrightarrow{\ \sigma\ } & T^* \\
{\scriptstyle \varphi}\downarrow & & \downarrow{\scriptstyle \psi} \\
M\varphi(c) \cup \{1\} & \xrightarrow[\ \tau\ ]{} & M_c
\end{array}
$$

The following lemma gives the remaining inclusion of $SF(A^*) = AP(A^*)$. Its proof is standard; it is presented here only to keep this paper self-contained.

Lemma 4. *For every language $L \in SF(A^*)$ there exists an integer $n(L) \in \mathbb{N}$ such that for all words $p, q, u, v \in A^*$ we have*

$$p\,u^{n(L)}q \in L \;\Leftrightarrow\; p\,u^{n(L)+1}q \in L.$$

Proof. For the languages A^* and $\{a\}$ with $a \in A$ we define $n(A^*) = 0$ and $n(\{a\}) = 2$. Let now $K, K' \in SF(A^*)$ such that $n(K)$ and $n(K')$ exist. We set

$$n(K \cup K') = n(K \setminus K') = \max\big(n(K), n(K')\big),$$
$$n(K \cdot K') = n(K) + n(K') + 1.$$

The correctness of the first two choices is straightforward. For the last equation, suppose $p\,u^{n(K)+n(K')+2}q \in K \cdot K'$. Then either $p\,u^{n(K)+1}q' \in K$ for some prefix q' of $u^{n(K')+1}q$ or $p'\,u^{n(K')+1}q \in K'$ for some suffix p' of $pu^{n(K)+1}$. By definition of $n(K)$ and $n(K')$ we have $p\,u^{n(K)}q' \in K$ or $p'\,u^{n(K')}q \in K'$, respectively. Thus $p\,u^{n(K)+n(K')+1}q \in K \cdot K'$. The other direction is similar: If $p\,u^{n(K)+n(K')+1}q \in K \cdot K'$, then $p\,u^{n(K)+n(K')+2}q \in K \cdot K'$. This completes the proof. $\square$

Theorem 1 (Schützenberger). *Let A be a finite alphabet and let $L \subseteq A^*$. The following conditions are equivalent:*

1. *L is star-free.*
2. *The syntactic monoid of L is finite and aperiodic.*
3. *L is recognized by a finite aperiodic monoid.*

Proof. "$1 \Rightarrow 2$": Every language $L \in SF(A^*)$ is regular. Thus $\mathrm{Synt}(L)$ is finite, *cf.* [5]. By Lemma 4, we see that $\mathrm{Synt}(L)$ is aperiodic. The implication "$2 \Rightarrow 3$" is trivial. If $\varphi^{-1}\big(\varphi(L)\big) = L$, then we can write $L = \bigcup_{p \in \varphi(L)} \varphi^{-1}(p)$. Therefore, "$3 \Rightarrow 1$" follows by Proposition 1. $\square$

The syntactic monoid of a regular language (for instance, given by a non-deterministic automaton) is effectively computable. Hence, from the equivalence of conditions "1" and "2" in Theorem 1 it follows that star-freeness is a decidable property of regular languages. The equivalence of "1" and "3" can be written as

$$SF(A^*) = AP(A^*).$$

The equivalence of "2" and "3" is rather trivial: The class of finite aperiodic monoids is closed under division, and the syntactic monoid of L divides any monoid that recognizes L, see *e.g.* [5].

Acknowlegdements. The author would like to thank Volker Diekert and Benjamin Steinberg for many interesting discussions on the proof method for Proposition 1.

References

1. Diekert, V., Gastin, P.: Pure future local temporal logics are expressively complete for Mazurkiewicz traces. Information and Computation 204, 1597–1619 (2006)
2. Fernández López, A., Tocón Barroso, M.: The local algebras of an associative algebra and their applications. In: Misra, J. (ed.) Applicable Mathematics in the Golden Age, pp. 254–275. Narosa (2002)
3. Kleene, S.C.: Representation of events in nerve nets and finite automata. In: Shannon, C.E., McCarthy, J. (eds.) Automata Studies. Annals of Mathematics Studies, vol. 34, pp. 3–40. Princeton University Press (1956)
4. Meyberg, K.: Lectures on algebras and triple systems. Technical report, University of Virginia, Charlottesville (1972)
5. Pin, J.-É.: Varieties of Formal Languages. North Oxford Academic, London (1986)
6. Rabin, M.O., Scott, D.: Finite automata and their decision problems. IBM Journal of Research and Development 3, 114–125 (1959); Reprinted in Moore, E.F. (ed.) Sequential Machines: Selected Papers. Addison-Wesley (1964)
7. Schützenberger, M.P.: On finite monoids having only trivial subgroups. Information and Control 8, 190–194 (1965)

Aperiodic Tilings by Right Triangles[*]

Nikolay Vereshchagin

Moscow State University, Leninskie gory 1, Moscow 119992, Russia
ver@mccme.ru

Abstract. Let ψ denote the square root of the golden ratio, $\psi = \sqrt{(\sqrt{5}-1)/2}$. A golden triangle is any right triangle with legs of lengths a, b where $a/b = \psi$. We consider tilings of the plane by two golden triangles: that with legs $1, \psi$ and that with legs ψ, ψ^2. Under some natural constrains all such tilings are aperiodic.

1 Introduction

1.1 Golden Triangles

The altitude of every right triangle cuts it into two similar triangles. Are there other polygons P that can be divided into two polygons each of which is similar to P? Any parallelogram whose width is $\sqrt{2}$ times bigger than its length has this property: its median cuts it into two equal such parallelograms. A more interesting example is the so called *Ammann hexagon*[1]:

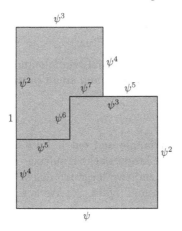

Here ψ stands for the square root of the golden ratio ($\psi^4 + \psi^2 = 1$). It turns out that that there are no other such hexagons. This was conjectured by Scherer in [6] and proved by Schmerl in [7].

---

[*] The work was in part supported by the RFBR grant 14-01-93107.
[1] This hexagon is attributed to Robert Ammann in [8]. Independently the hexagon was discovered by Scherer [6] who called it the *Golden Bee*.

H. Jürgensen et al. (Eds.): DCFS 2014, LNCS 8614, pp. 29–41, 2014.

Let us go back to the right triangle, whose altitude cuts it into two similar triangles. Assume that its hypotenuse and legs are proportional to 1, ψ and ψ^2, respectively, so that Pythagorean theorem hold:

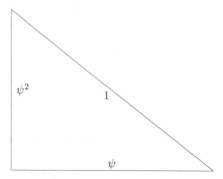

Call any such triangle a *golden triangle*. The altitude of the golden triangle cuts it into a *large* and *small* golden triangles, labeled by letters "L" and "S":

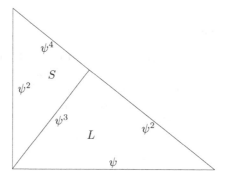

The ratio between the sizes of the initial triangle and its large part is equal to the ratio between the sizes of its large and small parts (hence the name).

1.2   Standard Tilings

Let us start with a golden triangle and cut it by its altitude into two smaller golden triangles. Then cut the larger of the resulting triangles into two triangles. We obtain one small triangle and two large triangles. Then again cut both large resulting triangles, then again and again .... We get the following tilings:

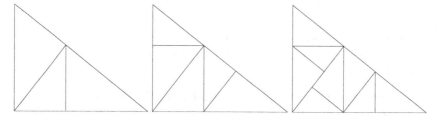

...

On each step we obtain a tiling of the original golden triangle by golden triangles of two sizes. We will call such tilings *standard*. The number of steps needed to obtain a standard tiling from the original golden triangle is called its *depth*. For example, the depth of the last tiling in the last picture is 4.

We can start from any golden triangle, so for each n and each d we can obtain a standard tiling of depth n consisting of triangles of sizes d and ψd. In this paper, we study tilings of the plane (or its parts) by golden triangles of two sizes d and ψd (where d is a fixed number, say, 1) that look locally like standard tilings. This means that for any circular window (of any diameter D) every pattern that we can observe in that tiling through such a window can be observed also in some standard tiling. Of course the depth of that tiling may depend on the diameter of the window. The larger the window is the larger the depth of the tiling may be. We will call such tilings *locally standard, LS*. In other words, a tiling is LS if each its finite subset is a subset of a standard tiling. (Throughout the paper we consider tiling as sets of triangles.)

Do locally standard tiling of the plane exist? This can be shown by well known arguments (used in the literature, for instance, for Berger's tilings [2]). Moreover, like Berger's tilings, all locally standard tilings of the plane are aperiodic.

Assume now that we bound the size of the window by some constant D. That is, consider only patterns of diameter at most D. Our Theorem 2 states that for any D there are finitely many patterns of diameter at most D that can be observed in standard tilings. (When counting patterns, we identify isometric ones.)

More specifically, we say that a finite tiling T is a *pattern* of a tiling T' if T is a subset of T'. For example, every standard tiling is a pattern of every standard tiling of larger depth, but not the other way around. A pattern is *standard* if it is a pattern of a standard tiling. Thus a tiling is LS iff all its patterns are standard. The *diameter of a pattern* is the maximal distance between two points lying in triangles of that pattern. Theorem 2 states that for every D there are finitely many standard patterns of diameter at most D.

Theorem 1 gives a hope to describe locally standard tilings by a finite number of patterns. This would be possible if there were D with the following property: if all patterns of diameter at most D of a tiling T are standard then T is locally standard (that is, all patterns of T are standard).

The main result of the paper, Theorem 3, states that this is not the case. In other words, for every D there is a tiling of the plane by golden triangles that is not LS and yet all its patterns of diameter at most D are standard. Speaking informally, locally standard tilings cannot be finitely presented, they cannot be defined by a finite set of local rules.

This result shows a crucial difference between tilings by Ammann hexagons and golden triangles. Recall that Ammann hexagon of size d can be also cut into two Ammann hexagons of sizes ψd and $\psi^2 d$ (see the picture on the first page). In the similar way one can define *Ammann standard tilings, Ammann locally standard tilings* etc. However, this time there is D such an Ammann tiling of the plane is locally standard iff all its patterns of diameter at most D

are standard [3]. Moreover, an Ammann tiling is locally standard iff all its pairs of adjacent hexagons form a standard pattern!

Let us return to tilings by golden triangles. There is yet another way to define what means that a tiling of a plane "looks like standard tilings". Let us call the operation used to define standard tiling *the refinement*. The refinement of a tiling T is the tiling obtained from T by cutting each large triangle from T by its altitude (and keeping all small triangles intact). It is not hard to see that different tilings have different refinements. Hence a reverse partial operation is well defined. That partial operation is called the *coarsening*. (Not every tiling has a coarsening: for example, a tiling consisting of one small triangle has no coarsening.) If a tiling admits n successive coarsenings, we call it n-*coarsenable*. For instance, any standard tiling of depth n is $n + 1$-coarsenable but not $n + 2$-coarsenable. If a tiling is n-coarsenable for all n we call it *infinitely coarsenable*, *IC*. One can show that every LS tiling is IC, but not the other way around. An example of IC tiling which is not LS will be given later. From this example it will be clear that the class of LS tilings is a more adequate formalization of tilings that "look like standard tilings" than the class of IC tilings.

Our main result applies to the class of IC tilings as well: IC tilings cannot be defined by a finite set of local rules. More specifically, for any D there is a non-IC tiling whose all patterns of diameter at most D are standard (and hence appear in an IC tiling, namely, in any LS tiling of the plane).

One can wonder if the class of locally standard tilings is "sofic". We say that a class $\mathcal{C}$ of tilings is *sofic* if the following holds. There are D, a finite set of colors and a finite set of patterns $\mathcal{P}$ of diameter at most D where each pattern consists of colored triangles (each triangle bears only one color) such that
(1) in every tiling from $\mathcal{C}$ each triangle can be colored so that all patterns of diameter at most D of the resulting colored tiling belong to $\mathcal{P}$, and the other way around:
(2) if every pattern of diameter at most D in a tiling of the plane by colored triangles belongs to $\mathcal{P}$, then after removing colors the resulting tiling belongs $\mathcal{C}$.

We do not know whether the families of LS tilings and of IC tilings are sofic. The Goodman-Strauss theorem [5], or its proof, might provide a positive answer to this question. The statement of that theorem itself does not imply the answer, as its conditions are not satisfied for the family of LS (or IC) tilings. The same applies to Fernique – Ollinger generalization of Goodman-Strauss theorem [4].

2   Preliminaries

The letter ψ denotes the square root of the golden ratio, $\psi = \sqrt{(\sqrt{5} - 1)/2}$. A *golden triangle* is any right triangle similar to that shown on the picture on page 30 (all points inside the triangle are considered as belonging to it). The *size* of a golden triangle is the length of its hypotenuse. A *d-tiling* is a non-empty set of golden triangles that pair wise have no common interior points and each of them is either of size d (such triangles are called *large*), or of size $d\psi$ (those are called *small*).

A *tiling* is a d-tiling for some d. A tiling T *tiles* A (where A is a subset of the plane), if A equals the union of all triangles in T. A tiling T is called *periodic*, if there is a nonzero vector v (called a *period*) such the result of transition of every triangle H in T by vector v belongs to T. Otherwise the tiling is called *aperiodic*.

The *refinement* of a d-tiling T is the ψd-tiling obtained from T by cutting each large triangle from T by its altitude. All small triangles remain intact and become large triangles of the refinement. It is easy to verify that the refinement is an injective operation. The reverse partial operation is called the *coarsening*. The k-*refinement* of a tiling is the result of applying k successive refinements to it. The partial operation of k-*coarsening* is defined in a similar way. If a tiling admits k successive coarsenings, that is, it is a k-refinement of some tiling, we call it k-*coarsenable*. If a tiling is k-coarsenable for all k we call it *infinitely coarsenable*.

A *standard d-tiling of depth n* is a d-tiling obtained from a single golden triangle H of size $d\psi^{-n}$ by n successive refinements.

A finite tiling P is a *pattern* of a tiling T if P is a subset of T. A finite tiling is a *standard pattern* if it is a subset of a standard tiling. A tiling T is *locally standard* all its patterns are standard, i.e. for all finite $W \subset T$ there is a standard tiling T' with $W \subset T'$.

The *diameter* of a finite tiling is the maximal distance between two points lying in triangles of that tiling. A tiling is called D-*locally standard* if all its patterns of diameter at most D are standard.

3   Results

Theorem 1. *(a) There are locally standard tilings of the plane. (b) Every locally standard tiling of the plane is infinitely coarsenable. (c) The converse is not true. (d) Every infinitely coarsenable tiling of the plane is aperiodic.*

Proof. (a) Let St_n denote the standard 1-tiling of depth n. Observe that St_7 has a large triangle that is located strictly inside the part of the plane tiled by St_7:

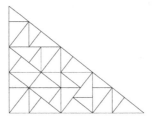

This implies that we can draw St_0 and St_7 on the plane so that St_0 is a subset of St_7 and, moreover, the triangle forming St_0 is strictly inside the part of the plane tiled by St_7. Similarly, we can draw St_{14} on the plane so that St_{14} includes St_7 as a subset and the part of the plane tiled by St_7 is strictly inside the part of the plane tiled by St_{14}.

In this way we can construct a sequence of tilings $\mathrm{St}_0, \mathrm{St}_7, \mathrm{St}_{14}, \ldots$ such that St_{7n} is a standard tiling of depth $7n$, St_{7n} is a subset of St_{7n+7} and the union $T = \bigcup_{n=0}^{\infty} \mathrm{St}_{7n}$ tiles the entire plane. On the other hand, the tiling T is locally standard by construction.

(c) In the same way as in item (a), we can construct a sequence of tilings $\mathrm{St}_0, \mathrm{St}_8, \mathrm{St}_{16}, \ldots$ such that the union $T = \bigcup_{n=0}^{\infty} \mathrm{St}_{8n}$ tiles a half-plane. This is because the tiling St_8

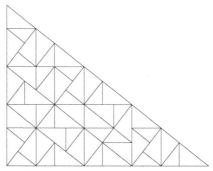

has a large triangle L such that the hypotenuse of L lies on the hypotenuse of the triangle tiled by St_8 and both legs of L are strictly inside the triangle tiled by St_8.

Let $\tilde{T}$ be the tiling obtained from T by applying the axial symmetry with the axis equal to the edge of the half plane tiled by T. Both T and $\tilde{T}$ are IS by construction. Then shift $\tilde{T}$ by a very small amount along the edge of the half plane. The shifted $\tilde{T}$ is IC as well. Hence the union of T and shifted $\tilde{T}$ is also IC. On the other hand, it is not LS, as all patterns along the edge of the half plane become non-standard after the shift.

(b) We will say that a small triangle S and a large triangle L form a *couple* if they are located as shown on the second picture on page 30. It is easy to see that in any standard tiling for any small triangle S there is a large triangle L forming a couple with S.

Consider any small triangle S in a locally standard d-tiling T of the plane. As T is locally standard, there is a large triangle $L \in T$ forming a couple with S. Replace in T the triangles L and S by their union $L \cup S$, for every small triangle S. We obtain a $\psi^{-1}d$-tiling T', whose refinement equals T. Thus T is coarsenable.

Let us show that T' is LS. Let W' be any finite subset of T'. We have to show that W' is a subset of a standard tiling. Let W stand for the refinement of W'. Then W is a subset of T. Since T is LS, W is a subset of a standard tiling, say, St_n. If $n = 0$ then W' is a small triangle and we are done. Otherwise W' is a subset St_{n-1}.

(d) Assume that an IC tiling T has a non-zero period v. Then v is also a period of the coarsening T' of T. Indeed, the refinement of $T' + v$ is equal to the tiling $T + v$, which equals T by the assumption; thus $T' + v$ and T' have the same refinement and hence coincide. Similarly, v is a period of the coarsening T'' of T' and so on. Note that the coarsening increases the sizes of triangles. Thus, on some step, v becomes smaller then the lengths of all sides of triangles and we get a contradiction.

Theorem 2. *For any D, d the family of patterns of diameter at most D of standard d-tilings is finite. (When counting patterns we identify isometric ones.)*

Proof. Call a tiling P a *simple* pattern of a tiling T if there is a node K of some triangle from T (called the *center* of the pattern) such that P consists of all the triangles from T whom K belongs to.

A *simple standard pattern* is a simple pattern of a standard tiling. An example of a simple standard pattern P is shown on the following picture (P consists of all triangles intersecting the circle, the center of the pattern is inside the circle):

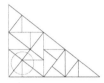

The proof is based on the following two lemmas.

Lemma 1. *Assume that a 1-tiling T tiles a convex set U. Assume further that S is a subset of U of diameter less than a certain positive constant ε. Then T has a simple pattern P that covers S.*

Lemma 2. *For every d the family of all simple patterns of d-tilings is finite and their number does not depend on d. (We identify here isometric patterns.)*

Both lemmas are quite technical and will be proved in the Appendix. Now we finish the proof of the theorem assuming the lemmas. Fix D and d. W.l.o.g. assume that $d = 1$. Consider a standard 1-tiling St_n of some depth n. Let W be any pattern of diameter at most D of the tiling St_n. We claim that there are a number k bounded by a function of D and a simple pattern P of a standard $(1/\psi)^k$-tiling such that k-refinement of P includes W.

Let k be the minimal integer such that $\psi^k D$ is less that the constant ε from Lemma 1. If it happens that $k > n$ then let $k = n$. Let St_{n-k} denote the k-coarsening of the tiling St_n. If $k = n$ we are done, as we can let $P = T_{n-k}$. Otherwise, $\psi^k D < \varepsilon$ and hence the diameter of W measured in units $(1/\psi)^k$ is less than ε. By Lemma 1 W is covered by a simple pattern P of T_{n-k}. Hence the k-refinement of P includes W.

By Lemma 2 the number of simple patterns of T_{n-k} is bounded by a constant and k is bounded by a function of D. For each k and each simple pattern P the number of subsets of the k-refinement of P is finite. This completes the proof of the theorem modulo the lemmas.

Theorem 1 gives a hope to describe LS tilings by a finite number of patterns. This would be possible if there were D such that every D-locally standard tiling is LS. The main result of this paper states that this is not the case.

Theorem 3. *For every D there is a D-locally standard tiling which is not locally standard and even not infinitely coarsenable.*

Proof. The proof of this theorem is fairly simple (but hard to find). Consider the following periodic 1-tiling U of the plane (the first configuration on the picture):

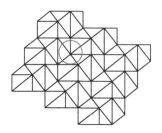 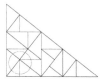

All its simple patterns are standard: they appear in the standard tiling of depth 6 (the second configuration on the picture).

Fix any D. Let ε be the constant from Lemma 1. Choose i so that $D\psi^i < \varepsilon$ and let U_i be the i-refinement of tiling U. Then U_i is the sought tiling.

Indeed, by Lemma 1 for every pattern of U_i of diameter less than D there is a simple pattern P of U such that W is a subset of the i-refinement of P. As we have seen, all simple patterns of U are standard and so does P. Hence W is a standard pattern as well.

On the other hand, being periodic, the tiling U_i is not infinitely coarsenable (actually, it admits only $i + 2$ coarsenings). ∎

Acknowledgments. The author is sincerely grateful to Alexander Shen, Andrey Romashchenko and Thomas Fernique for useful comments.

References

1. Ammann, R., Grünbaum, B., Shephard, G.C.: Aperiodic tiles. Discrete and Computational Geometry 8, 1–25 (1992)
2. Berger, R.: The undecidability of the domino problem. Memoirs of the American Mathematical Society 66 (1966)
3. Durand, B., Shen, A., Vereshchagin, N.: Ammann tilings: a classification and an application. CoRR abs/1112.2896 (2012)
4. Fernique, T., Ollinger, N.: Combinatorial substitutions and sofic tilings, http://arxiv.org/abs/1009.5167
5. Goodman-Strauss, C.: Matching rules and substitution tilings. Ann. of Math. 147(1), 181–223 (1998)
6. Scherer, K.: A puzzling journey to the reptiles and related animals. Privately published (1987)
7. Schmerl, J.: Dividing a polygon into two similar polygons. Discrete Math. 311(4), 220–231 (2011)
8. Senechal, M.: The mysterious Mr. Ammann. Math. Intelligencer 26, 10–21 (2004)

A   Proof of the Lemma 1

Assume that ε is small enough (in the end we will see how small it should be).

Let S denote the convex closure of W. Then U includes S. As convex closure has the same diameter as the set itself, the diameter of S is at most ε. Assuming that ε is less than the lengths of all sides of triangles from T we conclude that S has at most one node of a triangle from T. If it has such a node then let K be that (unique) node. In this case the simple pattern P of T with the center K covers S (and hence W). Indeed, if ε is less than the altitude of a small triangle then all the points of all the triangles from the simple pattern P are at the distance at least $\varepsilon > D$ from K. As the diameter of S is at most D, this implies all the points of S are at distance at most D from K and are thus covered by the pattern.

Assume now that S has no node of a triangle from T. If S covered by only one triangle we are done — any its node can be taken as the center of the sought simple pattern.

Otherwise S has no nodes of triangles from T and cannot be covered by one triangle from T. Let A be any triangle intersecting S, say in point C, and let D be any point from $S \setminus A$. Consider the segment $[C, D]$. At some point E that segment leaves the triangle A. The points of $[E, D]$ that lie very close to E belong to S and hence to some triangle B from T. That triangle includes the point E.

If E is close to a node of A or a node of B we can let K be that node. Indeed, all points in S are close to E and E is close to K. Hence all points from S are close to K and are thus covered by the simple pattern with center K. Otherwise E is the internal node of a leg of A and an internal node of a leg of B and thus A and B share a common segment. All nodes of A and B are far from E and hence from S. This implies that S is covered by $A \cup B$. It remains to notice that A and B belong to a simple pattern of T: indeed, both ends of the line segment shared by A, B can be chosen as the center of that simple pattern.

A calculation shows that ε equal to the half of altitude of the small triangle times ψ^2 will do.

B   Proof of the Lemma 2

The second statement of the theorem (the number of simple patterns of standard d-tilings does not depend on d) is obvious. So we will assume that $d = 1$.

Consider standard 1-tilings of depth 9 and 10 (Fig. 1). A careful examination reveals that every simple pattern of the second tiling is isometric to a simple pattern of the first tiling. This implies that every simple pattern of standard 1-tiling of depth 11 is isometric to a simple pattern of the standard 1-tilings of depth 10. Indeed, let K be a node of a triangle of St_{11} and P the simple pattern of St_{11} with center K.

Assume first that K is also a node of a triangle from the coarsening St_{10} of St_{11} (an "old" node). Let P' be the simple pattern of St_{10} with center K. Then P is a subset of the refinement of P'.

As every simple pattern of St_{10} is isomorphic to a simple pattern of St_9, so does P'. Let Q' be the simple pattern of St_9 that is isomorphic to P' and L its center. Then the set of all triangles in St_{10} that include the point L forms a simple pattern of St_{10} isomorphic to P.

Otherwise K is a "new node", it is not a node of a triangle from the coarsening St_{10} of St_{11}. Distinguish two cases (1) K is on the border of the area tiled by St_{11} and (2) K is an inner node of the area tiled by St_{11} (examples of such nodes are inside two circles on Fig. 1). The fist case is easy, as there are plenty such simple patterns of St_{10}.

The second case is more delicate. We claim that in this case K is always the center of a simple pattern isomorphic to that whose center lies inside the circle on Fig. 1. It is enough to prove that claim, as there are plenty such patterns in St_{10}.

Any new inner node K lies of the hypotenuse of a large triangle of St_{10}, say A, and also belongs to a triangle B, adjacent to A. Therefore we have to analyze the standard patterns consisting of two adjacent triangles. It turns out that there are 10 such patterns and we are going to find all of them.

Let T be a standard pattern consisting of adjacent triangles A, B. Let n be the minimal n such that T is a subset of a standard tiling St_n. Call such n the *index* of T. Let us show first that any pattern of index $n > 1$ is a subset of the refinement of a pattern of index $n - 1$.

For $n = 2$ this is easily verified ad hoc. Assume that $n > 2$. Then St_n is a disjoint union of two standard tilings T_{n-1} and T_{n-2}. Let I stand for the common segment of indexes of A, B. Coarsen T_{n-1} and T_{n-2} and denote the resulting tilings by T'_{n-2} and T'_{n-3}, respectively. Let A', B' be the triangles from T'_{n-2}, T'_{n-1} including A, B. Then $A'.B'$ also share the segment I and hence form a pattern of a smaller index than n.

This observation allows to find all standard patterns consisting of two adjacent triangles in the order of increasing its indexes. The only such pattern of index 1 is the pair of small and large triangles that is obtained by refining T_0 (a couple of a small and large triangles). All patterns of index 2 are subsets of T_2. There is only one such pattern (a large triangle whose large leg is the hypotenuse of a small triangle, see the third picture on page 30). All patterns of index 3 (or less) are subsets of the refinement of this pattern. As one can see on the third picture on page 30 the only pattern of index 3 is the pair of large triangles sharing the hypotenuse. All patterns of index 4 (or less) are subsets of the refinement of this pattern. Again on the third picture on page 30 we can see that there are two new different patterns of index 4 (two large triangles sharing a small part of their large legs and a small triangle whose small leg is a part of the large leg of a large triangle). Each of the patterns of index 4 produces one pattern of index 5:

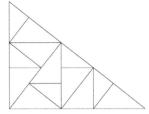

Thus we have two patterns of index 5: two large triangles sharing parts of their hypotenuses and a large triangle whose small leg is a part of the hypotenuse of a large triangle). The second pattern produces now new patterns. The first on one produces one pattern of index 6 (two small triangles sharing the small leg):

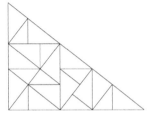

This pattern produces one pattern (two large triangles sharing the small leg) of index 7:

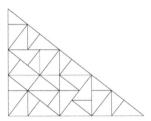

This pattern produces one pattern (two small triangles sharing the hypotenuse) of index 8:

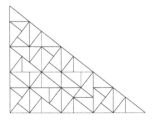

Finally, this pattern produces no new patterns:

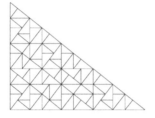

Now we can see that the only standard pair of adjacent triangles whose refinement yields a new inner node is the pattern of index 3 (the pair of large triangles sharing the hypotenuse). And both new inner nodes are centers of simple patterns isomorphic to that on Fig. 1.

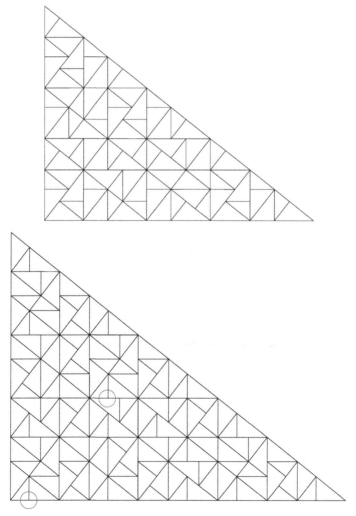

Fig. 1. Standard tilings of depths 9 and 10

In the similar way we can show that every simple pattern of standard 1-tiling of depth 12 is isometric to a simple pattern of the standard 1-tilings of depth 11 etc. Thus any simple pattern of any standard 1-tiling is isometric to a simple pattern of St_9.

Quantum Hashing via ε-Universal Hashing Constructions and Freivalds' Fingerprinting Schemas

Farid Ablayev and Marat Ablayev

Kazan Federal University, Kazan, Russia

Abstract. We define the concept of a quantum hash generator and offer a design, which allows one to build a large number of different quantum hash functions. The construction is based on composition of a classical ε-universal hash family and a given family of functions – quantum hash generators.

In particular, using the relationship between ε-universal hash families and Freivalds' fingerprinting schemas we present explicit quantum hash function and prove that this construction is optimal with respect to the number of qubits needed for the construction.

Keywords: quantum hashing, quantum hash function, ε-universal hashing, error-correcting codes.

1 Introduction

Quantum computing is inherently a very mathematical subject, and the discussions of how quantum computers can be more efficient than classical computers in breaking encryption algorithms started since Shor invented his famous quantum algorithm. The answer of the cryptography community is "Post-quantum cryptography", which refers to research on problems (usually public-key cryptosystems) that are no more efficiently breakable using quantum computers than by classical computer architectures. Currently post-quantum cryptography includes several approaches, in particular, hash-based signature schemes such as Lamport signatures and Merkle signature schemes.

Hashing itself is an important basic concept for the organization transformation and reliable transmission of information. The concept known as "universal hashing" was invented by Carter and Wegman [7] in 1979. In 1994 a relationship was discovered between ε-universal hash families and error-correcting codes [5]. In [16] Wigderson characterizes universal hashing as being a tool which "should belong to the fundamental bag of tricks of every computer scientist".

Gottesman and Chuang proposed a quantum digital system [9], based on quantum mechanics. Their results are based on quantum a fingerprinting technique and add "quantum direction" for post-quantum cryptography. Quantum fingerprints have been introduced by Buhrman, Cleve, Watrous and de Wolf in [6]. Gavinsky and Ito [8] viewed quantum fingerprints as cryptographic primitives.

H. Jürgensen et al. (Eds.): DCFS 2014, LNCS 8614, pp. 42–52, 2014.

In [2,3] we considered quantum fingerprinting as a construction for binary hash functions and introduced a non-binary hash function. The quantum hashing proposed a suitable one-way function for quantum digital signature protocol from [9]. For more introductory information we refer to [2].

In this paper, we define the concept of a quantum hash generator and offer a design, which allows one to build different quantum hash functions. The construction is based on the composition of classical ϵ-universal hash family with a given family of functions – quantum hash generator.

The construction proposed combines the properties of robust presentation of information by classical error-correcting codes together with the possibility of highly compressed presentation of information by quantum systems. In particular, using the relationship between ϵ-universal hash families and Freivalds' fingerprinting schemas we present an explicit quantum hash function and prove that this construction is optimal with respect to of number of qubits needed for the construction.

1.1   Definitions and Notations

We begin by recalling some definitions of classical hash families from [13]. Given a domain $\mathbb{X}$, $|\mathbb{X}| = K$, and a range $\mathbb{Y}$, $|\mathbb{Y}| = M$, (typically with $K \geq M$), a hash function f is a map

$$f : \mathbb{X} \to \mathbb{Y},$$

that hash *long* inputs to *short* outputs.

We let q to be a prime power and $\mathbb{F}_q$ be a finite field of order q. Let Σ^k be a set of words of length k over a finite alphabet Σ. In the paper we let $\mathbb{X} = \Sigma^k$, or $\mathbb{X} = \mathbb{F}_q$, or $\mathbb{X} = (\mathbb{F}_q)^k$, and $\mathbb{Y} = \mathbb{F}_q$. A hash family is a set $F = \{f_1, \ldots, f_N\}$ of hash functions $f_i : \mathbb{X} \to \mathbb{Y}$.

ϵ universal hash family. A hash family F is called an ϵ-universal hash family if for any two distinct elements $w, w' \in \mathbb{X}$, there exist at most ϵN functions $f \in F$ such that $f(w) = f(w')$. We will use the notation ϵ-U $(N; K, M)$ as an abbreviation for ϵ-universal hash family.

Clearly we have, that if function the f is chosen uniformly at random from a given ϵ-U $(N; K, M)$ hash family F, then the probability that any two distinct words collide under f is at most ϵ.

The case of $\epsilon = 1/N$ is known as universal hashing.

Classical-quantum function. The notion of a quantum function was considered in [11]. In this paper we use the following variant of a quantum function. First recall that mathematically a qubit $|\psi\rangle$ is described as $|\psi\rangle = \alpha|0\rangle + \beta|1\rangle$, where α and β are complex numbers, satisfying $|\alpha|^2 + |\beta|^2 = 1$. So, a qubit may be presented as a unit vector in the two-dimensional Hilbert complex space $\mathcal{H}^2$. Let $s \geq 1$. Let $(\mathcal{H}^2)^{\otimes s}$ be the 2^s-dimensional Hilbert space, describing the states of s qubits, i.e. $(\mathcal{H}^2)^{\otimes s}$ is made up of s copies of a single qubit space $\mathcal{H}^2$

$$(\mathcal{H}^2)^{\otimes s} = \mathcal{H}^2 \otimes \ldots \otimes \mathcal{H}^2 = \mathcal{H}^{2^s}.$$

For $K = |\mathbb{X}|$ and integer $s \geq 1$ we define a $(K; s)$ classical-quantum function to be a map of the elements $w \in \mathbb{X}$ to quantum states $|\psi(w)\rangle \in (\mathcal{H}^2)^{\otimes s}$

$$\psi : \mathbb{X} \to (\mathcal{H}^2)^{\otimes s}. \tag{1}$$

We will also use the notation $\psi : w \mapsto |\psi(w)\rangle$ for ψ.

2   Quantum Hashing

What we need to define for quantum hashing and what is implicitly assumed in various papers (see for example [2] for more information) is a collision resistance property. However, there is still no such notion as *quantum collision*. The reason why we need to define it is the observation that in quantum hashing there might be no collisions in the classical sense: since quantum hashes are quantum states they can store an arbitrary amount of data and can be different for different messages. But the procedure of comparing those quantum states implies measurement, which can lead to collision-type errors.

So, a *quantum collision* is a situation when a procedure that tests the equality of quantum hashes and outputs "true", while hashes are different. This procedure can be a well-known SWAP-test (see for example [2] for more information and citations) or something that is adapted for specific hashing function. Anyway, it deals with the notion of distinguishability of quantum states. Since non-orthogonal quantum states cannot be perfectly distinguished, we require them to be "nearly orthogonal".

– For $\delta \in (0, 1/2)$ we call a function

$$\psi : \mathbb{X} \to (\mathcal{H}^2)^{\otimes s}$$

δ-resistant, if for any pair w, w' of different elements,

$$|\langle \psi(w) \,|\, \psi(w') \rangle| \leq \delta.$$

Theorem 1. *Let* $\psi : \mathbb{X} \to (\mathcal{H}^2)^{\otimes s}$ *be a* δ-*resistant function. Then*

$$s \geq \log \log |\mathbb{X}| - \log \log \left(1 + \sqrt{2/(1 - \delta)}\right) - 1.$$

Proof. First we observe, that from the definition $\||\psi\rangle\| = \sqrt{\langle \psi | \psi \rangle}$ of the norm it follows that

$$\||\psi\rangle - |\psi'\rangle\|^2 = \||\psi\rangle\|^2 + \||\psi'\rangle\|^2 - 2\langle \psi | \psi' \rangle.$$

Hence for an arbitrary pair w, w' of different elements from $\mathbb{X}$ we have that

$$\||\psi(w)\rangle - |\psi(w')\rangle\| \geq \sqrt{2(1 - \delta)}.$$

We let $\Delta = \sqrt{2(1 - \delta)}$. For short we let $(\mathcal{H}^2)^{\otimes s} = V$ in this proof. Consider a set $\Phi = \{|\psi(w)\rangle : w \in \mathbb{X}\}$. If we draw spheres of radius $\Delta/2$ with centers

$|\psi\rangle \in \Phi$ then spheres do not pairwise intersect. All these K spheres are in a large sphere of radius $1 + \Delta/2$. The volume of a sphere of radius r in V is $cr^{2^{s+1}}$ for the complex space V. The constant c depends on the metric of V. From this we have, that the number K is bonded by the number of "small spheres" in the "large sphere"

$$K \leq \frac{c(1 + \Delta/2)^{2^{s+1}}}{c(\Delta/2)^{2^{s+1}}}.$$

Hence

$$s \geq \log\log K - \log\log\left(1 + \sqrt{2/(1-\delta)}\right) - 1.$$

$\square$

The notion of δ-resistance naturally leads to the following notion of quantum hash function.

Definition 1 (Quantum hash function). *Let K, s be positive integers and $K = |\mathbb{X}|$. We call a map*

$$\psi : \mathbb{X} \to (\mathcal{H}^2)^{\otimes s}$$

an δ-resistant $(K; s)$ quantum hash function if ψ is a δ-resistant function.

We use the notation δ-R $(K; s)$ as an abbreviation for δ-resistant $(K; s)$ quantum hash functions.

3   Generator for Quantum Hash Functions

In this section we present two constructions of quantum hash functions and define notion of quantum hash function generator, which generalizes these constructions.

3.1   Binary Quantum Hashing

One of the first explicit quantum hash functions was defined in [6]. Originally the authors invented a construction called "quantum fingerprinting" for testing the equality of two words for a quantum communication model. The cryptographical aspects of quantum fingerprinting are presented in [8]. The quantum fingerprinting technique is based on binary error-correcting codes. Later this construction was adopted for cryptographic purposes. Here we present the quantum fingerprinting construction from the quantum hashing point of view.

An (n, k, d) *error-correcting code* is a map

$$C : \Sigma^k \to \Sigma^n$$

such that, for any two distinct words $w, w' \in \Sigma^k$, the Hamming distance between code words $C(w)$ and $C(w')$ is at least d. The code is binary if $\Sigma = \{0, 1\}$.

The construction of a quantum hash function based on quantum fingerprinting in as follows.

- Let $c > 1$ and $\delta < 1$. Let k be a positive integer and $n > k$. Let $E : \{0,1\}^k \to \{0,1\}^n$ be an (n, k, d) binary error-correcting code with Hamming distance $d \geq (1 - \delta)n$.
- Define a family of functions $F_E = \{E_1, \ldots, E_n\}$, where $E_i : \{0,1\}^k \to \mathbb{F}_2$ is defined by the rule: $E_i(w)$ is the i-th bit of the code word $E(w)$.
- Let $s = \log n + 1$. Define the classical-quantum function $\psi_{F_E} : \{0,1\}^k \to (\mathcal{H}^2)^{\otimes s}$, determined by a word w as

$$\psi_{F_E}(w) = \frac{1}{\sqrt{n}} \sum_{i=1}^{n} |i\rangle |E_i(w)\rangle = \frac{1}{\sqrt{n}} \sum_{i=1}^{n} |i\rangle \left(\cos \frac{\pi E_i(w)}{2} |0\rangle + \sin \frac{\pi E_i(w)}{2} |1\rangle \right),$$

For $s = \log n + 1$, the function ψ_{F_E} is an δ-R $(2^k; s)$ quantum hash function, that is, for two different words w, w' we have

$$|\langle \psi_{F_E}(w) | \psi_{F_E}(w') \rangle| \leq \delta n/n = \delta.$$

Observe, that the authors in [6] propose, for the first choice of such binary codes, Justesen codes with $n = ck$, which give $\delta < 9/10 + 1/(15c)$ for any chosen $c > 2$. Next we observe, that the above construction of a quantum hash function needs $\log n + 1$ qubits for the fixed $\delta \approx 9/10 + 1/(15c)$. This number of qubits is good enough in the sense of the lower bound of Theorem 1.

A non-binary quantum hash function is presented in [2] and is based on the construction from [1].

3.2 Non-binary Quantum Hashing

We present the non-binary quantum hash function from [2] in the following form. For a field $\mathbb{F}_q$, let $B = \{b_1, \ldots, b_T\} \subseteq \mathbb{F}_q$. For every $b_j \in B$ and $w \in \mathbb{F}_q$, define a function $h_j : \mathbb{F}_q \to \mathbb{F}_q$ by the rule

$$h_j(w) = b_j w \pmod{q}.$$

Let $H = \{h_1, \ldots h_T\}$ and $t = \log T$. We define the classical-quantum function

$$\psi_H : \mathbb{F}_q \to (\mathcal{H}^2)^{\otimes(t+1)}$$

by the rule

$$|\psi_H(w)\rangle = \frac{1}{\sqrt{T}} \sum_{j=1}^{T} |j\rangle \left(\cos \frac{2\pi h_j(w)}{q} |0\rangle + \sin \frac{2\pi h_j(w)}{q} |1\rangle \right).$$

The following is proved in [2].

Theorem 2. *Let q be a prime power and $\mathbb{F}_q$ be a field. Then, for arbitrary $\delta > 0$, there exists a set $B = \{b_1, \ldots, b_T\} \subseteq \mathbb{F}_q$ (and, therefore, a corresponding family $H = \{h_1, \ldots, h_T\}$ of functions) with $T = \lceil (2/\delta^2) \ln(2q) \rceil$, such that the quantum function ψ_H is a δ-R $(q; t+1)$ quantum hash function.*

In the rest of the paper we use the notation $H_{\delta,q}$ to denote this family of functions from Theorem 2 and the notation $\psi_{H_{\delta,q}}$ to denote the corresponding quantum function.

Observe, that the above construction of the quantum hash function $\psi_{H_{\delta,q}}$ needs $t + 1 \leq \log\log 2q + 2\log 1/\delta + 3$ qubits. This number of qubits is good enough in the sense of the lower bound of Theorem 1.

Numerical results on $\psi_{H_{\delta,q}}$ are presented in [2].

3.3   Quantum Hash Generator

The above two constructions of quantum hash functions are using certain controlled rotations of target qubits. These transformations are generated by the corresponding discrete functions from a specific family of functions (F_E and $H_{\delta,q}$ respectively).

These constructions lead to the following definition.

Definition 2 (Quantum hash generator). *Let $K = |\mathbb{X}|$ and let $G = \{g_1, \ldots, g_D\}$ be a family of functions $g_j : \mathbb{X} \to \mathbb{F}_q$. Let $\ell \geq 1$ be an integer. For $g \in G$ let ψ_g be a classical-quantum function $\psi_g : \mathbb{X} \to (\mathcal{H}^2)^{\otimes\ell}$ determined by the rule*

$$\psi_g : w \mapsto |\psi_g(w)\rangle = \sum_{i=1}^{2^\ell} \alpha_i(g(w))|i\rangle, \qquad (2)$$

where the amplitudes $\alpha_i(g(w))$, $i \in \{1, \ldots, 2^\ell\}$, of the state $|\psi_g(w)\rangle$ are determined by $g(w)$. Let $d = \log D$. We define a classical-quantum function $\psi_G : \mathbb{X} \to (\mathcal{H}^2)^{\otimes(d+\ell)}$ by the rule

$$\psi_G : w \mapsto |\psi_G(w)\rangle = \frac{1}{\sqrt{D}} \sum_{j=1}^{D} |j\rangle|\psi_{g_j}(w)\rangle. \qquad (3)$$

We say that the family G generates the δ-R $(K; d+\ell)$ quantum hash function ψ_G and we call G a δ-R $(K; d+\ell)$ quantum hash generator, if ψ_G is a δ-R $(K; d+\ell)$ quantum hash function.

According to Definition 2 the family $F_E = \{E_1, \ldots, E_n\}$ from Section 3.1 is a δ-R $(2^k; \log n + 1)$ quantum hash generator and the family $H_{\delta,q}$ from Section 3.2 is δ-R $(q; t + 1)$ quantum hash generator.

4   Quantum Hashing via Classical ϵ-Universal Hashing Constructions

In this section we present a construction of a quantum hash generator based on the composition of an ϵ-universal hash family with a given quantum hash generator. We begin with the definitions and notation that we use in the rest of the paper.

Let $K = |\mathbb{X}|$, $M = |\mathbb{Y}|$. Let $F = \{f_1, \ldots, f_N\}$ be a family of functions, where

$$f_i : \mathbb{X} \to \mathbb{Y}.$$

Let q be a prime power and $\mathbb{F}_q$ be a field. Let $H = \{h_1, \ldots, h_T\}$ be a family of functions, where

$$h_j : \mathbb{Y} \to \mathbb{F}_q.$$

For $f \in F$ and $h \in H_B$, define composition $g = f \circ h$,

$$g : \mathbb{X} \to \mathbb{F}_q,$$

by the rule

$$g(w) = (f \circ h)(w) = h(f(w)).$$

Define composition $G = F \circ H$ of two families F and H as follows.

$$G = \{g_{ij} = f_i \circ h_j : i \in I, j \in J\},$$

where $I = \{1, \ldots, N\}$, $J = \{1, \ldots, T\}$.

Theorem 3. *Let $F = \{f_1, \ldots, f_N\}$ be an ϵ-U $(N; K, M)$ hash family. Let $\ell \geq 1$. Let $H = \{h_1, \ldots h_T\}$ be a δ-R $(M; \log T + \ell)$ quantum hash generator. Let $\log K > \log N + \log T + \ell$.*

Then the composition $G = F \circ H$ is an Δ-R $(K; s)$ quantum hash generator, where

$$s = \log N + \log T + \ell \tag{4}$$

and

$$\Delta \leq \epsilon + \delta. \tag{5}$$

The proof of Theorem 3 is presented in the next section.

4.1   Proof of Theorem 3

The δ-R $(M; \log T + \ell)$ quantum hash generator H generates the δ-R $(M; \log T + \ell)$ quantum hash function

$$\psi_H : v \mapsto \frac{1}{\sqrt{T}} \sum_{j \in J} |j\rangle |\psi_{h_j}(v)\rangle. \tag{6}$$

For $s = \log N + \log T + \ell$, using the family G, define the map

$$\psi_G : \mathbb{X} \to (\mathcal{H}^2)^{\otimes s}$$

by the rule

$$|\psi_G(w)\rangle = \frac{1}{\sqrt{N}} \sum_{i \in I} |i\rangle \otimes |\psi_H(f_i(w))\rangle. \tag{7}$$

We show the Δ resistance of ψ_G.

Consider a pair w, w' of different elements from $\mathbb{X}$ and their inner product $\langle \psi_G(w) | \psi_G(w') \rangle$. Using the linearity of the inner product we have that

$$\langle \psi_G(w) | \psi_G(w') \rangle = \frac{1}{N} \sum_{i \in I} \langle \psi_H(f_i(w)) | \psi_H(f_i(w')) \rangle.$$

We define two sets of indexes I_{bad} and I_{good}:

$$I_{bad} = \{ i \in I : f_i(w) = f_i(w') \}, \quad I_{good} = \{ i \in I : f_i(w) \neq f_i(w') \}.$$

Then we have

$$
\begin{aligned}
|\langle \psi_G(w) | \psi_G(w') \rangle| &\leq \frac{1}{N} \sum_{i \in I_{bad}} |\langle \psi_H(f_i(w)) | \psi_H(f_i(w')) \rangle| \\
&+ \frac{1}{N} \sum_{i \in I_{good}} |\langle \psi_H(f_i(w)) | \psi_H(f_i(w')) \rangle|.
\end{aligned}
\tag{8}
$$

The hash family F is ϵ-universal, hence

$$|I_{bad}| \leq \epsilon N.$$

The quantum function $\psi_H : \mathbb{Y} \to (\mathcal{H}^2)^{\log T + \ell}$ is δ-resistant, hence for an arbitrary pair v, v' of different elements from $\mathbb{Y}$ one has

$$|\langle \psi_H(v) | \psi_H(v') \rangle| \leq \delta.$$

Finally from (8) and the above two inequalities we have that

$$|\langle \psi_G(w) | \psi_G(w') \rangle| \leq \epsilon + \frac{|I_{good}|}{N} \delta \leq \epsilon + \delta.$$

The last inequality proves Δ-resistance of $\psi_G(w)$ (say for $\Delta = \epsilon + \delta(|I_{good}|)/N$) and proves the inequality (5).

To finish the proof of the theorem it remains to show that the function ψ_G can be presented in the form displayed in (3). From (6) and (7) we have that

$$|\psi_G(w)\rangle = \frac{1}{\sqrt{N}} \sum_{i \in I} |i\rangle \otimes \left(\frac{1}{\sqrt{T}} \sum_{j \in J} |j\rangle | \psi_{h_j}(f_i(w)) \rangle \right).$$

Using the notation from (2) the above expression can be presented in the following form (3).

$$|\psi_G(w)\rangle = \frac{1}{\sqrt{NT}} \sum_{i \in I, j \in J} |ij\rangle | \psi_{g_{ij}}(w) \rangle,$$

here $|ij\rangle$ denotes a basis quantum state, where ij is treated as a concatenation of the binary representations of i and j.

5 Constructions of Quantum Hashing Based on Classical Universal Hashing

The following statement is a corollary of Theorem 3 and a basis for explicit constructions of quantum hash functions in this section. Let q be a prime power and $\mathbb{F}_q$ be a field. Let $\delta \in (0,1)$. Let $H_{\delta,q}$ be the family of functions from Theorem 2. Let $|\mathbb{X}| = K$.

Theorem 4. *Let $F = \{f_1, \ldots, f_N\}$ be an ϵ-U $(N; K, q)$ hash family, where $f_i : \mathbb{X} \to \mathbb{F}_q$. Then for arbitrary $\delta > 0$, family $G = F \circ H_{\delta,q}$ is a Δ-R $(K; s)$ quantum hash generator, where*

$$s \leq \log N + \log \log q + 2 \log 1/\delta + 3$$

and

$$\Delta \leq \epsilon + \delta.$$

Proof. The family $H_{\delta,q} = \{h_1, \ldots, h_T\}$, where $h_i : \mathbb{F}_q \to \mathbb{F}_q$, $T = \lceil (2/\delta^2) \ln(2q) \rceil$, $\ell = 1$, and $s = \log T + 1 \leq \log n + \log \log q + 2 \log 1/\delta + 3$ is δ-R $(q; s)$ quantum hash generator. □

The next section presents an example of quantum hash functions based on Freivalds' fingerprinting construction.

5.1 Quantum Hashing Based on Freivalds' Fingerprinting

For a fixed positive constant k let $\mathbb{X} = \{0,1\}^k$. Let $c > 1$ be a positive integer and let $M = ck \ln k$. Let $\mathbb{Y} = \{0, 1, \ldots, M-1\}$.

For the i-th prime $p_i \in \mathbb{Y}$ define a function (fingerprint)

$$f_i : \mathbb{X} \to \mathbb{Y}$$

by the rule

$$f_i(w) = w \pmod{p_i}.$$

Here we treat a word $w = w_0 w_1 \ldots w_{k-1}$ also as an integer $w = w_0 + w_1 2 + \cdots + w_{k-1} 2^{k-1}$. Consider the set

$$F_M = \{f_1, \ldots, f_{\pi(M)}\}$$

of fingerprints. Here $\pi(M)$ denotes the number of primes less than or equal to M. Note that then $\pi(M) \sim M/\ln M$ as $M \to \infty$. Moreover,

$$\frac{M}{\ln M} \leq \pi(M) \leq 1.26 \frac{M}{\ln M} \qquad \text{for} \quad M \geq 17.$$

The following fact is based on a construction, "Freivalds' fingerprinting method", due to Freivalds [10].

Property 1. The set F_M of fingerprints is a $(1/c)$-U $(\pi(M); 2^k, M)$ hash family.

Proof (sketch). For any pair w, w' of distinct words from $\{0,1\}^k$ the number $N(w, w') = |\{f_i \in F_M : f_i(w) = f_i(w')\}|$ is bounded from above by k. Thus, if we pick a prime p_i (uniformly at random) from $\mathbb{Y}$ then

$$Pr[f_i(w) = f_i(w')] \leq \frac{k}{\pi(M)} \leq \frac{k \ln M}{M}.$$

Picking $M = ck \ln k$ for a constant c gives $Pr[f_i(w) = f_i(w')] \leq \frac{1}{c} + o(1)$.   □

Theorem 4 and Property 1 provide the following statement.

Theorem 5. *Let $c > 1$ be a positive integer and let $M = ck \ln k$. Let $q \in \{M, \ldots, 2M\}$ be a prime. Then, for arbitrary $\delta > 0$, family $G = F_M \circ H_{\delta, q}$ is a Δ-R $(2^k; s)$ quantum hash generator, where*

$$s \leq \log ck + \log \log k + \log \log q + 2 \log 1/\delta + 3$$

and

$$\Delta \leq \frac{1}{c} + \delta.$$

Proof. From Theorem 4 we have that

$$s \leq \log \pi(M) + \log \log q + 2 \log 1/\delta + 3.$$

From the choice of c above we have that $M = ck \ln k$. Thus

$$s \leq \log ck + \log \log k + \log \log q + 2 \log 1/\delta + 3.$$

□

Remark 1. Note that from Theorem 1 we have

$$s \geq \log k + \log \log q - \log \log \left(1 + \sqrt{2/(1 - \delta)}\right) - 1.$$

This lower bound shows that the quantum hash function ψ_{F_M} is good enough in the sense of the number of qubits used for the construction.

Acknowledgement. We are grateful to Professor Helmut Jürgensen and anonymous referee for their valuable suggestions, that improved the presentation of the paper.

References

1. Ablayev, F., Vasiliev, A.: Algorithms for quantum branching programs based on fingerprinting. In: Proceedings Fifth Workshop on Developments in Computational Models–Computational Models From Nature, DCM 2009, Rhodes, Greece, vol. 9, pp. 1–11 (July 11, 2009)

2. Ablayev, F., Vasiliev, A.: Quantum Hashing, arXiv:1310.4922 (quant-ph) (2013)
3. Ablayev, F., Vasiliev, A.: Cryptographic quantum hashing. Laser Physics Letters 11(2), 025202 (2014)
4. Ablayev, F., Ablayev, M.: Quantum Hashing via Classical ϵ-universal Hashing Constructions, arXiv:1404.1503 (quant-ph) (2014)
5. Bierbrauer, J., Johansson, T., Kabatianskii, G.A., Smeets, B.J.M.: On Families of Hash Functions via Geometric Codes and Concatenation. In: Stinson, D.R. (ed.) Advances in Cryptology - CRYPTO 1993. LNCS, vol. 773, pp. 331–342. Springer, Heidelberg (1994)
6. Buhrman, H., Cleve, R., Watrous, J., de Wolf, R.: Quantum fingerprinting. Phys. Rev. Lett. 87, 167902 (2001)
7. Carter, J., Wegman, M.: Universal Classes of Hash Functions. J. Computer and System Sciences 18, 143–154 (1979)
8. Gavinsky, D., Ito, T.: Quantum fingerprints that keep secrets. Quantum Information & Computation 13(7-8), 583–606 (2013)
9. Gottesman, D., Chuang, I.: Quantum digital signatures, T echnical report (2001), http://arxiv.org/abs/quant-ph/0105032
10. Freivalds, R.: Probabilistic Machines Can Use Less Running Time. In: Proceedings of the IFIP Congress 1977, Toronto, Canada, vol. 1977, pp. 839–842. North-Holland (1977)
11. Montanaro, A., Osborne, T.: Quantum Boolean functions. Chicago Journal of Theoretical Computer Science 1, arXiv:0810.2435 (2010)
12. Razborov, A., Szemeredi, E., Wigderson, A.: Constructing small sets that are uniform in arithmetic progressions. Combinatorics, Probability & Computing 2, 513–518 (1993)
13. Stinson, D.R.: On the connections between universal ϵ-hashing, combinatorial designs and error-correcting codes. Congressus Numerantium 114, 7–27 (1996)
14. Stinson, D.R.: Universal hash families and the leftover hash lemma, and applications to cryptography and computing. Journal of Combinatorial Mathematics and Combinatorial Computing 42, 3–31 (2002)
15. Stinson, D.R.: Cryptography: Theory and Practice, 3rd edn. Discrete Mathematics and Its Applications. CRC Press (2005)
16. Wigderson, A.: Lectures on the Fusion Method and Derandomization. Technical Report SOCS-95. 2, School of Computer Science, McGill University (file/pub/tech-reports/library/reports/95/TR95.2.ps.gz at the anonymousftpsiteftp.cs.mcgill.ca

Very Narrow Quantum OBDDs and Width Hierarchies for Classical OBDDs

Farid Ablayev[1,*], Aida Gainutdinova[1,*], Kamil Khadiev[1,**], and Abuzer Yakaryılmaz[2,3,***]

[1] Kazan Federal University, Kazan, Russia
[2] University of Latvia, Faculty of Computing, Raina bulv. 19 Riga, LV-1586, Latvia
[3] National Laboratory for Scientific Computing, Petrópolis, RJ, 25651-075, Brazil
{fablayev,aida.ksu,kamilhadi}@gmail.com, abuzer@lncc.br

Abstract. In the paper we investigate a model for computing of Boolean functions – Ordered Binary Decision Diagrams (OBDDs), which is a restricted version of Branching Programs. We present several results on the comparative complexity for several variants of OBDD models.

- We present some results on the comparative complexity of classical and quantum OBDDs. We consider a partial function depending on a parameter k such that for any $k > 0$ this function is computed by an exact quantum OBDD of width 2, but any classical OBDD (deterministic or stable bounded-error probabilistic) needs width 2^{k+1}.
- We consider quantum and classical nondeterminism. We show that quantum nondeterminism can be more efficient than classical nondeterminism. In particular, an explicit function is presented which is computed by a quantum nondeterministic OBDD with constant width, but any classical nondeterministic OBDD for this function needs non-constant width.
- We also present new hierarchies on widths of deterministic and nondeterministic OBDDs. We focus both on small and large widths.

1 Introduction

Branching programs are one of the well known models of computation. These models have been shown useful in a variety of domains, such as hardware verification, model checking, and other applications (see for example the book by Wegener [20]). It is known that the class of Boolean functions computed by polynomial size branching programs coincides with the class of functions computed by non-uniform log-space machines. Moreover branching programs are a convenient model for considering their various (natural) restrictive variants and various complexity measures such as size (number of inner nodes), length, and width.

---

* Partially supported by RFBR Grant 14-01-06036.
** Partially supported by RFBR Grant 14-07-00557.
*** Work done in part while visiting the Kazan Federal University. Partially supported by CAPES, ERC Advanced Grant MQC, and FP7 FET project QALGO.

H. Jürgensen et al. (Eds.): DCFS 2014, LNCS 8614, pp. 53–64, 2014.

One important class of restrictive branching programs is that of oblivious read-once branching programs, also known in applied computer science as Ordered Binary Decision Diagrams (OBDD) [20]. Since the length of an OBDD is at most linear (in the length of the input), the main complexity measure is "width".

OBDDs can also be seen as nonuniform automata (see for example [2]). Different variants of OBDDs were considered, i.e. deterministic, nondeterministic, probabilistic, and quantum, and many results have been proved on the comparative power of deterministic, nondeterministic, and randomized OBDDs [20]. For example, Ablayev and Karpinski [6] presented the first function that is polynomially easy for randomized OBDDs and exponentially hard for deterministic and even nondeterministic OBDDs. More specifically, it was proven that the OBDD variants of coRP and NP are different.

In the last decade the quantum model of OBDD came into play [3],[14],[17]. It was proven that quantum OBDDs can be exponentially cheaper than classical ones and it was shown that this bound is tight [5].

In this paper we present the first results on the comparative complexity for classical and quantum OBDDs computing partial functions. Then, we focus on the width complexity of deterministic and nondeterministic OBDDs, which have been investigated in several papers (see for more information and citations [11], [12]). Here we present very strict hierarchies for the classes of Boolean functions computed by deterministic and nondeterministic OBDDs.

The paper is organized as follows. Section 2 contains the definitions and notation used in the paper. In Section 3, we compare classical and exact quantum OBDDs. We consider a partial function depending on a parameter k such that, for any $k > 0$, this function is computed by an exact quantum OBDD of width 2 but deterministic or bounded-error probabilistic OBDDs need width 2^{k+1}. Also it is easy to show that nondeterministic OBDDs need width $k + 1$. In Section 4, we consider quantum and classical nondeterminism. We show that quantum nondeterministic OBDDs can be more efficient than their classical counterparts. We present an explicit function which is computed by a quantum nondeterministic OBDD with constant width, but any classical nondeterministic OBDD needs non-constant width. Section 5 contains our results on hierarchies on the sublinear (5.1) and larger (5.2) widths of deterministic and nondeterministic OBDDs.

The proofs of lower bound results (Theorem 2 and Lemma 3) are based on pigeonhole principle. The lower bound of Theorem 4 uses the technique of Markov chains. For the full proofs see the full version of the paper [4].

2   Preliminaries

We refer to [20] for more information on branching programs. The main model investigated throughout the paper is OBDD (Ordered Binary Decision Diagram), a restricted version of branching programs.

In this paper we use the following notation for vectors. We use subscripts for enumerating the elements of vectors and strings and superscripts for enumerating vectors and strings. For a binary string ν, $\#_1(\nu)$ and $\#_0(\nu)$ are the numbers of

1's and 0's in ν, respectively. We denote $\#_0^k(\nu)$ and $\#_1^k(\nu)$ to be the numbers of 1's and 0's in the first k elements of string ν, respectively.

For a given $n > 0$, a probabilistic OBDD P_n with width d, defined on $\{0,1\}^n$, is a 4-tuple $P_n = (T, v^0, Accept, \pi)$, where

- $T = \{T_j : 1 \leq j \leq n\}$ such that $T_j = (A_j(0), A_j(1))$ is an ordered pair of (left) stochastic matrices representing the transitions, where, at the j-th step, $A_j(0)$ or $A_j(1)$, determined by the corresponding input bit, is applied.
- v^0 is a zero-one initial stochastic vector (initial state of P_n).
- $Accept \subseteq \{1, \ldots, d\}$ is the set of accepting nodes.
- π is a permutation of $\{1, \ldots, n\}$ defining the order of testing the input bits.

For any given input $\nu \in \{0,1\}^n$, the computation of P_n on ν can be traced by a stochastic vector which is initially v^0. In each step j, $1 \leq j \leq n$, the input bit $x_{\pi(j)}$ is tested and then the corresponding stochastic operator is applied:

$$v^j = A_j(x_{\pi(j)})v^{j-1},$$

where v^j represents the probability distribution vector of nodes after the j-th step, $1 \leq j \leq n$. The accepting probability of P_n on ν is

$$\sum_{i \in Accept} v_i^n.$$

We say that a function f is computed by P_n with bounded error if there exists an $\varepsilon \in (0, \frac{1}{2}]$ such that P_n accepts all inputs from $f^{-1}(1)$ with a probability at least $\frac{1}{2} + \varepsilon$ and P_n accepts all inputs from $f^{-1}(0)$ with a probability at most $\frac{1}{2} - \varepsilon$. We say that P_n computes f *exactly* if $\varepsilon = 1/2$.

A deterministic OBDD is a probabilistic OBDD restricted to use only 0-1 transition matrices. In other words, the system is always in a single node and, from each node, there is exactly one outgoing transition for each tested input bit.

A nondeterministic OBDD (NOBDD) can have the ability of making more than one outgoing transition for each tested input bit from each node and so the program can follow more than one computational path and if one of the path ends with an accepting node, then the input is accepted (rejected, otherwise).

- An OBDD is called *stable* if each transition set T_j is identical for each level.
- An OBDD is called ID (ID-OBDD) if the input bits are tested in the order $\pi = (1, 2, \ldots, n)$. If a *stable* ID-OBDD has a fixed width and transition rules for each n, then it can be considered as a realtime finite automaton.

Quantum computation is a generalization of classical computation [19]. Therefore, each quantum model can simulate its probabilistic counterparts. In some cases, on the other hand, the quantum models are defined in a restricted way, e.g., using only unitary operators during the computation followed by a single measurement at the end, and so they may not simulate their probabilistic counterparts. The literature on quantum automata contains many results of this kind such as [13,7,9]. A similar result was also given for OBDDs in [17], in which a function with a small size of deterministic OBDD was given but the quantum OBDD defined in a restricted way needs exponential size to solve this function.

Quantum OBDDs that are defined with the general quantum operators, i.e. superoperator [18,19,22], followed by a measurement on the computational basis at the end can simulate their classical counterpart with the same size and width. So we can always conclude that any quantum class contains its classical counterpart.

A quantum OBDD is the same as a probabilistic OBDD with the following modifications:

- The state set is represented by a d-dimensional Hilbert space over the field of complex numbers. The initial state is $|\psi\rangle_0 = |q_0\rangle$ where q_0 corresponds to the initial node.
- Instead of a stochastic matrix, we apply a unitary matrix in each step. That is, $T = \{T_j : 1 \leq j \leq n$ and $T_j = (U_j^0, U_j^1)\}$, where, at the j-th step, U_j^0 or U_j^1, determined by the corresponding input bit, is applied,
- At the end, we take a measurement on the computational basis.

The state of the system is updated as follows after the j-th step:

$$|\psi\rangle_j = U_j^{x_{\pi(j)}}(|\psi\rangle_{j-1}),$$

where $|\psi\rangle_{j-1}$ and $|\psi\rangle_j$ represent the state of the system after the $(j-1)$-th and j-th steps, respectively, where $1 \leq j \leq n$.

The accepting probability of the quantum program on ν is calculated from $|\psi\rangle_n = (z_1, \ldots, z_d)$ as

$$\sum_{i \in Accept} |z_i|^2.$$

3   Exact Quantum OBDDs

In [8], Ambainis and Yakaryılmaz defined a new family of unary promise problems: For any $k > 0$, $A^k = (A_{yes}^k, A_{no}^k)$ such that $A_{yes}^k = \{a^{(2i)2^k} : i \geq 0\}$ and $A_{no}^k = \{a^{(2i+1)2^k} : i \geq 0\}$. They showed that each member of this family (A^k) can be solved exactly by a 2-state realtime quantum finite automaton (QFA), but any exact probabilistic finite automaton (PFA) needs at least 2^{k+1} states. Recently, Rashid and Yakaryılmaz [16] showed that bounded-error realtime PFAs also need at least 2^{k+1} states for solving A^k.[1] Based on this promise problem, we define a partial function:

$$\texttt{PartialMOD}_n^k(\nu) = \begin{cases} 1 \text{ , if } \#_1(\nu) = 0 \pmod{2^{k+1}}, \\ 0 \text{ , if } \#_1(\nu) = 2^k \pmod{2^{k+1}}, \\ * \text{ , otherwise,} \end{cases}$$

---

[1] The same result is also proved for two-way nondeterministic finite automata by Geffert and Yakaryılmaz [10].

where the function is not defined for the inputs mapping to "*". We call the inputs where the function takes the value of 1 (0) as yes-instances (no-instances).

Theorem 1. *For any $k \geq 0$,* PartialMOD$_n^k$ *can be solved by a stable quantum ID-OBDD with width 2 exactly.*

The OBDD can be constructed in the same way as a QFA, which solves problem A^k [8].

We show that the width of deterministic or bounded-error stable probabilistic OBDDs that solve PartialMOD$_n^k$ cannot be less than 2^{k+1}.

Remark 1. Note that, the proof for deterministic OBDDs is not similar to the proof for automata because potentially the nonstability can give profit. Also this proof is different from proofs for total functions (for example, MOD_p) due to the existence of incomparable inputs. Note that, classical one-way communication complexity techniques also fail for partial functions (for example, it can be shown that the communication complexity of PartialMOD$_n^k$ is 1), and we need to use a more careful analysis in the proof.

A deterministic stable ID-OBDD with width 2^{k+1} for PartialMOD$_n^k$ can be easily constructed. We left open the case of bounded-error non-stable probabilistic OBDDs.

Theorem 2. *For any $k \geq 0$, there are infinitely many n such that any deterministic OBDD computing the partial function* PartialMOD$_n^k$ *has width at least 2^{k+1}.*

Proof. Let $\nu \in \{0,1\}^n, \nu = \sigma\gamma$. We call γ valid for σ if $\nu \in ($PartialMOD$_n^k)^{-1}(0) \cup ($PartialMOD$_n^k)^{-1}(1)$. We call two substrings σ' and σ'' comparable if for all γ it holds that γ is valid for σ' iff γ is valid for σ''. We call two substrings σ' and σ'' nonequivalent if they are comparable and there exists a valid substring γ such that PartialMOD$_n^k(\sigma'\gamma) \neq$ PartialMOD$_n^k(\sigma''\gamma)$.

Let P be a deterministic OBDD computing the partial function PartialMOD$_n^k$. Note that paths associated with nonequivalent strings must lead to different nodes. Otherwise, if σ and σ' are nonequivalent, there exists a valid string γ such that PartialMOD$_n^k(\sigma\gamma) \neq$ PartialMOD$_n^k(\sigma'\gamma)$ and computations on these inputs lead to the same final node.

Let $N = 2^k$ and $\Gamma = \{\gamma : \gamma \in \{0,1\}^{2N-1}, \gamma = 0 \cdots 01 \cdots 1\}$. We will naturally identify any string ν with the element $a = \#_1(\nu) \pmod{2N}$ of the additive group $\mathbb{Z}_{2N}$. We call two strings of the same length different if the numbers of ones modulo $2N$ in them are different. We denote by $\rho(\gamma^1, \gamma^2) = \gamma^1 - \gamma^2$ the distance between numbers γ^1, γ^2.

Let the width of P be $t < 2N$. At each step i ($i = 1, 2, \dots$) of the proof we will count the number of different strings, which lead to the same node (denote this node v_i). At the i-th step we consider the $(2N-1)i$-th level of P.

Let $i = 1$. By the pigeonhole principle there exist two different strings σ^1 and σ^2 from the set Γ such that the corresponding paths lead to the same node v_1

of the $(2N-1)$-th level of P. Note that $\rho(\sigma^1, \sigma^2) \neq N$, because in this case σ^1 and σ^2 are nonequivalent and cannot lead to the same node.

We will show by induction that in each step of the proof the number of different strings which lead to the same node increases.

Step 2. By the pigeonhole principle there exist two different strings γ^1 and γ^2 from the set Γ such that corresponding paths from the node v_1 lead to the same node v_2 of the $(2N-1)2$-th level of P. In this case, the strings $\sigma^1\gamma^1, \sigma^2\gamma^1, \sigma^1\gamma^2$, and $\sigma^2\gamma^2$ lead to the node v_2. Note that $\rho(\gamma^1, \gamma^2) \neq N$, because in this case $\sigma^1\gamma^1$ and $\sigma^1\gamma^2$ are nonequivalent and cannot lead to the same node.

Adding the same number does not change the distance between the numbers, so we have

$$\rho(\sigma^1 + \gamma^1, \sigma^2 + \gamma^1) = \rho(\sigma^1, \sigma^2)$$

and

$$\rho(\sigma^1 + \gamma^2, \sigma^2 + \gamma^2) = \rho(\sigma^1, \sigma^2).$$

Let $\gamma^2 > \gamma^1$. Denote $\Delta = \gamma^2 - \gamma^1$. Let us count the number of different numbers among $\sigma^1 + \gamma^1$, $\sigma^2 + \gamma^1$, $\sigma^1 + \gamma^1 + \Delta$, and $\sigma^2 + \gamma^1 + \Delta$. Because σ^1 and σ^2 are different and $\rho(\sigma^1, \sigma^2) \neq N$, the numbers from the pair $\sigma^1 + \gamma^1$, and $\sigma^2 + \gamma^1$ coincide with corresponding numbers from the pair $\sigma^1 + \gamma^1 + \Delta$ and $\sigma^2 + \gamma^1 + \Delta$ iff $\Delta = 0 \pmod{2N}$. But $\Delta \neq 0 \pmod{2N}$ since the numbers γ^1 and γ^2 are different and $\gamma^1, \gamma^2 < 2N$. The numbers $\sigma^1 + \gamma^1 + \Delta$ and $\sigma^2 + \gamma^1 + \Delta$ cannot be a permutation of numbers $\sigma^1 + \gamma^1$ and $\sigma^2 + \gamma^1$ since $\rho(\gamma^1, \gamma^2) \neq N$ and $\rho(\sigma^1, \sigma^2) \neq N$. In this case, at least 3 numbers from $\sigma^1 + \gamma^1$, $\sigma^2 + \gamma^1$, $\sigma^1 + \gamma^2$, and $\sigma^2 + \gamma^2$ are different.

Step of induction. Let the numbers $\sigma^1, \ldots, \sigma^i$ be different on the step $i-1$ and the corresponding paths lead to the same node v_{i-1} of the $(2N-1)(i-1)$-th level of P.

By the pigeonhole principle there exist two different strings γ^1 and γ^2 from the set Γ such that the corresponding paths from the node v_{i-1} lead to the same node v_i of the $(2N-1)i$-th level of P. So the paths $\sigma^1\gamma^1, \ldots, \sigma^i\gamma^1, \sigma^1\gamma^2, \ldots, \sigma^i\gamma^2$ lead to the same node v_i. Let us estimate the number of different strings among them. Note that $\rho(\gamma^1, \gamma^2) \neq N$, because in this case the strings $\sigma^1\gamma^1$ and $\sigma^1\gamma^2$ are nonequivalent but lead to the same node.

The numbers $\sigma^1, \ldots, \sigma^i$ are different and $\rho(\sigma^l, \sigma^j) \neq N$ for each pair (l, j) such that $l \neq j$. Let $\sigma^1 < \cdots < \sigma^i$. We will show that among $\sigma^1 + \gamma^1, \ldots, \sigma^i + \gamma^1$ and $\sigma^1 + \gamma^1 + \Delta, \ldots, \sigma^i + \gamma^1 + \Delta$ at least $i+1$ numbers are different.

The sequence of numbers $\sigma^1 + \gamma^1, \ldots, \sigma^i + \gamma^1$ coincides with the sequence $\sigma^1 + \gamma^1 + \Delta, \ldots, \sigma^i + \gamma^1 + \Delta$ iff $\Delta = 0 \pmod{2N}$. But $\Delta \neq 0 \pmod{2N}$ since γ^1 and γ^2 are different and $\gamma^1, \gamma^2 < 2N$.

Suppose that the sequence $\sigma^1 + \gamma^1 + \Delta, \ldots, \sigma^i + \gamma^1 + \Delta$ is a permutation of the sequence $\sigma^1 + \gamma^1, \ldots, \sigma^i + \gamma^1$. In this case, we have numbers $a_0, \ldots, a_r$ from $\mathbb{Z}_{2N}$ such that all a_j are from the sequence $\sigma^1 + \gamma^1, \ldots, \sigma^i + \gamma^1$, $a_0 = a_r = \sigma^1 + \gamma^1$, and $a_j = a_{j-1} + \Delta$, where $j = 1, \ldots, r$. In this case, $r\Delta = 2Nm$. Because $N = 2^k$, $\Delta < 2N$, and $\Delta \neq N$ we have that r is even. For $z = r/2$ we have $z\Delta = Nm$. Since all numbers from $\sigma^1 + \gamma^1, \ldots, \sigma^i + \gamma^1$ are different, we have that $\rho(a_0, a_z) = N$. So we have that a_0 and a_z are nonequivalent, but the corresponding strings lead to the same node v_i. So after the i-th step, we have that at least $i + 1$ different strings lead to the same node v_i.

On the N-th step, we have that $N + 1$ different strings lead to the same node v_N. Among these strings, there must be at least two nonequivalent strings. Thus we can conclude that P cannot compute the function $\mathtt{PartialMOD_n^k}$ correctly. $\square$

Theorem 3. *For any $k \geq 0$, there are infinitly many n such that any nondeterministic OBDD computing the partial function $\mathtt{PartialMOD_n^k}$ has width at least $k + 1$.*

The proof is based on the well-known correspondence between nondeterministic and deterministic space complexity. That is, if the Boolean function $f(X)$ is computed by an NOBDD P of width d, then there is a deterministic OBDD P' computing f which has width 2^d. More precise lower bound for width of an NOBDD computing $\mathtt{PartialMOD_n^k}$ is presented in the full version of the paper [4].

Theorem 4. *For any $k \geq 0$, there are infinitely many n such that any stable probabilistic OBDD computing $\mathtt{PartialMOD_n^k}$ with bounded error has width at least 2^{k+1}.*

The proof of the Theorem is based on the technique of Markov chains and the details are given in [4].

4   Nondeterministic Quantum and Classical OBDDs

In [21], Yakaryılmaz and Say showed that nondeterministic QFAs define a superset of regular languages, called exclusive stochastic languages [15]. This class contains the *complements* of some interesting languages: $PAL = \{w \in \{0,1\}^* : w = w^r\}$, where w^r is the reverse of w, $O = \{w \in \{0,1\}^* : \#_1(w) = \#_0(w)\}$, $SQUARE = \{w \in \{0,1\}^* : \#_1(w) = (\#_0(w))^2\}$, and $POWER = \{w \in \{0,1\}^* : \#_1(w) = 2^{\#_0(w)}\}$.

Based on these languages, we define three symmetric functions for any input $\nu \in \{0,1\}^n$:

$$\mathtt{NotO_n}(\nu) = \begin{cases} 0 \text{ , if } \#_0(\nu) = \#_1(\nu) \\ 1 \text{ , otherwise} \end{cases},$$

$$\mathtt{NotSQUARE_n}(\nu) = \begin{cases} 0 \text{ , if } (\#_0(\nu))^2 = \#_1(\nu) \\ 1 \text{ , otherwise} \end{cases},$$

$$\mathtt{NotPOWER_n}(\nu) = \begin{cases} 0 \text{ , if } 2^{\#_0(\nu)} = \#_1(\nu) \\ 1 \text{ , otherwise} \end{cases}.$$

Theorem 5. *The Boolean functions $\mathtt{NotO_n}$, $\mathtt{NotSQUARE_n}$, and $\mathtt{NotPOWER_n}$ can be computed by a nondeterministic quantum OBDD with constant width.*

For each of these three functions, we can define a nondeterministic quantum (stable ID-) OBDD with constant width based on nondeterministic QFAs for the languages O, $SQUARE$, and $POWER$, respectively [21].

The complements of $PAL, O, SQUARE$ and $POWER$ cannot be recognized by classical nondeterministic finite automata. But, for example, the function version of the complement of PAL, $\mathtt{NotPAL_n}$, which returns 1 only for the non-palindrome inputs, is quite easy since it can be computed by a deterministic OBDD with width 3. Note that the order of such an OBDD is not the natural $(1, \ldots, n)$. However, as will be shown soon, this is not the case for the function versions of the complements of the other three languages.

Theorem 6. *There are infinitely many n such that any NOBDD P_n computing* $\mathtt{NotO_n}$ *has width at least* $\lfloor \log n \rfloor - 1$.

The proof of the theorem is based on the complexity properties of the Boolean function $\mathtt{NotO_n}$. At first we will discuss complexity properties of this function in Lemma 1. After that we will prove claim of the theorem.

Lemma 1. *There are infinitely many n such that any OBDD computing* $\mathtt{NotO_n}$ *has width at least $n/2 + 1$. (For the proof see [4]).*

Proof of Theorem 6. Let function $\mathtt{NotO_n}$ be computed by $NOBDD$ P_n of width d, then in the same way as in the proof of Theorem 3 we have $d \geq \log(n/2 + 1) > \log n - 1$.   □

In the same way we can show that there are infinitely many n such that any NOBDD P_n computing the function $\mathtt{NotSQUARE_n}$ has width at least $\Omega(\log(n))$ and any NOBDD P'_n computing the function $\mathtt{NotPOWER_n}$ has width at least $\Omega(\log \log(n))$.

5   Hierarchies for Deterministic and Nondeterministic OBDDs

We denote $\mathsf{OBDD^d}$ and $\mathsf{NOBDD^d}$ to be the sets of Boolean functions that can be computed by $OBDDs$ and $NOBDDs$ of width $d = d(n)$, respectively, where n is the number of variables. In this section, we present some width hierarchies for $\mathsf{OBDD^d}$ and $\mathsf{NOBDD^d}$. Moreover, we discuss relations between these classes. We consider $\mathsf{OBDD^d}$ and $\mathsf{NOBDD^d}$ with small (sublinear) widths and large widths.

5.1   Hierarchies and Relations for Small Width OBDDs

We have the following width hierarchy for the deterministic and nondeterministic models.

Theorem 7. *For any integer n, $d = d(n)$, and $1 < d \leq n/2$, we have*

$$\mathsf{OBDD^{d-1}} \subsetneq \mathsf{OBDD^d} \ \textit{and} \tag{1}$$
$$\mathsf{NOBDD^{d-1}} \subsetneq \mathsf{NOBDD^d}. \tag{2}$$

Proof of Theorem 7. It is obvious that $\mathsf{OBDD}^{d-1} \subseteq \mathsf{OBDD}^d$ and $\mathsf{NOBDD}^{d-1} \subseteq \mathsf{NOBDD}^d$. Let us show the inequalities of these classes. For this purpose we use the complexity properties of the Boolean function MOD_n^k.

Let k be a number such that $1 < k \leq n/2$. For any given input $\nu \in \{0,1\}^n$,

$$\mathsf{MOD}_n^k(\nu) = \begin{cases} 1, \text{ if } \#_1(\nu) = 0 \ (mod \ k), \\ 0, \text{ otherwise} \end{cases}.$$

Lemma 2. *There is an OBDD (and so a NOBDD) P_n of width d which computes the Boolean function MOD_n^k and $d = k$.*

Proof. At each level, P_n counts the number of 1's by modulo k. P_n answers 1 iff the number in the last step is zero. It is clear that the width of P_n is k. □

Lemma 3. *Any OBDD and NOBDD computing MOD_n^k has width at least k.*

Proof. The proof is based on the pigeonhole principle. Let P be a deterministic OBDD computing the function MOD_n^k. For each input ν from $(\mathsf{MOD}_n^k)^{-1}(1)$ there must be exactly one path in P leading from source node to accepting node. Let us consider k inputs $\{\nu^1, \nu^2, \dots, \nu^k\}$ from this set such that the last k bits in $\nu^j (j = 1, \dots, k)$ contain exactly j 1's and $(k-j)$ 0's. Let us consider the $(n-k)$-th level of P. The acceptance paths for different inputs from $\{\nu^1, \nu^2, \dots, \nu^k\}$ must pass trough different nodes of the $(n-k)$-th level of P. So the width of the $(n-k)$-th level of P is at least k.

The proof for the nondeterministic case is similar to the deterministic one. For each input from $(\mathsf{MOD}_n^k)^{-1}(1)$ for the function MOD_n^k there must be at least one path in P leading from the source node to an accepting node labelling this input. The accepting paths for different inputs from the set $\{\nu^1, \nu^2, \dots, \nu^k\}$ must go through different nodes of the $(n-k)$-th level of P. □

Boolean function $\mathsf{MOD}_n^d \in \mathsf{OBDD}^d$ and $\mathsf{MOD}_n^d \in \mathsf{NOBDD}^d$ due to Lemma 2 and Boolean function $\mathsf{MOD}_n^d \notin \mathsf{OBDD}^{d-1}$ and $\mathsf{MOD}_n^d \notin \mathsf{NOBDD}^{d-1}$ due to Lemma 3. This completes the proof of the Theorem 7. □

We have the following relationships between the deterministic and nondeterministic models.

Theorem 8. *For any integer n, $d = d(n)$, and $d' = d'(n)$ such that $d \leq n/2$ and $O(\log^2 d \log \log d) < d' \leq d - 1$, we have*

$$\mathsf{NOBDD}^{\lfloor \log(d) \rfloor} \subsetneq \mathsf{OBDD}^d \ and \tag{3}$$

$$\mathsf{OBDD}^d \ and \ \mathsf{NOBDD}^{d'} \ are \ not \ comparable. \tag{4}$$

Proof of Theorem 8. We start with (3). In the same way as in the proof of Theorem 3, we can show that $\mathsf{NOBDD}^{\lfloor \log(d) \rfloor} \subseteq \mathsf{OBDD}^d$ and, from Lemma 3, we know that $\mathsf{MOD}_n^d \notin \mathsf{NOBDD}^{\lfloor \log(d) \rfloor}$. Then we have $\mathsf{OBDD}^d \neq \mathsf{NOBDD}^{\lfloor \log(d) \rfloor}$.

We continue with (4). Let k be even and $1 < k \leq n$. For any given input $\nu \in \{0,1\}^n$,

$$\texttt{NotO}_n^k(\nu) = \begin{cases} 0, \text{ if } \#_0^k(\nu) = \#_1^k(\nu) = k/2, \\ 1, \text{ otherwise.} \end{cases}$$

Note that function $\texttt{NotO}_n^n$ is identical to $\texttt{NotO}_n$.

Lemma 4. *Any OBDD computing* $\texttt{NotO}_n^k$ *has width at least* $k/2 + 1$.

Proof. The proof can be obtained by the same technique as given in the proof of Lemma 1.   □

Lemma 5. *There is an NOBDD* P_n *of width* d *that computes the Boolean function* $\texttt{NotO}_n^k$ *and* $d \leq O(\log^2 k \log \log k)$. *(For the proof see [4]).*

Recall that $O(\log^2 d \log \log d) \leq d' \leq d - 1$, and, by Lemma 2 and Lemma 3, we have $\texttt{MOD}_n^d \in \texttt{OBDD}^d$ and $\texttt{MOD}_n^d \notin \texttt{NOBDD}^{d'}$; by Lemma 5, we have $\texttt{NotO}_n^{2d-1} \in \texttt{NOBDD}^{d'}$; and, by Lemma 4, we have $\texttt{NotO}_n^{2d-1} \notin \texttt{OBDD}^d$. Therefore, we cannot compare these classes and so we can conclude Theorem 8.   □

5.2 Hierarchies and Relations for Large Width OBDDs

In this section we consider OBDDs of large width. We obtain some hierarchies which are different from the ones in the previous section (Theorem 7).

Theorem 9. *For any integer* n, $d = d(n)$, $16 \leq d \leq 2^{n/4}$, *we have*

$$\texttt{OBDD}^{\lfloor d/8 \rfloor - 1} \subsetneq \texttt{OBDD}^d \text{ and} \tag{5}$$

$$\texttt{NOBDD}^{\lfloor d/8 \rfloor - 1} \subsetneq \texttt{NOBDD}^d, \tag{6}$$

Proof of Theorem 9. It is obvious that $\texttt{OBDD}^{\lfloor d/8 \rfloor - 1} \subseteq \texttt{OBDD}^d$ and $\texttt{NOBDD}^{\lfloor d/8 \rfloor - 1} \subseteq \texttt{NOBDD}^d$.

We define the Boolean function $\texttt{EQS}_n^k$ as a modification of the Boolean function *Shuffled Equality* which was defined in [6] and [1]. The proofs of the inequalities are based on the complexity properties of $\texttt{EQS}_n^k$.

Let k be a multiple of 4 such that $4 \leq k \leq 2^{n/4}$. The Boolean function $\texttt{EQS}_n$ depends only on the first k bits.

For any given input $\nu \in \{0,1\}^n$, we can define two binary strings $\alpha(\nu)$ and $\beta(\nu)$ in the following way. We call the odd bits of the input *marker bits* and the even bits *value bits*. For any i satisfying $1 \leq i \leq k/2$, the value bit ν_{2i} belongs to $\alpha(\nu)$ if the corresponding marker bit ν_{2i-1} is 0, and ν_{2i} belongs to $\beta(\nu)$ otherwise.

$$\texttt{EQS}_n^k(\nu) = \begin{cases} 1, \text{ if } \alpha(\nu) = \beta(\nu) \\ 0, \text{ otherwise} \end{cases}.$$

Lemma 6. *There is an OBDD* P_n *of width* $8 \cdot 2^{k/4} - 5$ *which computes the Boolean function* $\texttt{EQS}_n^k$. *(For the proof see [4]).*

Lemma 7. *There are infinitely many n such that any OBDD and NOBDD P_n computing EQS_n^k has width at least $2^{k/4}$. (For the proof see [4]).*

Boolean function $\text{EQS}_n^{4\lceil \log(d+5)\rceil - 12} \in \text{OBDD}^d$ and $\text{EQS}_n^{4\lceil \log(d+5)\rceil - 12} \in \text{NOBDD}^d$ due to Lemma 6.

Boolean function $\text{EQS}_n^{4\lceil \log(d+5)\rceil - 12} \notin \text{OBDD}^{\lfloor d/8\rfloor - 1}$ and $\text{EQS}_n^{4\lceil \log(d+5)\rceil - 12} \notin \text{NOBDD}^{\lfloor d/8\rfloor - 1}$ due to Lemma 7. So $\text{OBDD}^{\lfloor d/8\rfloor - 1} \neq \text{OBDD}^d$ and $\text{NOBDD}^{\lfloor d/8\rfloor - 1} \neq \text{NOBDD}^d$. These inequalities prove Statements (5) and (6) and complete the proof of Theorem 9. $\qquad\square$

In the following theorem, we present a relationship between the deterministic and nondeterministic models.

Theorem 10. *For any integer n, $d = d(n)$, and $d' = d'(n)$ satisfying $d \leq 2^{n/4}$ and $O(\log^4(d+1)\log\log(d+1)) < d' < d/8 - 1$, we have*

$$\text{NOBDD}^{\lfloor \log(d)\rfloor} \subsetneq \text{OBDD}^d \ and \tag{7}$$

$$\text{OBDD}^d \ and \ \text{NOBDD}^{d'} \ are \ not \ comparable. \tag{8}$$

Proof of Theorem 10. We start with (7). In the same way as in the proof of Theorem 3, we can show that $\text{NOBDD}^{\lfloor \log(d)\rfloor} \subseteq \text{OBDD}^d$. By Lemma 7, we have $\text{EQS}_n^{4\lceil \log(d+5)\rceil - 12} \notin \text{NOBDD}^{\lfloor \log(d)\rfloor}$ which means $\text{OBDD}^d \neq \text{NOBDD}^{\lfloor \log(d)\rfloor}$.

Now we continue with (8). We use the complexity properties of the Boolean function NotEQS_n^k, which is the negation of EQS_n^k.

Lemma 8. *There are infinitely many n such that any OBDD P_n computing NotEQS_n^k has width at least $2^{k/4}$.*

Proof. We can prove it in the same way as Lemma 7. $\qquad\square$

Lemma 9. *There is a NOBDD P_n of width d computing the Boolean function NotEQS_n^k where $d \leq O(k^4 \log k)$. (For the proof see [4]).*

Recall that $O(\log^4(d+1)\log\log(d+1)) \leq d' \leq \lfloor d/8\rfloor - 1$, and, by Lemma 7 and Lemma 6, we have $\text{EQS}_n^{4\lceil \log(d+5)\rceil - 12} \in \text{OBDD}^d$ and $\text{EQS}_n^{4\lceil \log(d+5)\rceil - 12} \notin \text{NOBDD}^{d'}$; by Lemma 9, we have $\text{NotEQS}_n^{4\lceil \log(d)\rceil + 4} \in \text{NOBDD}^{d'}$; and, by Lemma 8, we have $\text{NotEQS}_n^{4\lceil \log(d)\rceil + 4} \notin \text{OBDD}^d$. Therefore we cannot compare these classes and so we can conclude Theorem 10. $\qquad\square$

References

1. Ablayev, F.: Randomization and nondeterminsm are incomparable for ordered read-once branching programs. Electronic Colloquium on Computational Complexity (ECCC) 4(21) (1997)
2. Ablayev, F., Gainutdinova, A.: Complexity of quantum uniform and nonuniform automata. In: De Felice, C., Restivo, A. (eds.) DLT 2005. LNCS, vol. 3572, pp. 78–87. Springer, Heidelberg (2005)

3. Ablayev, F., Gainutdinova, A., Karpinski, M.: On computational power of quantum branching programs. In: Freivalds, R. (ed.) FCT 2001. LNCS, vol. 2138, pp. 59–70. Springer, Heidelberg (2001)

4. Ablayev, F., Gainutdinova, A., Khadiev, K., Yakaryılmaz, A.: Very narrow quantum OBDDs and width hierarchies for classical OBDDs. Technical Report arXiv:1405.7849, arXiv (2014)

5. Ablayev, F.M., Gainutdinova, A., Karpinski, M., Moore, C., Pollett, C.: On the computational power of probabilistic and quantum branching program. Information Computation 203(2), 145–162 (2005)

6. Ablayev, F.M., Karpinski, M.: On the power of randomized branching programs. In: Meyer auf der Heide, F., Monien, B. (eds.) ICALP 1996. LNCS, vol. 1099, pp. 348–356. Springer, Heidelberg (1996)

7. Ambainis, A., Freivalds, R.: 1-way quantum finite automata: strengths, weaknesses and generalizations. In: FOCS, pp. 332–341. IEEE Computer Society (1998), http://arxiv.org/abs/quant-ph/9802062

8. Ambainis, A., Yakaryılmaz, A.: Superiority of exact quantum automata for promise problems. Information Processing Letters 112(7), 289–291 (2012)

9. Bertoni, A., Carpentieri, M.: Analogies and differences between quantum and stochastic automata. Theoretical Computer Science 262(1-2), 69–81 (2001)

10. Geffert, V., Yakaryılmaz, A.: Classical automata on promise problems. In: DCFS 2014. LNCS, vol. 8614, pp. 125–136. Springer, Heidelberg (2014)

11. Hromkovič, J., Sauerhoff, M.: Tradeoffs between nondeterminism and complexity for communication protocols and branching programs. In: Reichel, H., Tison, S. (eds.) STACS 2000. LNCS, vol. 1770, pp. 145–156. Springer, Heidelberg (2000)

12. Hromkovic, J., Sauerhoff, M.: The power of nondeterminism and randomness for oblivious branching programs. Theory of Computing Systems 36(2), 159–182 (2003)

13. Kondacs, A., Watrous, J.: On the power of quantum finite state automata. In: FOCS, pp. 66–75. IEEE Computer Society (1997)

14. Nakanishi, M., Hamaguchi, K., Kashiwabara, T.: Ordered quantum branching programs are more powerful than ordered probabilistic branching programs under a bounded-width restriction. In: Du, D.-Z., Eades, P., Sharma, A.K., Lin, X., Estivill-Castro, V. (eds.) COCOON 2000. LNCS, vol. 1858, pp. 467–476. Springer, Heidelberg (2000)

15. Paz, A.: Introduction to Probabilistic Automata. Academic Press, New York (1971)

16. Rashid, J., Yakaryılmaz, A.: Implications of quantum automata for contextuality. In: Holzer, M., Kutrib, M. (eds.) CIAA 2014. LNCS, vol. 8587, pp. 318–331. Springer, Heidelberg (2014)

17. Sauerhoff, M., Sieling, D.: Quantum branching programs and space-bounded nonuniform quantum complexity. Theoretical Computer Science 334(1-3), 177–225 (2005)

18. Watrous, J.: On the complexity of simulating space-bounded quantum computations. Computational Complexity 12(1-2), 48–84 (2003)

19. Watrous, J.: Quantum computational complexity. In: Encyclopedia of Complexity and System Science. Springer arXiv:0804.3401 (2009)

20. Wegener, I.: Branching Programs and Binary Decision Diagrams. SIAM (2000)

21. Yakaryılmaz, A., Say, A.C.C.: Languages recognized by nondeterministic quantum finite automata. Quantum Information and Computation 10(9-10), 747–770 (2010)

22. Yakaryılmaz, A., Say, A.C.C.: Unbounded-error quantum computation with small space bounds. Information and Computation 279(6), 873–892 (2011)

Matter and Anti-Matter in Membrane Systems

Artiom Alhazov[1], Bogdan Aman[2], Rudolf Freund[3], and Gheorghe Păun[4]

[1] Institute of Mathematics and Computer Science, Academy of Sciences of Moldova,
Academiei 5, MD-2028, Chişinău, Moldova
artiom@math.md
[2] Institute of Computer Science, Romanian Academy, Iaşi, Romania
bogdan.aman@iit.academiaromana-is.ro
[3] Faculty of Informatics, Vienna University of Technology,
Favoritenstr. 9, 1040 Vienna, Austria
rudi@emcc.at
[4] Institute of Mathematics, Romanian Academy, Bucharest, Romania
gpaun@us.es

Abstract. The concept of a matter object being annihilated when meeting its corresponding anti-matter object is investigated in the context of membrane systems, i.e., of (distributed) multiset rewriting systems applying rules in the maximally parallel way. Computational completeness can be obtained with using only non-cooperative rules besides these matter/anti-matter annihilation rules if these annihilation rules have priority over the other rules. Without this priority condition, in addition catalytic rules with one single catalyst are needed to get computational completeness. Even deterministic systems are obtained in the accepting case. Universal P systems with a rather small number of rules – 57 for computing systems, 59 for generating and 53 for accepting systems – can be constructed when using non-cooperative rules together with matter/anti-matter annihilation rules having weak priority.

1 Introduction

Membrane systems (usually called *P systems*) are distributed multiset rewriting systems, where all objects – if possible – evolve in parallel in the membrane regions and may be communicated through the membranes. Membrane systems were introduced in [11] and since then have become an emerging research field. Overviews can be found in the monograph [12] and the handbook of membrane systems [13]; for actual news and results we refer to the P systems webpage [15]. Computational completeness (computing any partial recursive relation on non-negative integers) can be obtained with using cooperative rules or with catalytic rules (eventually) together with non-cooperative rules.

In this paper, we look for very weak forms of object interactions: for any object a *(matter)*, we consider its anti-object *(anti-matter)* a^- as well as the corresponding *annihilation rule* $aa^- \to \lambda$, which is assumed to exist in all membranes; this annihilation rule could be assumed to remove a pair a, a^- in zero time, but here we use these annihilation rules as special non-cooperative rules

H. Jürgensen et al. (Eds.): DCFS 2014, LNCS 8614, pp. 65–76, 2014.

having priority over all other rules in the sense of weak priority (e.g., see [1], i.e., other rules then also may be applied if objects cannot be bound by some annihilation rule any more). The idea of anti-matter has already been considered in another special variant of P systems with motivation coming from modeling neural activities, which are known as spiking neural P systems. For example, spiking neural P systems with anti-matter *(anti-spikes)* were already investigated in [10].

As expected (for example, compare with the Geffert normal forms, see [14]), the annihilation rules are rather powerful. Yet it is still surprising that using matter/anti-matter annihilation rules as the only non-cooperative rules, with the annihilation rules having priority, we already get computational completeness without using any catalyst; without giving the annihilation rules priority, we need one single catalyst. Even more surprising is the result that with priorities we obtain *deterministic* systems in the case of accepting P systems. Finally, we show how rather small universal P systems with anti-matter can be obtained based on the universal register machine U_{32} constructed by Korec, see [8].

2 Prerequisites

The set of integers is denoted by $\mathbb{Z}$, the set of non-negative integers by $\mathbb{N}$. Given an alphabet V, a finite non-empty set of abstract symbols, the free monoid generated by V under the operation of concatenation is denoted by V^*. The elements of V^* are called strings, the empty string is denoted by λ, and $V^*\backslash\{\lambda\}$ is denoted by V^+. For an arbitrary alphabet $\{a_1, \ldots, a_n\}$, the number of occurrences of a symbol a_i in a string x is denoted by $|x|_{a_i}$, while the length of a string x is denoted by $|x| = \Sigma_{a_i}|x|_{a_i}$. The Parikh vector associated with x with respect to $a_1, \ldots, a_n$ is $(|x|_{a_1}, \ldots, |x|_{a_n})$. The Parikh image of an arbitrary language L over $\{a_1, \ldots, a_n\}$ is the set of all Parikh vectors of strings in L, and is denoted by $Ps(L)$. For a family of languages FL, the family of Parikh images of languages in FL is denoted by $PsFL$, while for families of languages over a one-letter (d-letter) alphabet, the corresponding sets of non-negative integers are denoted by NFL (N^dFL).

A (finite) multiset over a (finite) alphabet $V = \{a_1, \ldots, a_n\}$, is a mapping $f : V \to \mathbb{N}$ and can be represented by $\langle a_1^{f(a_1)}, \ldots, a_n^{f(a_n)} \rangle$ or by any string x for which $(|x|_{a_1}, \ldots, |x|_{a_n}) = (f(a_1), \ldots, f(a_n))$. In the following we will not distinguish between a vector $(m_1, \ldots, m_n)$, a multiset $\langle a_1^{m_1}, \ldots, a_n^{m_n} \rangle$ or a string x having $(|x|_{a_1}, \ldots, |x|_{a_n}) = (m_1, \ldots, m_n)$.

The family of regular and recursively enumerable string languages is denoted by REG and RE, respectively. For more details of formal language theory the reader is referred to the monographs and handbooks in this area as [3] and [14].

Register Machines. A *register machine* is a tuple $M = (m, B, l_0, l_h, P)$, where m is the number of registers, B is a set of labels, $l_0 \in B$ is the initial label, $l_h \in B$ is the final label, and P is the set of instructions bijectively labeled by elements of B. The instructions of M can be of the following forms:

- $l_1 : (ADD(j), l_2, l_3)$, with $l_1 \in B \backslash \{l_h\}$, $l_2, l_3 \in B$, $1 \leq j \leq m$.
  Increases the value of register j by one, followed by a non-deterministic jump to instruction l_2 or l_3. This instruction is usually called *increment*.
- $l_1 : (SUB(j), l_2, l_3)$, with $l_1 \in B \backslash \{l_h\}$, $l_2, l_3 \in B$, $1 \leq j \leq m$.
  If the value of register j is zero then jump to instruction l_3; otherwise, the value of register j is decreased by one, followed by a jump to instruction l_2. The two cases of this instruction are usually called *zero-test* and *decrement*, respectively.
- $l_h : HALT$. Stops the execution of the register machine.

A *configuration* of a register machine is described by the contents of each register and by the value of the current label, which indicates the next instruction to be executed. Computations start by executing the instruction l_0 of P, and terminate with reaching the HALT-instruction l_h.

3   P Systems

The basic ingredients of a (cell-like) P system are the membrane structure, the multisets of objects placed in the membrane regions, and the evolution rules. The *membrane structure* is a hierarchical arrangement of membranes, in which the space between a membrane and the immediately inner membranes defines a *region/compartment*. The outermost membrane is called the *skin membrane*, the region outside is the *environment*. Each membrane can be labeled, and the label (from a set *Lab*) will identify both the membrane and its region; the skin membrane is identified by (the label) 0. The membrane structure can be represented by an expression of correctly nested labeled parentheses, and also by a rooted tree (with the label of a membrane in each node and the skin in the root). The *multisets of objects* are placed in the compartments of the membrane structure and usually represented by strings of the form $a_1^{m_1} \ldots a_n^{m_n}$.

The *evolution rules* are multiset rewriting rules of the form $u \to v$, where $u \in O^*$ and $v = (b_1, tar_1) \ldots (b_k, tar_k)$ with $b_i \in O^*$ and $tar_i \in \{here, out, in\}$. Using such a rule means "consuming" the objects of u and "producing" the objects from $b_1, \ldots, b_k$ of v, where the target *here* means that the objects remain in the same region where the rule is applied, *out* means that they are sent out of the respective membrane (in this way, objects can also be sent to the environment, when the rule is applied in the skin region), and *in* means that they are sent to one of the immediately inner membranes, chosen in a non-deterministic way; in general, the target indication *here* is omitted.

Formally, a (cell-like) *P system* is a construct

$$\Pi = (O, \mu, w_1, \ldots, w_m, R_1, \ldots, R_m, l_{in}, l_{out})$$

where O is the alphabet of objects, μ is the membrane structure (with m membranes), $w_1, \ldots, w_m$ are multisets of objects present in the m regions of μ at the beginning of a computation, $R_1, \ldots, R_m$ are finite sets of evolution rules, associated with the regions of μ, l_{in} is the label of the membrane region where the inputs are put at the beginning of a computation, and l_{out} indicates the region from which the outputs are taken; l_{out}/l_{in} being 0 indicates that the output/input is taken from the environment.

If a rule $u \to v$ has $|u| > 1$, then it is called *cooperative* (abbreviated *coo*); otherwise, it is called *non-cooperative* (abbreviated *ncoo*). In *catalytic P systems* non-cooperative as well as *catalytic* rules of the form $ca \to cv$ are used, where c is a *catalyst* – a special object that never evolves and never passes through a membrane, but it just assists object a to evolve to the multiset v. In a *purely catalytic P system* only catalytic rules are allowed. In both catalytic and purely catalytic P systems, in their description O is replaced by O, C in order to specify those objects from O that are the catalysts in the set C.

The evolution rules are used in the non-deterministic maximally parallel way, i.e., in any computation step of Π a multiset of rules is chosen from the sets $R_1, \ldots, R_m$ in such a way that no further rule can be added to it so that the obtained multiset would still be applicable to the existing objects in the membrane regions $1, \ldots, m$. A *configuration* of a system is given by the membranes and the objects present in the compartments of the system. Starting from a given *initial configuration* and applying evolution rules as described above, we get *transitions* among configurations; a sequence of transitions forms a computation. A computation is *halting* if it reaches a configuration where no rule can be applied any more.

In the *generative* case, a halting computation has associated a result, in the form of the number of objects present in membrane l_{out} in the halting configuration (l_{in} can be omitted). In the *accepting* case, for $l_{in} \neq 0$, we accept all (vectors of) non-negative integers whose input, given as the corresponding numbers of objects in membrane l_{in}, leads to a halting computation (l_{out} can be omitted). For the input being taken from the environment, i.e., for $l_{in} = 0$, we need an additional target indication *come*; $(a, come)$ means that the object a is taken into the skin from the environment (all objects there are assumed to be available in an unbounded number). The multiset of all objects taken from the environment during a halting computation then is the multiset accepted by this accepting P system, which in this case we shall call a *P automaton*, see [2]. The set of non-negative integers and the set of (Parikh) vectors of non-negative integers obtained as results of halting computations in Π in case α, $\alpha \in \{gen, acc, aut\}$, are denoted by $N_\alpha(\Pi)$ and $Ps_\alpha(\Pi)$, respectively. A P system Π can also be considered as a system *computing* a partial recursive function (in the deterministic case) or even a partial recursive relation (in the non-deterministic case), with the input being given in a membrane region $l_{in} \neq 0$ as in the accepting case or being taken from the environment as in the automaton case. The corresponding functions/relations computed by halting computations in Π are denoted by $ZY_\alpha(\Pi)$, $Z \in \{Fun, Rel\}$, $Y \in \{N, Ps\}$, $\alpha \in \{acc, aut\}$.

Computational completeness for (generating) catalytic P systems can be achieved when using two catalysts or with three catalysts in purely catalytic P systems, and the same number of catalysts is needed for P automata; in accepting P systems, the number of catalysts increases with the number of components in the vectors of natural numbers to be analyzed, see [4]. It is a long-time open problem how to characterize the families of sets of (vectors of) natural numbers generated by (purely) catalytic P systems with only one (two) catalysts. Using additional control mechanisms as, for example, priorities or promoters/inhibitors, P systems with only one (two) catalyst(s) can be shown to be computationally complete, e.g., see Chapter 4 in [13]. Last year several other variants of control mechanism have been shown to lead to computational completeness in (purely) catalytic P systems using only one (two) catalyst(s), e.g., see [6], and [7]. In this paper we are going to investigate the power of using matter/antimatter annihilation rules.

The family of sets $Y_\delta(\Pi)$, $Y \in \{N, Ps\}$, $\delta \in \{gen, acc, aut\}$, computed by P systems with at most m membranes and cooperative rules and with non-cooperative rules is denoted by $Y_\delta OP_m(coo)$ and $Y_\delta OP_m(ncoo)$, respectively. The family of sets $Y_\delta(\Pi)$, $Y \in \{N, Ps\}$, $\delta \in \{gen, acc, aut\}$, computed by (purely) catalytic P systems with at most m membranes and at most k catalysts is denoted by $Y_\delta OP_m(cat_k)$ ($Y_\delta OP_m(pcat_k)$). It is well known that, for any $m \geq 1$ and any $Y \in \{N, Ps\}$, $YREG = Y_{gen}OP_m(ncoo) \subset Y_{gen}OP_m(coo) = YRE$, as well as, for any $m \geq 1$, $d \geq 1$, $\delta \in \{gen, aut\}$,

$$Ps_{acc}OP_m(cat_{d+2}) = Ps_{acc}OP_m(pcat_{d+3}) = N^d RE;$$
$$Ps_\delta OP_m(cat_2) = Ps_\delta OP_m(pcat_3) = PsRE.$$

4 Using Matter and Anti-Matter

This concept to be used in (catalytic) P systems is a direct generalization of the idea of anti-spikes from spiking neural P systems (see [10]): for each object a we may introduce the anti-matter object a^-. We here look at these anti-matter objects a^- as objects of their own. Both objects and anti-objects are handled by usual evolution rules, but whenever related matter a and anti-matter a^- meet, they may annihilate each other by an application of the (non-context-free!) rule $aa^- \to \lambda$. If all these annihilation rules are given weak priority over all other rules, immediate annihilation is guaranteed.

We also consider catalytic P systems extended by allowing annihilation rules $aa^- \to \lambda$, with these rules having weak priority over all other rules, i.e., other rules can only be applied if no annihilation rule could still bind the corresponding objects. The families of sets $Y_\delta(\Pi)$, $Y \in \{N, Ps\}$, $\delta \in \{gen, acc, aut\}$, and the families of functions/relations $ZY_\alpha(\Pi)$, $Z \in \{Fun, Rel\}$, $\alpha \in \{acc, aut\}$, computed by such extended P systems with at most m membranes and k catalysts are denoted by $Y_\delta OP_m(cat(k), antim/pri)$ and $ZY_\alpha OP_m(cat(k), antim/pri)$; we omit $/pri$ for the families without priorities.

The matter/anti-matter annihilation rules, when having weak priority over the non-cooperative rules, are so powerful that we do not need catalysts at all:

Theorem 1. *For any $n \geq 1$, $k \geq 0$, $Y \in \{N, Ps\}$, $\delta \in \{gen, acc, aut\}$, $\alpha \in \{acc, aut\}$, and $Z \in \{Fun, Rel\}$,*

$$Y_\delta OP_n\left(cat(k), antim/pri\right) = Y_\delta OP_1\left(ncoo, antim/pri\right) \quad = YRE \text{ and}$$
$$ZY_\alpha OP_n\left(cat(k), antim/pri\right) = ZY_\alpha OP_1\left(ncoo, antim/pri\right) = ZYRE.$$

Proof. Let $M = (m, B, l_0, l_h, P)$ be a register machine. We now construct a one-membrane P system $\Pi = (O, [\ ]_1, l_0, R_1, l_{in}, 1)$ which simulates M. The contents of register r is represented by the number of copies of the object a_r, $1 \leq r \leq m$, and for each object a_r we also consider the corresponding anti-object a_r^-; in sum, we take $O = \{a_r, a_r^- \mid 1 \leq r \leq m\} \cup \{l, l' \mid l \in B\} \cup \{\#, \#^-\}$. The instructions of M are simulated by rules in R_1 as follows:

- $l_1 : (ADD\,(j), l_2, l_3)$, with $l_1 \in B \setminus \{l_h\}$, $l_2, l_3 \in B$, $1 \leq j \leq m$.
  Simulated by the rules $l_1 \rightarrow a_r l_2$ and $l_1 \rightarrow a_r l_3$.
  For $l_{in} = 0$, a_r is replaced by $(a_r, come)$ in case r is an input register.
- $l_1 : (SUB\,(r), l_2, l_3)$, with $l_1 \in B \setminus \{l_h\}$, $l_2, l_3 \in B$, $1 \leq r \leq m$.
  As rules common for all simulations of SUB-instructions, we have $a_r^- \rightarrow \#^-$, $a_r a_r^- \rightarrow \lambda$, $\#\#^- \rightarrow \lambda$, $\#^- \rightarrow \#\#$, and $\# \rightarrow \#\#$ (which last two rules lead the system into an infinite computation whenever a trap symbol is left without being annihilated).
  The *zero test* for instruction l_1 is simulated by the rules $l_1 \rightarrow l_1' a_r^-$ and $l_1' \rightarrow \#l_3$: the symbol $\#$ generated by the second rule $l_1' \rightarrow \#l_3$ can only be eliminated if the anti-matter a_r^- generated by the first rule $l_1 \rightarrow l_1' a_r^-$ is not annihilated by a_r, i.e., only if register r is empty.
  The *decrement case* for instruction l_1 is simulated by the rule $l_1 \rightarrow l_2 a_r^-$: the anti-matter a_r^- either correctly annihilates one matter a_r thus decrementing the register r or else traps an incorrect guess by forcing the symbol a_r^- to evolve to $\#^-$ and then to $\#\#$ in the next two steps in case register r is empty.
- $l_h : HALT$. Simulated by $l_h \rightarrow \lambda$.
  When the computation in M halts, the object l_h is removed, and no further rules can be applied provided the simulation has been carried out correctly, i.e., if no trap symbols $\#$ are present in this situation. The remaining objects in the system represent the result computed by M. □

Without this priority of the annihilation rules, the construction does not work, hence, a characterization of the families $Y_\delta OP_n\,(ncoo, antim)$ as well as $ZY_\alpha OP_n\,(ncoo, antim)$ remains as an open problem. Yet using catalytic rules with one catalyst again allows us to obtain computational completeness:

Theorem 2. *For any $n \geq 1$, $k \geq 1$, $Y \in \{N, Ps\}$, $\delta \in \{gen, acc, aut\}$, $\alpha \in \{acc, aut\}$, and $Z \in \{Fun, Rel\}$,*

$$Y_\delta OP_n\,(cat(k), antim) = YRE \quad and$$
$$ZY_\alpha OP_n\,(cat(k), antim) = ZYRE.$$

Proof. We again consider a register machine $M = (m, B, l_0, l_h, P)$ as in the previous proof, and construct the catalytic P system $\Pi = (O, \{c\}, [\ ]_1, cl_0, R_1, l_{in}, 1)$ with the single catalyst c and with $O = \{a_r, a_r^- \mid 1 \le r \le m\} \cup \{l, l', l'' \mid l \in B\} \cup \{\#, \#^-, d\}$. The results now are sent to the environment, in order not to have to count the catalyst in the skin membrane; for that purpose, we simply use the rule $a_i \to (a_i, out)$ for the output symbols a_i (we assume that output registers of M are only incremented).

For each ADD-instruction $l_1 : (ADD\,(j), l_2, l_3)$ in P, we again take the rules $l_1 \to a_r l_2$ and $l_1 \to a_r l_3$, and $l_h : HALT$ is simulated by $l_h \to \lambda$.

For each SUB-instruction $l_1 : (SUB\,(r), l_2, l_3)$, we now consider the four rules $l_1 \to l_2 a_r^-$, $l_1 \to l_1'' d a_r^-$, $l_1'' \to l_1'$, and $l_1' \to \#l_k$. As rules common for all SUB-instructions, we again add the matter/antimatter annihilation rules $a_r a_r^- \to \lambda$ and $\#\#^- \to \lambda$ as well as the trap rules $\# \to \#\#$ and $\#^- \to \#\#$, but in addition, also $d \to \#\#$ as well as the catalytic rules $cd \to c$ and $ca_r^- \to c\#^-$, $1 \le r \le m$. The decrement case is simulated as in the previous proof, by using the rule $l_1 \to l_2 a_r^-$ and then applying the annihilation rule $a_r a_r^- \to \lambda$. The zero-test now is initiated with the rule $l_i \to l_i'' d a_r^-$; the symbol d keeps the catalyst busy for one step with $cd \to c$, and only then, rule $ca_r^- \to c\#^-$ produces a symbol $\#^-$ which afterwards annihilates the symbol $\#$ generated by the rule $l_i' \to \#l_k$. □

In the accepting case, with priorities, we can even simulate the actions of a deterministic register machine in a deterministic way, i.e., for each configuration of the system, there can be at most one multiset of rules applicable to it.

Theorem 3. *For any $n \ge 1$ and $Y \in \{N, Ps\}$,*

$$Y_{dacc}OP_n\,(ncoo, antim/pri) = YRE \text{ and}$$
$$FunY_{dacc}OP_n\,(ncoo, antim/pri) = FunYRE.$$

Proof. We only show how the SUB-instructions of a register machine $M = (m, B', l_0, l_h, P)$ can be simulated in a deterministic way without introducing a trap symbol and therefore causing infinite loops by them:

Let $B = \{l \mid l : (SUB\,(r), l', l'') \in P\}$ and, for every register r,

$$\tilde{M}_r = \left\{\tilde{l} \mid l : (SUB\,(r), l', l'') \in P\right\}, \quad \hat{M}_r = \left\{\hat{l} \mid l : (SUB\,(r), l', l'') \in P\right\},$$
$$\tilde{M}_r^- = \left\{\tilde{l}^- \mid l : (SUB\,(r), l', l'') \in P\right\}, \hat{M}_r^- = \left\{\hat{l}^- \mid l : (SUB\,(r), l', l'') \in P\right\}.$$

Moreover, we take the rules $a_r^- \to \tilde{M}_r^- \hat{M}_r$ and the annihilation rules $a_r a_r^- \to \lambda$ for every register r as well as $\hat{l}\hat{l}^- \to \lambda$ and $\tilde{l}\tilde{l}^- \to \lambda$ for all $l \in B$. Then a SUB-instruction $l_1 : (SUB\,(r), l_2, l_3)$, with $l_1 \in B$, $l_2, l_3 \in B'$, $1 \le r \le m$, is simulated by the rules $l_1 \to \bar{l}_1 a_r^-$, $\bar{l}_1 \to \hat{l}_1^-(\tilde{M}_r \setminus \{\tilde{l}_1\})$, $\hat{l}_1^- \to l_2(\tilde{M}_r^- \setminus \{\tilde{l}_1^-\})$, and $\hat{l}_1^- \to l_3(\hat{M}_r^- \setminus \{\hat{l}_1^-\})$. The symbol $\hat{l}_1^-$ generated by the second rule is eliminated again and replaced by $\tilde{l}_1^-$ if a_r^- is not annihilated (which indicates that the register is empty). □

5   Small Universal P Systems with Anti-Matter

In [8], several variants of universal register machines were exploited. The main interesting variant for the results presented in this paper is shown in Figure 1.

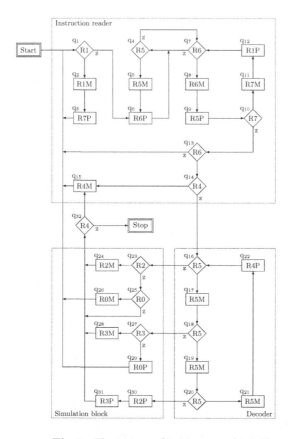

Fig. 1. The universal register machine U_{32}

In the diagram of the universal register machine U_{32} in Figure 1, the operations used on the registers are: the *zero-test* on register i is indicated by a rhomboid inclosing the encryption Ri, and in the case that the contents of register i is zero, the next operation is the one to be reached with the arc labeled by z; the *increment* operation is depicted by a rectangle with the encryption RiP, and the *decrement* operation by a rectangle with the encryption RiM (as the decrement operation RiM is always preceded by the corresponding zero-test, it can always be carried out). The states are depicted directly at the corresponding operations; q_1 is the initial state, and the state where the U_{32} stops is indicated by $STOP$ in Figure 1. A thorough analysis of the universal register machine U_{32} shows that when it halts not only register 2 as the output register, but also register 6 and register 1 (still containing the code of the register machine to be simulated) may be non-empty; for more details we refer to [5].

In order to produce better descriptional complexity results with respect to the number of rules than those we would immediately get when applying the constructions given in the proof of Theorem 1, we introduce a generalization of register machines or counter automata.

Generalized Counter Automata. For a register machine $M = (m, B, l_0, l_h, P)$ consider the more general type of instructions $i : (q, M_-, N, M_+, q')$, where $q, q' \in Q$ are states, $N \subseteq R$ is a set of registers, and M_-, M_+ are multisets of registers. Such a register machine applies instruction i as follows: first, multiset M_- is subtracted from the register values (i.e., for each register $j \in R$, $M_-(j)$ is subtracted from the contents of register j; if at least one resulting value would be negative, the machine is blocked without producing any result); second, the subset N of registers is checked to be zero (if at least one of them is found to be non-zero, the machine is blocked without producing any result); third, the multiset M_+ is added to the register values (i.e., for each register $j \in R$, $M_+(j)$ is added to the contents of register j), and finally the state changes to q'.

The work of such a register machine, now also called a *generalized counter automaton* and written $M = (m, B, l_0, q_h, P)$, consists of derivation steps applying instructions, chosen in a non-deterministic way, associated with the current state. The computation starts in the initial state q_0, and we say that it halts if the final state q_h has been reached (which replaces the condition of reaching the final HALT-instruction labeled by l_h).

Theorem 4. *There exist small universal P systems with non-cooperative rules and matter/anti-matter annihilation rules – with 9 annihilation rules and, in total, 53 rules in the accepting case, 59 rules in the generating case, and 57 rules in the computing case.*

Proof. We start with a slightly changed variant of the P system from Theorem 4 in [5] (obtained from the universal register machine U_{32} machine in [8]). This modified sequential antiport P system with forbidden contexts can be written with the instructions of a generalized counter machine as follows:

$$
\begin{array}{ll}
1 : (q_1, \langle 1 \rangle, \{\}, \langle 7 \rangle, q_1), & 10 : (q_{18}, \langle 5^3 \rangle, \{\}, \langle 4 \rangle, q_{18}), \\
2 : (q_1, \langle \rangle, \{1\}, \langle 6 \rangle, q_4), & 11 : (q_{18}, \langle \rangle, \{5,3\}, \langle 0 \rangle, q_1), \\
3 : (q_4, \langle 5 \rangle, \{\}, \langle 6 \rangle, q_4), & 12 : (q_{18}, \langle 5^2, 0 \rangle, \{5,2\}, \langle \rangle, q_1), \\
4 : (q_4, \langle 6 \rangle, \{5\}, \langle 5 \rangle, q_{10}), & 13 : (q_{18}, \langle 5^2, 2 \rangle, \{5\}, \langle \rangle, q_1), \\
5 : (q_{10}, \langle 7,6 \rangle, \{\}, \langle 1,5 \rangle, q_{10}), & 14 : (q_{18}, \langle 5^2 \rangle, \{5,2,0\}, \langle \rangle, q_1) \\
6 : (q_{10}, \langle 7 \rangle, \{6\}, \langle 1 \rangle, q_4), & 15 : (q_{18}, \langle 3 \rangle, \{5\}, \langle \rangle, q_{32}), \\
7 : (q_{10}, \langle \rangle, \{6,7\}, \langle \rangle, q_1), & 16 : (q_{18}, \langle 5 \rangle, \{5\}, \langle 2,3 \rangle, q_{32}), \\
8 : (q_{10}, \langle 6,4 \rangle, \{7\}, \langle \rangle, q_1), & 17 : (q_{32}, \langle 4 \rangle, \{\}, \langle \rangle, q_1), \\
9 : (q_{10}, \langle 6,5 \rangle, \{7,4\}, \langle \rangle, q_{18}), & 18 : (q_{32}, \langle \rangle, \{4\}, \langle \rangle, q_h).
\end{array}
$$

For a generalized counter automaton $M = (m, B, l_0, q_h, P)$, let

$$
k = 1 + \max_{i:(q, M_-, N, M_+, q') \in P} (|M_-|, |N|).
$$

We consider the following rules (common for different instructions of M):

$$\#^- \to \#^k, \# \to \#^k, \#\#^- \to \lambda, a_r \to \#^-, a_r a_r^- \to \lambda, \ r \in R.$$

Now we present the simulation of instruction $i : (q, M_-, N, M_+, q') \in P$. First we consider the case when M_- and N have no common elements, and moreover, we also assume that M_- does not overlap with M_+ (otherwise such an instruction can be split into two instructions; notice that this condition is already satisfied in the rules given above).

$$q \to l_i \prod_{r \in N} a_r^-, \ l_i \to q'(\prod_{r \in N} \#)(\prod_{r \in M_-} a_r^-) \prod_{r \in M_+} a_r.$$

Indeed, the zero-test is successful if *none* of the objects a_r^- generated in the first step annihilates with the corresponding register symbols a_r; they have to change into objects $\#^-$ to annihilate with the same number of objects $\#$ produced in the next step. The decrement is successful if *all* objects a_r^- generated in the second step annihilate with the corresponding register symbols a_r. If either decrement or zero-test fail, then at least either one $\#$ or one $\#^-$ will be produced without its annihilation partner, leading to producing objects $\#$ in a geometric progression, ensuring that such computations do not produce any result (notice that no objects $\#$ or $\#^-$ are produced in the first step of the simulation of any instruction).

If the zero-test set N is empty, then the first step is a simple renaming, and thus can be combined with the second step, yielding just one rule

$$q \to q'(\prod_{r \in M_-} a_r^-) \prod_{r \in M_+} a_r.$$

Clearly, if M_- and N overlap, such an instruction can be broken down into two subsequent instructions of the generalized counter automaton. However, a more efficient solution with only three rules exists:

$$q \to l_i \prod_{r \in M_-} a_r^-, \ l_i \to l_i' \prod_{r \in N} a_r^-, \ l_i' \to q(\prod_{r \in N} \#^-) \prod_{r \in M_+} a_r.$$

The generalized counter automaton obtained by rewriting the sequential antiport P system with inhibitors from [5] (with the modifications described above) has 18 instructions, out of which only 4 have overlaps between the decrement multiset and the zero-test set, and other 5 have empty zero-test sets. Hence, applying the constructions described above we get a universal P system with anti-matter having $(18 \times 2 + 4 - 5) + 8 + 2 + (8 + 1) = 54$ rules, i.e., 45 non-cooperative rules and 9 model-defined annihilation rules:

$$\Pi = (O, [\ ]_1, q_1, R_1, 1, 1) \text{ where}$$
$$O = \{l_2, l_4, l_6, l_7, l_8, l_9, l_{11}, l_{12}, l_{12}', l_{13}, l_{13}', l_{14}, l_{14}', l_{15}, l_{16}, l_{16}', l_{18}\}$$
$$\cup \{q_1, q_4, q_{10}, q_{18}, q_{32}, q_h\} \cup \{a, a^- \mid a \in \{a_j \mid 0 \le j \le 7\} \cup \{\#\}\}$$

and R_1 contains the following rules:

$$q_1 \to q_1 a_1{}^- a_7,$$
$$q_1 \to l_2 a_1{}^-, \qquad\qquad l_2 \to q_4 \# a_6,$$
$$q_4 \to q_4 a_5{}^- a_6,$$
$$q_4 \to l_4 a_5{}^-, \qquad\qquad l_4 \to q_{10} \# a_6{}^- a_5,$$
$$q_{10} \to q_{10} a_7{}^- a_6{}^- a_1 a_5,$$
$$q_{10} \to l_6 a_6{}^-, \qquad\qquad l_6 \to q_4 \# a_7{}^- a_1,$$
$$q_{10} \to l_7 a_6{}^- a_7{}^-, \qquad\qquad l_7 \to q_1 \#\#,$$
$$q_{10} \to l_8 a_7{}^-, \qquad\qquad l_8 \to q_1 \# a_6{}^- a_4{}^-,$$
$$q_{10} \to l_9 a_7{}^- a_4{}^-, \qquad\qquad l_9 \to q_{18} \#\# a_6{}^- a_5{}^-,$$
$$q_{18} \to q_{18} a_5{}^- a_5{}^- a_5{}^- a_4,$$
$$q_{18} \to l_{11} a_5{}^- a_3{}^-, \qquad\qquad l_{11} \to q_1 \#\# a_0,$$
$$q_{18} \to l_{12} a_5{}^- a_5{}^- a_0{}^-, \quad l_{12} \to l'_{12} a_5{}^- a_2{}^-, \quad l'_{12} \to q_1 \#\#,$$
$$q_{18} \to l_{13} a_5{}^- a_5{}^- a_2{}^-, \quad l_{13} \to l'_{13} a_5{}^-, \qquad l'_{13} \to q_1 \#,$$
$$q_{18} \to l_{14} a_5{}^- a_5{}^-, \qquad\quad l_{14} \to l'_{14} a_5{}^- a_2{}^- a_0{}^-, \; l'_{14} \to q_1 \#\#\#,$$
$$q_{18} \to l_{15} a_5{}^-, \qquad\qquad l_{15} \to q_{32} \# a_3{}^-,$$
$$q_{18} \to l_{16} a_5{}^-, \qquad\qquad l_{16} \to l'_{16} a_5{}^-, \qquad l'_{16} \to q_{32} \# a_2 a_3,$$
$$q_{32} \to q_1 a_4{}^-,$$
$$q_{32} \to l_{18} a_4{}^-, \qquad\qquad l_{18} \to q_h \#,$$
$$\#^- \to \#^4, \qquad\qquad \# \to \#^4, \qquad\qquad (\#\#^- \to \lambda),$$
$$a_r \to \#^-, \qquad\qquad (a_r a_r{}^- \to \lambda), \qquad 0 \le r \le 7.$$

As the rules with l_7 and l'_{12} on the left side have the same right side, we can replace l'_{12} by l_7, thus decreasing the number of non-cooperative rules down to 44. In sum, we finish with 53 rules in the *accepting case*. In the *computing case*, we have to "clean" registers 1 and 6 (see Section 5) and add the following four rules and the state q'_h:

$$q_h \to q_h a_1{}^-, \; q_h \to q_h a_6{}^-, \; q_h \to q'_h a_1{}^- a_6{}^-, \; q'_h \to \#\#.$$

The P system now halts with the skin membrane only containing copies of the symbol a_2 representing the output value. Finally, in the *generating case*, we start with the new initial state q_0 and add the two rules $q_0 \to a_2 q_0$ and $q_0 \to q_1$, which allows us to produce, in a non-deterministic way, an input for U_{32} simulating the identity function on the domain of the set to be generated by the P system. □

6   Conclusions

We have shown that only non-cooperative rules together with matter/anti-matter annihilation rules are needed to obtain computational completeness in P systems working in the maximally parallel derivation mode if annihilation rules have weak priority; without priorities, one catalyst is needed. In the case of accepting P systems we can even get deterministic systems.

There may be a lot of other interesting models of P systems allowing for introducing anti-matter objects and matter/anti-matter annihilation rules. Several problems remain open even for the models presented here, for example, can we avoid both catalysts and priorities. Moreover, the number of rules needed for universal P systems with anti-matter might still be reduced.

Acknowledgements. Artiom Alhazov acknowledges project STCU-5384 *Models of high performance computations based on biological and quantum approaches* awarded by the Science and Technology Center in the Ukraine.

References

1. Alhazov, A., Sburlan, D.: Static Sorting P Systems. In: Ciobanu, G., Păun, G., Pérez-Jiménez, M.J. (eds.) Applications of Membrane Computing. Natural Computing Series, pp. 215–252. Springer (2005)
2. Csuhaj-Varjú, E., Vaszil, G.: P Automata or Purely Communicating Accepting P Systems. In: Păun, G., Rozenberg, G., Salomaa, A., Zandron, C. (eds.) WMC-CdeA 2002. LNCS, vol. 2597, pp. 219–233. Springer, Heidelberg (2003)
3. Dassow, J., Păun, G.: Regulated Rewriting in Formal Language Theory. Springer (1989)
4. Freund, R., Kari, L., Oswald, M., Sosík, P.: Computationally Universal P Systems without Priorities: Two Catalysts Are Sufficient. Theoretical Computer Science 330, 251–266 (2005)
5. Freund, R., Oswald, M.: A Small Universal Antiport P System with Forbidden Context. In: Leung, H., Pighizzini, G. (eds.) 8th International Workshop on Descriptional Complexity of Formal Systems, DCFS 2006, June 21-23. Proceedings DCFS, pp. 259–266. New Mexico State University, Las Cruces (2006)
6. Freund, R., Oswald, M.: Catalytic and Purely Catalytic P Automata: Control Mechanisms for Obtaining Computational Completeness. In: Bensch, S., Drewes, F., Freund, R., Otto, F. (eds.) Fifth Workshop on Non-Classical Models of Automata and Applications (NCMA 2013), pp. 133–150. OCG, Wien (2013)
7. Freund, R.: Gh. Păun: How to Obtain Computational Completeness in P Systems with One Catalyst. In: Proceedings Machines, Computations and Universality, MCU 2013, Zürich, Switzerland, September 9-11. EPTCS, vol. 128, pp. 47–61 (2013)
8. Korec, I.: Small Universal Register Machines. Theoretical Computer Science 168, 267–301 (1996)
9. Minsky, M.L.: Computation: Finite and Infinite Machines. Prentice Hall, Englewood Cliffs (1967)
10. Pan, L., Păun, G.: Spiking Neural P Systems with Anti-Matter. International Journal of Computers, Communications & Control 4(3), 273–282 (2009)
11. Păun, G.: Computing with Membranes. Journal of Computer and System Sciences 61(1), 108–143 (2000); (Turku Center for Computer Science-TUCS Report 208 (November 1998), www.tucs.fi)
12. Păun, G.: Membrane Computing. An Introduction. Springer (2002)
13. Păun, G., Rozenberg, G., Salomaa, A. (eds.): The Oxford Handbook of Membrane Computing. Oxford University Press (2010)
14. Rozenberg, G., Salomaa, A. (eds.): Handbook of Formal Languages, vol. 3. Springer (1997)
15. The P Systems Website, www.ppage.psystems.eu

Complexity of Extended vs. Classic LR Parsers*

Angelo Borsotti, Luca Breveglieri, Stefano Crespi Reghizzi,
and Angelo Morzenti

Politecnico di Milano, *DEIB*, Piazza Leonardo da Vinci n. 32, I-20133, Milano, Italy
angelo.borsotti@mail.polimi.it,
{luca.breveglieri,stefano.crespireghizzi,angelo.morzenti}@polimi.it

Abstract. For the deterministic context-free languages, we compare the space and time complexity of their $LR(1)$ parsers, constructed in two different ways: the classic method by Knuth [7] for *BNF* grammars, and the recent one by the authors [2], which directly builds the parser from *EBNF* grammars represented as transition networks. For the *EBNF* grammars, the classic Knuth's method is indirect as it needs to convert them to *BNF*. We describe two parametric families of formal languages indexed by the number of stars, which exhibit a linear growth of the parser size (number of states) in passing from the classic to our novel direct methods. Experimental measurements of the number of parser states and of the parsing speed for two real languages (*Java* and *JSON*) confirm the advantage of the new direct parser model for *EBNF* grammars.

Keywords: Extended *BNF* grammar, *EBNF*, $LR(1)$, $ELR(1)$, transition network, *TN*, shift-reduce, bottom-up parser, parsing performance.

1   Introduction

The grammars in Extended Backus-Naur Form (*EBNF*), also known as grammars with regular right parts, are more readable than the pure context-free (*BNF*) ones and are widely used for language specifications. An *EBNF* grammar is a set of regular expressions (*RE*), one per nonterminal. Since each *RE* is recognizable by a deterministic finite-state automaton (*DFA*) or *machine*, the grammar can be viewed as a set of machines, called a *Transition Network* (*TN*), e.g., pioneered in [3]. A *TN* may contain cycles, which correspond to the (Kleene) stars in the *RE*'s, instead absent in the *DFA*'s associated to a *BNF* grammar.

Given an *EBNF* grammar, we recap the *classic method* for constructing the $LR(1)$ language parser, to be abstractly viewed as a *DPDA*. First, the grammar is transformed into an equivalent *BNF* one by converting every sub-*RE* into a set of left- (or right-) recursive grammar rules; for that, new nonterminals are needed. Second, the well known Knuth's algorithm [7] is used to construct the language parser. Provided *conflicts* (shift-reduce or reduce-reduce) do not occur, the resulting *PDA* is deterministic. It is straightforward to adjust the Knuth's construction to let it work on the acyclic *TN* that represents a *BNF* grammar.

---

* Partially supported by *PRIN* "Automi e Linguaggi Formali" and by *CNR - IEIIT*.

H. Jürgensen et al. (Eds.): DCFS 2014, LNCS 8614, pp. 77–89, 2014.

In contrast with the classic method, which is indirect, the new *direct* (or *ELR* (1)) *method* recently presented in [2] (more details are in [4]), constructs the graph of the *DPDA* directly from the *TN*. There, the new method is compared with former unsatisfactory solutions, a formal simple condition for the parser to be deterministic is stated and proved correct, but a quantitative analysis of the classic vs direct method is missing. Our present work optimizes the direct method of [2]. We add some static pre-computed information (so-called keys) to the *DPDA* graph, to increase the parsing performance by a limited overhead.

Clearly, the (optimized) direct method has the advantage that any source *EBNF* grammar or *TN* can be fed unchanged to the parser generator, whereas the grammar transformations used by the classic generators, e.g., *YACC* or *Bison*, are tedious and, what is more annoying, impose to maintain two different grammars: one for language specification, the other for compiler implementation.

However, such advantages of the direct method would not suffice to qualify it as a replacement of the classic method, which has a use record consolidated over 50 years, without a thorough realistic comparison of the size and performance of classic and direct parsers. Such a comparison is the subject of this paper: it is based on a formal analysis of the parser size for parametrized grammar families, and on experiments and measurements of both size and speed for real programming languages. To be convincing, the experimental comparison should focus on real languages rather than on small formal languages, but the latter help to understand the origin of size differences. Our benchmarks cover two important cases, the *Java* and *JSON* languages. To generate parsers, we used both our new parser generator *E-Bison* and the widespread *GNU Bison* tool.

To mention some results, we found the size of our new parser model is consistently smaller for a parametrized language family, while it is comparable in the experiments on real languages. Concerning the run-time performance of our model, we achieve a better execution time by anticipating some computations at parser generation-time and by storing them into the *DPDA* transition function.

There is little related work: in recent years, the research on deterministic parsers has been rare, and to find examples of complexity studies for parsing algorithms, one has to go back to when a limited memory capacity motivated the invention of the *LALR* (1) and similar methods, e.g., Pager [8], that use small parsing tables. In those years, also the size and speed of top-down *LL* (1) parsers and bottom-up *LR* (1) parsers were compared. Since it is easy to directly write *LL* (recursive descent) parsers for *EBNF* grammars, this has often been put forward as a major advantage of top-down over bottom-up parsers. Despite many attempts and suggestions to extend *LR* methods to *EBNF* grammars, no proposal has gained consensus. The recent critical survey [6] says:

> "It is a striking phenomenon that the ideas behind the recursive descent parsing of *ECFG*'s [i.e., *EBNF*] can be grasped and applied immediately, whereas most literature on the *LR*-like parsing of *RRPG*'s [i.e., *TN*] is very difficult to access. Given the developments in computing power and software engineering, and the practical importance of *ECFG*'s and *RRPG*'s, a uniform and coherent treatment of the subject seems in order."

Our present work shows that now a direct construction method for *ELR* parsers of *EBNF* grammars is available and competitive with the classic *LR* method.

Sect. 2 sets terminology and notation. Sect. 3 shows the unified construction of the classic and direct parsers. Sect. 4 shows the quantitative analysis of simple language families, reports experiments and discusses findings. Sect. 5 concludes.

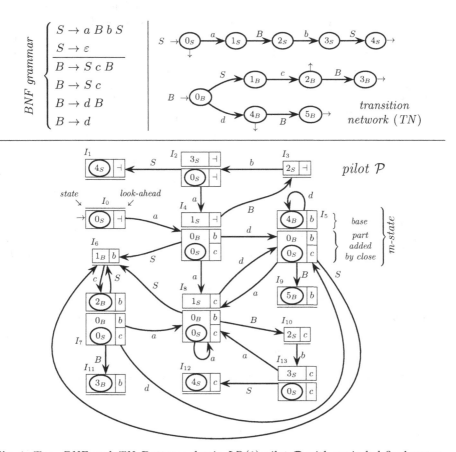

Fig. 1. Top: *BNF* and *TN*. Bottom: classic *LR* (1) pilot $\mathcal{P}$ with encircled final states.

2   Preliminaries

The *terminal* alphabet is Σ, the empty string is ε, the end-of-text is '$\dashv$', and $\Sigma_\dashv = \Sigma \cup \{\dashv\}$. A context-free *grammar* G is a 4-tuple (Σ, V, P, S), with V the *nonterminal* alphabet ($\Sigma \cap V = \emptyset$), P the set of *rules* and $S \in V$ the *axiom*. A *grammar symbol* is an element of the alphabet *symbols* $= \Sigma \cup V$.

An *EBNF* grammar $G = (\Sigma, V, P, S)$ has exactly one rule $A \to \alpha$ for each nonterminal $A \in V$. The rule right part α is a *regular expression* (*RE*) over the alphabet *symbols*, and it may contain concatenation, union and star "$*$" (or cross "$+$"). The regular language, over the alphabet *symbols*, that is associated to A, is defined by the *RE* α of rule $A \to \alpha$ and is denoted R_A or also $R(\alpha)$. A *BNF* rule is an *EBNF* one where the rule right part α is a sequence of symbols, or a finite union of such sequences if alternative *BNF* rules are grouped. It is well known that *EBNF* and *BNF* grammars have the same generative capacity.

An *immediate derivation* is a relation such as $u A v \Rightarrow u w v$, where the strings u, v and w are defined over the alphabet *symbols*, and it holds $w \in R_A$. A *derivation* is its reflexive and transitive closure, and is denoted $\overset{*}{\Rightarrow}$. The reverse relation of a derivation $w \overset{*}{\Rightarrow} z$ is named *reduction* and is denoted $z \rightsquigarrow w$. The context-free language generated by grammar G is $L(G) = \{ x \in \Sigma^* \mid S \overset{*}{\Rightarrow} x \}$.

We represent an *EBNF* rule $A \to \alpha$ as a deterministic *FA* (*DFA*) that recognizes the language R_A. Such a *DFA* is denoted M_A and is called a *machine*. A set $\mathcal{M}$ of machines for all the nonterminals represents an *EBNF* grammar and is named a *transition net* (*TN*), see Def. 1. Such a *TN* represents any other equivalent grammar that has the same nonterminals and regular languages.

Definition 1. *Let G be an EBNF grammar $S \to \sigma$, $A \to \alpha$, ... The transition network $\mathcal{M} = \{ M_S, M_A, \ldots \}$ or TN, is a set of DFA's (machines) that accept the regular languages R_S, R_A, ... To prevent confusion, the state set of a machine M_A is denoted $Q_A = \{ 0_A, \ldots, q_A, \ldots \}$, with initial state 0_A and final state set $F_A \subseteq Q_A$. The state set of $\mathcal{M}$ is $Q = \bigcup_{M_A \in \mathcal{M}} Q_A$. The transition function of $\mathcal{M}$ is denoted δ, with no clashes since all the machine state sets are disjoint. Every machine is assumed to be non-reentrant, i.e., there is no edge that enters its unique initial state (but the initial state may be final). The regular language over the grammar symbols accepted by a machine M_A starting from state q_A and ending to some final state is denoted $R(M_A, q_A)$. The context-free language over the alphabet Σ defined by a machine M_A starting from any state q_A is $L(M_A, q_A) = \{ x \in \Sigma^* \mid \eta \in R(M_A, q_A) \land \eta \overset{*}{\Rightarrow} x \} \neq \emptyset$. It is simply denoted $L(q_A)$ or $L(q)$ if machine M_A is understood or indifferent.* □

A path in the graph of a non-reentrant machine never revisits the initial state. To simplify the parser reduction moves, any machine can be so normalized by adding a new initial state and a few outgoing edges with a negligible overhead.

Example 2. Running example started.

Fig. 2, top, shows an *EBNF* grammar, which has only one nonterminal; its only rule features two nested iterations. The equivalent *TN* is made of one machine, the graph of which is cyclic. Fig. 1, top, shows the equivalent *BNF* grammar, obtained by transforming the iterations into right-recursive rules; for that, two nonterminals are needed. The equivalent *TN* consists of two machines, the graphs of which are trees. The bottom parts of both figures contain the control graphs of the related parsers, also called *pilots*, to be discussed later. □

3   *ELR*(1) Parsers

The classic $LR(1)$ theory is well known, for instance see [1,5]. Thus for brevity the construction of the $LR(1)$ parsers is not repeated here; instead, their application is shown on the running example 2. The acronyms $LR(1)$ and $ELR(1)$ refer to the classic theory and to the present development for *TN*, respectively.

We use a *TN* instead of a grammar. Thus a *dotted rule* becomes a machine state. For instance, in Fig. 1 the rule $S \to a\,B\,b\,S$ corresponds to (a path of) machine M_S, and the dotted rule $S \to a \bullet B\,b\,S$ is encoded by the *TN* state 1_S.

A *classic* $LR(1)$ parser is a *DPDA* that uses the so-called *item sets* as push-down stack symbols. Basically, an item consists of a *state* and a *look-ahead*. Next, we define the item and the extended version needed by our parsers (Def. 3).

Definition 3. *Item (and tuple).*

An "item" is a 2-tuple $\langle q, \lambda \rangle \in Q \times \wp(\Sigma_\dashv)$. The fields q and λ are the state and the (non-empty) look-ahead set. Two or more items $\langle q, \lambda_i \rangle$ with the same state q are written as the equivalent unified item $\langle q, \lambda_1 \cup \ldots \cup \lambda_k \rangle$ $(1 \le i \le k)$.

An "item extended with a key" is a 3-tuple $\langle q, \lambda, k \rangle$, where the fields q and λ are as before, and the field k is a key named PIK (predecessor item key) that identifies another 3-tuple. The "nil" key value is denoted "$\perp$". The key value i that identifies a 3-tuple t is written as a prefix, in the format "$i: t$". Two or more 3-tuples with the same state q and PIK k are written unified as before.

A generic item (with key) is a "tuple", and the set of all TN tuples is named "tuples". A tuple with $q = 0_S$ and $\lambda = \{ \dashv \}$ $(PIK = \perp)$ is said "axiomatic". □

Specifying how to compute the keys is unnecessary, provided the keys of any two tuples in the same set are distinct. For any *TN*, the set *tuples* is finite.

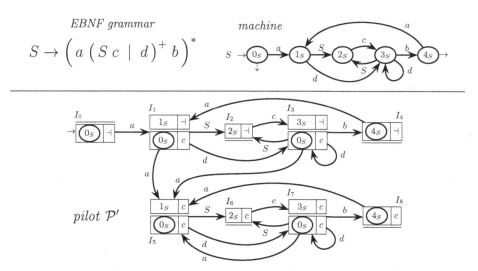

Fig. 2. Top: *EBNF* and *TN* with cycles. Bottom: direct $ELR(1)$ (key-less) pilot $\mathcal{P}'$.

Given a *TN*, the classic $LR(1)$ parser generator builds the state transition function ϑ of a *DFA*, originally named "recognizer of viable $LR(1)$ prefixes" [7], here renamed *pilot* for brevity. A pilot state is a set of tuples (Def. 3), and it is named a *macro-state* (*m-state*) to avoid confusion with the states of the *TN*.

The *shift* function [7] builds the pilot transitions between m-states for all the three pilot types we have (Def. 6 and Alg. 7): the *classic* pilot $\mathcal{P}$, the *direct (key-less) pilot* $\mathcal{P}'$ already defined in [2], and the novel *direct pilot with keys*, named $\mathcal{P}''$. Here (Def. 4) shift is formulated for a *TN*, in such three versions.

Definition 4. *Shift function (total).*
For the pilot $\mathcal{P}''$, function shift: $\wp\,(tuples) \times symbols \rightarrow \wp\,(tuples)$ is specified as follows, on a 3-tuple $i\colon \langle r, \rho, k \rangle$ with key i and a grammar symbol X:

$$shift\left(\{\, i\colon \langle r, \rho, k \rangle \,\}, X\right) = \left\{\, \langle s, \rho, i \rangle \,\middle|\, arc\left(r \xrightarrow{X \in symbols} s\right) \in \delta \,\right\} \quad (1)$$

For the two (key-less) pilots $\mathcal{P}$ and $\mathcal{P}'$, the shift function is defined as in (1) but without item key and PIK field. Shift is extended to sets of items I, i.e., m-states, in the obvious way. If it holds shift $(I, X) = \emptyset$, i.e., no item of I shifts, then the pilot transition from I by grammar symbol X is undefined. □

The *closure* function [7] creates initial tuples. Here (Def. 5) it is formulated for a *TN*, in the two versions (2) for tuples without/with key. For a language L over the alphabet Σ, define the set $First\,(L) = \{\, a \in \Sigma_{\dashv} \mid\ a\,x \in L\,\dashv\ \wedge\ |x| \geq 0 \,\}$.

Definition 5. *Closure function (total).*
Let I_{base} be a tuple set to be closed and let 0_X be a machine initial state. Function close: $\wp\,(tuples) \rightarrow \wp\,(tuples)$ computes the smallest set close (I_{base}) such that:

$$close\,(I_{base}) = I_{base} \cup \left\{\, \langle 0_X, \lambda, \bot \rangle \,\middle|\, \begin{matrix} \langle q, \rho, k \rangle \in close\,(I_{base}) \\ \wedge\ arc\left(q \xrightarrow{X \in V} r\right) \in \delta \\ \wedge\ \lambda = First\left(L\,(r) \cdot \rho\right) \end{matrix} \,\right\} \quad (2)$$

All the initial 3-tuples created and added to I_{base} have a field PIK $= \bot$. For the key-less pilots, the closure function is defined as in (2) without PIK field. □

Definition 6. *Pilot (i.e., parser control unit).*
Given a TN, each of the three pilot models $\mathcal{P}$, $\mathcal{P}'$ or $\mathcal{P}''$ is a DFA defined by a 4-tuple (symbols, $\mathcal{I}$, I_0, ϑ), where $\mathcal{I} = \{\, I_i \mid\ i \geq 0\ \wedge\ I_i \subseteq tuples \,\}$ is the (finite) set of pilot m-states, I_0 is the initial m-state, all the m-states are final, and $\vartheta\colon \mathcal{I} \times symbols \rightarrow \mathcal{I}$ is the pilot transition function computed by Alg. 7. □

Algorithm 7 *Pilot; input: TN (i.e., Q and δ); output: $\mathcal{I}\,(= \mathcal{I}_{pilot})$ and ϑ.*

$I_0 := close\,(\,\{\, \textbf{axiomatic tuple} \,\}\,)$	// pilot initial m-state
$\mathcal{I}_{pilot} := \{\, I_0 \,\}\quad \mathcal{I}_{new} := \{\, I_0 \,\}\quad \vartheta := \emptyset$	// global initializations
repeat	// loop till fixed point
$\quad \mathcal{I}_{add} := \mathcal{I}_{new}\quad \mathcal{I}_{new} := \emptyset$	// loop initializations
$\quad$ **for each** $I_{add} \in \mathcal{I}_{add}$ **and** $X \in symbols$ **do**	// scan m-states/sym.s
$\qquad I_{base} := shift\,(I_{add}, X)$	// shift m-state by sym.
$\qquad$ **if** $I_{base} \neq \emptyset$ **then**	// if transition defined
$\qquad\quad \mathcal{I}_{new} := close\,(I_{base})$	// close shifted m-state
$\qquad\quad \vartheta := \vartheta \cup \{arc\,(I_{add} \xrightarrow{X} I_{new})\}$	// add to transition set
$\qquad\quad$ **if** $I_{new} \notin \mathcal{I}_{pilot}$ **then**	// if shifted m-state new

$$\mathcal{I}_{pilot} := \mathcal{I}_{pilot} \cup \{\, I_{new}\,\} \qquad \text{// add to pilot m-states}$$
$$\mathcal{I}_{new} := \mathcal{I}_{new} \cup \{\, I_{new}\,\} \qquad \text{// add to new m-states}$$

until $\mathcal{I}_{new} = \emptyset$ // no new m-states   □

Every m-state is divided into a *base* part created by *shift* and a part added by *close*, with non-initial and initial tuples, respectively; the base of I_0 is empty.

Example 8. Running example 2 continued.

Fig. 1, bottom, shows the classic pilot $\mathcal{P}$. A m-state such as I_5 consists of an item set. The divider splits it into the *base* $\{\, \langle 4_B, \{\, b\,\}\,\rangle\,\}$ and the part added by *close* $\{\, \langle 0_B, \{\, b\,\}\,\rangle, \langle 0_S, \{\, c\,\}\,\rangle\,\}$. The first and the third item have a final state. Fig. 2, bottom, depicts the direct (key-less) pilot $\mathcal{P}'$ in a similar way. Notice pilot $\mathcal{P}'$ has fewer m-states than the classic pilot $\mathcal{P}$; but more of that later.   □

It is well known that the classic pilot $\mathcal{P}$ serves these two different purposes:

- to check that the grammar or the *TN* has the *LR* (1) property
- as state transition function ϑ of the parser *DPDA* (recognizer)

A *TN* satisfies the classic *LR* (1) condition if in the classic pilot $\mathcal{P}$ there is not a m-state I that has either one of the two following *conflicts* (3):

$$\left. \begin{array}{ll} \textit{shift-reduce}: & \text{set } I \text{ contains an item } \langle r, \rho\rangle \text{ with state } r \text{ final and the} \\ & \text{function } \vartheta\,(\,I, a\,) \text{ is } \neq \emptyset \text{ for some } a \in \rho \\ \textit{reduce-reduce}: & \text{set } I \text{ contains two items } \langle r, \rho\rangle \text{ and } \langle s, \sigma\rangle \text{ with states} \\ & r \text{ and } s \text{ final, and it holds } \rho \cap \sigma \neq \emptyset \end{array} \right\} \quad (3)$$

All the m-states of the classic pilot $\mathcal{P}$ shown in Fig. 1 meet the *LR* (1) condition.

Dropping the actions for building the syntax tree, the classic parser is equivalent to a *DPDA* that has the m-states of the classic pilot $\mathcal{P}$ as its stack alphabet. Why the new direct parser has two pilot versions instead of one, is purely contingent. Pilot $\mathcal{P}'$ is the formal model defined in [2], which allows to check the new *ELR* (1) conditions (4). But the parser of [2] is (equivalent to) a *DPDA* that needs additional information to link the stack top tuples to those of the m-state underneath in the stack. It runs faster if the same linkage information is precomputed and stored in its pilot, which then becomes a novel pilot model with keys, named $\mathcal{P}''$. With this provision, the m-states of $\mathcal{P}''$ are the stack symbols of the *DPDA*. Since pilot $\mathcal{P}''$ derives from $\mathcal{P}'$ by just adding the keys, the new direct parser behaves equivalently to the one already proved correct in [2].

The direct parser satisfies the *ELR* (1) *condition* if the direct pilot $\mathcal{P}'$ is free from conflicts (3), i.e., it satisfies the *LR* (1) condition, and additionally if its transition function ϑ does not have any *convergence conflicts* [2], defined as (4):

$$\langle r, \rho\rangle, \langle s, \sigma\rangle \in I, \quad \delta\,(r, X) = \delta\,(s, X) \quad \text{and} \quad \rho \cap \sigma \neq \emptyset \qquad (4)$$

where I is a m-state and X is a grammar symbol. Clearly, if clause (4) holds, then the next m-state $\vartheta\,(I, X)$ contains the item $\langle q, \rho \cup \sigma\rangle$ with state $q = \delta\,(r, X) = \delta\,(s, X)$. We refer to [2] for an explanation and a few examples. All the arcs of the direct pilot $\mathcal{P}'$ in Fig. 2 are free from convergence conflicts (4).

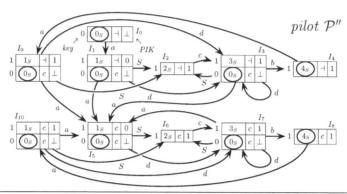

pilot $\mathcal{P}''$

Fig. 3. Top: direct $ELR(1)$ pilot with keys $\mathcal{P}''$. Bottom: simulation of recognition.

input (valid) string	$a\,c\,b\,a\,a\,d\,d\,b\,c\,b$ ⊣
initial stack	I_0
shifted a	$I_0\,a\,I_1$
reduced $\varepsilon \leadsto S$ (null red.)	$I_0\,a\,I_1$
shifted S (nonterm. shift)	$I_0\,a\,I_1\,S\,I_2$
shifted $c\,b\,a\,a\,d\,d\,b$	$I_0\,a\,I_1\,S\,I_2\,c\,I_3\,b\,I_4\,a\,I_9\,a\,I_5\,d\,I_7\,d\,I_7\,b\,I_8$
reduced $a\,d\,d\,b \leadsto S$	$I_0\,a\,I_1\,S\,I_2\,c\,I_3\,b\,I_4\,a\,I_9$
shifted S (nonterm. shift)	$I_0\,a\,I_1\,S\,I_2\,c\,I_3\,b\,I_4\,a\,I_9\,S\,I_2$
shifted $c\,b$	$I_0\,a\,I_1\,S\,I_2\,c\,I_3\,b\,I_4\,a\,I_9\,S\,I_2\,c\,I_3\,b\,I_4$
reduced $a\,S\,c\,b\,a\,S\,c\,b \leadsto S$	I_0
acceptance	

Next, we examine how the direct parser with keys $\mathcal{P}''$ differs from the one without keys $\mathcal{P}'$. The difference comes from using different shift and closure functions (Def.s 4 and 5) in the Alg. 7. If the PIK field of the 3-tuple is deleted in (1), then some m-states of $\mathcal{P}''$ may become identical and coalesce into the same m-state of $\mathcal{P}'$. Thus pilots $\mathcal{P}'$ and $\mathcal{P}''$ are equivalent, viewed as DFA's.

We describe the parser controlled by the direct pilot with keys $\mathcal{P}''$, and so we demonstrate it is equivalent to the (key-less) parser presented in [2]. The parser stack alternates m-states and grammar symbols; initially it contains only I_0. The *shift* operation is identical to that of the classic LR parser. A difference occurs in the *reduction* operation. The running example helps to understand.

Example 9. Running example 2 finished.

Fig. 3, top, shows the direct pilot with keys $\mathcal{P}''$, which has more m-states than the direct key-less pilot $\mathcal{P}'$ of Fig. 2. For instance, the m-states I_1 and I_9 of $\mathcal{P}''$ only differ in their PIK fields and coalesce into the m-state I_1 of $\mathcal{P}'$. Fig. 3, bottom, shows the parsing trace of a valid string, with three reductions. We refer to [2] for a formal presentation. Here is how each reduction traces back its stack *handle* by means of the chain of tuple links based on the PIK field:

"$\varepsilon \leadsto S$" The final tuple 0: $\langle 0_S, \{\, c \,\}, \bot \rangle$ in I_1 matches look-ahead and input, and it has $PIK = \bot$ (created by closure); so the reduction handle is null.

"$a\,d\,d\,b \leadsto S$" The final tuple 1: $\langle 4_S, \{\, c \,\}, 1 \rangle$ in I_8 matches look-ahead and input, and it has $PIK = 1 \neq \bot$; so it points back to tuple 1: $\langle 3_S, \{\, c \,\}, 1 \rangle$

in I_7; then to itself in I_7; then to 1: $\langle 1_S, \{\,c\,\}, 0\rangle$ in I_5; and finally to 0: $\langle 0_S, \{\,c\,\}, \bot\rangle$ in I_9, with $PIK = \bot$; so the reduction handle is complete. "$a\,S\,c\,b\,a\,S\,c\,b \rightsquigarrow S$" The final tuple 1: $\langle 4_S, \{\dashv\}, 1\rangle$ in I_4 matches look-ahead and input, and it has $PIK = 1 \neq \bot$; so it points back to tuple 1: $\langle 3_S, \{\dashv\}, 1\rangle$ in I_3; and so on through I_2, I_9, I_4, I_3, I_2 and I_1, until tuple 0: $\langle 0_S, \{\dashv\}, \bot\rangle$ is found in I_0, with $PIK = \bot$; so the reduction handle is complete.

The string is eventually accepted because the stack contents reduce to I_0, there is not any input left to analyze and the last reduction is to the axiom S.   □

4   Quantitative Measurements and Comparisons

We make a theoretical and experimental comparison between classic and direct parsers. By showing the parser size and speed, first for formal parameterized language families and then for real programming languages such as *Java*, we provide consistent evidence that our novel parsers have a lower or comparable cost and a quite competitive performance with respect to the classic ones.

The code of an $LR(1)$ parser consists of a fixed language-independent part and of a language-dependent part, which are data structures or tables that represent the state transition function ϑ of the parser pilot. Traditionally, the descriptive complexity of LR parsers is measured as the number $\#ms$ of pilot m-states. As said, early research introduced some simplified algorithms, e.g., $LALR(1)$ and Simple $LR(1)$, to reduce the number $\#ms$ and the memory for the tables.

4.1   Formal Analysis of the Pilot Size

We compute the parameter $\#ms$ for the classic $LR(1)$ parser, i.e., model $\mathcal{P}$, and for the direct $ELR(1)$ one enriched with keys, i.e., model $\mathcal{P}''$, starting with a family of regular languages parameterized by star height (Def. 10 and Th. 11).

Definition 10. *Let* $\Sigma = \{\,a_1, a_2, \ldots, a_k, \ldots\,\}$ *be a countable alphabet. The language family* $\mathcal{L} = \{\,L_1, L_2, \ldots, L_k, \ldots\,\}$ *is defined as follows* $(k \geq 1)$:

$$L_1 = (a_1)^*, \ L_2 = \big((a_1)^*\, a_2\big)^*, \ \ldots, \ L_k = \big((\ldots (a_1)^* \ldots a_{k-1})^*\, a_k\big)^*, \ \ldots \ \square$$

Theorem 11. *Let* $\#ms\_classic_k$ *and* $\#ms\_direct_k$ *be the numbers of m-states of the classic* $LR(1)$ *pilot and of the direct* $ELR(1)$ *one with keys, for* $L_k \in \mathcal{L}$ $(k \geq 1$, *see Def. 10). It holds* $\#ms\_classic_k = 3k$ *and* $\#ms\_direct_k = k + 1$.   □

Proof. For the classic $LR(1)$ pilot, i.e., model $\mathcal{P}$, consider the relation between languages L_{k-1} and L_k. The pilots for L_1 and L_2 can be drawn directly, and they have 3 and 6 m-states, respectively. Fig. 4 shows the *BNF* grammar and the acyclic *TN* of L_k, and it sketches the pilot, which has a modular structure.

The pilot for L_k can be inductively obtained from that for L_{k-1} (with $k \geq 3$) by updating the initial m-state I_0 with one more item for nonterminal A_k, namely $\langle 0_{A_k}, \{\dashv\}\rangle$, plus three more m-states I_{3k-3}, I_{3k-2} and I_{3k-1} similarly updated, which are modeled and connected to the (sub-)pilots for L_{k-1} and L_1 as shown

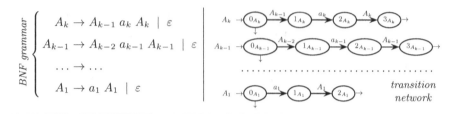

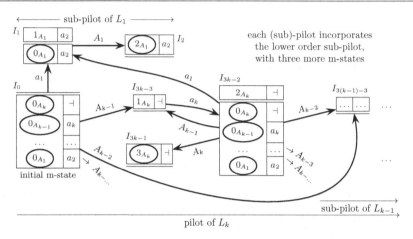

Fig. 4. Top: *BNF* grammar and *TN*. Bottom: classic $LR(1)$ pilot for L_k.

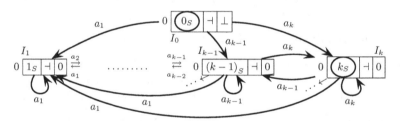

Fig. 5. Direct $ELR(1)$ pilot with keys for L_k, isomorphic to the *DFA* of the rule

in Fig. 4. Since the pilot for L_1 has 3 m-states (I_0 considering only $\langle 0_{A_1}, \{ a_2 \} \rangle$, I_1 and I_2), the thesis follows from induction on index k, for all values $k \geq 1$.

Concerning the direct $ELR(1)$ pilot with keys, i.e., model $\mathcal{P}''$, the one-rule *EBNF* grammar $S \to \Big(\big(\dots \big((a_1)^* a_2 \big)^* \dots a_{k-1} \big)^* a_k \Big)^*$ defines language L_k (with $k \geq 1$). Since language L_k is local, it is easy to obtain a local (and minimal) *DFA* that recognizes L_k, for any $k \geq 1$. Such a *DFA* can be turned into a direct $ELR(1)$ pilot with (trivial) keys, as each *DFA* state becomes a pilot m-state that contains only one 3-tuple. See the pilot for L_k sketched in Fig. 5, valid for any $k \geq 1$. The number of m-states of this pilot for language L_k is $k + 1$.   □

Similarly, we can define another language family for $k \geq 1$: $L_1 = a_1^*$, $L_2 = a_1^* a_2^*$, ..., $L_k = a_1^* a_2^* \ldots a_{k-1}^* a_k^*$, ... Also this family exhibits a linear separation of pilot complexities (we omit the analysis), and it is more representative of the typical iterative structures that occur in the grammars of programming languages.

Last and foremost, the analysis of such language families shows the growth of parameter #*ms_classic* is mainly due to the stars in the rule *RE*'s, e.g., Def. 10 and Th. 11, yet their significance for real parsing applications is questionable because no ordinary *EBNF* grammar has rules with *RE*'s of a large star height. Thus, to practically assess the complexity we resort to experimentation.

4.2 Experimental Analysis

We developed a parser generator tool, named *E-Bison*, that translates the *EBNF* grammar to a *TN*, constructs the pilot and generates the classic or direct parser encoded in *Java*. We can generate several parser variants: the classic $LR(1)$ parser, i.e., $\mathcal{P}$, if the grammar is (converted to) *BNF*; or the direct (key-less) $ELR(1)$ parser of [2], i.e., $\mathcal{P}'$; or the direct $ELR(1)$ parser with keys of Sect. 3, i.e., $\mathcal{P}''$. We used our tool for a static analysis of the descriptive complexity: we generated pilots for various grammars and obtained their numbers of m-states.

Table 1. Characteristics of the benchmarks and platform used for the experiments. Sources: for *Java JDK*, for *JSON* several files (Gospel of John, statistics on the n-grams of English in Google Books, the *UK* Comprehensive Knowledge Archive Network, etc.).

benchmark language	total size (MByte)	number of files	number of code lines	number of tokens	execution platform (for both benchmarks)
Java	16	2 743	500 000	1 103 799	AMD Athlon 64X2, 2.2 GHz,
JSON	1.9	5	1 per file	239 288	2 GB RAM, Win. 7 32 bit

To measure the parsing speed in a realistic and unbiased way, we compared a few classic and direct parsers generated by *E-Bison* against the classic ones generated as *Java* code by *GNU Bison*, which is the most popular generator. We remark however, that *GNU Bison* is of $LALR(1)$ type [1] and so accepts a strict subfamily of deterministic context-free languages. Pilots of $LALR(1)$ type have fewer m-states than the $LR(1)$ ones, but the difference is uninfluential for speed. For comparison, in all cases we used the scanners generated by *E-Bison*.

Two popular source languages are the corpora of our experiments: *Java* and *JSON*. Table 1 shows the sizes of the two benchmarks and the execution platform we used. Table 2 reports the grammar, *TN* and pilot sizes for *Java* and *JSON*, and it shows the m-state number #*ms* and the average parsing speed measured.

The *BNF* grammars have approximately twice as many rules as the *EBNF* ones. The average number of machine states per *EBNF* rule is ≈ 5 for *Java* and ≈ 4 for *JSON*, which has simpler syntactic constructs. The pilot generated by *GNU Bison* is small, but it is $LALR(1)$ and is incomparable with the $LR(1)$ pilot generated by *E-Bison*. The size ratio of the direct $ELR(1)$ pilot $\mathcal{P}''$ vs the classic $LR(1)$ pilot $\mathcal{P}$ is 1.07 ($Java$) and ≈ 1 ($JSON$). In both cases the figure is opposite to the result of Th. 11. The reason is twofold: the $EBNF$ grammars of both languages have few stars in their RE's, and the PIK field in the pilot tuples diversifies the m-states, a fact also visible by comparing Fig.s 2 and 3.

Of course, speed is the most important property to make a parser competitive. Table 2 reports the parsing speeds for $Java$ and $JSON$, measured as the number of tokens processed per millisecond. Our novel direct $ELR(1)$ parsers with keys are significantly faster than the classic $LR(1)$ ones, and the speedup is even larger if we consider the $LALR(1)$ parsers generated by GNU $Bison$. The larger speedup is for the simpler language, $JSON$. In fact the direct parsers can use longer rules than the classic ones do, thus they save 16.0% ($Java$) and 34.6% ($JSON$) reduction moves w.r.t. the latter, and the saving is larger for $JSON$.

Table 2. Experimental results: speed is measured in thousand tokens per millisecond

grammar and TN size			Bison LALR(1)		E-Bison LR(1)		E-Bison ELR(1)	
grammar of lang.	BNF rules	EBNF rules	TN states	pilot #ms	parser speed	pilot $\mathcal{P}$ #ms	parser speed	pilot $\mathcal{P}''$ #ms	parser speed
Java	263	133	660	655	2.20	2946	2.39	3153	2.46
JSON	25	11	43	46	4.13	70	4.97	71	6.13

We quote also the size of the direct $ELR(1)$ (key-less) pilots, i.e., $\mathcal{P}'$: they have 1973 ($Java$) and 50 ($JSON$) m-states. Notice they are smaller than the classic $LR(1)$ pilots, i.e., $\mathcal{P}$, in line with Th. 11. Their parsers are less efficient, so their speeds are not shown. As said, such pilots serve to prove determinism.

It would be out of scope to discuss programming details. It suffices to say we made a few optimizations in the direct $ELR(1)$ parser with keys, i.e., $\mathcal{P}''$, which concern the reduction move and aim at attaining the same speed as the classic $LR(1)$ parser, i.e., $\mathcal{P}$, whenever the grammar rules applied are non-extended.

5   Conclusion

The formal analysis and experimental measurements reported indicate that our novel direct ELR parser model (in particular with keys) for the languages defined by Extended BNF grammars [2], is competitive with those that have been in use for half a century for the less expressive BNF grammars: an almost negligible increase of parser size is offset by a more significant increase of parsing speed.

Sharing: ftp://ftp.elet.polimi.it/outgoing/Luca.Breveglieri/ebison; code: binary; data: access; source: free (information for reproducing our experiments).

References

1. Aho, A., Lam, M., Sethi, R., Ullman, J.: Compilers: principles, techniques and tools. Prentice-Hall, Englewood Cliffs (2006)
2. Breveglieri, L., Crespi Reghizzi, S., Morzenti, A.: Shift-reduce parsers for transition networks. In: Dediu, A.-H., Martín-Vide, C., Sierra-Rodríguez, J.-L., Truthe, B. (eds.) LATA 2014. LNCS, vol. 8370, pp. 222–235. Springer, Heidelberg (2014)
3. Conway, M.E.: Design of a separable transition-diagram compiler. Comm. ACM 6(7), 396–408 (1963)
4. Crespi Reghizzi, S., Breveglieri, L., Morzenti, A.: Formal languages and compilation, 2nd edn. Springer, London (2013)
5. Grune, D., Jacobs, C.: Parsing techniques: a practical guide, 2nd edn. Springer, London (2009)
6. Hemerik, K.: Towards a taxonomy for ECFG and RRPG parsing. In: Dediu, A.H., Ionescu, A.M., Martín-Vide, C. (eds.) LATA 2009. LNCS, vol. 5457, pp. 410–421. Springer, Heidelberg (2009)
7. Knuth, D.E.: On the translation of languages from left to right. Information and Control 8, 607–639 (1965)
8. Pager, D.: A practical general method for constructing LR(k) parsers. Acta Inf. 7, 249–268 (1977)

Most Complex Regular Right-Ideal Languages[*]

Janusz Brzozowski[1] and Gareth Davies[2]

[1] David R. Cheriton School of Computer Science, University of Waterloo,
Waterloo, ON, Canada N2L 3G1
brzozo@uwaterloo.ca
[2] Department of Pure Mathematics, University of Waterloo,
Waterloo, ON, Canada N2L 3G1
gdavies@uwaterloo.ca

Abstract. A right ideal is a language L over an alphabet Σ that satisfies the equation $L = L\Sigma^*$. We show that there exists a sequence $(R_n \mid n \geqslant 3)$ of regular right-ideal languages, where R_n has n left quotients and is most complex among regular right ideals under the following measures of complexity: the state complexities of the left quotients, the number of atoms (intersections of complemented and uncomplemented left quotients), the state complexities of the atoms, the size of the syntactic semigroup, the state complexities of reversal, star, product, and all binary boolean operations that depend on both arguments. Thus $(R_n \mid n \geqslant 3)$ is a universal witness reaching the upper bounds for these measures.

Keywords: atom, operation, quotient, regular language, right ideal, state complexity, syntactic semigroup, universal witness.

1 Introduction

Brzozowski [3] called a regular language *most complex* if it meets the upper bounds for a large set of commonly used language properties and operations, and found a single *witness* language of state complexity n for each $n \geqslant 3$ that meets all these bounds. In particular, this language has the maximal number of atoms and the state complexities of these atoms are maximal. Moreover, it meets the upper bounds for the state complexities of all the basic operations: reverse, Kleene star, boolean operations, product (also known as concatenation or catenation), as well as a large number of combined operations. In view of this, such a witness has been called *universal*.

If we restrict our attention to some subclass of regular languages, then the universal witness mentioned above no longer works because it lacks the properties of the subclass. In this paper we ask whether the approach used for general regular languages can be extended to subclasses. We answer this question positively for regular right ideals by presenting a universal right-ideal witness.

[*] This work was supported by the Natural Sciences and Engineering Research Council of Canada under grant No. OGP0000871.

H. Jürgensen et al. (Eds.): DCFS 2014, LNCS 8614, pp. 90–101, 2014.
© Springer International Publishing Switzerland 2014

For a further discussion of regular right ideals see [5,8]. It was pointed out in [5] that right ideals deserve to be studied for several reasons: They are fundamental objects in semigroup theory, they appear in the theoretical computer science literature as early as 1965, and they continue to be of interest in the present. Right ideal languages are complements of prefix-closed languages. Besides being of theoretical interest, right ideals also play a role in algorithms for pattern matching: When searching for all words beginning in a word from some set L, one is looking for all the words of the right ideal $L\Sigma^*$.

2   Background

A *deterministic finite automaton (DFA)* $\mathcal{D} = (Q, \Sigma, \delta, q_1, F)$ consists of a finite non-empty set Q of *states*, a finite non-empty *alphabet* Σ, a *transition function* $\delta \colon Q \times \Sigma \to Q$, an *initial state* $q_1 \in Q$, and a set $F \subseteq Q$ of *final* states. The transition function is extended to functions $\delta' \colon Q \times \Sigma^* \to Q$ and $\delta'' \colon 2^Q \times \Sigma^* \to 2^Q$ as usual, and these extensions are also denoted by δ. A state q of a DFA is *reachable* if there is a word $w \in \Sigma^*$ such that $\delta(q_1, w) = q$. The *language accepted* by $\mathcal{D}$ is $L(\mathcal{D}) = \{w \in \Sigma^* \mid \delta(q_1, w) \in F\}$. The *language of a state* q is the language accepted by the DFA $\mathcal{D}_q = (Q, \Sigma, \delta, q, F)$. A state is *empty* if its language is empty. Two DFAs are *equivalent* if their languages are the same. Two states are *equivalent* if their languages are equal; otherwise, they are *distinguishable* by some word that is in the language of one of the states, but not of the other. If $S \subseteq Q$, two states $p, q \in Q$ are *distinguishable with respect to* S if there is a word w such that $\delta(p, w) \in S$ if and only if $\delta(q, w) \notin S$. A DFA is *minimal* if all of its states are reachable and no two states are equivalent.

A *nondeterministic finite automaton (NFA)* is a tuple $\mathcal{N} = (Q, \Sigma, \eta, Q_1, F)$, where Q, Σ, and F are as in a DFA, $\eta \colon Q \times \Sigma \to 2^Q$ is the transition function and $Q_1 \subseteq Q$ is the *set of initial states*. An ε-*NFA* has all the features of an NFA but its transition function $\eta \colon Q \times (\Sigma \cup \{\varepsilon\}) \to 2^Q$ allows also transitions under the empty word ε. The *language accepted* by an NFA or an ε-NFA is the set of words w for which there exists a sequence of transitions such that the concatenation of the symbols inducing the transitions is w, and this sequence leads from a state in Q_1 to a state in F. Two NFAs are *equivalent* if they accept the same language.

We use the following operations on automata:

1. The *determinization* operation D applied to an NFA $\mathcal{N}$ yields a DFA $\mathcal{N}^D$ obtained by the subset construction, where only subsets reachable from the initial subset of $\mathcal{N}^D$ are used and the empty subset, if present, is included.

2. The *reversal* operation R applied to an NFA $\mathcal{N}$ yields an NFA $\mathcal{N}^R$, where sets of initial and final states of $\mathcal{N}$ are interchanged and transitions are reversed.

Let $\mathcal{D} = (Q, \Sigma, \delta, q_1, F)$ be a DFA. For each word $w \in \Sigma^*$, the transition function induces a transformation t_w of Q by w: for all $q \in Q$, $qt_w \overset{\text{def}}{=} \delta(q, w)$. The set $T_\mathcal{D}$ of all such transformations by non-empty words forms a semigroup of transformations called the *transition semigroup* of $\mathcal{D}$ [11]. Conversely, we can use a set $\{t_a \mid a \in \Sigma\}$ of transformations to define δ, and so also the DFA $\mathcal{D}$. We also write $a \colon t$ to mean that the transformation induced by $a \in \Sigma$ is t.

The *syntactic congruence* $\leftrightarrow_L$ of a language $L \subseteq \Sigma^*$ is defined on Σ^+: For $x, y \in \Sigma^+$, $x \leftrightarrow_L y$ if and only if $uxv \in L \Leftrightarrow uyv \in L$ for all $u, v \in \Sigma^*$. The quotient set $\Sigma^+ / \leftrightarrow_L$ of equivalence classes of the relation $\leftrightarrow_L$ is a semigroup called the *syntactic semigroup* of L. If $\mathcal{D}$ is the minimal DFA of L, then $T_{\mathcal{D}}$ is isomorphic to the syntactic semigroup T_L of L [11], and we represent elements of T_L by transformations in $T_{\mathcal{D}}$.

A *permutation* of Q is a mapping of Q *onto* itself. The *identity* transformation $\mathbf{1}$ maps each element to itself, that is, $q\mathbf{1} = q$ for $q \in Q$. A transformation t is a *cycle* of length k if there exist pairwise different elements $p_1, \ldots, p_k$ such that $p_1 t = p_2, p_2 t = p_3, \ldots, p_{k-1} t = p_k, p_k t = p_1$, and other elements of Q are mapped to themselves. A cycle is denoted by $(p_1, p_2, \ldots, p_k)$. A *transposition* is a cycle (p, q). For $p \neq q$, a *unitary* transformation $t \colon (p \to q)$, has $pt = q$ and $rt = r$ for all $r \neq p$.

The set of all permutations of a set Q of n elements is a group, called the *symmetric group* of degree n. Without loss of generality, from now on we assume that $Q = \{1, 2, \ldots, n\}$. It is well known that the symmetric group of degree n can be generated by any cyclic permutation of n elements together with any transposition. In particular, it can be generated by $(1, 2, \ldots, n)$ and $(1, 2)$.

The set of all transformations of a set Q, denoted by $\mathcal{T}_Q$, is a monoid with $\mathbf{1}$ as the identity. It is well known that the transformation monoid $\mathcal{T}_Q$ of size n^n can be generated by any cyclic permutation of n elements together with any transposition and any unitary transformation. In particular, $\mathcal{T}_Q$ can be generated by $(1, 2, \ldots, n)$, $(1, 2)$ and $(n \to 1)$.

The *state complexity* [12] *of a regular language* L over an alphabet Σ is the number of states in any minimal DFA recognizing L. An equivalent notion is that of *quotient complexity* [2], which is the number of distinct left quotients of L, where the left quotient of $L \subseteq \Sigma^*$ by a word $w \in \Sigma^*$ is the language $w^{-1}L = \{x \in \Sigma^* \mid wx \in L\}$. This paper uses *complexity* for both of these equivalent notions, and this term will not be used for any other property here.

The *(state/quotient) complexity of an operation* [12] on regular languages is the maximal complexity of the language resulting from the operation as a function of the complexities of the arguments. For example, for $L \subseteq \Sigma^*$, the complexity of the reverse L^R of L is 2^n if the complexity of L is n, since a minimal DFA for L^R can have at most 2^n states and there exist languages meeting this bound [9].

There are two parts to the process of establishing the complexity of an operation. First, one must find an *upper bound* on the complexity of the result of the operation by using quotient computations or automaton constructions. Second, one must find *witnesses* that meet this upper bound. One usually defines a sequence $(L_n \mid n \geqslant k)$ of languages, where k is some small positive integer. This sequence will be called a *stream*. The languages in a stream differ only in the parameter n. For example, one might study unary languages $(\{a^n\}^* \mid n \geqslant 1)$ that have zero occurrences of the letter a modulo n. A unary operation takes its argument from a stream $(L_n \mid n \geqslant k)$. For a binary operation, one adds a stream $(K_n \mid n \geqslant k)$ as the second argument. While the witness streams are normally different for different operations, our main result shows that a single stream can meet the complexity bounds for all operations in the case of right ideals.

Atoms of regular languages were studied in [7], and their complexities, in [6]. Let L be a regular language with quotients $K = \{K_1, \ldots, K_n\}$. Each subset S of K defines an *atomic intersection* $A_S = \widetilde{K_1} \cap \cdots \cap \widetilde{K_n}$, where $\widetilde{K_i}$ is K_i if $K_i \in S$ and $\overline{K_i}$ otherwise. An *atom* of L is a non-empty atomic intersection. Since non-empty atomic intersections are pairwise disjoint, every atom A has a unique atomic intersection associated with it, and this atomic intersection has a unique subset S of K associated with it. This set S is called the *basis* of A and is denoted by $\mathcal{B}(A)$. The *cobasis* of A is $\overline{\mathcal{B}}(A) = K \setminus \mathcal{B}(A)$. The basis of an atom is the set of quotients of L that occur uncomplemented as terms of the corresponding intersection, and the cobasis is the set of quotients that occur complemented.

It was proven in [7] that each regular language L defines a unique set of atoms, that every quotient of L (including L itself) is a union of atoms, and that every quotient of every atom of L is a union of atoms. Thus the atoms of L are its basic building blocks. It was argued in [3] that the complexity of the atoms of a language should be considered when searching for "most complex" regular languages, since a complex language should have complex building blocks. We shall show that – as was the case for arbitrary regular languages – for right ideals there is a tight upper bound on the complexity of any atom with a basis of a given size.

3   Main Results

The stream of right ideals that turns out to be most complex is defined as follows:

Definition 1. *For $n \geqslant 3$, let $\mathcal{R}_n = \mathcal{R}_n(a, b, c, d) = (Q, \Sigma, \delta, 1, \{n\})$, where $Q = \{1, \ldots, n\}$ is the set of states[1], $\Sigma = \{a, b, c, d\}$ is the alphabet, the transformations defined by δ are $a\colon (1, \ldots, n-1)$, $b\colon (2, \ldots, n-1)$, $c\colon (n-1 \to 1)$ and $d\colon (n-1 \to n)$, 1 is the initial state, and $\{n\}$ is the set of final states. Let $R_n = R_n(a, b, c, d)$ be the language accepted by $\mathcal{R}_n$.*

The structure of the DFA $\mathcal{R}_n(a, b, c, d)$ is shown in Figure 1. Note that input b induces the identity transformation in $\mathcal{R}_n$ for $n = 3$.

The stream of languages of Definition 1 is very similar to the stream $(L_n \mid n \geqslant 2)$ shown to be a universal witness for regular languages in [3,6]. In that stream, L_n is defined by the DFA $\mathcal{D}_n = \mathcal{D}_n(a, b, c) = (Q, \Sigma, \delta, 1, \{n\})$, where $Q = \{1, \ldots, n\}$, $\Sigma = \{a, b, c\}$, and δ is defined by $a\colon (1, \ldots, n)$, $b\colon (1, 2)$, and $c\colon (n \to 1)$. The automaton $\mathcal{R}_n$ can be constructed by taking $\mathcal{D}_{n-1}$, adding a new state n and a new input $d\colon (n-1 \to n)$, making n the only final state, and having b induce the cyclic permutation $(2, \ldots, n-1)$, rather than the transposition $(1, 2)$. The new state and input are necessary to ensure that R_n is a right ideal for all n.

---

[1] Although Q and δ depend on n, this dependence is usually not shown to keep the notation as simple as possible.

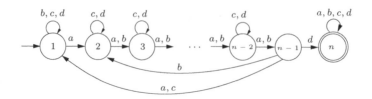

Fig. 1. Automaton $\mathcal{R}_n$ of a most complex right ideal R_n

Changing the transformation induced by b is necessary since, if b induces $(1, 2)$ in $\mathcal{R}_n$, then R_n does not meet the bound for product.

We can generalize this definition to a stream $(R_n \mid n \geqslant 1)$ by noting that when $n = 1$, all four inputs induce the identity transformation, and when $n = 2$, a, b and c induce the identity transformation, while d induces $(1 \to 2)$. Hence $R_1 = \{a, b, c, d\}^*$ and $R_2 = \{a, b, c\}^* d \{a, b, c, d\}^*$. However, the complexity bound for star is not reached by R_1, and the complexity bounds for boolean operations are not reached when one of the operands is R_1 or R_2. Thus we require $n \geqslant 3$.

In some cases, the complexity bounds can be reached even when the alphabet size is reduced. If c is not needed, let $\mathcal{R}_n(a, b, d)$ be the DFA of Definition 1 restricted to inputs a, b and d, and let $R_n(a, b, d)$ be the language recognized by this DFA. If both b and c are not needed, we use $\mathcal{R}_n(a, d)$ and $R_n(a, d)$. We also define $\mathcal{R}_n(b, a, d)$ to be the DFA obtained from $\mathcal{R}_n(a, b, d)$ by interchanging the roles of the inputs a and b, and let $R_n(b, a, d)$ be the corresponding language.

Theorem 1 (Main Results). *The language $R_n = R_n(a, b, c, d)$ has the properties listed below. Moreover, all the complexities of R_n are the maximal possible for right ideals. The results hold for all $n \geqslant 1$ unless otherwise specified.*

1. *$R_n(a, d)$ has n quotients, that is, its (state/quotient) complexity is n.*
2. *The syntactic semigroup of $R_n(a, b, c, d)$ has cardinality n^{n-1}.*
3. *Quotients of $R_n(a, d)$ have complexity n, except for the quotient $\{a, d\}^*$, which has complexity 1.*
4. *$R_n(a, b, c, d)$ has 2^{n-1} atoms.*
5. *The atom of $R_n(a, b, c, d)$ with the empty cobasis has complexity 2^{n-1}. If an atom of $R_n(a, b, c, d)$ has a cobasis of size r, $1 \leqslant r \leqslant n-1$, its complexity is*

$$1 + \sum_{k=1}^{r} \sum_{h=k+1}^{k+n-r} \binom{n-1}{h-1} \binom{h-1}{k}.$$

6. *The reverse of $R_n(a, d)$ has complexity 2^{n-1}.*
7. *For $n \geqslant 2$, the star of $R_n(a, d)$ has complexity $n + 1$.*
8. *For $m, n \geqslant 3$, the complexity of $R_m(a, b, d) \cap R_n(b, a, d)$ is mn.*
9. *For $m, n \geqslant 3$, the complexity of $R_m(a, b, d) \oplus R_n(b, a, d)$ is mn.*
10. *For $m, n \geqslant 3$, the complexity of $R_m(a, b, d) \setminus R_n(b, a, d)$ is $mn - (m-1)$.*
11. *For $m, n \geqslant 3$, the complexity of $R_m(a, b, d) \cup R_n(b, a, d)$ is $mn - (m+n-2)$.*

12. For $m, n \geqslant 3$, since any binary boolean operation can be expressed as a combination of the four operations above (and complement, which does not affect complexity), the complexity of $R_m(a, b, d) \circ R_n(b, a, d)$ is maximal for all binary boolean operations $\circ$.
13. For $m, n \geqslant 3$, if $m \neq n$, then the complexity of $R_m(a, b, d) \circ R_n(a, b, d)$ is maximal for all binary boolean operations $\circ$.
14. The complexity of $R_m(a, b, d) \cdot R_n(a, b, d)$ is $m + 2^{n-2}$.

The proof of Theorem 1 is the topic of the remainder of the paper.

4 Conditions for the Complexity of Right Ideals

1. Complexity of the Language: $R_n(a, d)$ has n quotients because the DFA $\mathcal{R}_n(a, d)$ is minimal. This holds since the non-final state i accepts $a^{n-1-i}d$ and no other non-final state accepts this word, for $1 \leqslant i \leqslant n - 1$, and all non-final states are distinguishable from the final state n by the empty word.

2. Cardinality of the Syntactic Semigroup: It was proved in [8] that the syntactic semigroup of a right ideal of complexity n has cardinality at most n^{n-1}. To show $R_n(a, b, c, d)$ meets this bound, one first verifies the following:

Remark 1. For $n \geqslant 3$, the transposition $(1, 2)$ in $\mathcal{R}_n$ is induced by $a^{n-2}b$.

Theorem 1 (2) The syntactic semigroup of $R_n(a, b, c, d)$ has cardinality n^{n-1}.

Proof. The cases $n \leqslant 3$ are easily checked. For $n \geqslant 4$, let the DFA $\mathcal{P}_n$ be $\mathcal{P}_n = (Q, \Sigma, \delta, 1, \{n\})$, where $Q = \{1, \ldots, n\}$, $\Sigma = \{a, b, c, d\}$, and $a \colon (1, \ldots, n - 1)$, $b \colon (1, 2)$, $c \colon (n - 1 \to 1)$ and $d \colon (n - 1 \to n)$. It was proved in [8] that the syntactic semigroup of $P_n(a, b, c, d)$ has cardinality n^{n-1}. Since words in Σ^* can induce all the transformations of $\mathcal{P}_n$ in $\mathcal{R}_n(a, b, c, d)$, the claim follows. $\square$

3. Complexity of Quotients: Each quotient of $R_n(a, d)$, except the quotient $\{a, d\}^*$, has complexity n, since states $1, \ldots, n - 1$ are strongly connected. So the complexities of the quotients are maximal for right ideals.

4. Number of Atoms: It was proved in [6] that the number of atoms of L is precisely the complexity of the reverse of L. It was shown in [5] that the maximal complexity of L^R for right ideals is 2^{n-1}. For $n \leqslant 3$ it is easily checked that our witness meets this bound. For $n > 3$, it was proved in [8] that the reverse of $R_n(a, d)$, and hence also of $R_n(a, b, c, d)$, reaches this bound.

5. Complexity of Atoms: This is the topic of Section 5.

6. Reversal: See **4. Number of Atoms**.

7. Star: The complexity of the star of a right ideal is at most $n + 1$ [5]. This follows because, if $\varepsilon \notin L$, we need to add ε to $L = L\Sigma^*$ to obtain L^*. Our witness meets this bound, as one can easily verify:

Remark 2 (Star). For $n \geqslant 2$, the complexity of $(R_n(a, d))^*$ is $n + 1$.

8.–14. Boolean Operations and **Product:** See Sections 6 and 7.

Table 1. Maximal complexity of atoms of right ideals

n	1	2	3	4	5	6	7	$\cdots$
r=0	1/1	2/3	4/7	8/15	16/31	32/63	64/127	$\cdots$
r=1		2/3	5/10	13/29	33/76	81/187	193/442	$\cdots$
r=2		*/3	4/10	16/43	53/141	156/406	427/1,086	$\cdots$
r=3			*/7	8/29	43/141	166/501	542/1,548	$\cdots$
r=4				*/15	16/76	106/406	462/1,548	$\cdots$
r=5					*/31	32/187	249/1,086	$\cdots$
r=6						*/63	64/442	$\cdots$
max	1/1	2/3	5/10	16/43	53/141	166/501	542/1,548	$\cdots$
ratio	–	2/3	2.50/3.33	3.20/4.30	3.31/3.28	3.13/3.55	3.27/3.09	$\cdots$

5   Complexity of Atoms

In [6], for the language stream $(L_n \mid n \geqslant 2)$ described after Definition 1, it was proved that the atoms of L_n have maximal complexity amongst all regular languages of complexity n. We want to prove that the atoms of $R_n(a, b, c, d)$ have maximal complexity amongst all right ideals of complexity n. We only give a high-level outline following the approach of [6].

1. Derive upper bounds for the complexities of atoms of right ideals.
The cobasis of an atom cannot contain Σ^*; if it did, then $\overline{\Sigma^*} = \emptyset$ would be a term in the corresponding atomic intersection and the intersection would be empty. Since all right ideals have Σ^* as a quotient, every atom of a right ideal must contain Σ^* in its basis. It follows the cobasis of an atom of a right ideal is either empty or contains r quotients, where $1 \leqslant r \leqslant n - 1$. Knowing this, we can derive the upper bounds by the same method as in [6].

2. Describe the transition function of the átomaton of $R_n(a, b, c, d)$.
Let $\mathbf{A} = \{A_1, \ldots, A_m\}$ be the set of atoms of L. The *átomaton*[2] of L is the NFA $\mathcal{A} = (\mathbf{A}, \Sigma, \eta, \mathbf{A}_I, A_f)$, where the *initial* atoms are $\mathbf{A}_I = \{A_i \mid L \in \mathcal{B}(A_i)\}$, the *final* atom A_f is the unique atom such that $K_i \in \mathcal{B}(A_f)$ if and only if $\varepsilon \in K_i$, and $A_j \in \eta(A_i, a)$ if and only if $aA_j \subseteq A_i$. In the átomaton the language of state A of $\mathcal{A}$ is the atom A of L. Since each regular language defines a unique set of atoms, each regular language also defines a unique átomaton.

3. Prove that certain strong connectedness and reachability results hold for states of minimal DFAs of atoms of $R_n(a, b, c, d)$.

4. Prove that the complexity of each atom of $R_n(a, b, c, d)$ meets the established bound.

Many steps of this proof are similar or identical to the proof for L_n given in [6]; for the details see [4]. Table 4 shows the bounds for right ideals (first entry) and compares them to those of regular languages (second entry). An asterisk indicates the case is impossible for right ideals. The *ratio* row shows the ratio m_n/m_{n-1} for $n \geqslant 2$, where m_i is the ith entry in the *max* row.

---

[2] The accent in *átomaton* avoids confusion with *automaton*, and suggests that the stress should be on the first syllable, since the word comes from *atom*.

6 Boolean Operations

Tight upper bounds for boolean operations on right ideals [5] are mn for intersection and symmetric difference, $mn - (m+n)$ for difference, and $mn - (m+n-2)$ for union. Since $L_n \cup L_n = L_n \cap L_n = L_n$, and $L_n \setminus L_n = L_n \oplus L_n = \emptyset$, two different languages must be used to reach the bounds if $m = n$. We use $R_m = R_m(a,b,d)$ and $R_n = R_n(b,a,d)$, shown in Figure 2 for $m = 4$ and $n = 5$.

Let $\mathcal{R}_{m,n} = \mathcal{R}_m \times \mathcal{R}_n = (Q_m \times Q_n, \Sigma, \delta, (1,1), F_{m,n})$ with $\delta((i,j),\sigma) = (\delta_m(i,\sigma), \delta_n(j,\sigma))$, where δ_m (δ_n) is the transition function of $\mathcal{R}_m$ ($\mathcal{R}_n$). Depending on $F_{m,n}$, this DFA recognizes different boolean operations on R_m and R_n. The direct product of $\mathcal{R}_4(a,b,d)$ and $\mathcal{R}_5(b,a,d)$ is in Figure 3.

In our proof that the bounds for boolean operations are reached, we use a result of Bell, Brzozowski, Moreira and Reis [1]. A binary boolean operation $\circ$ on regular languages is a mapping $\circ : 2^{\Sigma^*} \times 2^{\Sigma^*} \to 2^{\Sigma^*}$. If $L, L' \subseteq \Sigma^*$, the result of the operation $\circ$ is denoted by $L \circ L'$. We say that such a boolean operation is *proper* if $\circ$ is not a constant, and not a function of one variable only, that is, it is not the identity or the complement of one of the variables.

Let S_n denote the symmetric group of degree n. A *basis* [10] of S_n is an ordered pair (s,t) of distinct transformations of $Q_n = \{1, \ldots, n\}$ that generate S_n. Two bases (s,t) and (s',t') of S_n are *conjugate* if there exists a transformation $r \in S_n$ such that $rsr^{-1} = s'$, and $rtr^{-1} = t'$. A DFA *has a basis* (t_a, t_b) *for* S_n if it has letters $a, b \in \Sigma$ such that a induces t_a and b induces t_b.

Proposition 1 (Symmetric Groups and Boolean Operations [1]). *Suppose that $m, n \geqslant 1$, L_m and L'_n are regular languages of complexity m and n respectively, and $\mathcal{D}_m = (Q_m, \Sigma, \delta, 1, F)$ and $\mathcal{D}'_n = (Q_n, \Sigma, \delta', 1, F')$ are minimal DFAs for L_m and L'_n, where $\emptyset \subsetneq F \subsetneq Q_m$ and $\emptyset \subsetneq F' \subsetneq Q_n$. Suppose further that $\mathcal{D}_m$ has a basis $B = (t_a, t_b)$ for S_m and $\mathcal{D}'_n$ has a basis $B' = (t'_a, t'_b)$ for S_n. Let $\circ$ be a proper binary boolean operation. Then the following hold:*

1. In the direct product $\mathcal{D}_m \times \mathcal{D}'_n$, all mn states are reachable if and only if $m \neq n$, or $m = n$ and the bases B and B' are not conjugate.

2. For $m, n \geqslant 2$, $(m,n) \notin \{(2,2), (3,4), (4,3), (4,4)\}$, $L_m \circ L'_n$ has complexity mn if and only if $m \neq n$, or $m = n$ and the bases B and B' are not conjugate.

This implies that if the conditions of the proposition hold, then no matter how we choose the sets F and F', as long as $\emptyset \subsetneq F \subsetneq Q_m$ and $\emptyset \subsetneq F' \subsetneq Q_n$,

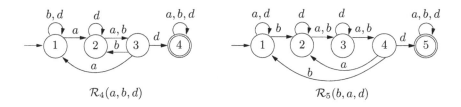

Fig. 2. Right-ideal witnesses for boolean operations

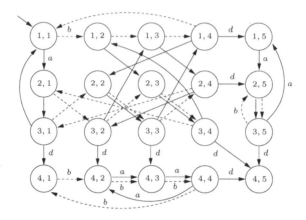

Fig. 3. Direct-product automaton for boolean operations, $m = 4, n = 5$. Transitions under a and d are in solid lines and under b, in dotted lines. Unlabelled solid transitions are under a. Self-loops are omitted.

and the boolean function $\circ$ is proper, the direct product DFA $\mathcal{D}_m \times \mathcal{D}_n$ has mn states and is minimal.

In the case of our right ideal $\mathcal{R}_m$ ($\mathcal{R}_n$), the transitions t_a and t_b (t_a' and t_b') restricted to $\{1, \ldots, n-1\}$, constitute a basis for S_{m-1} (S_{n-1}). This implies that in the direct product $\mathcal{R}_{m,n}$, all states in the set $S = \{(i,j) \mid 1 \leqslant i \leqslant m-1, 1 \leqslant j \leqslant n-1\}$ are reachable by words in $\{a,b\}^*$. Furthermore, if $m, n \geqslant 3$ and $(m,n) \notin \{(3,3),(4,5),(5,4),(5,5)\}$, then every pair of states in S is distinguishable with respect to $F \circ F'$, the set of final states of the direct product.

Theorem 1 (8–11) (Boolean Operations) If $m, n \geqslant 3$, then
- The complexity of $R_m(a, b, d) \cap R_n(b, a, d)$ is mn.
- The complexity of $R_m(a, b, d) \oplus R_n(b, a, d)$ is mn.
- The complexity of $R_m(a, b, d) \setminus R_n(b, a, d)$ is $mn - (m - 1)$.
- The complexity of $R_m(a, b, d) \cup R_n(b, a, d)$ is $mn - (m + n - 2)$.

Proof. In the cases where $(m,n) \in \{(3,3),(4,5),(5,4),(5,5)\}$, we cannot apply Proposition 1, but we have verified computationally that the bounds are met. For the remainder of the proof we assume $(m,n) \notin \{(3,3),(4,5),(5,4),(5,5)\}$.

Our first task is to show that all mn states of $\mathcal{R}_{m,n}$ are reachable. By Proposition 1, all states in the set $S = \{(i,j) \mid 1 \leqslant i \leqslant m-1, 1 \leqslant j \leqslant n-1\}$ are reachable. The remaining states are the ones in the last row or last column (that is, row m or column n) of the direct product.

For $1 \leqslant j \leqslant n-2$, from state $(m-1, j)$ we can reach (m, j) by d. From state $(m, n-2)$ we can reach $(m, n-1)$ by a. From state $(m-1, n-1)$ we can reach (m, n) by d. Hence all states in row m are reachable.

For $1 \leqslant i \leqslant m-2$, from state $(i, n-1)$ we can reach (i, n) by d. From state $(m-2, n)$ we can reach $(m-1, n)$ by a. Hence all states in column n are reachable, and thus all mn states are reachable.

We now count the number of distinguishable states for each operation. Let $H = \{(m, j) \mid 1 \leqslant j \leqslant n\}$ be the set of states in the last row and let $V = \{(i, n) \mid 1 \leqslant i \leqslant m\}$ be the set of states in the last column. If $\circ \in \{\cap, \oplus, \setminus, \cup\}$, then $R_m(a, b, d) \circ R_n(b, a, d)$ is recognized by $\mathcal{R}_{m,n}$, where the set of final states is taken to be $H \circ V$.

Let $H' = \{(m-1, j) \mid 1 \leqslant j \leqslant n-1\}$ and let $V' = \{(i, n-1) \mid 1 \leqslant i \leqslant m-1\}$. By Proposition 1, all states in S are distinguishable with respect to $H' \cap V' = \{(m-1, n-1)\}$. We claim that they are also distinguishable with respect to $H \circ V$ for $\circ \in \{\cap, \oplus, \setminus, \cup\}$.

Distinguishability with respect to $H' \cap V'$ implies that for all pairs of states $(i, j), (k, \ell) \in S$, there exists a word w that sends (i, j) to $(m-1, n-1)$ and sends (k, ℓ) to some other state in S. It follows that the word wd sends (i, j) to (m, n) (which is in $H \cap V$), while (k, ℓ) is sent to a state outside of $H \cap V$. Hence all states in S are distinguishable with respect to $H \cap V$. The same argument works for $H \oplus V$, $H \setminus V$, and $H \cup V$.

Thus for each boolean operation $\circ$, all $(m-1)(n-1) = mn - m - n + 1$ states in S are distinguishable with respect to the final state set $H \circ V$. To show that the complexity bounds are reached by $R_m(a, b, d) \circ R_n(b, a, d)$, it suffices to consider how many of the $m + n - 1$ states in $H \cup V$ are distinguishable with respect to $H \circ V$.

Intersection: Here the set of final states is $H \cap V = \{(m, n)\}$. State (m, n) is the only final state and hence is distinguishable from all the other states. Any two states in H (V) are distinguished by words in b^*d (a^*d). State $(m, 1)$ accepts $b^{n-2}d$, while $(1, n)$ rejects it. For $2 \leqslant i \leqslant n-1$, (m, i) is sent to $(m, 1)$ by b^{n-1-i}, while state $(1, n)$ is not changed by that word. Hence (m, i) is distinguishable from $(1, n)$. By a symmetric argument, (j, n) is distinguishable from $(m, 1)$ for $2 \leqslant j \leqslant m-1$. For $2 \leqslant i \leqslant n-1$ and $2 \leqslant j \leqslant m-1$, (m, i) is distinguished from (j, n) because b^{n-i} sends the former to $(m, 1)$ and the latter to a state of the form (k, n), where $2 \leqslant k \leqslant m-1$. Hence all pairs of states from $H \cup V$ are distinguishable. Siince there are $m + n - 1$ states in $H \cup V$, it follows there are $(mn - m - n + 1) + (m + n - 1) = mn$ distinguishable states.

Symmetric Difference: Here the set of final states is $H \oplus V$, that is, all states in the last row and column except (m, n), which is the only empty state. This situation is complementary to that for intersection. Thus every two states from $H \cup V$ are distinguishable by the same word as for intersection. Hence there are mn distinguishable states.

Difference: Here the set of final states is $H \setminus V$, that is, all states in the last row H except (m, n), which is empty. All other states in the last column V are also empty. The m empty states in V are all equivalent, and the $n-1$ final states in $H \setminus V$ are distinguished in the same way as for intersection. Hence there are $(n-1) + 1 = n$ distinguishable states in $H \setminus V$. It follows there are $(mn - m - n + 1) + n = mn - (m - 1)$ distinguishable states.

Union: Here the set of final states is $H \cup V$. From a state in $H \cup V$ it is only possible to reach other states in $H \cup V$, and all these states are final; so every

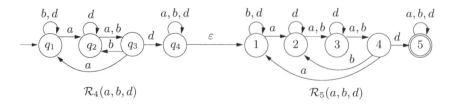

Fig. 4. Right-ideal witnesses for product

state in $H \cup V$ accepts Σ^*. Thus all the states in $H \cup V$ are equivalent, and so there are $(mn - m - n + 1) + 1 = mn - (m + n - 2)$ distinguishable states.   □

Although it is impossible for the stream $(R_n(a, b, d) \mid n \geqslant 3)$ to meet the bound for boolean operations when $m = n$, this stream is as complex as it could possibly be in view of the following theorem proved in [4]:

Theorem 1 (13) (Boolean Operations, $m \neq n$) If $m, n \geqslant 3$ and $m \neq n$,
- The complexity of $R_m(a, b, d) \cap R_n(a, b, d)$ is mn.
- The complexity of $R_m(a, b, d) \oplus R_n(a, b, d)$ is mn.
- The complexity of $R_m(a, b, d) \setminus R_n(a, b, d)$ is $mn - (m - 1)$.
- The complexity of $R_m(a, b, d) \cup R_n(a, b, d)$ is $mn - (m + n - 2)$.

7   Product

We show that the complexity of the product of $R_m(a, b, d)$ with $R_n(a, b, d)$ reaches the maximum possible bound derived in [5]. To avoid confusing states of the two DFAs, we label their states differently. Let $\mathcal{R}_m = \mathcal{R}_m(a, b, d) = (Q'_m, \Sigma, \delta', q_1, \{q_m\})$, where $Q'_m = \{q_1, \ldots, q_m\}$, and let $\mathcal{R}_n = \mathcal{R}_n(a, b, d)$, as in Definition 1. Define the ε-NFA $\mathcal{P} = (Q'_m \cup Q_n, \Sigma, \delta_{\mathcal{P}}, \{q_1\}, \{n\})$, where $\delta_{\mathcal{P}}(q, a) = \{\delta'(q, a)\}$ if $q \in Q'_m$, $a \in \Sigma$, $\delta_{\mathcal{P}}(q, a) = \{\delta(q, a)\}$ if $q \in Q_n$, $a \in \Sigma$, and $\delta_{\mathcal{P}}(q_m, \varepsilon) = \{1\}$. This ε-NFA accepts $R_m R_n$, and is illustrated in Figure 4.

Theorem 1 (14) (Product) For $m \geqslant 1$, $n \geqslant 2$, the complexity of the product $R_m(a, b, d) \cdot R_n(a, b, d)$ is $m + 2^{n-2}$.

Proof. It was shown in [5] that $m + 2^{n-2}$ is an upper bound on the complexity of the product of two right ideals. To prove this bound is met, we apply the subset construction to $\mathcal{P}$ to obtain a DFA $\mathcal{D}$ for $R_m R_n$. The states of $\mathcal{D}$ are subsets of $Q'_m \cup Q_n$. We prove that all states of the form $\{q_i\}$, $i = 1, \ldots, m - 1$ and all states of the form $\{q_m, 1\} \cup S$, where $S \subseteq Q_n \setminus \{1, n\}$, and state $\{q_m, 1, n\}$ are reachable, for a total of $m + 2^{n-2}$ states.

State $\{q_1\}$ is the initial state, and $\{q_i\}$ is reached by a^{i-1} for $i = 2, \ldots, m - 1$. Also, $\{q_m, 1\}$ is reached by $a^{m-2}d$, and states q_m and 1 are present in every subset reachable from $\{q_m, 1\}$. By applying ab^{j-1} to $\{q_m, 1\}$ we reach $\{q_m, 1, j\}$; hence all subsets $\{q_m, 1\} \cup S$ with $|S| = 1$ are reachable. Assume now that we can reach all sets $\{q_m, 1\} \cup S$ with $|S| = k$, and suppose that we want to reach

$\{q_m, 1\} \cup T$ with $T = \{i_0, i_1, \ldots, i_k\}$ with $2 \leqslant i_0 < i_1 < \cdots < i_k \leqslant n - 1$. This can be done by starting with $S = \{i_1 - i_0 + 1, \ldots, i_k - i_0 + 1\}$ and applying ab^{i_0-2}. Finally, to reach $\{q_m, 1, n\}$, start with $\{q_m, 1, n - 1\}$ and apply d.

If $1 \leqslant i < j \leqslant m - 1$, then state $\{q_i\}$ is distinguishable from $\{q_j\}$ by $a^{m-1-j}da^{n-1}d$. Also, state $i \in Q_n$ with $2 \leqslant i \leqslant n - 1$ accepts $a^{n-1-i}d$ and no other state $j \in Q_n$ with $2 \leqslant j \leqslant n - 1$ accepts this word. Hence, if $S, T \subseteq Q_n \setminus \{1, n\}$ and $S \neq T$, then $\{q_m, 1\} \cup S$ and $\{q_m, 1\} \cup T$ are distinguishable. State $\{q_k\}$ with $2 \leqslant k \leqslant m - 1$ is distinguishable from state $\{q_m, 1\} \cup S$ because there is a word with a single d that is accepted from $\{q_m, 1\} \cup S$ but no such word is accepted by $\{q_k\}$. Hence all the non-final states are distinguishable, and $\{q_m, 1, n\}$ is the only final state. $\Box$

8 Conclusion

Our stream of right ideals is a universal witness for all common operations.

References

1. Bell, J., Brzozowski, J., Moreira, N., Reis, R.: Symmetric groups and quotient complexity of boolean operations. In: Esparza, J., Fraigniaud, P., Husfeldt, T., Koutsoupias, E. (eds.) ICALP 2014, Part II. LNCS, vol. 8573, pp. 1–12. Springer, Heidelberg (2014)
2. Brzozowski, J.: Quotient complexity of regular languages. J. Autom. Lang. Comb. 15(1/2), 71–89 (2010)
3. Brzozowski, J.: In search of the most complex regular languages. Internat. J. Found. Comput. Sci. 24(6), 691–708 (2013)
4. Brzozowski, J., Davies, G.: Most complex regular right-ideal languages (2013), http://arxiv.org/abs/1311.4448
5. Brzozowski, J., Jirásková, G., Li, B.: Quotient complexity of ideal languages. Theoret. Comput. Sci. 470, 36–52 (2013)
6. Brzozowski, J., Tamm, H.: Complexity of atoms of regular languages. Int. J. Found. Comput. Sci. 24(7), 1009–1027 (2013)
7. Brzozowski, J., Tamm, H.: Theory of átomata. Theoret. Comput. Sci. 539, 13–27 (2014)
8. Brzozowski, J., Ye, Y.: Syntactic complexity of ideal and closed languages. In: Mauri, G., Leporati, A. (eds.) DLT 2011. LNCS, vol. 6795, pp. 117–128. Springer, Heidelberg (2011)
9. Mirkin, B.G.: On dual automata. Kibernetika (Kiev) 2, 7–10 (1970) (Russian); English translation: Cybernetics 2, 6–9 (1966)
10. Piccard, S.: Sur les bases du groupe symétrique. Časopis Pro Pěstování Matematiky a Fysiky 68(1), 15–30 (1939)
11. Pin, J.E.: Syntactic semigroups. In: Handbook of Formal Languages. Word, Language, Grammar, vol. 1, pp. 679–746. Springer, New York (1997)
12. Yu, S.: State complexity of regular languages. J. Autom. Lang. Comb. 6, 221–234 (2001)

State Complexity of Inversion Operations

Da-Jung Cho[1], Yo-Sub Han[1], Sang-Ki Ko[1], and Kai Salomaa[2]

[1] Department of Computer Science, Yonsei University,
50, Yonsei-Ro, Seodaemun-Gu, Seoul 120-749, Republic of Korea
{dajung,emmous,narame7}@cs.yonsei.ac.kr
[2] School of Computing, Queen's University,
Kingston, Ontario K7L 3N6, Canada
ksalomaa@cs.queensu.ca

Abstract. The reversal operation is well-studied in literature and the deterministic (respectively, nondeterministic) state complexity of reversal is known to be 2^n (respectively, n). We consider the inversion operation where some substring of the given string is reversed. Formally, the inversion of a language L consists of all strings $ux^R v$ such that $uxv \in L$. We show that the nondeterministic state complexity of inversion is in $\Theta(n^3)$. We establish that the deterministic state complexity of the inversion is $2^{\Omega(n \cdot \log n)}$, which is strictly worse than the worst case state complexity of the reversal operation. We also study the state complexity of different variants of the inversion operation, including prefix-, suffix-, and pseudo-inversion.

Keywords: State complexity, Inversion operations, Regular languages.

1 Introduction

Questions of descriptional complexity belong to the very foundations of automata and formal language theory [10, 12, 23, 27]. The state complexity of finite automata has been studied since the 60's [13, 16, 17]. Maslov [15] originated the study of operational state complexity and Yu et al. [27] investigated the state complexity for basic operations. Later, Yu and his co-authors [7, 8, 20, 21] initiated the study on the state complexity of combined operations such as star-of-union, star-of-intersection and so on.

In biology, a *chromosomal inversion* occurs when a segment of a single chromosome breaks and rearranges within itself in reverse order [18]. It is known that the chromosomal inversion often causes genetic diseases [14]. Informally, the inversion operation reverses an infix of a given string. This can be viewed as a generalization of the reversal operation which reverses the whole string. The inversion of a language L is defined as the union of all inversions of strings in L. Therefore, the inversion of L always contains the reversal of L since a string is always an infix of itself.

Many researchers [2, 4–6, 11, 24] have considered the inversion of DNA sequences in terms of formal language theory. Searls [22] considered closure properties of languages under various bio-inspired operations including inversion. Later,

H. Jürgensen et al. (Eds.): DCFS 2014, LNCS 8614, pp. 102–113, 2014.
© Springer International Publishing Switzerland 2014

Yokomori and Kobayashi [26] showed that inversion can be simulated by the set of primitive operations and languages. Dassow et al. [6] investigated a generative mechanism based on some operations inspired by mutations in genomes such as deletion, transposition, duplication and inversion. Daley et al. [4] considered a hairpin inverse operation, which replaces the hairpin part of a string with the inversion of the hairpin part. Note that the hairpin inversion operation is a variation of the inversion operation that reverses substrings of a string. Recently, Cho et al. [3] defined the pseudo-inversion operation and examined closure properties and decidability problems regarding the operation. Moreover, several string matching problems allowing inversions have been studied [2, 24].

Reversal is an "easy" operation for NFAs. The reversal of a regular language L can be, roughly speaking, recognized by an NFA that is obtained by reversing the transitions of an NFA for L and, consequently, the nondeterministic state complexity of the reversal operation is n for NFAs that allow multiple initial states [9][1]. However, a corresponding simple NFA construction does not work for inversion and here we show that the nondeterministic state complexity of inversion is $\Theta(n^3)$. Also we show that the nondeterministic state complexity of prefix- and suffix-inversion is $\Theta(n^2)$. Moreover, we establish the nondeterministic complexity of the pseudo-inversion, which is defined as the reversal of inversion, and the pseudo-prefix- and pseudo-suffix-inversion operations.

It is known that the deterministic state complexity of the reversal operation is 2^n [19]. The inversion operation is, in some sense, an extension of the reversal operation and using this correspondence it is easy to verify that the state complexity of inversion is at least exponential. Based on their nondeterministic state complexity we establish an upper bound 2^{n^3+2n} for the deterministic state complexity of inversion and an upper bound 2^{n^2+n} for the deterministic state complexity of prefix- and suffix-inversion. Also using a non-constant alphabet (of exponential size) we give a lower bound $2^{\Omega(n \cdot \log n)}$ for inversion and prefix-inversion. This establishes that the deterministic state complexity of these operations is strictly worse than the deterministic state complexity of ordinary reversal. For the nondeterministic and deterministic state complexity of pseudo-inversion we establish exactly the same bounds as for inversion.

There remains a possibility that there could be a more efficient DFA construction for the inversion of a language recognized by a given DFA A than first constructing an NFA for the inversion of $L(A)$ and then determinizing the NFA. The precise deterministic state complexity of inversion remains open.

We give the basic notations and definitions in Section 2. We introduce the inversion and related operations in Section 3 and present the state complexity results in Section 4 and in Section 5. In Section 6, we conclude the paper.

---

[1] The result stated in [9] is $n + 1$ because the NFA model used there allows only one initial state.

2 Preliminaries

We briefly present definitions and notations used throughout the paper. The reader may refer to the books [23, 25] for more details on language theory.

Let Σ be a finite alphabet and Σ^* be a set of all strings over Σ. A language over Σ is any subset of Σ^*. The symbol λ denotes the null string and Σ^+ denotes $\Sigma^* \setminus \{\lambda\}$. Given a string $w = z_1 z_2 \cdots z_m$, $z_i \in \Sigma$, $1 \leq i \leq m$, we denote the reversal of w by $w^R = z_m z_{m-1} \cdots z_1$.

A *nondeterministic finite automaton with λ-transitions* (λ-NFA) is a five-tuple $A = (\Sigma, Q, Q_0, F, \delta)$ where Σ is a finite alphabet, Q is a finite set of states, $Q_0 \subseteq Q$ is the set of initial states, $F \subseteq Q$ is the set of final states and δ is a multi-valued transition function from $Q \times (\Sigma \cup \{\lambda\})$ into 2^Q. By an NFA we mean a nondeterministic automaton without λ-transitions, that is, A is an NFA if δ is a function from $Q \times \Sigma$ into 2^Q. The automaton A is *deterministic* (a DFA) if Q_0 is a singleton set and δ is a (total single-valued) function $Q \times \Sigma \rightarrow Q$. It is well known that the λ-NFAs, NFAs and DFAs all recognize the regular languages [19, 23, 25]. Moreover, the language recognized by a λ-NFA A can be recognized also by an NFA (without λ-transitions) of the same size as A [25].

The (right) *Kleene congruence* of a language $L \subseteq \Sigma^*$ is the relation $\equiv_L \subseteq \Sigma^* \times \Sigma^*$ defined by setting, for $x, y \in \Sigma^*$,

$$x \equiv_L y \text{ iff } [(\forall z \in \Sigma^*)\, xz \in L \Leftrightarrow yz \in L].$$

It is well known that L is regular if and only if the index of $\equiv_L$ is finite and, in this case, the number of classes of $\equiv_L$ is equal to the size of the minimal DFA for L [19, 23, 25].

The deterministic (respectively, nondeterministic) state complexity of a regular language L, $\mathrm{sc}(L)$ (respectively, $\mathrm{nsc}(L)$) is the size of the minimal DFA (respectively, the size of a minimal NFA) recognizing L. Thus, $\mathrm{sc}(L)$ is equal to the number of classes of $\equiv_L$.

For the nondeterministic state complexity problem, we show a technique called the *fooling set technique* that gives a lower bound for the size of NFAs.

Proposition 1 (Fooling set technique [1]). *Let $L \subseteq \Sigma^*$ be a regular language. Suppose that there exists a set $P = \{(x_i, w_i) \mid 1 \leq i \leq n\}$ of pairs such that*

(i) $x_i w_i \in L$ for $1 \leq i \leq n$;
(ii) if $i \neq j$, then $x_i w_j \notin L$ or $x_j w_i \notin L$, $1 \leq i, j \leq n$.

Then, a minimal NFA for L has at least n states.

The set P satisfying the conditions of Proposition 1 is called a *fooling set* for the language L.

3 Inversion Operations

We give the formal definition of the inversion as follows:

Definition 1 (Yokomori and Kobayashi [26]). *The inversion of a string w is defined as the set*

$$\mathbb{INV}(w) = \{ux^R v \mid w = uxv, \ u, x, v \in \Sigma^*\}.$$

For instance, given a string $w = abcd$, we have

$$\mathbb{INV}(w) = \{abcd, bacd, cbad, dcba, acbd, abdc, adcb\}.$$

Note that $\mathbb{INV}(\lambda) = \{\lambda\}$. The inversion operation is extended to the languages in the natural way:

$$\mathbb{INV}(L) = \bigcup_{w \in L} \mathbb{INV}(w).$$

We define the *prefix-inversion* that reverses a prefix of a given string and the *suffix-inversion* that reverses a suffix of a given string.

Definition 2. *For a string w, we define the* prefix-inversion *of w as*

$$\mathrm{Pref}\mathbb{INV}(w) = \{u^R x \mid w = ux, \ u, x \in \Sigma^*\}.$$

Definition 3. *For a string w, we define the* suffix-inversion *of w as*

$$\mathrm{Suf}\mathbb{INV}(w) = \{ux^R \mid w = ux, \ u, x \in \Sigma^*\}.$$

See Fig. 1 for examples. Note that the sets $\mathrm{Pref}\mathbb{INV}(L)$ and $\mathrm{Suf}\mathbb{INV}(L)$ are always included in the set $\mathbb{INV}(L)$.

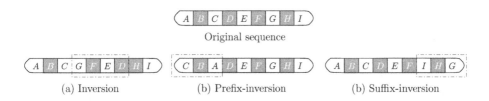

Original sequence

(a) Inversion   (b) Prefix-inversion   (b) Suffix-inversion

Fig. 1. Examples of the inversion operations

As variants of the inversion operations, we consider the *pseudo-inversion* operations [3] which are defined as the reversal of the inversion operations. Informally, the pseudo-inversion of a given string is defined as a set of strings that are obtained by reversing the given string while maintaining a central substring.

Definition 4. *For a string w, we define the* pseudo-inversion *of w as*

$$\mathbb{PI}(w) = \{v^R x u^R \mid w = uxv, \ u, x, v \in \Sigma^*\}.$$

Furthermore, given a set L of strings, $\mathbb{PI}(L) = \bigcup_{w \in L} \mathbb{PI}(w).$

We call the operation the *pseudo-inversion* in the sense that the inversion is not properly performed. Note that $\mathbb{PI}(L) = \mathbb{INV}(L)^R$. We also define similar operations called the *prefix-pseudo-inversion* and *suffix-pseudo-inversion* as follows:

Definition 5. *We define the* prefix-pseudo-inversion *of a string w as*

$$\mathrm{Pref}\mathbb{PI}(w) = \{xu^R \mid w = ux, \; u, x \in \Sigma^*\}.$$

Definition 6. *We define the* suffix-pseudo-inversion *of a string w as*

$$\mathrm{Suf}\mathbb{PI}(w) = \{x^R u \mid w = ux, \; u, x \in \Sigma^*\}.$$

See Fig. 2 for examples of the pseudo-inversion operations.

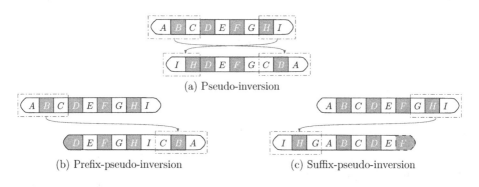

(a) Pseudo-inversion

(b) Prefix-pseudo-inversion   (c) Suffix-pseudo-inversion

Fig. 2. Examples of the pseudo-inversion operations

Lastly, we consider one more non-trivial inversion operation called the *non-overlapping-inversion*. The non-overlapping-inversion operation allows any character in the string to be involved in at most one inversion operation.

Definition 7. *For a string w, we define the* non-overlapping-inversion *of w as*

$$\mathrm{NonO}\mathbb{INV}(w) =$$
$$\{w'_1 w'_2 \cdots w'_n \mid w = w_1 w_2 \cdots w_n, \; w_i \in \Sigma^*, \; w'_i = w_i \text{ or } w'_i = w_i^R \text{ for } 1 \leq i \leq n\}.$$

4   Nondeterministic State Complexity

We establish upper and lower bounds for the nondeterministic state complexity of inversion, prefix-inversion and suffix-inversion. We begin with the upper bound construction of an NFA for $\mathbb{INV}(L)$ when we are given an NFA for a regular language L.

Lemma 1. *Let L be a regular language recognized by an NFA with n states. Then, $\mathbb{INV}(L)$ is recognized by an NFA with $n^3 + 2n$ states.*

Proof. Let $A = (\Sigma, Q, Q_0, F_A, \delta)$ be an NFA for L. We define a λ-NFA $B = (\Sigma, P, P_0, F_B, \gamma)$ for the language $\mathbb{INV}(L)$ where

$$P = Q^3 \cup Q \cup \overline{Q},$$

$\overline{Q} = \{\overline{q} \mid q \in Q\}, P_0 = Q_0, F_B = F_A \cup \overline{F_A}$ and the transition function $\gamma : P \times (\Sigma \cup \{\lambda\}) \to 2^P$ is defined as follows:

(i) For all $q, p \in Q$ and $a \in \Sigma$, if $p \in \delta(q, a)$, then $p \in \gamma(q, a)$ and $\overline{p} \in \gamma(\overline{q}, a)$.

(ii) For all $q, p \in Q$, $(p, q, q) \in \gamma(p, \lambda)$.

(iii) For all $q, p, r_1, r_2 \in Q$ and $a \in \Sigma$, if $r_2 \in \delta(r_1, a)$, then $(p, r_1, q) \in \gamma((p, r_2, q), a)$.

(iv) For all $q, p \in Q, \overline{q} \in \gamma((p, p, q), \lambda)$.

The automaton B operates as follows. The transitions in (i) simulate the original computation of A. For any state $p \in Q$, we choose a state q nondeterministically using a λ-transition, and we reach a state (p, q, q) according to the transitions in (ii). The transitions in (iii) allow B to simulate the computation of A in reverse. Note that the first and third elements in Q^3 remember the start and ending positions of the reversed part, while the second element simulatTes the computation of A in reverse. After B reaches the state (p, p, q), it can make a λ-transition to the state $\overline{q}$, and B continues the original computation of A following the transition (i). Fig. 3 shows the computation of B as an illustrative example. As a consequence of the transitions, B recognizes a string $ux^R v$ if A has an accepting computation for uxv.   □

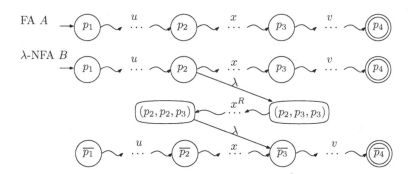

Fig. 3. An illustrative example of constructing an NFA B recognizning $\mathbb{INV}(L(A))$. Note that if A accepts a string uxv, then B accepts $ux^R v \in \mathbb{INV}(L(A))$.

We present the following lower bound using the fooling set technique.

Lemma 2. *For every $n_0 \in \mathbb{N}$, there exists an NFA $A = (Q, \Sigma, Q_0, F_A, \delta)$ with $n \geq n_0$ states over an alphabet of size 4 such that $\mathrm{nsc}(\mathbb{INV}(L(A))) \geq \frac{1}{8}n^3 - f(n)$, where $f(n) \in O(n^2)$.*

Proof. Let $m \geq 1$ be an integer and consider the language $L = \{\#a^m\$, \#b^m\$\}^*$ over the alphabet $\Sigma = \{a, b, \#, \$\}$. We construct a fooling set P for the language $\mathbb{INV}(L(A))$.

Take the set of pairs P to be the set

$$P = \{(\#a^i b^j \# \$ b^k, b^{m-k} \# \$ a^{m-i} b^{m-j} \$) \mid 1 \leq i, j, k \leq m\}.$$

Consider

$$(x, w) = (\#a^i b^j \# \$ b^k, b^{m-k} \# \$ a^{m-i} b^{m-j} \$) \in P.$$

Now the string xw is in $\mathbb{INV}(L)$ because we can write

$$xw = \#a^i (a^{m-i} \$ \# b^m \$ \# b^j)^R b^{m-j} \$.$$

On the other hand, consider another pair

$$(x', w') = (\#a^{i'} b^{j'} \# \$ b^{k'}, b^{m-k'} \# \$ a^{m-i'} b^{m-j'} \$) \in P,$$

where $(i, j, k) \neq (i', j', k')$. Now

$$x \cdot w' = \#a^i b^j \# \$ b^k \cdot b^{m-k'} \# \$ a^{m-i'} b^{m-j'} \$.$$

Now if $xw' \in \mathbb{INV}(L)$ it must be obtained from a string of L by inverting one substring. Since in strings of L the markers $\#$ and $\$$ alternate (when we disregard symbols a and b), the only way we could obtain a string of L from $x \cdot w'$ is to invert a substring z that begins between the first two markers $\#$ and ends between the last two markers $\$$. If $k \neq k'$, the resulting string is not in L. If $k = k'$, necessarily we have $i \neq i'$ or $j \neq j'$ which means again that inverting z cannot produce a string in L.

Hence there are at least $|P| = m^3$ states for any NFA accepting $\mathbb{INV}(L)$ by Proposition 1. It is easy to verify that $n = 2m + 2$ states are sufficient for an NFA that accepts L. Therefore, we have the lower bound $\frac{1}{8}n^3 - f(n)$ for the nondeterministic state complexity of $\mathbb{INV}(L)$, where $f(n) \in O(n^2)$.   □

As a consequence of Lemma 1 and Lemma 2, we have:

Theorem 1. *The nondeterministic state complexity of inversion is in $\Theta(n^3)$.*

Now we consider the upper bound construction for the prefix-inversion which is a restricted variant of the general inversion in the sense that only the prefixes of the given string can be reversed.

Lemma 3. *Let L be a regular language recognized by an NFA with n states. Then, $\text{Pref}\mathbb{INV}(L)$ is recognized by an NFA with $n^2 + n$ states.*

Proof. Let $A = (\Sigma, Q, Q_0, F_A, \delta)$ be an NFA for L. We define a λ-NFA $B = (\Sigma, P, P_0, F_B, \gamma)$ for the language $\text{Pref}\mathbb{INV}(L)$. We choose

$$P = Q^2 \cup Q,$$

where $P_0 = \{(q, q) \mid q \in Q\}$, $F_B = F_A$ and the transition function $\gamma : P \times (\Sigma \cup \{\lambda\}) \to 2^P$ is defined as follows:

(i) For all $p, q \in Q$ and $a \in \Sigma$, if $p \in \delta(q, a)$, then $p \in \gamma(q, a)$.
(ii) For all $p, r_1, r_2 \in Q$ and $a \in \Sigma$, if $r_2 \in \delta(r_1, a)$, then $(p, r_1) \in \gamma((p, r_2), a)$.
(iii) For all $q_0 \in Q_0$ and $r \in Q$, $r \in \gamma((r, q_0), \lambda)$.

The simulation begins in an arbitrary state (q, q), $q \in Q$. The transitions (ii) simulate a computation of A in reverse in the second component of the state, while the first component of the state pair remembers the state where B starts reverse computation of A. After B reaches a state (q, q_0), where $q_0 \in Q_0$, it can make a λ-transition to state q using the rule (iii). The transitions (i) allow B to simulate the original computation of A from q to a final state. Therefore, B accepts exactly all strings $u^R x$ where A has an accepting computation for ux.

$\square$

We also establish that the nondeterministic state complexity of the suffix-inversion coincides with that of the prefix-inversion:

Lemma 4. *Let L be a regular language recognized by an NFA with n states. Then, $\mathrm{SufINV}(L)$ is recognized by an NFA with $n^2 + n$ states.*

Next we give the following lower bound for the nondeterministic state complexity of the prefix- and suffix-inversion. Using an analogous fooling set construction as in the proof of Lemma 2 we can easily get an $\Omega(n^2)$ lower bound for the nondeterministic state complexity of suffix-inversion.

Lemma 5. *For every $n_0 \in \mathbb{N}$, there exists an NFA A with $n \geq n_0$ states over an alphabet Σ of size 4 such that $\mathrm{nsc}(\mathrm{PrefINV}(L(A))) \geq \frac{1}{4}n^2 - f(n)$, where $f(n) \in O(n)$.*

We have the following statement based on Lemma 3, Lemma 4 and Lemma 5.

Theorem 2. *The nondeterministic state complexity of prefix- and suffix-inversion is in $\Theta(n^2)$.*

The following Observation 3 is now immediate since the state complexity of the reversal operation is n.

Observation 3. *The following statements hold:*

(i) $\mathrm{nsc}(\mathrm{SufINV}(L)) = \mathrm{nsc}(\mathrm{PrefPI}(L))$,
(ii) $\mathrm{nsc}(\mathrm{PrefINV}(L)) = \mathrm{nsc}(\mathrm{SufPI}(L))$, *and*
(iii) $\mathrm{nsc}(\mathrm{INV}(L)) = \mathrm{nsc}(\mathrm{PI}(L))$.

Based on Observation 3, we establish the following results.

Corollary 1. *The nondeterministic state complexity of prefix- and suffix-pseudo-inversion is in $\Theta(n^2)$.*

Corollary 2. *The nondeterministic state complexity of pseudo-inversion is in $\Theta(n^3)$.*

Now we discuss the nondeterministic state complexity of non-overlapping-inversion. Interestingly, we have slightly smaller upper bound for the non-overlapping-inversion than the upper bound for the general inversion.

Lemma 6. *Let L be a regular language recognized by an NFA with n states. Then, $\text{NonO}\mathbb{I}\text{NV}(L)$ is recognized by an NFA with $n^3 + n$ states.*

Proof. Let a λ-NFA B be an automaton for $\text{NonO}\mathbb{I}\text{NV}(L)$. The computation of the automaton B is similar to the computation in the proof of Lemma 1. But, the set of states $\overline{Q}$ and the transitions in (iv) of the proof of Lemma 1 are useless for the language $\text{NonO}\mathbb{I}\text{NV}(L)$ since the non-overlapping-inversion $\text{NonO}\mathbb{I}\text{NV}$ allows more than one inversion without overlap. Note that the automaton B can make a λ-transition to the state q if B reaches the state (p, p, q).   □

5   Deterministic State Complexity

We first consider the deterministic state complexity of $\text{Pref}\mathbb{I}\text{NV}(L)$. Recall that if A is an NFA with n states, Lemma 3 gives a construction of an NFA with $n^2 + n$ states for the language $\text{Pref}\mathbb{I}\text{NV}(L(A))$. This implies an upper bound for the deterministic state complexity $2^{n^2 + n}$ of prefix-inversion. Next we present a lower bound $2^{\Omega(n \cdot \log n)}$ using an alphabet of size 5.

Lemma 7. *For $n \in \mathbb{N}$ there exists an alphabet of size 5 and a DFA A with $2n + 3$ states such that the minimal DFA for $\text{Pref}\mathbb{I}\text{NV}(L(A))$ has size at least $2^{n \cdot \log n}$.*

Proof. We define $L_n \subseteq \Sigma_n^3$ by setting

$$L_n = \{1^j \cdot [1^{i_0}, \ldots, 1^{i_m}] \cdot 1^{i_j} \mid j < n, m \ge j\}.$$

Note that all strings of L_n have length exactly three.
    Consider a DFA $A = (\Sigma_n, Q, q_0, \{q_{\text{acc}}\}, \delta)$, where

$$Q = \{q_0, q_1, \ldots, q_n, p_1, \ldots, p_n\} \cup \{q_{\text{acc}}, q_{\text{dead}}\},$$

and the transitions of δ are defined by setting

  (i) $\delta(q_0, i) = q_i$, $i \in [n]$,
  (ii) $\delta(q_i, f) = p_{f(i)}$, $i \in [n]$, $f \in \text{func}_n$,
  (iii) $\delta(p_i, i) = q_{\text{acc}}$, $i \in [n]$,
  (iv) all transitions not defined above go to the dead state q_{dead}.

It is clear that $L(A) = L_n$. We show that any distinct alphabet symbols $f_1, f_2 \in \text{func}_n$ belong to distinct classes of $\equiv_{\text{Pref}\mathbb{I}\text{NV}(L_n)}$ which gives a lower bound for the size of a minimal DFA for $\text{Pref}\mathbb{I}\text{NV}(L_n)$.
    If $f_1 \ne f_2$, there exists $i \in [n]$ such that $f_1(i) \ne f_2(i)$. Now $i \cdot f_1 \cdot f_1(i) \in L_n$ and hence $f_1 \cdot i \cdot f_1(i) \in \text{Pref}\mathbb{I}\text{NV}(L_n)$.
    On the other hand, since all words of L_n have length three and have an element of func_n in the middle position, the only way $f_2 \cdot i \cdot f_1(i)$ could be in $\text{Pref}\mathbb{I}\text{NV}(L_n)$

is that $i \cdot f_2$ would be the prefix of a word of L_n that has been reversed. However, since $f_2(i) \neq f_1(i)$, $i \cdot f_2 \cdot f_1(i) \notin L_n$ and hence $f_2 \cdot i \cdot f_1(i) \notin \text{PrefINV}(L_n)$ and, consequently, $f_1 \not\equiv_{\text{PrefINV}(L_n)} f_2$.

Thus, $\equiv_{\text{PrefINV}(L_n)}$ has at least

$$|\text{func}_n| = n^n = 2^{n \cdot \log n}$$

equivalence classes. $\square$

Note that the deterministic state complexity of the prefix-inversion is strictly worse than the deterministic state complexity of reversal since the latter is known to be 2^n. Moreover, it is easily verified that the same lower bound as in the proof of Lemma 7 applies to the inversion and non-overlapping-inversion operations.

Lemma 8. *A lower bound for the deterministic state complexity of inversion and of non-overlapping-inversion is $2^{\Omega(n \cdot \log n)}$.*

As a consequence of Lemma 1,3 and 7 we have:

Theorem 4. *Let L be a regular language having a DFA with n states. Then $sc(\text{PrefINV}(L)) \leq 2^{n^2+n}$. There exist languages $L(n)$ defined over an alphabet depending on n such that $sc(L(n)) \in O(n)$, $sc(\text{PrefINV}(L(n))) \geq 2^{n \cdot \log n}$.*

Theorem 5. *Let L be a regular language having a DFA with n states. Then $sc(\text{INV}(L)) \leq 2^{n^3+2n}$. There exist languages $L(n)$ defined over an alphabet depending on n such that $sc(L(n)) \in O(n)$ and $sc(\text{INV}(L(n))) \geq 2^{n \cdot \log n}$.*

Theorem 5 leaves a larger gap between the state complexity upper and lower bounds for inversion than was the case for prefix-inversion.

The method of Lemma 7 is based on the idea that, in order to force the DFA to remember more information, we want to move the function symbols to the beginning of the string, and this construction does not seem to work for suffix-inversion. For the state complexity of suffix-inversion, the best immediate lower bound is 2^n. The same situation applies for suffix-pseudo-inversion.

Lastly, we observe that similar state complexity lower bounds apply for pseudo-inversion and prefix-pseudo-inversion. The below corollary again follows from the proof of Lemma 7 in the same way as Lemma 8.

Lemma 9. *A lower bound for the deterministic state complexity of pseudo-inversion and prefix-pseudo-inversion is $2^{\Omega(n \cdot \log n)}$.*

6 Conclusion

We have considered the (non)deterministic state complexity of inversion operations that are motivated by evolutionary operations on DNA sequences. While the reversal operation completely reverses the whole string, the inversion operation reverses any infix of a string. Initially, one might think that the state complexity of inversion operations should be similar to that of the reversal operation. However, both the nondeterministic and deterministic state complexity of the inversion have turned out to be strictly worse than the known bounds for the reversal operation. The prefix- and suffix-inversions which are simplified variants of inversion were also considered. We have shown that the nondeterministic state complexity of prefix- and suffix-inversion is $\Theta(n^2)$ while that of the inversion operation is $\Theta(n^3)$.

We have also obtained a deterministic state complexity lower bound $2^{\Omega(n \cdot \log n)}$ for inversion and prefix-inversion operations using an exponential-size alphabet. This is strictly worse than the deterministic state complexity of reversal, however, it does not match the corresponding upper bounds. It is possible that given a DFA A for L, there is a more efficient construction of a DFA for $\mathbb{INV}(L)$ than first constructing an NFA and then determinizing it. However, when working on this question it seems that a DFA for $\mathbb{INV}(L)$ needs to remember sets of triples of states of A (and similarly a DFA for the prefix-inversion or suffix-inversion needs to remember sets of pairs of states of A). The main open question is to determine the precise deterministic state complexity for inversion and its variants.

References

1. Birget, J.-C.: Intersection and union of regular languages and state complexity. Information Processing Letters 43(4), 185–190 (1992)
2. Cantone, D., Cristofaro, S., Faro, S.: Efficient string-matching allowing for non-overlapping inversions. Theoretical Computer Science 483, 85–95 (2013)
3. Cho, D.-J., Han, Y.-S., Kang, S.-D., Kim, H., Ko, S.-K., Salomaa, K.: Pseudo-inversion on formal languages. In: Proceeding of the 13th International Conference on Unconventional and Natural Computation (to appear)
4. Daley, M., Ibarra, O.H., Kari, L.: Closure and decidability properties of some language classes with respect to ciliate bio-operations. Theoretical Computer Science 306(1-3), 19–38 (2003)
5. Dassow, J., Mitrana, V.: Operations and language generating devices suggested by the genome evolution. Theoretical Computer Science 270(12), 701–738 (2002)
6. Dassow, J., Mitrana, V., Salomaa, A.: Context-free evolutionary grammars and the structural language of nucleic acids. Biosystems 43(3), 169–177 (1997)
7. Ésik, Z., Gao, Y., Liu, G., Yu, S.: Estimation of state complexity of combined operations. Theoretical Computer Science 410(35), 3272–3280 (2009)
8. Gao, Y., Salomaa, K., Yu, S.: The state complexity of two combined operations: Star of catenation and star of reversal. Fundamenta Informaticae 83(1-2), 75–89 (2008)
9. Holzer, M., Kutrib, M.: Nondeterministic descriptional complexity of regular languages. International Journal of Foundations of Computer Science 14(6), 1087–1102 (2003)
10. Holzer, M., Kutrib, M.: Descriptional and computational complexity of finite automata – a survey. Information and Computation 209, 456–470 (2011)
11. Kececioglu, J.D., Sankoff, D.: Exact and approximation algorithms for the inversion distance between two chromosomes. In: Apostolico, A., Crochemore, M., Galil, Z., Manber, U. (eds.) CPM 1993. LNCS, vol. 684, pp. 87–105. Springer, Heidelberg (1993)

12. Kutrib, M., Pighizzini, G.: Recent trends in descriptional complexity of formal languages. Bulletin of the EATCS 111, 70–86 (2013)
13. Lupanov, O.: A comparison of two types of finite sources. Problemy Kibernetiki 9, 328–335 (1963)
14. Lupski, J.R.: Genomic disorders: structural features of the genome can lead to DNA rearrangements and human disease traits. Trends in Genetics 14(10), 417–422 (1998)
15. Maslov, A.: Estimates of the number of states of finite automata. Soviet Mathematics Doklady 11, 1373–1375 (1970)
16. Meyer, A., Fisher, M.: Economy of description by automata, grammars and formal systems. In: Proceedings of the 12th Annual Symposium on Switching and Automata Theory, pp. 188–191 (1971)
17. Moore, F.: On the bounds for state-set size in the proofs of equivalence between deterministic, nondeterministic and two-way finite automata. IEEE Transactions on Computers C-20, 1211–1214 (1971)
18. Painter, T.S.: A New Method for the Study of Chromosome Rearrangements and the Plotting of Chromosome Maps. Science 78, 585–586 (1933)
19. Rozenberg, G., Salomaa, A. (eds.): Handbook of Formal Languages. Beyond Words, vol. 3. Springer-Verlag New York, Inc. (1997)
20. Salomaa, A., Salomaa, K., Yu, S.: State complexity of combined operations. Theoretical Computer Science 383(2-3), 140–152 (2007)
21. Salomaa, K., Yu, S.: On the state complexity of combined operations and their estimation. International Journal of Foundations of Computer Science 18, 683–698 (2007)
22. Searls, D.B.: The Computational Linguistics of Biological Sequences. In: Artificial Intelligence and Molecular Biology, pp. 47–120 (1993)
23. Shallit, J.: A Second Course in Formal Languages and Automata Theory, 1st edn. Cambridge University Press, New York (2008)
24. Vellozo, A.F., Alves, C.E.R., do Lago, A.P.: Alignment with non-overlapping inversions in $O(n^3)$-time. In: Bücher, P., Moret, B.M.E. (eds.) WABI 2006. LNCS (LNBI), vol. 4175, pp. 186–196. Springer, Heidelberg (2006)
25. Wood, D.: Theory of Computation. Harper & Row (1986)
26. Yokomori, T., Kobayashi, S.: DNA evolutionary linguistics and RNA structure modeling: A computational approach. In: Proceedings of INBS 1995, pp. 38–45. IEEE Computer Society (1995)
27. Yu, S., Zhuang, Q., Salomaa, K.: The state complexities of some basic operations on regular languages. Theoretical Computer Science 125(2), 315–328 (1994)

Cycles and Global Attractors
of Reaction Systems*

Enrico Formenti[1], Luca Manzoni[1], and Antonio E. Porreca[2]

[1] Univ. Nice Sophia Antipolis, CNRS, I3S, UMR 7271
06900 Sophia Antipolis, France
formenti@unice.fr, luca.manzoni@i3s.unice.fr
[2] Dipartimento di Informatica, Sistemistica e Comunicazione
Università degli Studi di Milano-Bicocca
Viale Sarca 336/14, 20126 Milano, Italy
porreca@disco.unimib.it

Abstract. Reaction systems are a recent formal model inspired by the chemical reactions that happen inside cells and possess many different dynamical behaviours. In this work we continue a recent investigation of the complexity of detecting some interesting dynamical behaviours in reaction system. We prove that detecting global behaviours such as the presence of global attractors is PSPACE-complete. Deciding the presence of cycles in the dynamics and many other related problems are also PSPACE-complete. Deciding bijectivity is, on the other hand, a coNP-complete problem.

1 Introduction

This paper completes the investigations started in [7], in which we studied the complexity of a collection of problems related to finding fixed points and local fixed points attractors in *reaction systems* (RS). Here we study the complexity of determining the existence of cycles and global attractors (either fixed points or cycles) in the dynamics of RS. In the first half of this investigation [7], the problems studied were either NP, coNP, or at most Π_2^P-complete. We show that moving from fixed points to cycles and from local to global attractors pushes the complexity to PSPACE in the majority of the cases. Recall that RS are a computational model, inspired by chemical reactions, recently introduced by Ehrenfeucht and Rozenberg [6]. After its introduction, many different aspects of the model were investigated [4,5,2,13]. Indeed, the success of these systems is essentially due to the fact that they can be used to study practical problems [3] and, at the same time, they are clean and formal enough to allow formal investigations. Roughly speaking, a reaction system is made of some finite set and

---

* This work has been supported by the French National Research Agency project EMC (ANR-09-BLAN-0164) and by Fondo d'Ateneo (FA) 2013 of Università degli Studi di Milano-Bicocca: "Complessità computazionale in modelli di calcolo bioispirati: Sistemi a membrana e sistemi a reazioni".

a list of generating rules. The finite set consists of entities, or chemical species, that are used as reactants, inhibitor and products. Finally, a generation rule is activated if the reactants are present and if the inhibitors are not present and then it replaces the current set with the set of products.

The study of the dynamical behaviour of RS is one of the current trends in this domain. For example, in [5], the authors analysed the dynamical behaviour that can be obtained under the constraint of limited resources. In [4], the author considered some particular state as a death state (the empty set in this case) and computed the probability of a system to reach the death state.

Notice that, from a certain point of view, the dynamics of reaction systems are pretty well-known. Indeed, these are finite systems and hence their dynamics is always ultimately periodic. However, there are very interesting questions which are both useful for practical applications and highly non-trivial. For example, one can ask if a particular product will appear at a certain point of the dynamics or not [15,14]. The present paper follows this trend by greatly extending the first results on complexity proved in [6,15,14], where the idea that RS can be used to evaluate Boolean formulae was introduced. In particular, we investigate the complexity of establishing if a RS admits a fixed point global attractor (PSPACE-complete). We also study the complexity of finding if two RS share all fixed points that are global attractors (PSPACE-complete). This is in some sense a concept of equivalence *w.r.t.* global attractors. We also explore the difficulty of finding if a state is part of cycle, a local attractor cycle, or a global attractor cycle, resulting in all cases in PSPACE-completeness. On the other hand, deciding if a RS admits a local attractor cycle is NP-complete. The other decision problems studied are about finding if two RS share one or all of their local attractor cycles (both PSPACE-complete).

The paper is structured as follows. Section 2 provides the basic notions on RS and two lemmata that will be used in the remaining part of the paper. Section 3 gives a description in logical terms of the problems we consider. The decision problems regarding global fixed points attractors are presented in Section 4. The decision problems regarding cycles are investigated in Section 5. A summary of the results and a hint at possible future developments is given in Section 6.

2   Basic Notions

This section provides a brief recollection of all the basic notions of RS and of dynamical systems necessary for the rest of the paper. Notations are taken from [6]. First of all, we recall the definitions of *reaction, reaction system,* and of their dynamics.

Definition 1. *Consider a finite set S, whose elements are called* entities. *A reaction a over S is a triple (R_a, I_a, P_a) of subsets of S. The set R_a is the set of* reactants, *I_a the set of* inhibitors, *and P_a is the set of* products. *The set of all reactions over S is denoted by* $\mathrm{rac}(S)$.

Definition 2. *A reaction system $\mathcal{A}$ is a pair (S, A) where S is a finite set, called the* background set, *and $A \subseteq \mathrm{rac}(S)$.*

Given a *state* $T \subseteq S$, a reaction a is said to be *enabled* in T when $R_a \subseteq T$ and $I_a \cap T = \varnothing$. The *result function* $\text{res}_a : 2^S \to 2^S$ of a, where 2^S denotes the power set of S, is defined as

$$\text{res}_a(T) = \begin{cases} P_a & \text{if } a \text{ is enabled in } T \\ \varnothing & \text{otherwise.} \end{cases}$$

The definition of res_a naturally extends to sets of reactions. Indeed, given $T \subseteq S$ and $A \subseteq \text{rac}(S)$, define $\text{res}_A(T) = \bigcup_{a \in A} \text{res}_a(T)$. The result function res_A of a RS $\mathcal{A} = (S, A)$ is res_A, *i.e.*, it is the result function on the whole set of reactions.

Example 1 (NAND gate). To implement a NAND gate using a RS we use as background set $S = \{0_a, 1_a, 0_b, 1_b, 0_{\text{out}}, 1_{\text{out}}\}$. The first four elements represent the two inputs (denoted by the subscripts a and b), the last two, on the other hand, denote the two possible outputs. The reactions used to model a NAND gate are the followings: $(\{0_a, 0_b\}, \varnothing, \{1_{\text{out}}\})$, $(\{0_a, 1_b\}, \varnothing, \{1_{\text{out}}\})$, $(\{1_a, 0_b\}, \varnothing, \{1_{\text{out}}\})$, and $(\{1_a, 1_b\}, \varnothing, \{0_{\text{out}}\})$. Similarly to NAND gates, others gates can be simulated and it is possible to build circuits with gates of limited fan-in using only a number of entities and reactions that is linear in the size of the modeled circuit.

After this brief introduction of RS, the reader may notice that there exist different bio-inspired models that are, in some sense, similar or related to RS. The most prominent are membrane systems [12], chemical reaction networks [17], and Boolean automata networks (BAN) [10,16]. In the first two cases the object of the evolution is a multiset of symbols. In the last case one can show that, for each RS, there exists a BAN which simulates it and which is polynomial in the size of the given RS. However, an RS simulating a BAN may be exponentially larger than the original BAN [7].

We can now proceed by recalling the necessary definitions of the dynamical properties investigated in this work. Given a set $T \subseteq S$, the set of states visited by T is $(T, \text{res}_A(T), \text{res}_A^2(T), \ldots)$, that is, the sequence of $\text{res}_A^i(T)$ for $i \in \mathbb{N}$. Since 2^S is finite, every sequence of states is ultimately periodic, *i.e.*, there exists $h, k \in \mathbb{N}$ such that for all $t \in \mathbb{N}$, $\text{res}_A^{h+kt}(T) = \text{res}_A^{h+k}(T)$. The integer h is usually called the *length of the transient* and k the *length of the period*. A state $T \subseteq S$ is in a cycle if there exists $k \in \mathbb{N}$ such that $\text{res}_A^k(T) = T$. The smallest such k is the length of the cycle. If T is in a cycle of length 1 we say that T is a *fixed point*.

The notion of attractor is a central concept in the study of dynamical systems. Recall that an invariant set for $\mathcal{A}$ is a set of states $\mathcal{U}$ with $\bigcup_{U \in \mathcal{U}} \{\text{res}_A(U)\} = \mathcal{U}$. For RS all invariant sets consist of cycles. A local attractor in a RS is an invariant set $\mathcal{U}$ such that there exists $T \notin \mathcal{U}$ with $\text{res}_A(T) \in \mathcal{U}$. Intuitively, a local attractor is a set of states $\mathcal{U}$ from which the dynamics never escapes and such that there exists at least one external state whose dynamics ends up in $\mathcal{U}$. A global attractor for an RS $\mathcal{A}$ is an invariant set of states $\mathcal{U}$ such that for all $T \in 2^S$ there exists $t \in \mathbb{N}$ such that $\text{res}_A^t(T) \in \mathcal{U}$. A global attractor $\mathcal{U}$ a *global fixed-point attractor* if $\mathcal{U} = \{T\}$ and hence, necessarily, T is a fixed point. Similarly, we call $\mathcal{U}$ a *global attractor cycle* if all the states in $\mathcal{U}$ belong to the same cycle.

2.1   Counters and Turing Machines

We conclude this section with two technical results that are useful in the sequel. The first one tells us that it is always possible to build a RS containing a cycle of a given period. In the constructions that will follow, this cycle will be used as a kind of internal counter. This construction is inspired by Brijder et al. [2]. The second lemma ensures that it is possible to simulate a Turing machine (TM) with a family of RS and fixes the bounds for some simulation parameters. For an introduction, basic results, and notation on Turing machines we refer the reader to [8].

Lemma 1. *For every integer $k \geq 1$ there exists a RS $C_k = (S, A)$ with background set of cardinality $n = |S| = \lceil \log_2 k \rceil$ having a cycle of period k reachable from any initial configuration after at most $2^n - k$ steps, i.e,, C_k is a binary counter of n bits overflowing at $k \leq n$.*

Proof. Let $S = \{b_0, \ldots, b_{n-1}\}$. A subset B of S is interpreted as an n-bit integer m defined as $\sum_{b_i \in B} 2^i$, i.e., the i-th bit of m is 1 iff $b_i \in B$. Let B_m be the subset of S representing the integer m. We shall design the reactions in A in order to obtain the following result function

$$\text{res}_{C_k}(B_m) = \begin{cases} B_{(m+1) \bmod k} & \text{if } m < k \\ B_{(m+1) \bmod 2^n} & \text{if } m \geq k \end{cases} \tag{1}$$

where $(x \bmod y)$ is the remainder of the integer division of x and y. The RS can thus be seen as a binary counter overflowing at k; if the state instead represents a number larger than $k - 1$, the counter is increased normally until it overflows at 2^n.

The set A contains the following reactions:

$$(\{b_i\} \cup \{b_\ell\}, \{b_j\}, \{b_i\}) \qquad \text{for } 0 \leq i < n,\, 0 \leq j < i,\, b_\ell \notin B_{k-1} \tag{2}$$

$$(\{b_i\}, \{b_j\} \cup \{b_\ell\}, \{b_i\}) \qquad \text{for } 0 \leq i < n,\, 0 \leq j < i,\, b_\ell \in B_{k-1} \tag{3}$$

$$(\{b_0, \ldots, b_{i-1}\} \cup \{b_\ell\}, \{b_i\}, \{b_i\}) \quad \text{for } 0 \leq i < n,\, b_\ell \notin B_{k-1} \tag{4}$$

$$(\{b_0, \ldots, b_{i-1}\}, \{b_i\} \cup \{b_\ell\}, \{b_i\}) \quad \text{for } 0 \leq i < n,\, b_\ell \in B_{k-1}. \tag{5}$$

Reactions of types (2) and (3) preserve the bits set to 1 if there is a less significant bit set to 0, since they are not going to be modified by the increment operation. Reactions of types (4) and (5) set a currently null bit to 1 when all the less significant bits are 1. Notice that the empty state (corresponding to a null value of the counter) is mapped to state $\{b_0\}$ by the reactions of type (5) having no reactants. By construction, none of these reactions are enabled when the current state of C_k is B_{k-1}, because reactions of types (2) and (4) require the missing b_ℓ, and those of types (3) and (5) are inhibited by b_ℓ occurring in B_{k-1}.

Hence, the RS C_k defines the result function of Equation 1, and has a cycle of period k with a pre-period of at most $2^n - k < k$ steps.   □

It is important to remark that the RS C_k of Lemma 1 can be constructed in polynomial time from the binary encoding of the integer k.

Theorem 1. *Let M be a Turing machine, $x \in \{0,1\}^*$, and $m \geq |x|$ an integer. Then, there exists a RS $\mathcal{M} = \mathcal{M}_{M,m} = (S, A)$ and a state $X \subseteq S$ such that M reaches its final state q_F in t steps on input x, using at most m tape cells if and only if $\text{res}^t_{\mathcal{M}}(X) = \{q_F\}$.*

Proof. Let M be a Turing machine $(Q, \Sigma, \Gamma, \delta, q_I, q_F)$, where Q is the set of states, $\Sigma = \{0,1\}$ is the input alphabet, $\Gamma = \Sigma \cup \{\triangleright\}$ is the tape alphabet, $q_I, q_F \in Q$ are the initial and final states, respectively (we assume $q_I \neq q_F$), and the transition function is $\delta \colon (Q - \{q_F\}) \times \Gamma \to Q \times \Gamma \times \{-1, 0, +1\}$. Without loss of generality, we assume that M accepts by final state and rejects by diverging and that the tape is delimited on the left by $\triangleright$, a symbol that is never written or overwritten by M, and that forces the tape head to move to the right.

The RS $\mathcal{M}$ has $S = \{q_i : q \in Q - \{q_F\}, -1 \leq i \leq m\} \cup \{q_F\} \cup \{w_0, \ldots, w_{m-1}\}$ as its set of entities. A configuration of M is encoded as a subset of S as follows: q_i is present when the machine is in state q and the head is positioned on the i-th tape cell (the final state q_F has no subscript); the presence (resp., the absence) of w_i indicates that the i-th tape cell of M contains the symbol 1 (resp., the symbol 0). The state of the Turing machine tape in the RS is preserved by means of reactions which rewrite the entities w_i into themselves when there is no object q_i, (*i.e.*, the head of the Turing machine is not on the i-th tape cell). The entity q_i rewrite itself by looking at the presence or absence of the entity w_i (*i.e.*, by reading the tape) and according to the transition function of the Turing machine.

A transition $\delta(q, \sigma) = (r, \tau, d)$ of M not entering the final state is implemented by the following reactions

$(\{q_i\}, \{w_i, q_F\}, \{r_{i+d}\})$	for $0 \leq i < m$, if $\sigma = 0$, $\tau = 0$, $r \neq q_F$
$(\{q_i, w_i\}, \{q_F\}, \{r_{i+d}\})$	for $0 \leq i < m$, if $\sigma = 1$, $\tau = 0$, $r \neq q_F$
$(\{q_i\}, \{w_i, q_F\}, \{r_{i+d}, w_i\})$	for $0 \leq i < m$, if $\sigma = 0$, $\tau = 1$, $r \neq q_F$
$(\{q_i, w_i\}, \{q_F\}, \{r_{i+d}, w_i\})$	for $0 \leq i < m$, if $\sigma = 1$, $\tau = 1$, $r \neq q_F$.

If the transition enters the final state, we produce the symbol q_F with no subscript, e.g., $\delta(q, 0) = (q_F, 1, -1)$ is simulated by $(\{q_i\}, \{w_i, q_F\}, \{q_F, w_i\})$ for $0 \leq i < m$. If the tape head exceeds the space constraint of m cells (*i.e.*, we have an item of the form q_m), we do not generate any other state, effectively halting the simulation.

If the tape head reaches the $\triangleright$ symbol (cell -1) in state q, the RS simulates the subsequent transition $\delta(q, \triangleright) = (r, \triangleright, +1)$ as $(\{q_{-1}\}, \{q_F\}, \{r_0\})$.

Finally, in order to preserve the portion of tape not currently scanned by M we use the reactions $(\{w_i\}, Q_i, \{w_i\})$ for $0 \leq i < m$, where $Q_i = \{q_i : q \in Q\}$.

Now let $X = \{q_{I,0}\} \cup \{w_i : x_i = 1\}$, where $x = x_1 \cdots x_n$ is the input string for M. Each iteration of $\text{res}_{\mathcal{M}}$ starting from X simulates a computation step of M, as long as M does not exceed m tape cells; if this happens, the simulation stops at a fixed point state not containing q_F. The statement follows. $\square$

Notice that the RS $\mathcal{M}$ of Theorem 1 can be built in polynomial time *w.r.t.* the size of the description of the Turing machine M, the length of the input x, and the integer m.

3   Logical Description

This section recalls a logical description of RS and formulae related to their dynamics (see [7] for its first introduction) that will be used in the rest of the paper. This description (or a slight adaptation) will usually be sufficient for proving membership in many complexity classes. For the background notions of logic and descriptive complexity we refer the reader to the classical book of Neil Immerman [9].

We describe a RS $\mathcal{A} = (S, A)$ where the background set is $S \subseteq \{0, \ldots, n-1\}$ and $|A| \leq n$ by the vocabulary $(\mathsf{S}, \mathsf{R}_{\mathcal{A}}, \mathsf{I}_{\mathcal{A}}, \mathsf{P}_{\mathcal{A}})$, where S is a unary relation symbol and $\mathsf{R}_{\mathcal{A}}$, $\mathsf{I}_{\mathcal{A}}$, and $\mathsf{P}_{\mathcal{A}}$ are binary relation symbols. The symbols have the following intended meaning: the background set is $S = \{i : \mathsf{S}(i)\}$ and each of reaction $a_j = (R_j, I_j, P_j) \in A$ is described by the three sets $R_j = \{i \in S : \mathsf{R}_{\mathcal{A}}(i,j)\}$, $I_j = \{i \in S : \mathsf{I}_{\mathcal{A}}(i,j)\}$, and $P_j = \{i \in S : \mathsf{P}_{\mathcal{A}}(i,j)\}$.

We will also use some additional vocabularies: $(\mathsf{S}, \mathsf{R}_{\mathcal{A}}, \mathsf{I}_{\mathcal{A}}, \mathsf{P}_{\mathcal{A}}, \mathsf{T})$, where T is a unary relation that represents a subset of S, $(\mathsf{S}, \mathsf{R}_{\mathcal{A}}, \mathsf{I}_{\mathcal{A}}, \mathsf{P}_{\mathcal{A}}, \mathsf{T}_1, \mathsf{T}_2)$ with two additional unary relations that represent sets, and $(\mathsf{S}, \mathsf{R}_{\mathcal{A}}, \mathsf{I}_{\mathcal{A}}, \mathsf{P}_{\mathcal{A}}, \mathsf{R}_{\mathcal{B}}, \mathsf{I}_{\mathcal{B}}, \mathsf{P}_{\mathcal{B}})$ that denotes two RS having the same background set. The following formulae describe basic properties of $\mathcal{A}$. The first is true if a reaction a_j is enabled in T: $\mathrm{EN}_{\mathcal{A}}(j, T) \equiv \forall i (\mathsf{S}(i) \Rightarrow (\mathsf{R}_{\mathcal{A}}(i,j) \Rightarrow T(j)) \wedge (\mathsf{I}_{\mathcal{A}}(i,j) \Rightarrow \neg T(j)))$ the latter is: $\mathrm{RES}_{\mathcal{A}}(T_1, T_2) \equiv \forall i (\mathsf{S}(i) \Rightarrow (T_2(i) \Leftrightarrow \exists j (\mathrm{EN}_{\mathcal{A}}(j, T_1) \wedge \mathsf{P}_{\mathcal{A}}(i,j))))$ and it is verified if $\mathrm{res}_{\mathcal{A}}(T_1) = T_2$ for $T_1, T_2 \subseteq S$. Notice that $\mathrm{EN}_{\mathcal{A}}$ and $\mathrm{RES}_{\mathcal{A}}$ are both first-order (FO) formulae.

Since FO logic is insufficient for our purposes, we will formulate some problems using stronger logics: existential second order logic SO∃ characterising NP (Fagin's theorem); universally quantified second order logic SO∀ giving coNP; second order logic with a transitive closure operator (SO(TC), characterizing PSPACE). We denote the transitive closure of a formula $\varphi(X, Y)$ with two free second-order variables by $\varphi^{\star}(X, Y)$. We define the bounded second order quantifiers $(\forall X \subseteq Y)\varphi$ and $(\exists X \subseteq Y)\varphi$ as shorthands for $\forall X (\forall i (X(i) \Rightarrow Y(i)) \Rightarrow \varphi)$ and $\exists X (\forall i (X(i) \Rightarrow Y(i)) \wedge \varphi)$, respectively. We say that a formula is SO∃ or SO∀ if it is logically equivalent to a formula in the required prenex normal form.

4   Global Attractors

The study of fixed points that are global attractors is closely related to the analysis of local fixed point attractors presented in [7]. However, the difficulty of the corresponding decision problems appears to be higher since we require that any point eventually evolves to the fixed point. Using SO(TC), one can define a formula $\mathrm{PATH}_{\mathcal{A}}(T_1, T_2) \equiv \mathrm{RES}_{\mathcal{A}}^{\star}(T_1, T_2)$ to denote the existence of a (possibly empty) path in $\mathcal{A}$ from state T_1 to T_2. A formula expressing that T is a global fixed point attractor of $\mathcal{A}$ is $\mathrm{GLOB}_{\mathcal{A}}(T) \equiv \mathrm{FIX}_{\mathcal{A}}(T) \wedge (\forall U \subseteq \mathsf{S}) \mathrm{PATH}_{\mathcal{A}}(U, T)$, where $\mathrm{FIX}_{\mathcal{A}}(T) \equiv \mathrm{RES}_{\mathcal{A}}(T, T)$.

The following theorem tells us that deciding if a given point is a global fixed point attractor is PSPACE-complete since it is possible to design a RS that simulates a polynomial-space bounded Turing machine such that the dynamics of the system ends in fixed point iff the Turing machine halts.

Theorem 2. *Given a RS $\mathcal{A} = (S, A)$ and a state $T \subseteq S$, it is* PSPACE-*complete to decide whether T is a global attractor of $\mathcal{A}$.*

Proof. The problem is in PSPACE, since $\mathrm{GLOB}_{\mathcal{A}}(T)$ is a SO(TC) formula. In order to show PSPACE-hardness we perform a reduction from the PSPACE-complete [11] problem

$$L = \{(M, x, 1^m) : M \text{ is a TM halting on } x \in \{0, 1\}^\star \text{ in space } m \geq |x|\}.$$

Given $(M, x, 1^m)$, there are $|Q| \geq 2$ states of M, $m+1$ possible head positions (including cell -1 containing the left delimiter $\triangleright$), and 2^m possible strings on the tape, for a total of $k = |Q| \times (m + 1) \times 2^m$ potential configurations.

We use a variant of the RS $\mathcal{C}_k = (S_{\mathcal{C}}, A_{\mathcal{C}})$ of Lemma 1 to count the configurations of M, which is simulated by means of a RS $\mathcal{M} = (S_{\mathcal{M}}, A_{\mathcal{M}})$, a modified version of that of Theorem 1. Denote by $\mathcal{A} = (S, A)$ the resulting RS.

We want the counter implemented by $\mathcal{C}_k$ to stop if and when the Turing machine simulated by $\mathcal{M}$ enters its final state q_{F}. Hence, we add q_{F} as an inhibitor to all reactions of $\mathcal{C}_k$.

Let $\blacktriangleleft \notin S_{\mathcal{C}_k} \cup S_{\mathcal{M}}$ be a new entity. The entity $\blacktriangleleft$ is used to reset the simulation of M to the initial configuration if the counter implemented by $\mathcal{C}_k$ reaches $k - 1$ (*i.e.*, if the configuration of M are exhausted and the machine has entered a loop). To this purpose, $\blacktriangleleft$ is added as an inhibitor to all reactions of $\mathcal{M}$, and we define the reaction $(B_{k-2}, (S_{\mathcal{C}} - B_{k-2}) \cup \{q_{\mathrm{F}}\}, \{\blacktriangleleft\})$, which is enabled when the state of $\mathcal{A}$ restricted to $S_{\mathcal{C}_k}$ is B_{k-2}, *i.e.*, the representation of $k - 2$ (see Lemma 1), and produces $\blacktriangleleft$ when the counter reaches $k - 1$. When $\blacktriangleleft$ appears, it causes the restoration of the initial state of M by means of the reaction $(\{\blacktriangleleft\}, \{q_{\mathrm{F}}\}, X)$, where $X = \{q_{\mathrm{I},0}\} \cup \{w_i : x_i = 1\}$ is the initial configuration of M on input $x = x_1 \cdots x_n$. One more reaction is needed to preserve the final state of M: $(\{q_{\mathrm{F}}\}, \varnothing, \{q_{\mathrm{F}}\})$. This is the only reaction enabled when the state of $\mathcal{A}$ contains q_{F}, ensuring that $\{q_{\mathrm{F}}\}$ is a fixed point.

Notice that $\{q_{\mathrm{F}}\}$ is the *only* fixed point, since in the absence of q_{F}, reactions incrementing the binary counter are always enabled (ensuring that the next state is different); on the other hand, if T contains q_{F} together with other entities it is never a fixed point, since $\mathrm{res}_{\mathcal{A}}(T) = \{q_{\mathrm{F}}\} \neq T$ for all $T \supsetneq \{q_{\mathrm{F}}\}$.

When the initial state of $\mathcal{A}$ is exactly X (hence, the binary counter is null) and M halts on x in space m, then, by Theorem 1, the RS reaches the state $\{q_{\mathrm{F}}\}$; on the other hand, if M does not halt or uses more than m tape cells, the binary counter eventually reaches $k - 1$, resetting the configuration of $\mathcal{A}$ to X in the next step.

Any other initial state of $\mathcal{A}$ either reaches $\{q_{\mathrm{F}}\}$ before the counter reaches $k-1$ (this requires less than $2k$ steps, depending on the initial value of the counter as in Lemma 1), or the counter reaches $k - 1$ and the state of $\mathcal{A}$ is set to X, once again eventually reaching $\{q_{\mathrm{F}}\}$ iff M halts on x in space m.

The mapping $(M, x, 1^m) \mapsto (\mathcal{A}, \{q_{\mathrm{F}}\})$ can be carried out in polynomial time and the PSPACE-hardness of the problem follows.   □

By slightly tuning the proof of Theorem 2, it is possible to establish that determining if there exists a global fixed point attractor in a RS or if two RS share the same global fixed-point attractor are both PSPACE-complete problems. Furthermore, it follows directly from the previous proof that the reachability problem for RS is PSPACE-complete.

Corollary 1. *Given a RS $\mathcal{A} = (S, A)$ and two states $T_1, T_2 \subseteq S$, it is PSPACE-complete to decide whether T_2 is reachable from T_1, i.e., PATH$_{\mathcal{A}}(T_1, T_2)$* $\square$

Corollary 2. *Given a RS $\mathcal{A} = (S, A)$, it is PSPACE-complete to decide whether $\mathcal{A}$ has a global fixed point attractor.*

Proof. The problem lies in PSPACE, since $(\exists T \subseteq S)$ GLOB$_{\mathcal{A}}(T)$ is a SO(TC) formula. The reduction $(M, x, 1^m) \mapsto (\mathcal{A}, \{q_F\})$ in the proof of Theorem 2 can be modified to obtain a reduction $(M, x, 1^m) \mapsto \mathcal{A}$. Notice that if $\mathcal{A}$ has a global fixed point attractor it is $\{q_F\}$. Hence the PSPACE-hardness follows. $\square$

Corollary 3. *Given two RS $\mathcal{A}$ and $\mathcal{B}$ over the same background set S, it is PSPACE-complete to decide whether they share a global fixed-point attractor.*

Proof. The problem lies in PSPACE since $(\exists T \subseteq S)($GLOB$_{\mathcal{A}}(T) \wedge$ GLOB$_{\mathcal{B}}(T))$ is a SO(TC) formula. The reduction $(M, x, 1^m) \mapsto (\mathcal{A}, \{q_F\})$ in the proof of Theorem 2 can be adapted to a reduction $(M, x, 1^m) \mapsto (\mathcal{A}, \mathcal{B})$ where $\mathcal{B}$ is the RS having $(\varnothing, \varnothing, \{q_F\})$ as its only reaction, *i.e.*, $\mathcal{B}$ has $\{q_F\}$ as global attractor. The two RS share $\{q_F\}$ as a global attractor iff $\{q_F\}$ is a global attractor also for $\mathcal{A}$, that is to say, iff M accepts x in space m; Hence, the PSPACE-hardness follows. $\square$

5   Cycles

We now turn our attention to the only other possible behaviour of RS, namely having cycles of length greater than 1. Notice that, while we only need to analyse the application of the result function when checking properties related to fixed points, dealing with cycles essentially involves reachability problems. This shifts the complexity of most of the decision problems to PSPACE.

Theorem 3. *Given a RS $\mathcal{A} = (S, A)$ and a state $T \subseteq S$, it is PSPACE-complete to decide whether T is part of a cycle, i.e., whether res$_{\mathcal{A}}^t(T) = T$ for some $t \in \mathbb{N}$.*

Proof. The problem lies in PSPACE, since it can be expressed by the following SO(TC) formula: CYCLE$_{\mathcal{A}}(T) \equiv \exists U(RES_{\mathcal{A}}(T, U) \wedge$ PATH$_{\mathcal{A}}(U, T))$. The reduction $(M, x, 1^m) \mapsto (\mathcal{A}, \{q_F\})$ of Theorem 2 can be adapted to a reduction $(M, x, 1^m) \mapsto (\mathcal{A}, X)$, where X is the encoding of the initial state of the Turing machine. If M accepts x using space m, then the dynamics of X eventually reaches the fixed point $\{q_F\}$, and X is not in a cycle; conversely, the state is eventually rewritten into X, *i.e.*, X belongs to a cycle. This proves that the problem is PSPACE-hard. Since the complement of a PSPACE-complete problem is also PSPACE-complete, the statement follows. $\square$

Since every RS is finite, a cycle always exists. Therefore, in this case, it turn out to be much more interesting to study comparison problems between RS. That is, if they share one (resp., every) cycle.

Theorem 4. *Given two RS $\mathcal{A}$ and $\mathcal{B}$ over the same background set S, it is* PSPACE-*complete to decide whether they share a common cycle.*

Proof. The problem lies in PSPACE since it can be expressed as a SO(TC) formula. Let $\text{RES}_{\mathcal{A},\mathcal{B}}(T,U) \equiv \text{RES}_{\mathcal{A}}(T,U) \wedge \text{RES}_{\mathcal{B}}(T,U)$ be the formula denoting the fact that T has the same image U under both $\text{res}_{\mathcal{A}}$ and $\text{res}_{\mathcal{B}}$. From it we define $\text{PATH}_{\mathcal{A},\mathcal{B}}(T,U) \equiv \text{RES}^{\star}_{\mathcal{A},\mathcal{B}}(T,U)$ to denote that the same path leads from T to U in $\mathcal{A}$ and $\mathcal{B}$. We then express the fact that T belongs to a shared cycle by means of the formula $\text{CYCLE}_{\mathcal{A},\mathcal{B}}(T) \equiv \exists U(\text{RES}_{\mathcal{A},\mathcal{B}}(T,U) \wedge \text{PATH}_{\mathcal{A},\mathcal{B}}(U,T))$. Finally, $(\exists T \subseteq S)\, \text{CYCLE}_{\mathcal{A},\mathcal{B}}(T)$ is a SO(TC) formula denoting the existence of a shared cycle between $\mathcal{A}$ and $\mathcal{B}$.

Consider the reduction $(M, x, 1^m) \mapsto (\mathcal{A}, \{q_F\})$ of Theorem 2, this can be transformed into a reduction $(M, x, 1^m) \mapsto (\mathcal{A}, \mathcal{B})$, where $\mathcal{B}$ is equal to $\mathcal{A}$ except that the reaction $(\{q_F\}, \varnothing, \{q_F\})$ is replaced by $(\{q_F\}, \varnothing, X)$, in which X represents the initial configuration of M on input x. The two RS behave as follows:

- If M halts on x within space m, then $\mathcal{A}$ has only the fixed point $\{q_F\}$ as a cycle, while $\mathcal{B}$ has a cycle going from X to $\{q_F\}$ and immediately back to X.
- Otherwise, $\mathcal{A}$ has two cycles: the fixed point $\{q_F\}$ and the cycle starting from X and going back to X (when the binary counter overflows). The latter also exists in $\mathcal{B}$, since the behaviours of the two systems only differ *w.r.t.* states containing q_F.

Hence, $\mathcal{A}$ and $\mathcal{B}$ share a cycle if and only if M does *not* halt on input x within space m. Hence, the PSPACE-hardness of the problem follows. $\qquad\square$

Theorem 5. *Given two RS $\mathcal{A}$ and $\mathcal{B}$ over the same background set S, it is* PSPACE-*complete to decide whether they share all their cycles.*

Proof. The problem is in PSPACE, since it can be expressed by the following SO(TC) formula: $(\forall T \subseteq S)(\text{CYCLE}_{\mathcal{A}}(T) \vee \text{CYCLE}_{\mathcal{B}}(T) \Rightarrow \text{CYCLE}_{\mathcal{A},\mathcal{B}}(T))$.

The reduction $(M, x, 1^m) \mapsto (\mathcal{A}, \{q_F\})$ of Theorem 2 can be transformed into the reduction $(M, x, 1^m) \mapsto (\mathcal{A}, \mathcal{B})$, where $\mathcal{B}$ is the RS having $(\varnothing, \varnothing, \{q_F\})$ as its only reaction. The fixed point $\{q_F\}$ is the only cycle of $\mathcal{B}$, and it is shared by both systems. Remark that $\{q_F\}$ is the only cycle of $\mathcal{A}$ only if M halts on x within space m. This reduction establishes the PSPACE-hardness. $\qquad\square$

Theorem 6. *Given a RS $\mathcal{A} = (S, A)$ and a state $T \subseteq S$, it is* PSPACE-*complete to decide whether T is part of a local attractor cycle.*

Proof. Since $\text{ATTC}_{\mathcal{A}}(T) \equiv \text{CYCLE}_{\mathcal{A}}(T) \wedge (\exists U \subseteq S)(\text{PATH}_{\mathcal{A}}(U,T) \wedge \neg\text{PATH}_{\mathcal{A}}(T,U))$ is a SO(TC) formula, the problem lies in PSPACE.

We can transform the reduction $(M, x, 1^m) \mapsto (\mathcal{A}, \{q_F\})$ of Theorem 2 into the reduction $(M, x, 1^m) \mapsto (\mathcal{A}, X)$, where X is the encoding of the initial state of the Turing machine. Notice that if X belongs to a cycle then this cycle is an attractor. Irrespective of the behaviour of M, state X has at least two preimages, namely $\{\blacktriangleleft\}$ and $B_{k-1} \cup \{\blacktriangleleft\}$ (notice that $B_{k-1} \neq \varnothing$ since $k \geq 2$). Finally, if (and only if) M does not halt on x in space m, starting from the state X we eventually reach X again when the binary counter overflows. This shows that the problem is PSPACE-hard. □

Theorem 7. *Given a RS $\mathcal{A} = (S, A)$, deciding if $\mathrm{res}_{\mathcal{A}}$ is a bijection is coNP-complete.*

Proof. It is possible to express bijectivity with the following SO∀ formula:

$$\mathrm{BIJ}_{\mathcal{A}} \equiv (\forall T \subseteq S)(\forall U \subseteq S)(\forall V \subseteq S)(\mathrm{RES}_{\mathcal{A}}(T, V) \wedge \mathrm{RES}_{\mathcal{A}}(U, V) \Rightarrow \mathrm{EQ}(T, U))$$

where $\mathrm{EQ}(T, U) \equiv \forall i(S(i) \Rightarrow (T(i) \Leftrightarrow U(i)))$. Hence the problem is in coNP.

We reduce TAUTOLOGY [6] (also known as VALIDITY [11]) to this problem. Let φ be a Boolean formula in DNF (*i.e.*, $\varphi = \varphi_1 \vee \ldots \vee \varphi_m$, where each φ_j, for $1 \leq j \leq m$, is a conjunctive clause) over the variables $V = \{x_1, \ldots, x_n\}$.

Let $\mathcal{C}_{2^n} = (V, A')$ be the RS of Lemma 1, implementing a binary counter ranging over the set $\{0, \ldots, 2^n - 1\}$, with the entities $b_0, \ldots, b_{n-1}$ renamed as $x_1, \ldots, x_n$. Let $\mathcal{A}$ be a RS with $S = V \cup \{\heartsuit\}$ and having all the reactions of A' plus the following ones:

$$(\mathrm{pos}(\varphi_j) \cup \{\heartsuit\}, \mathrm{neg}(\varphi_j), \{\heartsuit\}) \qquad \text{for } 1 \leq j \leq m \qquad (6)$$

where $\mathrm{pos}(\varphi_j)$ and $\mathrm{neg}(\varphi_j)$ denote the variables appearing, respectively, as positive and negative literals in φ_j. A set $X_i \subseteq V$ represents both the integer i, as in the proof of Lemma 1, and a truth assignment of φ, where the variables having a true value are those in X_i. If the state of $\mathcal{A}$ has the form $X_i \cup \{\heartsuit\}$, then φ is evaluated under X_i, preserving $\heartsuit$ if satisfied, by reaction (6); in any case, the subset of the state representing i is incremented modulo 2^n by the reactions in A'. Hence, the result function of $\mathcal{A}$ is

$$\mathrm{res}_{\mathcal{A}}(T) = \begin{cases} \{\heartsuit\} \cup X_{(i+1) \bmod 2^n} & \text{if } T = \{\heartsuit\} \cup X_i \text{ and } X_i \vDash \varphi \\ X_{(i+1) \bmod 2^n} & \text{if } T = X_i \text{ or if } T = \{\heartsuit\} \cup X_i \text{ and } X_i \nvDash \varphi. \end{cases}$$

If φ is not a tautology, then there exists an assignment X_i such that $X_i \nvDash \varphi$, thus $\mathrm{res}_{\mathcal{A}}(X_i)$ is equal to $\mathrm{res}_{\mathcal{A}}(X_i \cup \{\heartsuit\})$, *i.e.*, $\mathrm{res}_{\mathcal{A}}$ is not bijective. On the other hand, if φ is a tautology, then $\heartsuit$ is always preserved when present, and the dynamics of $\mathcal{A}$ consists of two disjoint cycles, namely $(X_0, \ldots, X_{2^n - 1})$ and $(X_0 \cup \{\heartsuit\}, \ldots, X_{2^n - 1} \cup \{\heartsuit\})$, *i.e.*, $\mathrm{res}_{\mathcal{A}}$ is a bijection. Since the mapping $\varphi \mapsto \mathcal{A}$ is computable in polynomial time, the problem is coNP-hard. □

Corollary 4. *Given a RS $\mathcal{A} = (S, A)$, it is NP-complete to decide whether $\mathcal{A}$ has a local attractor cycle.*

Proof. A finite system has an attractor cycle if and only if the next-state function is not a bijection. Hence, this is the complement of the coNP-complete problem of Theorem 7. □

Theorem 8. *Given two RS $\mathcal{A}$ and $\mathcal{B}$ over the same background set S, it is* PSPACE-*complete to decide whether $\mathcal{A}$ and $\mathcal{B}$ have a common local attractor cycle.*

Proof. The problem is in PSPACE since $(\exists T \subseteq S)$ ATTC$_{\mathcal{A},\mathcal{B}}(T)$ is a SO(TC) formula, where

$$\text{ATTC}_{\mathcal{A},\mathcal{B}}(T) \equiv \text{CYCLE}_{\mathcal{A},\mathcal{B}}(T) \wedge (\exists U \subseteq S)(\text{PATH}_{\mathcal{A}}(U,T) \wedge \neg\text{PATH}_{\mathcal{A}}(T,U)) \wedge$$
$$(\exists V \subseteq S)(\text{PATH}_{\mathcal{B}}(V,T) \wedge \neg\text{PATH}_{\mathcal{B}}(T,V)).$$

Consider the reduction $(M, x, 1^m) \mapsto (\mathcal{A}, \mathcal{B})$ of Theorem 4, and notice that X, the state encoding the initial configuration of M, has at least two distinct preimages (see the proof of Theorem 6). Hence, when $\mathcal{A}$ and $\mathcal{B}$ share a common cycle, it is always an attractor cycle. The PSPACE-hardness follows. $\square$

Theorem 9. *Given two RS $\mathcal{A}$ and $\mathcal{B}$ over the same background set S, it is* PSPACE-*complete to decide whether $\mathcal{A}$ and $\mathcal{B}$ share all their local attractor cycles.*

Proof. The problem is in PSPACE since it can be expressed by the following SO(TC) formula: $(\forall T \subseteq S)(\text{ATTC}_{\mathcal{A}}(T) \vee \text{ATTC}_{\mathcal{B}}(T) \Rightarrow \text{ATTC}_{\mathcal{A},\mathcal{B}}(T))$.

Consider the reduction $(M, x, 1^m) \mapsto (\mathcal{A}, \mathcal{B})$ of the proof of Theorem 5; both RS share the fixed point $\{q_F\}$, which is an attractor since every state containing q_F is mapped to $\{q_F\}$. When M does *not* halt on x in space m, $\mathcal{A}$ has another cycle containing X, which is an attractor as shown in the proof of Theorem 6. This is enough to establish the PSPACE-hardness. $\square$

Since a global attractor state is a special case of global attractor cycle, and the corresponding decision problems remain in PSPACE, we immediately have the following statement:

Corollary 5. *Given two RS $\mathcal{A}$ and $\mathcal{B}$ over the same background set S and a state $T \subseteq S$, it is* PSPACE-*complete to decide if (i) T is a part of a global attractor cycle in $\mathcal{A}$, (ii) $\mathcal{A}$ has a global attractor cycle, (iii) $\mathcal{A}$ and $\mathcal{B}$ have a common global attractor cycle.* $\square$

6   Conclusions

In this paper we have studied the complexity of checking the presence of many different dynamical behaviours of a RS, extending the work started in [7]. When global fixed point attractors are considered, all problems are PSPACE-complete, differently from the case of local fixed points attractors, where all the problems lied in the polynomial hierarchy. We proved that PSPACE-completeness remains the most common complexity class for the decision problems regarding cycles that we analysed. While some PSPACE-completeness results are known for more expressive or different computational models [1], it is interesting that, even if they are quite simple, RS exhibit difficult decision problems.

The paper, while closing some open question, still discloses many interesting research directions. In particular, we have only studied deterministic RS, *i.e.*, the next-state is uniquely determined. However many significant modelling questions involve non-deterministic RS where at every time step an external device inserts some entities in the state of the RS (these kind of RS are called *RS with context* in the literature). It is interesting to understand how the complexity of decision problems about dynamics behaves in this case. Does everything shift to PSPACE? Is PSPACE the upper bound?

References

1. Barrett, C.L., Hunt III, H.B., Marathe, M.V., Ravi, S., Rosenkrantz, D.J., Stearns, R.E.: Complexity of reachability problems for finite discrete dynamical systems. Int. J. Found. Comput. Sci. 72(8), 1317–1345 (2006)
2. Brijder, R., Ehrenfeucht, A., Rozenberg, G.: Reaction systems with duration. In: Kelemen, J., Kelemenová, A. (eds.) Păun Festschrif. LNCS, vol. 6610, pp. 191–202. Springer, Heidelberg (2011)
3. Corolli, L., Maj, C., Marini, F., Besozzi, D., Mauri, G.: An excursion in reaction systems: From computer science to biology. Theor. Comp. Sci. 454, 95–108 (2012)
4. Ehrenfeucht, A., Main, M., Rozenberg, G.: Combinatorics of life and death for reaction systems. Int. J. Found. Comput. Sci. 21(03), 345–356 (2010)
5. Ehrenfeucht, A., Main, M., Rozenberg, G.: Functions defined by reaction systems. Int. J. Found. Comput. Sci. 22(1), 167–168 (2011)
6. Ehrenfeucht, A., Rozenberg, G.: Reaction systems. Fundam. Inform. 75, 263–280 (2007)
7. Formenti, E., Manzoni, L., Porreca, A.E.: Fixed points and attractors of reaction systems. In: Beckmann, A., Csuhaj-Varjú, E., Meer, K. (eds.) CiE 2014. LNCS, vol. 8493, pp. 194–203. Springer, Heidelberg (2014)
8. Hopcroft, J.E., Motwani, R., Ullman, J.D.: Introduction to Automata Theory, Languages, and Computation. Addison-Wesley (1979)
9. Immerman, N.: Descriptive Complexity. Graduate Texts in Computer Science. Springer (1999)
10. Kauffman, S.A.: Metabolic stability and epigenesis in randomly constructed genetic nets. J. Theor. Biol. 22(3), 437–467 (1969)
11. Papadimitriou, C.H.: Computational Complexity. Addison-Wesley (1993)
12. Păun, G.: Computing with membranes. Journal of Computer and System Sciences 61(1), 108–143 (2000)
13. Salomaa, A.: Functions and sequences generated by reaction systems. Theoretical Computer Science 466, 87–96 (2012)
14. Salomaa, A.: Functional constructions between reaction systems and propositional logic. Int. J. Found. Comput. Sci. 24(1), 147–159 (2013)
15. Salomaa, A.: Minimal and almost minimal reaction systems. Natural Computing 12(3), 369–376 (2013)
16. Shmulevich, I., Dougherty, E.R.: Probabilistic boolean networks: the modeling and control of gene regulatory networks. SIAM (2010)
17. Soloveichik, D., Cook, M., Winfree, E., Bruck, J.: Computation with finite stochastic chemical reaction networks. Natural Computing 7, 615–633 (2008)

Classical Automata on Promise Problems[*]

Viliam Geffert[1],[**] and Abuzer Yakaryılmaz[2],[3],[***]

[1] Dept. Comput. Sci., P.J. Šafárik University Jesenná 5, 04154 Košice, Slovakia
[2] University of Latvia, Faculty of Computing, Raina bulv. 19, Rīga, LV-1586, Latvia
[3] National Laboratory for Scientific Computing, Petrópolis, RJ, 25651-075, Brazil
viliam.geffert@upjs.sk, abuzer@lncc.br

Abstract. In automata theory, promise problems have been mainly examined for quantum automata. In this paper, we focus on classical automata and obtain some new results regarding the succinctness of models
and their computational powers. We start with a negative result. Recently, Ambainis and Yakaryılmaz (2012) introduced a quantumly very
cheap family of unary promise problems, i.e. solvable exactly by using
only a single qubit. We show that two-way nondeterminism does not
have any advantage over realtime determinism for this family of promise
problems. Secondly, we present some basic facts for classical models:
The computational powers of deterministic, nondeterministic, alternating, and Las Vegas probabilistic automata are the same. Then, we show
that any gap of succinctness between any two of deterministic, nondeterministic, and alternating automata models for language recognition
cannot be violated on promise problems. On the other hand, we show
that the tight quadratic gap between Las Vegas realtime probabilistic
automata and realtime deterministic automata given for language recognition can be replaced with a tight exponential gap on promise problems.
Lastly, we show how the situation can be different when considering two-
sided bounded-error. Similar to quantum case, we present a probabilistically very cheap family of unary promise problems, i.e. solvable by a
2-state automaton with bounded-error. Then, we show that this family
is not cheap for any of the aforementioned classical models. Moreover,
we show that bounded-error probabilistic automata are more powerful
than any other classical model on promise problems.

1 Introduction

Promise problem is a generalization of language recognition. Instead of
considering all strings, we focus on a subset of strings and the input is promised
to be only from this subset. Thus, the language under consideration and its

---

[*] ArXiv version is at http://arxiv.org/abs/1405.6671.
[**] Geffert was partially supported by the Slovak grant contracts VEGA 1/0479/12
and APVV-0035-10. 1.
[***] Yakaryılmaz was partially supported by CAPES, ERC Advanced Grant MQC, and
FP7 FET project QALGO. Moreover, the part of the research work was done while
Yakaryılmaz was visiting Kazan Federal University.

H. Jürgensen et al. (Eds.): DCFS 2014, LNCS 8614, pp. 126–137, 2014.
© Springer International Publishing Switzerland 2014

complement must form this subset (instead of the set of all strings). Promise problems have provided many different perspectives in computational complexity (see the survey by Goldreich [9]). For example, it is not known whether the class of languages recognized by bounded-error probabilistic polynomial-time algorithms has a complete problem, but, the class of promise problems solvable by the same type algorithms has some complete problems. A similar tendency also exists for quantum complexity classes [20]. The first known result we are aware on promise problems for the restricted computational models was given by Condon and Lipton in 1989 [4]. In the literature, some separation results regarding restricted models have also been given in the form of promise problems (e.g. [7,6]). The first result regarding restricted quantum models was given by Murakami et. al. [15]: There is a promise problem solvable by quantum pushdown automata exactly but not by any deterministic pushdown automaton. Recently, Ambainis and Yakaryılmaz [3] showed that there is an infinite family of promise problems which can be solved exactly by just tuning transition amplitudes of a realtime two-state quantum finite automaton (QFA), whereas the size of the corresponding classical automata grows without bound. Moreover, Rashid and Yakaryılmaz [18] presented many superiority results of QFAs over their classical counterparts. There are also some new results on succinctness of QFAs [23,10,11,22] and a new result on quantum pushdown automata [16].

In this paper, we turn our attention to classical models and obtain some new results. We give the preliminaries to follow the rest of the paper in the next section. We show that two-way nondeterminism is useless for the family of promise problems given by Ambainis and Yakaryılmaz [3] in Section 3.1. Then, we present the basic facts for classical models in Section 3.2. That is, (i) the computational powers of deterministic, nondeterministic, alternating, and Las Vegas probabilistic automata are the same; and, (ii) any gap of succinctness between any two of deterministic, nondeterministic, and alternating automata models for language recognition cannot be violated on promise problems. In Section 3.3, we focus on Las Vegas probabilistic automata and we show that the tight quadratic gap between them and realtime deterministic automata given for language recognition can be replaced with a tight exponential gap on promise problems. In Section 3.4, we show how the situation can be different when considering two-sided bounded-error. Firstly, we present a probabilistically very cheap family of unary promise problems, i.e. solvable by a 2-state automaton with bounded-error. Then, we show that this family is not cheap for any of the aforementioned classical models. And, lastly, we prove that bounded-error probabilistic automata are more powerful than any other classical model on promise problems.

2   Preliminaries

We represent any automaton model in the paper as $\mathtt{xY}FA$, where $\mathtt{x}$ can be "2" or "rt" that stands for *two-way* or *realtime*, respectively; and, $\mathtt{Y}$ can be "D", "N", "A", or "P" that stands for *deterministic*, *nondeterministic*, *alternating*, or *probabilistic*, respectively. For two-way models, we assume that the input is placed between a left and a right end-markers on a tape and the input head can move to the left, move to the right, or stay on the same square. For realtime models, we assume that the input is not stored on a tape, there is no end-markers, and the input is fed for the automaton from left to the right symbol by symbol. A two-way automaton is called sweeping if the input head is allowed to change its direction only on the end-markers [19,13]. A very restricted version of sweeping automaton called restarting realtime automaton runs a realtime algorithm in an infinite loop, [21], i.e. if the computation is not terminated on the right end-marker, the same realtime algorithm is executed again.

We assume that the reader knows the definitions of 2DFA, 2NFA, 2AFA, and their realtime versions. A rtPFA [17] is a 5-tuple $\mathcal{P} = (S, \Sigma, \{A_\sigma \mid \sigma \in \Sigma\}, s_1, S_a)$, where Σ is the input alphabet, S is the set of internal states, s_1 is the initial state, $S_a \subseteq S$ is the set of accepting states, and A_σ is a stochastic transition matrix whose $(i, j)^{th}$ entry represent the probability of going from the ith state to j^{th} state when reading symbol σ. The computation starts in state s_1 and the input is accepted if it ends in a state belonging to S_a. The overall accepting probability of $\mathcal{P}$ on a given input $w \in \Sigma^*$, denoted $f_\mathcal{P}(w)$, can be calculated over all accepting paths. So, the overall rejecting probability is $1 - f_\mathcal{P}(w)$. A Las Vegas rtPFA is 7-tuple and additionally has S_r and S_n, the sets of rejecting states and neutral ("don't know") states, respectively, such that the union of S_a, S_r, and S_n is S and they are pairwise disjoint; and, the automaton gives the decision of "acceptance", "rejection", or "don't know" if it is in a state in S_a, S_r, or S_n, respectively, at the end of the computation. A 2PFA (firstly defined in [14]) is a two-way version of rtPFA such that during the computation the symbol under the head is read and the automaton also updates the head position after each transition. It has also some specified rejecting states and the computation is terminated with the decision of "acceptance" or "rejection" if it enters an accepting or a rejecting state, respectively, during the computation.

A promise problem is a pair $\mathrm{P} = (\mathrm{P_{yes}}, \mathrm{P_{no}})$, where $\mathrm{P_{yes}}, \mathrm{P_{no}} \subseteq \Sigma^*$ and $\mathrm{P_{yes}} \cap \mathrm{P_{no}} = \emptyset$ [20]. Here, the members of $\mathrm{P_{yes}}$ ($\mathrm{P_{no}}$) are called yes-instances (no-instances). P is said to be solved by a machine $\mathcal{M}$ with error bound $\epsilon \in (0, \frac{1}{2})$ if any member of $\mathrm{P_{yes}}$ is accepted with a probability at least $1 - \epsilon$ and any member of $\mathrm{P_{no}}$ is rejected by $\mathcal{M}$ with a probability at least $1 - \epsilon$. P is said to be solved by $\mathcal{M}$ with bounded-error if it is solved by $\mathcal{M}$ with an error bound. If $\epsilon = 0$, then it is said that the problem is solved by $\mathcal{M}$ *exactly*. A special case of bounded-error is one-sided bounded-error where either all members of $\mathrm{P_{yes}}$ are accepted with probability 1 or all members of $\mathrm{P_{no}}$ are rejected with probability 1. $\mathcal{M}$ is said to be Las Vegas with a success probability $p \in (0, 1]$ [12] if

- for a member of $\mathrm{P_{yes}}$, $\mathcal{M}$ gives the decision of "acceptance" with a probability at least p and gives the decision of "don't know" with the remaining probability; and,

- for a member of $\mathrm{P_{no}}$, $\mathcal{M}$ gives the decision of "rejection" with a probability at least p and gives the decision of "don't know" with the remaining probability.

If P satisfies $P_{yes} \cup P_{no} = \Sigma^*$ and P is *solvable* by $\mathcal{M}$, then it is conventional said that P_{yes} is *recognized* by $\mathcal{M}$.

3   Main Results

3.1   Classically Expensive Promise Problem

Recently, Ambainis and Yakaryılmaz [3] presented a family of promise problems $\{EVENODD^k \mid k \in \mathbb{Z}^+\}$ such that each member of the family, say $EVENODD^k$, can be solved by a 2-state realtime quantum finite automaton exactly but any rtDFA needs at least 2^{k+1} states, where

$$EVENODD^k_{yes} = \{a^{m2^k} \mid m \text{ is even}\}$$
$$EVENODD^k_{no} = \{a^{m2^k} \mid m \text{ is odd}\}$$

Later, it was shown that any bounded-error rtPFAs can also need at least 2^{k+1} states [18].[1] Here we show that two-way nondeterminism does not help us to save some states. Our method is not new but it needs some modifications on promise problems. Currently we do not know whether bounded-error 2PFAs can do better but we know that realtime alternation can help us to save some states (we give the proof in an expanded version of this paper).

Theorem 1. *2NFAs need at least 2^{k+1} states to solve* $EVENODD^k$, *where* $k \in \mathbb{Z}^+$.

Proof. Suppose, for contradiction, that the promise problem can be solved by a 2NFA $\mathcal{N}$ with less than 2^{k+1} states.

Consider now the input $a^{2^{k+1}}$. Clearly, it must be accepted by $\mathcal{N}$, and hence there must exist at least one accepting computation path, so we can fix one such path. (Besides, being nondeterministic, $\mathcal{N}$ can have also other paths, not all of them necessarily accepting, but we do not care for them.)

Along this fixed path, take the sequence of states

$$q_0, p_1, q_1, p_2, q_2, p_3, q_3, \ldots, p_r, q_r, p_{r+1},$$

where q_0 is the initial state, p_{r+1} is an accepting state, and all the other states ($\{p_i \cup q_i \mid 1 \le i \le r\}$) are at the left/right end-markers of the input 2^{k+1} such that:

- If p_i is at the left (resp., right) end-marker, then q_i is at the opposite end-marker, and the path from p_i to q_i traverses the entire input from left to right (resp., right to left), not visiting any of end-markers in the meantime.
- The path between q_i and p_{i+1} starts and ends at the same end-marker, possible visiting this end-marker several times, but not visiting the opposite end-marker. Such a path is called a "U-turn". This covers also the case of $q_i = p_{i+1}$ with zero number of executed steps in between.

---

[1] Recently, it is also shown that the width of any nondeterministic or bounded-error stable OBDD cannot be less than 2^{k+1} for a function version of $EVENODD^k$ [1].

Now, since the path connecting p_i with q_i must visit all input tape positions in the middle of the input $a^{2^{k+1}}$, $\mathcal{N}$ enters 2^{k+1} different states, so some state r_i must be repeated. That is, between p_i and q_i, there must exists a loop, starting and ending in the same state r_i, travelling some l_i positions to the right, not visiting any of the end-markers. For each traversal, we fix one such loop (even though the path from p_i to q_i may contain many such loops). Note that the argument for travelling across the input from right to left is symmetrical. Since this loop is in the middle of the traversal, we have that $l_i < 2^{k+1}$. But then l_i can be expressed in the form

$$l_i = 2^{\alpha_i} \gamma_i,$$

where $\gamma_i > 0$ is odd (not excluding $\gamma_i = 1$) and $\alpha_i \in \{0, 1, \ldots, k\}$. Remark that if $\alpha_i \geq k + 1$, we would have a contradiction with $l_i < 2^{k+1}$.

Now, consider the value

$$l = lcm\{l_1, l_2, \ldots, l_r\},$$

the least common multiple of all fixed loops, for all traversals of the input $a^{2^{k+1}}$. Clearly, l can also be expressed in the form

$$l = 2^{\alpha} \gamma,$$

where $\gamma > 0$ is odd (actually $\gamma = lcm\{\gamma_1, \ldots, \gamma_r\}$) and $\alpha \in \{0, 1, \ldots, k\}$ (actually $\alpha = \max\{\alpha_1, \ldots, \alpha_r\}$).

We shall now show that the machine $\mathcal{N}$ must also accept the input

$$a^{2^{k+1} + l2^{k-\alpha}}.$$

First, if q_i and p_{i+1} are connected by a U-turn path on the input $a^{2^{k+1}}$, they will be connected also on the input $a^{2^{k+1} + l2^{k-\alpha}}$ since such path does not visit the opposite end-marker and $2^{k+1} + l2^{k-\alpha} \geq 2^{k+1}$. Second, if p_i and q_i are connected by a left-to-right traversal along $a^{2^{k+1}}$, they will stay connected also along the input $a^{2^{k+1} + l2^{k-\alpha}}$. This only requires to repeat the loop of the length l_i beginning and ending in the state r_i. Namely, we make $2^{k-\alpha} \frac{l}{l_i}$ more iterations. Note that $l = lcm\{l_1, \ldots, l_r\}$ and hence l is divisible by l_i, for each $i = 1, \ldots, r$, which gives that $2^{k-\alpha} \frac{l}{l_i}$ is an integer. Note also that these $2^{k-\alpha} \frac{l}{l_i}$ additional iterations of the loop of the length l_i travel exactly $(2^{k-\alpha} \frac{l}{l_i}) l_i = 2^{k-\alpha} l$ additional positions to the right. Thus, if $\mathcal{N}$ has an accepting path for $a^{2^{k+1}}$, it must also have an accepting path for $a^{2^{k+1} + 2^{k-\alpha}}$ (just a straightforward induction on r). Therefore, $\mathcal{N}$ accepts $a^{2^{k+1} l2^{k-\alpha}}$. (Actually, $\mathcal{N}$ can have many other paths for this longer input, but they cannot rule out accepting decision of the path constructed above.) However, the input $a^{2^{k+1} l2^{k-\alpha}}$ should be rejected, since $2^{k+1} + l2^{k-\alpha} = 2^k(2 + \gamma)$, where γ is odd. This is a contradiction. So, $\mathcal{N}$ must have at least 2^{k+1} states.   □

3.2 Basic Facts on Classical Automata

We continue with some basic facts regarding classical automata. We show that the class of promise problems solvable by deterministic, nondeterministic, alternating, and Las Vegas probabilistic finite automata are identical.

Theorem 2. *Any promise problem* P *defined on* Σ *solvable by a 2AFA* $\mathcal{A}$ *can be solved by a rtDFA.*

Proof. For any given string w in Σ, $\mathcal{A}$ can give a single decision and all strings accepted by $\mathcal{A}$ form a regular language, say R. Note that all string rejected by $\mathcal{A}$ form $\bar{\text{R}}$. Since P is solvable by $\mathcal{A}$, P_{yes} is a subset of R and P_{no} is a subset of $\bar{\text{R}}$. R can also be recognized by a rtDFA, say $\mathcal{D}$, i.e. any member of R are accepted by $\mathcal{D}$ and any member of $\bar{\text{R}}$ is rejected by $\mathcal{D}$. Therefore, $\mathcal{D}$ can solve P, too.   □

Based on this proof, we can also obtain the following corollary.

Corollary 1. *Any gap of succinctness between any two of deterministic, nondeterministic, and alternating automata models for language recognition cannot be violated on promise problems.*

Now, we show that Las Vegas PFAs cannot gain any computational power.

Theorem 3. *Any promise problem* P *solvable by a Las Vegas 2PFA* $\mathcal{P}$ *can be solved by a rtDFA.*

Proof. For each member of P_{yes} (P_{no}), $\mathcal{P}$ ends its computation with the decision of "acceptance" ("rejection") in one of the paths and there is no path ending with the decision of "rejection" ("acceptance"). So, by removing the probabilities and converting neutral states into rejecting states, we obtain a 2NFA $\mathcal{N}$ recognizing regular language R such that $\text{P}_{\text{yes}} \subseteq \text{R}$ and $\text{P}_{\text{no}} \cap \text{R} = \emptyset$. Therefore, $\text{P}_{\text{no}} \subseteq \bar{\text{R}}$. That is, $\mathcal{N}$ can solve R, too. Due to the previous theorem, we can follow that there is also a rtDFA solving P.   □

We follow the same result for Las Vegas rtPFA by also giving a state bound.

Corollary 2. *Any promise problem* P *solvable by an n-state Las Vegas rtPFA can be solved by a $(2^n - 1)$-state rtDFA.*

Proof. The proof is the same. Moreover, n-state rtPFA can be converted to a n-state rtNFA which can be simulated by a $(2^n - 1)$-state rtDFA by using standard subset construction.   □

In the next section, we show that this bound is actually asymptotically tight.

3.3   Las Vegas Probabilistic Finite Automata

In the case of language recognition, the gap of succinctness between Las Vegas rtPFAs and rtDFAs can be at most quadratic and this gap is tight [5,12]. In the case of promise problems, we show that quadratic gap can be replaced with an exponential tight gap. For this purpose, we start with a promise problem recently defined by Gruska et. al. [10]: GQZ-EQ(n, k) ($n \in \mathbb{Z}^+$ and $k \geq \frac{n}{2}$) such that

$$\text{GQZ-EQ}_{\text{yes}}(n, k) = \{x \# x \mid x \in \{0,1\}^n\}$$
$$\text{GQZ-EQ}_{\text{no}}(n, k) = \{x \# y \mid x, y \in \{0,1\}^n \text{ and } H(x,y) = k\},$$

where H returns the Hamming distance between two strings. They also showed that GQZ-EQ(n, k) can be solved by an exact realtime finite automaton with quantum and classical states (rtQCFA)[2] with $O(n^2)$ states ($n+1$ quantum and $O(n)$ classical states (the number of quantum states is n if $k = \frac{n}{2}$ [22])) but the size of states of the corresponding rtDFA is $2^{\Omega(n)}$. We present an $O(n)$-state rtPFA, say $\mathcal{P}$, solving GQZ-EQ(n, k) with one-sided error bound $1 - \frac{k}{n}$.

Let $x\#y$ be the input as promised. The automaton has $3n$ states, i.e.

$$\{s_i \mid 1 \leq i \leq n\} \cup \{c_{i,0}, c_{i,1} \mid 1 \leq i \leq n-1\} \cup \{s_a, s_r\},$$

where s_1 and s_a are the initial and accepting states, respectively. The automaton starts its computation with state s_1. When reading x_i and being in state s_i ($1 \leq i < n$), $\mathcal{P}$ switches to states c_{1,x_i} and s_{i+1} with probability p_i and $1 - p_i$, respectively; and, when reading x_n and being in state s_n, it switches to c_{1,x_n} with probability 1. By carefully setting the values of p_i's, we can guarantee that $\mathcal{P}$ switches to $c_{1,*}$-states with equal probability ($\frac{1}{n}$). That is, it switches to state c_{1,x_i} on symbol x_i with probability $\frac{1}{n}$. State c_{1,x_i} is responsible to compare x_i with y_i. The automaton is in this state on the $(i+1)$th symbol of the input. After n transitions, $\mathcal{P}$ will be on symbol y_i by changing the states as

$$c_{1,x_i} \rightarrow c_{2,x_i} \rightarrow \cdots \rightarrow c_{n-1,x_i}$$

where $\mathcal{P}$ does not change its state on symbol $\#$. If $x_i = y_i$, then $\mathcal{P}$ enters s_a, and s_r otherwise. It does not change its state once entering s_a or s_r. So, $\mathcal{P}$ detects any index j with $x_j \neq y_j$ with probability $\frac{1}{n}$. Therefore, each member of GQZ-EQ$_{\text{yes}}$(n, k) is accepted exactly and each member of GQZ-EQ$_{\text{no}}$(n, k) is rejected with probability $\frac{k}{n}$. Therefore, the error bound is $1 - \frac{k}{n}$ which can be arbitrary small by choosing a convenient (n, k)-pair. Remark that the same algorithm can also be implemented by a $O(n)$-state one-sided bounded-error rtQCFA, where the automaton needs only a qubit.

Now, by using an idea given by Rashid and Yakaryılmaz [18], we define a new promise problem based on GQZ-EQ(n, k) which can be solved by a Las Vegas rtPFA, say $\mathcal{P}'$, with $O(n)$ states:

$$\text{GQZ-EQ}'_{\text{yes}}(\mathbf{n}) = \{x\#x\#y \mid x, y \in \{0,1\}^n \text{ and } H(x, y) = \frac{n}{2}\}$$
$$\text{GQZ-EQ}'_{\text{no}}(\mathbf{n}) = \{x\#y\#y \mid x, y \in \{0,1\}^n \text{ and } H(x, y) = \frac{n}{2}\}.$$

Lemma 1. GQZ-EQ'(n) *can be solved by a Las Vegas rtPFA with $O(n)$ states with probability $\frac{1}{4}$.*

Proof. Let $x\#y\#z$ be the input as promised. At the beginning of the computation, $\mathcal{P}'$ branches the computation into two branches with equal probability. In the first branch, $\mathcal{P}'$ executes $\mathcal{P}$ on $y\#z$ and it accepts the input if $\mathcal{P}$ gives the decision of "rejection" and says "don't know" if $\mathcal{P}$ gives the

---

[2] It is a realtime version of the two-way quantum finite automaton model defined by Ambainis and Watrous [2] and its formal definition was given in [24].

decision of "acceptance". And so, in this branch, the members of $\text{GQZ-EQ}'_{\text{yes}}(n)$ are accepted with probability $\frac{1}{2}$ and the decision of "rejection" is never given by $\mathcal{P}'$. In the second branch, $\mathcal{P}'$ executes $\mathcal{P}$ on $x\#y$ and it rejects the input if $\mathcal{P}$ gives the decision of "rejection" and says "don't know" if $\mathcal{P}$ gives the decision of "acceptance". And so, in this branch, the members of $\text{GQZ-EQ}'_{\text{no}}(n)$ are rejected with probability $\frac{1}{2}$ and the decision of "acceptance" is never given by $\mathcal{P}'$. Therefore, any member of $\text{GQZ-EQ}'_{\text{yes}}(n)$ $(\text{GQZ-EQ}'_{\text{no}}(n))$ is accepted (rejected) with probability $\frac{1}{4}$ and the decision of "don't know" is given with the remaining probability. □

It is quite clear that any 2PFA (restarting rtPFA) can exactly solve this problem in linear expected time by restarting the computation when the decision of "don't know" is obtained.

Corollary 3. $\text{GQZ-EQ}'(n)$ *can be solved exactly by a $O(n)$-state restarting rtPFA in linear expected time.*

To get a higher probability for Las Vegas rtPFAs, we define a new promise problem $\text{GQZ-EQ}'(n,r)$ by modifying $\text{GQZ-EQ}'(n)$:

$$\text{GQZ-EQ}'_{\text{yes}}(n,r) = \{(x\#x\#y\#)^r \mid x,y \in \{0,1\}^n \text{ and } H(x,y) = \frac{1}{2}\}$$
$$\text{GQZ-EQ}'_{\text{no}}(n,r) = \{(x\#y\#y\#)^r \mid x,y \in \{0,1\}^n \text{ and } H(x,y) = \frac{1}{2}\},$$

where $n, r > 0$.

Theorem 4. $\text{GQZ-EQ}'(n,r)$ *can be solved by a Las Vegas rtPFA with $O(n)$ states with probability $1 - (\frac{3}{4})^r$.*

Proof. Let w be the string $x\#y\#z\#$. We know the behaviour of $\mathcal{P}'$ on w. We can design a new Las Vegas rtPFA which reads w r times. Then, the probability of saying "don't know" can be reduced to $(\frac{3}{4})^r$. □

By modifying the proof given for $\text{GQZ-EQ}(n,k)$ in [10,22], we can follow that any rtDFA needs $2^{\Omega(n)}$ states to solve $\text{GQZ-EQ}'(n)$ and $\text{GQZ-EQ}'(n,r)$.

Corollary 4. *The tight gap of succinctness between rtDFA and Las Vegas rtPFA is asymptotically exponential on promise problems.*

3.4 Two-Sided Bounded-Error Probabilistic Finite Automata

We show the limitations of Las Vegas rtPFAs on the previous sections. One-sided error PFAs[3] have also similar limitations since they can be simulated by a nondeterministic or a universal finite automota by removing the probabilities: If any no-instance is rejected exactly (only yes-instances are accepted), then we obtain a NFA from the one-sided error PFA; and, if any yes-instance is accepted exactly (only non-instances are rejected), then we obtain a universal

---

[3] We also cover the case of *unbounded-error*, i.e., there might be no bound on the error and so it can be arbitrarily close to 0 for some inputs.

finite automaton from the one-sided error PFA. In this subsection, we show that we have a different picture in the case of two-sided bounded-error.

Firstly, we show that bounded-error rtPFAs can be very succinct compared to DFAs, NFAs, AFAs, and Las Vegas rtPFAs. In fact, we present a parallel result to that of Ambainis and Yakaryılmaz [3] but with bounded error instead of exact acceptance. Let $\mathcal{UP}_p$ be a 2-state unary rtPFA such that

- $\Sigma = \{a\}$ is the input alphabet,
- s_1 is the initial and accepting state, and s_2 is the rejecting state, and,
- $\mathcal{UP}_p$ stays in s_1 with probability p and switches to s_2 with the remaining probability for each reading symbol.

It is straightforward that the accepting probability of a^j is p^j, where $j \geq 0$. Let $\mathrm{UP}(\mathrm{p})$ be a promise problem such that

$$\mathrm{UP}_{\mathrm{yes}}(\mathrm{p}) = \{a^j \mid f_{\mathcal{UP}_p}(a^j) \geq \tfrac{3}{4}\}$$
$$\mathrm{UP}_{\mathrm{no}}(\mathrm{p}) = \{a^j \mid f_{\mathcal{UP}_p}(a^j) \leq \tfrac{1}{4}\}.$$

It is clear that, for any $p < 1$, $\mathrm{UP}(\mathrm{p})$ can be solved by a 2-state bounded-error rtPFA. On the other hand, the number of states required by a rtDFA increases when p approaches to 1. We define $A_p = \lfloor \log_p(\tfrac{3}{4}) \rfloor$ and $R_p = \lceil \log_p(\tfrac{1}{4}) \rceil$. Note that $A_p \to \infty$ as $p \to 1$. It is clear that $\mathrm{UP}_{\mathrm{yes}}(\mathrm{p})$ contains only the first j^{th} strings satisfying $j \leq A_p$ and $\mathrm{UP}_{\mathrm{no}}(\mathrm{p})$ contains infinitely many strings, i.e. any string with length at least R_p.

Theorem 5. *For any $p < 0$, $\mathcal{UP}_p$ can be solved by a 2-state rtPFA with error bound $\tfrac{1}{4}$ but any rtDFA (or rtNFA) needs at least $A_p + 1$ states.*

Proof. We need to prove the second statement. Suppose that there is a rtDFA, say $\mathcal{D}$, solving $\mathcal{UP}_p$ has less than $A_p + 1$ states. Since $\mathcal{D}$ must accept all strings in $\{a^j \mid j \leq A_p\}$, all states visited by $\mathcal{D}$ when reading a^{A_p} must be accepting states including the initial one. Moreover, since the number of states is less than $A_p + 1$, then there must be a cycle containing an accepting state. This means there must be infinitely many accepted strings but any string with length at least R_p must be rejected. This is a contradiction. Therefore, any such rtDFA must have at least $A_p + 1$ states. The same proof also works for rtNFAs by only considering a single accepting path. □

We can easily design an (A_p+1)-state rtDFA solving $\mathcal{UP}_p$, i.e. after A_p accepting states, the automaton enters a rejecting state and stays there. Therefore, the bound $A_p + 1$ is tight. Moreover, the error bound ($\tfrac{1}{4}$) in the definition of $\mathrm{UP}(\mathrm{p})$ can be replaced with an arbitrary error and the same arguments can still be obtained.

Corollary 5. *There is a family of unary promise problems such that 2-state bounded-error rtPFAs are sufficient to solve all family but the state number required by 2AFAs to solve all family cannot be bounded.*

Proof. The second part is due to Theorem 5 and Corollary 1. □

Now, we present a separation result between bounded-error rtPFAs and rtDFAs (and so 2AFAs). For this purpose, we again use an idea given by Jibran and Yakaryılmaz [18]. 2PFAs can recognize some non-regular languages, e.g. $EQ = \{a^n b^n \mid n \geq 1\}$ [8], with bounded error but it requires exponential time [7]. It was also shown that EQ can be recognized by a restarting rtPFA for any error bound [21]. If it is promised that any string $a^m b^n$ $(m, n \geq 1)$ is given exponentially many times, then the algorithm given in [21] for EQ can distinguish the cases where $m = n$ and $m \neq n$ with high probability. We define a new promise problem $EXPPromiseEQ$:

$$EXPPromiseEQ_{yes} = \{(a^n b^n)^t \mid n \geq 1 \text{ and } t \geq T\}$$
$$EXPPromiseEQ_{no} = \{(a^m b^n)^t \mid m \neq n \geq 1 \text{ and } t \geq T\}\text{'}$$

where the value of T will be given later. Let $\mathcal{P}_{\frac{1}{3}}$ be a restarting rtPFA recognizing EQ with error bound $\frac{1}{3}$ [21]. Let $w = u^t = (a^m b^n)^t$ be a promised input as described above, where $m, n \geq 1$ and $t \geq T$. A rtPFA, say $\mathcal{P}$, can execute $\mathcal{P}_{\frac{1}{3}}$ on u at least t times. From [21], we know that, in a single round of $\mathcal{P}_{\frac{1}{3}}$, u is accepted with probability

$$p_{acc} = \frac{1}{3}\left(\frac{1}{18}\right)^{m+n},$$

and the rejecting probability, say p_{rej}, is exactly $\frac{1}{3}$ of p_{acc} if $m = n$ and at least 3 times p_{acc} if $m \neq n$. The accepting and rejecting probabilities of $\mathcal{P}$ on w can be respectively calculated as

$$ACC = \sum_{i=1}^{t} p_{acc}(1 - p_{acc} - p_{rej})^{i-1} = p_{acc}\frac{1 - (1 - p_{acc} - p_{rej})^t}{p_{acc} + p_{rej}}$$

and

$$REJ = \sum_{i=1}^{t} p_{rej}(1 - p_{acc} - p_{rej})^{i-1} = p_{rej}\frac{1 - (1 - p_{acc} - p_{rej})^t}{p_{acc} + p_{rej}}.$$

If $m = n$, then $p_{acc} = 3p_{rej}$ and so ACC can be at least

$$\frac{3}{4}\left[1 - \left(1 - \frac{4}{27}\left(\frac{1}{18}\right)^{2n}\right)^t\right]$$

If $m \neq n$, then $p_{rej} \geq 3p_{acc}$ and so REJ can be at least

$$\frac{3}{4}\left[1 - \left(1 - \frac{4}{27}\left(\frac{1}{18}\right)^{m+n}\right)^t\right]$$

It is a well-known fact that $\left(1 - \frac{1}{x}\right)^x$ is less than $\frac{1}{e}$ for every $x \in \mathbb{Z}^+$. If T is $\frac{27}{2}18^{|u|}$, then the terms

$$\left(1 - \frac{4}{27}\left(\frac{1}{18}\right)^{2n}\right)^t \quad \text{or} \quad \left(1 - \frac{4}{27}\left(\frac{1}{18}\right)^{m+n}\right)^t$$

in ACC and REJ, respectively, can be at most $\frac{1}{e^2}$. Therefore both ACC and REJ can be at least

$$\frac{3}{4}\left(1 - \frac{1}{e^2}\right) \geq 0.64.$$

By using conventional amplification techniques and/or picking bigger T's, this value can be made arbitrarily close to 1.

Theorem 6. *For any $p > \frac{1}{2}$, there is a promise problem solvable by a bounded-error rtPFA with probability p but there is no rtDFA solving the same problem.*

Proof. We need to show that EXPPromiseEQ cannot be recognized by any rtDFA. Suppose that there is such a rtDFA, say $\mathcal{D}$, with n states $\{s_1, \ldots, s_n\}$. Any rtDFAs enters a cycle and stays there on sufficiently long unary input strings. Let c_i be the length of this cycle for $\mathcal{D}$ on symbol a when starting in state s_i, where $1 \leq i \leq n$. We define $c = lcm(c_1, \ldots, c_n)$. Let a^m be a string such that $m \gg n$. For any state of $\mathcal{D}$, say s, we can follow that if $\mathcal{D}$ is in s on the first symbols of a^m and a^{m+c}, then $\mathcal{D}$ leaves a^m and a^{m+c} in the same state. Therefore, the behaviours of $\mathcal{D}$ on $(a^m b^m)^t$ and $(a^{m+c} b^m)^t$ are the same for any $t > 0$. But, this is a contradiction since, for sufficiently long t's, the former strings must be accepted by $\mathcal{D}$ and the latter ones must be rejected by $\mathcal{D}$. Thus, we can follow that there is no rtDFA solving EXPPromiseEQ. □

Acknowledgements. We thank the anonymous reviewers for their helpful comments.

References

1. Ablayev, F., Gainutdinova, A., Khadiev, K., Yakaryılmaz, A.: Very narrow quantum OBDDs and width hierarchies for classical OBDDs. In: DCFS 2014. LNCS, vol. 8614, pp. 53–64. Springer, Heidelberg (2014)
2. Ambainis, A., Watrous, J.: Two–way finite automata with quantum and classical states. Theoretical Computer Science 287(1), 299–311 (2002)
3. Ambainis, A., Yakaryılmaz, A.: Superiority of exact quantum automata for promise problems. Information Processing Letters 112(7), 289–291 (2012)
4. Condon, A., Lipton, R.J.: On the complexity of space bounded interactive proofs (extended abstract). In: FOCS 1989: Proceedings of the 30th Annual Symposium on Foundations of Computer Science, pp. 462–467 (1989)
5. Duris, P., Hromkovic, J., Rolim, J.D.P., Schnitger, G.: Las Vegas versus determinism for one-way communication complexity, finite automata, and polynomial-time computations. In: Reischuk, R., Morvan, M. (eds.) STACS 1997. LNCS, vol. 1200, pp. 117–128. Springer, Heidelberg (1997)
6. Dwork, C., Stockmeyer, L.: Finite state verifiers I: The power of interaction. Journal of the ACM 39(4), 800–828 (1992)
7. Dwork, C., Stockmeyer, L.J.: A time complexity gap for two-way probabilistic finite-state automata. SIAM Journal on Computing 19(6), 1011–1123 (1990)
8. Freivalds, R.: Probabilistic two-way machines. In: Gruska, J., Chytil, M.P. (eds.) MFCS 1981. LNCS, vol. 118, pp. 33–45. Springer, Heidelberg (1981)

9. Goldreich, O.: On promise problems: A survey. In: Goldreich, O., Rosenberg, A.L., Selman, A.L. (eds.) Shimon Even Festschrift. LNCS, vol. 3895, pp. 254–290. Springer, Heidelberg (2006)
10. Gruska, J., Qiu, D., Zheng, S.: Generalizations of the distributed Deutsch-Jozsa promise problem. Technical report, arXiv:1402.7254 (2014)
11. Gruska, J., Qiu, D., Zheng, S.: Potential of quantum finite automata with exact acceptance. Technical Report arXiv:1404.1689 (2014)
12. Hromkovic, J., Schnitger, G.: On the power of Las Vegas for one-way communication complexity, OBDDs, and finite automata. Information and Computation 169(2), 284–296 (2001)
13. Kapoutsis, C.A., Královic, R., Mömke, T.: Size complexity of rotating and sweeping automata. Journal of Computer and System Sciences 78(2), 537–558 (2012)
14. Kuklin, Y.I.: Two-way probabilistic automata. Avtomatika i Vyčlstitelnaja Tekhnika 5, 36–36 (1973) (Russian)
15. Murakami, Y., Nakanishi, M., Yamashita, S., Watanabe, K.: Quantum versus classical pushdown automata in exact computation. IPSJ Digital Courier 1, 426–435 (2005)
16. Nakanishi, M.: Quantum pushdown automata with a garbage tape. Technical Report arXiv:1402.3449 (2014)
17. Rabin, M.O.: Probabilistic automata. Information and Control 6, 230–243 (1963)
18. Rashid, J., Yakaryılmaz, A.: Implications of quantum automata for contextuality. In: CIAA. LNCS. Springer (to appear, 2014); arXiv:1404.2761
19. Sipser, M.: Lower bounds on the size of sweeping automata. Journal of Computer and System Sciences 21(2), 195–202 (1980)
20. Watrous, J.: Quantum computational complexity. In: Meyers, R.A. (ed.) Encyclopedia of Complexity and Systems Science, pp. 7174–7201. Springer (2009)
21. Yakaryılmaz, A., Say, A.C.C.: Succinctness of two-way probabilistic and quantum finite automata. Discrete Mathematics and Theoretical Computer Science 12(2), 19–40 (2010)
22. Zheng, S., Gruska, J., Qiu, D.: On the state complexity of semi-quantum finite automata. In: Dediu, A.-H., Martín-Vide, C., Sierra-Rodríguez, J.-L., Truthe, B. (eds.) LATA 2014. LNCS, vol. 8370, pp. 601–612. Springer, Heidelberg (2014)
23. Zheng, S., Qiu, D., Gruska, J., Li, L., Mateus, P.: State succinctness of two-way finite automata with quantum and classical states. Theor. Comput. Sci. 499, 98–112 (2013)
24. Zheng, S., Qiu, D., Li, L., Gruska, J.: One-way finite automata with quantum and classical states. In: Bordihn, H., Kutrib, M., Truthe, B. (eds.) Dassow Festschrift 2012. LNCS, vol. 7300, pp. 273–290. Springer, Heidelberg (2012)

From Ultrafilters on Words to the Expressive Power of a Fragment of Logic*

Mai Gehrke[1], Andreas Krebs[2], and Jean-Éric Pin[1]

[1] LIAFA, CNRS and Univ. Paris-Diderot, Case 7014, 75205 Paris Cedex 13, France
[2] Wilhelm-Schickard-Institut für Informatik, Universität Tübingen, Germany

Abstract. We give a method for specifying ultrafilter equations and identify their projections on the set of profinite words. Let $\mathcal{B}$ be the set of languages captured by first-order sentences using unary predicates for each letter, arbitrary uniform unary numerical predicates and a predicate for the length of a word. We illustrate our methods by giving profinite equations characterizing $\mathcal{B} \cap \mathrm{Reg}$ via ultrafilter equations satisfied by $\mathcal{B}$. This suffices to establish the decidability of the membership problem for $\mathcal{B} \cap \mathrm{Reg}$.

In two earlier papers, Gehrke, Grigorieff, and Pin proved the following results:

Result 1. [5] *Any Boolean algebra of regular languages can be defined by a set of equations of the form $u = v$, where u and v are profinite words.*[1]

Result 2. [6] *Any Boolean algebra of languages can be defined by a set of equations of the form $u = v$, where u and v are ultrafilters on the set of words.*

These two results can be summarized by saying that Boolean algebras of languages can be defined by *ultrafilter equations* and by *profinite equations* in the regular case. Restricted instances of Result 1 have proved to be very successful long before the result was stated in full generality. It is in particular a powerful tool for characterizing classes of regular languages or for determining the expressive power of various fragments of logic, see the book of Almeida [2] or the survey [9] for more information.

Result 2 however is still awaiting convincing applications and even an idea of how to apply it in a concrete situation. The main problem in putting it into practice is to cope with ultrafilters, a difficulty nicely illustrated by Jan van Mill, who cooked up the nickname *three headed monster* for the set of ultrafilters on $\mathbb{N}$. Facing this obstacle, the authors thought of using Results 1 and 2 simultaneously to obtain a new proof of the equality

$$\mathbf{FO}[\mathcal{N}] \cap \mathrm{Reg} = [\![\ (x^{\omega-1}y)^{\omega+1} = (x^{\omega-1}y)^{\omega}$$

$$\text{for } x, y \text{ words of the same length}\ ]\!] \quad (1)$$

---

* Work supported by the project ANR 2010 BLAN 0202 02 FREC.
[1] In [5], these were denoted by $u \leftrightarrow v$.

H. Jürgensen et al. (Eds.): DCFS 2014, LNCS 8614, pp. 138–149, 2014.
© Springer International Publishing Switzerland 2014

This formula gives the profinite equations characterizing the regular languages captured by $\textbf{FO}[\mathcal{N}]$, the first order logic using arbitrary numerical predicates and the usual letter predicates. This result follows from the work of Barrington, Straubing and Thérien [3] and Straubing [10] and is strongly related to circuit complexity. Indeed its proof makes use of the equality between $\textbf{FO}[\mathcal{N}]$ and AC^0, the class of languages accepted by unbounded fan-in, polynomial size, constant-depth Boolean circuits [11, Theorem IX.2.1, p. 161]. See also [7] for similar results and problems.

However, before attacking this problem in earnest we have to tackle the following questions: how does one get hold of an ultrafilter equation given the non-constructibility of each one of them (save the trivial ones given by pairs of words)? In particular, how does one generalize the powerful use in the regular setting of x^ω? And how does one project such ultrafilter equations to the regular fragment? In answering these questions and facing these challenges, we have chosen to consider a smaller and simpler logic fragment first. Our choice was dictated by two parameters: we wanted to be able to handle the corresponding ultrafilters and we wished to obtain a reasonably understandable list of profinite equations. Finally, we opted for $\textbf{FO}[\mathcal{N}_0, \mathcal{N}_1^u]$, the restriction of $\textbf{FO}[\mathcal{N}]$ to constant numerical predicates and to uniform unary numerical predicates. Here we obtain the following result (Theorem 4.7)

$$\textbf{FO}[\mathcal{N}_0, \mathcal{N}_1^u] \cap \mathrm{Reg} = [\![(x^{\omega-1}s)(x^{\omega-1}t) = (x^{\omega-1}t)(x^{\omega-1}s),$$
$$(x^{\omega-1}s)^2 = (x^{\omega-1}s) \text{ for } x, s, t \text{ words of the same length}]\!] \quad (2)$$

which shows in particular that membership in $\textbf{FO}[\mathcal{N}_0, \mathcal{N}_1^u]$ is decidable for regular languages.

Although this result is of interest in itself, we claim that our *proof method* is more important than the result. Indeed, this case study demonstrates for the first time the workability of the ultrafilter approach.

This method can be summarized as follows. First we find a set of ultrafilter equations satisfied by $\textbf{FO}[\mathcal{N}_0, \mathcal{N}_1^u]$ (Theorem 3.2). These equations do not necessarily suffice to characterize $\textbf{FO}[\mathcal{N}_0, \mathcal{N}_1^u]$ [2], but projecting ultrafilters onto profinite words, we convert our ultrafilter equations to profinite equations for $\textbf{FO}[\mathcal{N}_0, \mathcal{N}_1^u] \cap \mathrm{Reg}$ (Theorems 3.3 and 3.4). The last step consists in verifying that the set of profinite equations thus obtained suffices to characterize $\textbf{FO}[\mathcal{N}_0, \mathcal{N}_1^u] \cap \mathrm{Reg}$ (Theorem 4.7).

Now, a closer look at our proof shows that we are far from making use of the potential power of ultrafilters. For instance, difficult combinatorial results like Szemeredi's theorem on arithmetic progressions can be formulated in terms of ultrafilters. Thus it is quite possible that more sophisticated arguments are required to extend our results to larger fragments of logic, including $\textbf{FO}[\mathcal{N}]$.

---

[2] We recently proved that these equations actually do suffice to characterize $\textbf{FO}[\mathcal{N}_0, \mathcal{N}_1^u]$, but this will be the topic of another paper.

1   Stone Duality and Equations

In this paper, we denote by S^c the complement of a subset S of a set E.

1.1   Stone Duality

Let A be a finite alphabet. A *Boolean algebra of languages* is a set $\mathcal{B}$ of languages of A^* closed under finite unions, finite intersections and complement. It is *closed under quotients* if, for each $L \in \mathcal{B}$ and $u \in A^*$, the languages $u^{-1}L$ and Lu^{-1} are also in $\mathcal{B}$. Recall that $u^{-1}L = \{x \in A^* \mid ux \in L\}$ and $Lu^{-1} = \{x \in A^* \mid xu \in L\}$.

Let $\mathcal{B}$ be a Boolean algebra of languages of A^*. An *ultrafilter* of $\mathcal{B}$ is a non-empty subset γ of $\mathcal{B}$ such that:

(1) the empty set does not belong to γ,

(2) if $K \in \gamma$ and $K \subseteq L$, then $L \in \gamma$ (closure under extension)[3],

(3) if $K, L \in \gamma$, then $K \cap L \in \gamma$ (closure under intersection),

(4) for every $L \in \mathcal{B}$, either $L \in \gamma$ or $L^c \in \gamma$ (ultrafilter condition).

Stone duality tells us that $\mathcal{B}$ has an associated compact Hausdorff space $S(\mathcal{B})$, called its *Stone space*. This space is given by the set of ultrafilters of $\mathcal{B}$ with the topology generated by the basis of clopen sets of the form $\{\gamma \in S(\mathcal{B}) \mid L \in \gamma\}$, where $L \in \mathcal{B}$.

Two Stone spaces are of special interest for this paper. The first one is the Stone space of the Boolean algebra of all the subsets of a set X. It is known as the *Stone-Čech compactification* of X and is usually denoted by βX. An important property is that every map $f : X \to Y$ has a unique continuous extension $\beta f : \beta X \to \beta Y$ defined by $L \in \beta f(\gamma)$ if and only if $f^{-1}(L) \in \gamma$ for each subset L of Y. Moreover, the map sending an element x of X to the principal ultrafilter generated by x defines an injective map from X into βX.

Our second example is the Stone space of the Boolean algebra Reg of all *regular* subsets of A^*. It was proved by Almeida [1] to be equal to the topological space underlying the free profinite monoid on A, denoted by $\widehat{A^*}$. We refer to [2,8,9] for more information on this space, but it can be seen as the completion of A^* for the *profinite metric d* defined as follows. A finite monoid M *separates* two words u and v of A^* if there is a monoid morphism $\varphi : A^* \to M$ such that $\varphi(u) \neq \varphi(v)$. We set

$$r(u, v) = \min\{|M| \mid M \text{ is a finite monoid that separates } u \text{ and } v\ \}$$

and $d(u, v) = 2^{-r(u,v)}$, with the usual conventions $\min \emptyset = +\infty$ and $2^{-\infty} = 0$. Then d is a *metric* on A^* and the completion of A^* for this metric is denoted by $\widehat{A^*}$. The product on A^* can be extended by continuity to $\widehat{A^*}$, making $\widehat{A^*}$ a compact topological monoid, called the *free profinite monoid*. Its elements are called *profinite words*. We will only use two types of profinite words in this paper. In a compact monoid, the smallest closed subsemigroup containing a given

---

[3] In other words, γ is an *upset*.

element x has a unique idempotent, denoted by x^ω. Thus if x is a (profinite) word, so is x^ω. In fact, one can show that x^ω is the limit of the converging sequence $x^{n!}$. Moreover, the sequence $x^{n!-1}$ is also converging to an element denoted by $x^{\omega-1}$. More details can be found in [2,8,9].

1.2   Equations

Assigning to a Boolean algebra its Stone space is a contravariant functor: if $\mathcal{B}'$ is a subalgebra of $\mathcal{B}$, then $S(\mathcal{B}')$ is a quotient of $S(\mathcal{B})$. More precisely, the function which maps an ultrafilter of $\mathcal{B}$ onto its trace on $\mathcal{B}'$ induces a surjective continuous map $\pi : S(\mathcal{B}) \to S(\mathcal{B}')$.

This leads to the notion of equation relative to $\mathcal{B}$ or $\mathcal{B}$-*equation*. Let γ_1, γ_2 be two ultrafilters on $\mathcal{B}$ and let $L \in \mathcal{B}$. We say that L satisfies the $\mathcal{B}$-equation $\gamma_1 = \gamma_2$ provided

$$L \in \gamma_1 \iff L \in \gamma_2. \tag{3}$$

By extension, we say that $\mathcal{B}'$ satisfies the $\mathcal{B}$-equation $\gamma_1 = \gamma_2$ provided (3) holds for all $L \in \mathcal{B}'$, or equivalently $\pi(\gamma_1) = \pi(\gamma_2)$. Note that if $\mathcal{B}'$ is generated as a Boolean algebra by a subset $\mathcal{C}$, then $\mathcal{B}'$ satisfies a $\mathcal{B}$-equation as soon as each $L \in \mathcal{C}$ does. Finally, we say that $\mathcal{B}'$ is defined by a set E of $\mathcal{B}$-equations if for each $L \in \mathcal{B}$, $L \in \mathcal{B}'$ if and only if L satisfies all the $\mathcal{B}$-equations in E. The following result is an immediate consequence of Stone duality.

Theorem 1.1. *Every subalgebra of a Boolean algebra $\mathcal{B}$ can be defined by a set of $\mathcal{B}$-equations.*

Specializing this result to $\mathcal{B} = \mathrm{Reg}$ and to $\mathcal{B} = \mathcal{P}(A^*)$ yields Results 1 and 2 of the introduction. Another case of interest for this paper is to take $\mathcal{B} = \mathrm{Reg}$ and for $\mathcal{B}'$ a Boolean algebra closed under quotients. In this case, it is easier to reformulate Result 1 in terms of syntactic morphisms. Let L be a regular language and $\eta : A^* \to M$ be its syntactic morphism. We say that η *satisfies the profinite equation* $u = v$ or that L *syntactically satisfies the profinite equation* $u = v$ if $\hat{\eta}(u) = \hat{\eta}(v)$, where $\hat{\eta} : \widehat{A^*} \to M$ is the unique continuous extension of η to $\widehat{A^*}$. It is easy to see that a regular language syntactically satisfies a profinite equation if and only if all of its quotients satisfy this equation. Therefore we have

Result 3. *Any Boolean algebra of regular languages closed under quotients can be syntactically defined by a set of profinite equations.*

When working with ultrafilter equations, the following two observations will be helpful. Let us denote by $K \triangle L$ the symmetric difference of the sets K and L.

Proposition 1.2. *Let γ be an ultrafilter of $\mathcal{B}$ and let $K, L \in \mathcal{B}$. Then the following statements are equivalent:*

(1) $K \in \gamma$ if and only if $L \in \gamma$,

(2) $K \triangle L \notin \gamma$.

Proposition 1.3. *Let $f : X \to Y$ be a map and let L be a subset of Y. If $f^{-1}(L)$ satisfies $u = v$ for some $u, v \in \beta X$, then L satisfies $\beta f(u) = \beta f(v)$.*

2   A Boolean Algebra and Its Logical Description

For each word $u = a_0 \ldots a_{n-1}$ where $a_0, \ldots, a_{n-1} \in A$ and each letter $a \in A$, let $\mathbf{a}_u = \{i \in \mathrm{Dom}(u) \mid a_i = a\}$. For instance, if $u = aabcbaba$, then $\mathbf{a}_u = \{0, 1, 5, 7\}$, $\mathbf{b}_u = \{2, 4, 6\}$, and $\mathbf{c}_u = \{3\}$. The *length* of u is denoted by $|u|$.

For each letter a in A and for each subset P of $\mathbb{N}$, let
$$L_P = \{u \in A^+ \mid |u| - 1 \in P\} \text{ and } L_{a,P} = \{u \in A^+ \mid \mathbf{a}_u \subseteq P\}.$$
In this paper, we are interested in the Boolean algebra $\mathcal{B}$ generated by the languages L_P and $L_{a,P}$ for $P \subseteq \mathbb{N}$ and $a \in A$. A little combinatorics on words leads to the following result:

Proposition 2.1. *The Boolean algebras $\mathcal{B}$ and $\mathcal{B} \cap \mathrm{Reg}$ are closed under quotients and under the operations $L \to uL$ for each word $u \in A^*$.*

Let us turn to the logical description of $\mathcal{B}$. Let $u = a_0 \ldots a_{n-1}$ be a nonempty word where $a_0, \ldots, a_{n-1}$ are letters of the alphabet A. Then u may be viewed as a first-order model whose *domain* is the set $\mathrm{Dom}(u) = \{0, \ldots, |u|-1\}$, carrying, for each letter $a \in A$, the unary predicate $\mathbf{a}_u$ as defined above. For each subset P of $\mathbb{N}$, we also define two predicates: a 0-ary predicate which is true on u if and only if $|u| - 1 \in P$ and a unary uniform predicate[4] defined by $P(n) = P \cap \{0, ..., n-1\}$. Its interpretation on a word u is the subset $P(|u|)$ of $\{0, ..., |u| - 1\}$.

We denote by $\mathbf{FO}[\mathcal{N}_0, \mathcal{N}_1^u]$ the set of first-order formulas built on these predicates. Note that we do not consider $=$ as a logical symbol so that each formula is equivalent to one of quantifier depth one. The language defined by a sentence φ is the set[5]
$$L(\varphi) = \{u \in A^+ \mid u \text{ satisfies } \varphi\}$$
For instance if $\varphi = \exists x \; \mathbf{a}x$, then $L(\varphi) = A^* a A^*$. Let P be a subset of $\mathbb{N}$. If P is considered as a 0-ary numerical relation, then $L(P) = L_P$. If P is interpreted as a unary uniform numerical relation, then the formula $\forall x \; (\mathbf{a}x \to Px)$ defines the language $L_{a,P}$ since P is interpreted as $P(|u|)$. This proves one direction of the following logical description of $\mathcal{B}$.

Theorem 2.2. *A language L of A^+ belongs to $\mathcal{B}$ if and only it can be defined by a sentence of $\mathbf{FO}[\mathcal{N}_0, \mathcal{N}_1^u]$.*

3   Some Equations of $\mathcal{B}$

For $1 \leqslant i \leqslant k$, let $\pi_i : A^* \times \mathbb{N}^k \to \mathbb{N}$ be the map defined by $\pi_i(u, n_1, \ldots, n_k) = n_i$.

The following proposition shows that the classes of equations we will define subsequently contain at least one non-trivial equation for each $\alpha \in \beta\mathbb{N} - \mathbb{N}$.

---

[4] Following the terminology of [11], a unary *numerical relation* R associates to each $n > 0$ a subset $R(n)$ of $\{0, ..., n - 1\}$. It is *uniform* if there exists a subset P of $\mathbb{N}$ such that, for all $n > 0$, $R(n) = P \cap \{0, ..., n - 1\}$. Not every numerical relation is uniform: for instance, the unary numerical relation R defined by $R(n) = \{n - 1\}$ is not uniform.

[5] The empty word is excluded to avoid any problem related to empty models.

Proposition 3.1. *Let* $\gamma \in \beta(A^* \times \mathbb{N}^k)$ *with* $k \geqslant 1$. *Then, for each* $\alpha \in \beta\mathbb{N}$, *the following conditions are equivalent:*

(1) $\beta\pi_i(\gamma) = \alpha$ *for each* $i \in \{1, \ldots, k\}$;

(2) $\{A^* \times P^k \mid P \in \alpha\} \subseteq \gamma$.

Furthermore, these conditions hold for γ *with respect to some* α *if and only if*

(3) *For each partition* $\{P_1, \ldots, P_n\}$ *of* $\mathbb{N}$, *we have* $\bigcup_{j=1}^n (A^* \times P_j^k) \in \gamma$.

Proof. (1) implies (2) since $A^* \times P^k = \bigcap_{i=1}^k \pi_i^{-1}(P)$ and γ is closed under finite intersections.

(2) implies (1). Let $P \in \alpha$ and $i \in \{1, \ldots, k\}$. Then by (2), $A^* \times P^k \in \gamma$ and thus $\pi_i^{-1}(\pi_i(A^* \times P^k)) \in \gamma$ so that $P = \pi_i(A^* \times P^k) \in \beta\pi_i(\gamma)$. It follows that $\alpha \subseteq \beta\pi_i(\gamma)$ and thus $\alpha = \beta\pi_i(\gamma)$ since ultrafilters are maximal.

For the second assertion, suppose there is an $\alpha \in \beta\mathbb{N}$ such that (1) and (2) hold and $\{P_1, \ldots, P_n\}$ is a partition of $\mathbb{N}$. Then $\bigcup_{j=1}^n P_j = \mathbb{N}$ implies $P_\ell \in \alpha$ for some ℓ and thus $A^* \times P_\ell^k \in \gamma$ by (2). Since γ is an upset, condition (3) holds.

Suppose now that γ satisfies (3) and let $\alpha = \{P \mid A^* \times P^k \in \gamma\}$. Then α is an upset closed under intersection. Furthermore, for each $P \subseteq \mathbb{N}$, the partition $\{P, P^c\}$ forces $A^* \times P^k \in \gamma$ or $A^* \times (P^c)^k \in \gamma$ so that α is an ultrafilter. It follows by the equivalence of (1) and (2) that $\beta\pi_i(\gamma) = \alpha$ for each $i \in \{1, \ldots, k\}$. □

We are now ready to introduce the first class of equations pertinent to the languages treated in this paper. For this purpose, given $u, s, t \in A^*$, where $u = u_0 \cdots u_{n-1}$ with each $u_k \in A$ and $|s| = |t| = \ell$, and $i, j \in \mathbb{N}$, define

$$u(s@i, t@j) = \begin{cases} u_0 \ldots u_{i-1} s u_{i+\ell} \ldots u_{j-1} t u_{j+\ell} \ldots u_{n-1} & \text{if } i + \ell \leqslant j \text{ and } j + \ell \leqslant n \\ u & \text{otherwise} \end{cases}$$

Informally, we put s at position i and t at position j.

| u_0 | $\cdots$ | u_{i-1} | u_i | $\cdots$ | $u_{i+\ell-1}$ | $u_{i+\ell}$ | $\cdots$ | u_{j-1} | u_j | $\cdots$ | $u_{j+\ell-1}$ | $u_{j+\ell}$ | $\cdots$ | u_{n-1} |

For each pair (s, t) of words of the same length, let $f_{s,t} : A^* \times \mathbb{N}^2 \to A^*$ be the function defined by $f_{s,t}(u, i, j) = u(s@i, t@j)$.

Theorem 3.2. *Let* $s, t \in A^*$ *with* $|s| = |t|$. *If* $\gamma \in \beta(A^* \times \mathbb{N}^2)$ *and* $\beta\pi_1(\gamma) = \beta\pi_2(\gamma)$, *then* $\mathcal{B}$ *satisfies the equation*

$$\beta f_{s,t}(\gamma) = \beta f_{t,s}(\gamma). \tag{4}$$

Proof. Let $a \in A$ and $P \subseteq \mathbb{N}$. We show that $L_{a,P}$ and L_P satisfy the equations (4). First we have

$$L_{a,P} \in \beta f(\gamma) \iff f^{-1}(L_{a,P}) \in \gamma.$$

Thus (4) holds for $L_{a,P}$ if and only if

$$f_{s,t}^{-1}(L_{a,P}) \in \gamma \iff f_{t,s}^{-1}(L_{a,P}) \in \gamma$$

and by Proposition 1.2 this is equivalent to $S \notin \gamma$, where

$$S = f_{s,t}^{-1}(L_{a,P}) \triangle f_{t,s}^{-1}(L_{a,P}).$$

Let ℓ be the common length of s and t. If an element $(u, n_1, n_2) \in A^* \times \mathbb{N}^2$ is in S then $n_1 + 2\ell \leqslant n_2 + \ell \leqslant |u|$ since otherwise $f_{s,t}(u, n_1, n_2) = f_{t,s}(u, n_1, n_2) = u$. Suppose that $(u, n_1, n_2) \in f_{s,t}^{-1}(L_{a,P}) \setminus f_{t,s}^{-1}(L_{a,P})$, that is, $f_{s,t}(u, n_1, n_2) \in L_{a,P}$ and $f_{t,s}(u, n_1, n_2) \notin L_{a,P}$. Then all the positions of a in $f_{s,t}(u, n_1, n_2)$ are in P and some position of a in $f_{t,s}(u, n_1, n_2)$ is not in P. This latter position necessarily occurs inside one of the factors s or t of $f_{s,t}(u, n_1, n_2)$. Consequently, there is an $i \in \{0, \ldots, \ell - 1\}$ such that one of the two following possibilities occurs:

(1) the letter in position $n_1 + i$ in $f_{t,s}(u, n_1, n_2)$ is an a but $n_1 + i \notin P$,

(2) the letter in position $n_2 + i$ in $f_{t,s}(u, n_1, n_2)$ is an a but $n_2 + i \notin P$.

Now, in the first case, the letter in position $n_2 + i$ in $f_{s,t}(u, n_1, n_2)$ is an a. Thus $n_2 + i \in P$ since $f_{s,t}(u, n_1, n_2) \in L_{a,P}$. Similarly, we conclude that $n_1 + i \in P$ in the second case. In summary, we have either $n_1 + i \notin P$ and $n_2 + i \in P$ (first case) or $n_1 + i \in P$ and $n_2 + i \notin P$ (second case). In both cases we conclude that

$$(u, n_1, n_2) \in \bigcup_{i=0}^{\ell-1} \left(\pi_1^{-1}(P - i) \triangle \pi_2^{-1}(P - i) \right).$$

The case $(u, n_1, n_2) \in f_{t,s}^{-1}(L_{a,P}) \setminus f_{s,t}^{-1}(L_{a,P})$ leads to the same conclusion and thus we have shown that

$$S \subseteq \bigcup_{i=0}^{\ell-1} \left(\pi_1^{-1}(P - i) \triangle \pi_2^{-1}(P - i) \right).$$

If $S \in \gamma$, then $\bigcup_{i=0}^{\ell-1} \left(\pi_1^{-1}(P - i) \triangle \pi_2^{-1}(P - i) \right) \in \gamma$ and since γ is an ultrafilter, $\pi_1^{-1}(P - i) \triangle \pi_2^{-1}(P - i) \in \gamma$ for some $i \in \{0, \ldots, \ell - 1\}$. We complete the proof that $S \notin \gamma$ by showing that, for every $Q \subseteq \mathbb{N}$ we have $\pi_1^{-1}(Q) \triangle \pi_2^{-1}(Q) \notin \gamma$, or equivalently, $(\pi_1^{-1}(Q) \triangle \pi_2^{-1}(Q))^c \in \gamma$. But this is a direct consequence of Proposition 3.1(3) since

$$(\pi_1^{-1}(Q) \triangle \pi_2^{-1}(Q))^c = A^* \times \left((Q \times Q) \cup (Q^c \times Q^c) \right).$$

Thus $S \notin \gamma$ and $L_{a,P}$ satisfies the equation $\beta f_{s,t}(\gamma) = \beta f_{t,s}(\gamma)$.

By the same argument as applied above, L_P satisfies the equations (4) if and only if $f_{s,t}^{-1}(L_P) \triangle f_{t,s}^{-1}(L_P) \notin \gamma$. However, since $|f_{s,t}(u, n_1, n_2)| = |f_{t,s}(u, n_1, n_2)|$ and since $x \in L_P$ implies $y \in L_P$ if $|y| = |x|$, we have $f_{s,t}^{-1}(L_P) = f_{t,s}^{-1}(L_P)$ and thus $f_{s,t}^{-1}(L_P) \triangle f_{s,t}^{-1}(L_P) = \emptyset$ and therefore it does not belong to γ. $\square$

We now consider the projection of the equations introduced above on the Stone space of the regular fragment of the Boolean algebra $\mathcal{B}$.

Theorem 3.3. *Let* $x, s, t, \in A^*$ *with* $|s| = |t| = |x|$. *Then* $\mathcal{B} \cap \mathrm{Reg}$ *satisfies the profinite equation* $x^{\omega-1}sx^{\omega-1}t = x^{\omega-1}tx^{\omega-1}s$.

Proof. It suffices to show that there is a $\gamma \in \beta(A^* \times \mathbb{N}^2)$ with $\beta\pi_1(\gamma) = \beta\pi_2(\gamma)$ such that the projection $\pi_{\mathrm{Reg}} : \beta A^* \to \widehat{A^*}$ defined by

$$\pi_{\mathrm{Reg}}(\gamma) = \gamma \cap \mathrm{Reg}$$

maps $\beta f_{s,t}(\gamma)$ to $x^{\omega-1}sx^{\omega-1}t$ and $\beta f_{t,s}(\gamma)$ to $x^{\omega-1}tx^{\omega-1}s$.

Proposition 3.1 shows that in order for γ to satisfy $\beta\pi_1(\gamma) = \beta\pi_2(\gamma)$, we just need γ to contain the down-directed filter base

$$\left\{ \bigcup_{j=1}^{n}(A^* \times P_j^2) \mid \{P_1, \ldots, P_n\} \text{ is a partition of } \mathbb{N} \right\}.$$

We now show that for $\ell = |x|$, adding the sets $W_N = \{(x^{m!}, (k!-1)\ell, (m!-1)\ell) \mid N \leqslant k < m\}$ for each $N \in \mathbb{N}$ to this filter base still yields a filter base. To this end we just need to show that for each partition $\{P_1, \ldots, P_n\}$ of $\mathbb{N}$ and $N \in \mathbb{N}$, the set

$$W_N \cap (\bigcup_{j=1}^{n}(A^* \times P_j^2))$$

is non-empty. But since $\{P_1, \ldots, P_n\}$ is a partition of $\mathbb{N}$, there is $j \in \{1, \ldots, n\}$ with $P_j \cap \{(k!-1)\ell \mid k \geqslant N\}$ infinite. It readily follows that $W_N \cap (A^* \times P_j^2)$ is infinite and thus the bigger set $W_N \cap (\bigcup_{j=1}^{n}(A^* \times P_j^2))$ is non-empty.

Let $\gamma \in \beta(A^* \times \mathbb{N}^2)$ be an ultrafilter containing the extended filter base. Then clearly $\beta\pi_1(\gamma) = \beta\pi_2(\gamma)$ so that, by Theorem 3.2, the Boolean algebra $\mathcal{B}$ satisfies the equation $\beta f_{s,t}(\gamma) = \beta f_{t,s}(\gamma)$.

Now let $L \in \beta f_{s,t}(\gamma) \cap \mathrm{Reg}$. Then $f_{s,t}^{-1}(L) \in \gamma$. Also, since $W_N \in \gamma$ for each $N \in \mathbb{N}$, it follows that $f_{s,t}^{-1}(L) \cap W_1$ is infinite, or equivalently $L \cap f_{s,t}(W_1)$ is infinite. But

$$f_{s,t}(W_1) = \{x^{n!-1}sx^{(m!-n!)-1}t \mid 1 \leqslant n < m\}$$

and $m! - n! = (m!/n! - 1)n!$ where $(m!/n! - 1) \geqslant 1$. Since any sequence in this set with $n \to \infty$ converges to $x^{\omega-1}sx^{\omega-1}t$ in $\widehat{A^*}$, and since $L \cap f_{s,t}(W_1) \subseteq \widehat{L}$ with the latter closed, we must have $x^{\omega-1}sx^{\omega-1}t \in \widehat{L}$. But as $\widehat{A^*}$ is Hausdorff,

$$\bigcap \{\widehat{L} \mid L \in \beta f_{s,t}(\gamma) \text{ and } L \in \mathrm{Reg}\} = \pi_{\mathrm{Reg}}(\beta f_{s,t}(\gamma))$$

so $x^{\omega-1}sx^{\omega-1}t = \pi_{\mathrm{Reg}}(\beta f_{s,t}(\gamma))$. Similarly $x^{\omega-1}tx^{\omega-1}s = \pi_{\mathrm{Reg}}(\beta f_{t,s}(\gamma))$.   □

A similar argument using the ultrafilter equations $\beta f_{tss}(\gamma) = \beta f_{tts}(\gamma)$ with $\beta\pi_1(\gamma) = \beta\pi_2(\gamma) = \beta\pi_3(\gamma)$ and projecting yields the profinite equation
$$(x^{\omega-1}t)(x^{\omega-1}s)(x^{\omega-1}s) = (x^{\omega-1}t)(x^{\omega-1}t)(x^{\omega-1}s).$$
Specializing to $x = t$ we get $(x^{\omega-1}s)(x^{\omega-1}s) = (x^{\omega-1}s)$, which proves the following result.

Theorem 3.4. *Let $x, s \in A^*$ with $|s| = |x|$. Then $\mathcal{B} \cap \mathrm{Reg}$ satisfies the profinite equation $(x^{\omega-1}s)(x^{\omega-1}s) = (x^{\omega-1}s)$.*

Applications of the ultrafilter equations introduced in this section are not limited to the interplay with regular languages and they can also be used to prove separation results for nonregular languages. For instance, it is easy to find an ultrafilter equation of $\mathcal{B}$ not satisfied by the language $\{uav \mid u, v \in \{a, b\}^* \text{ and } |u| = |v|\}$.

4   The Regular Case

Consider the two profinite equations introduced in the previous sections, where x, s and t are words of the same length

$$(x^{\omega-1}s)(x^{\omega-1}t) = (x^{\omega-1}t)(x^{\omega-1}s) \tag{5}$$

$$(x^{\omega-1}s)(x^{\omega-1}s) = (x^{\omega-1}s) \tag{6}$$

We will show that the regular languages of our class $\mathcal{B}$ are exactly the languages whose syntactic morphism satisfies the equations (5) and (6) for all words x, s and t of the same length. Before we do this, it is useful to introduce some further notation.

Let $k, r, d \in \mathbb{N}$ with $d > 0$. Given a word $u = a_0 \cdots a_n \in A^*$ where $a_i \in A$, let $p_k(u) = a_0 \cdots a_{k-1}$ be the prefix of length k of u and let

$$C_{d,r}(u) = \{a_i \mid i \geqslant d \text{ and } i \equiv r \bmod d\}$$

be the *content* of u at r modulo d. For instance, if $u = ccbbacabac$, then $p_5(u) = ccbba$, $C_{3,0} = \{a, b, c\}$, $C_{3,1} = \{a, b\}$ and $C_{3,2} = \{a, c\}$.

For each positive integer d, let $\sim_d$ be the equivalence on A^* defined as follows. Given $u, v \in A^*$, $u \sim_d v$ if and only if the three following conditions are satisfied:

(1) for $0 < k \leqslant d$, $p_k(u) = p_k(v)$,

(2) $|u| \equiv |v| \bmod d$,

(3) for $0 \leqslant r < d$, $C_{d,r}(u) = C_{d,r}(v)$.

Proposition 4.1. *The relation $\sim_d$ is a congruence of finite index on A^*.*

We now consider a regular language L and we denote by $\eta : A^* \to M$ its syntactic morphism. We also let $d = |M|!$. It is well known that, for each $x \in M$, x^d is idempotent, that is, $x^{2d} = x^d$. For the remainder of the paper, we use the notation $u =_\eta v$ for $\eta(u) = \eta(v)$, and, for any $r \in \mathbb{N}$, we denote by $[r]$ the remainder after division of r by d. We will need a small combinatorial lemma:

Lemma 4.2. *Let u be a word of length at least $|M|$. Then there exist a prefix p of u of length lesser than $|M|$ and a word x of length $|M|!$ such that $px =_\eta p$.*

Proof. For each $k \geqslant 0$, let $s_k = \eta(p_k(u))$. If $s_0, \ldots, s_{|M|-1}$ are all distinct, one of them, say s_i, is idempotent. Then $p = p_i(u)$ and $x = p^{|M|!/(i+1)}$ give the result. On the other hand, if $s_i = s_j$ with $i < j < |M|$, then $p = p_i(u)$ and $x = z^{|M|!/|z|}$ where $p_j(u) = pz$ yield the result.   □

Let $\mathcal{B}_{\text{Reg}}$ be the Boolean algebra generated by the languages L_P or $L_{a,P}$ where P is a regular subset of $\mathbb{N}$, that is, a finite union of languages of the form $r + d\mathbb{N}$ for $r, d \in \mathbb{N}$. Clearly $\mathcal{B}_{\text{Reg}} \subseteq \mathcal{B} \cap \text{Reg}$. Our aim is to show that if η satisfies the equations (5) and (6), then L is a union of $\sim_d$-classes. In view of the following proposition, it then follows that $L \in \mathcal{B}_{\text{Reg}}$.

Proposition 4.3. *For every $d \geqslant 1$, every $\sim_d$-class is a language of $\mathcal{B}_{\text{Reg}}$.*

We now suppose that η satisfies equations (5) and (6) for all words x, s and t of the same length.

Lemma 4.4. *Let $a_0, a_1, \ldots, a_r$ be letters and let p and x be two words such that $|x| = d$ and $px =_\eta p$. Setting $x = b_0 \cdots b_{d-1}$ where $b_0, b_1, \ldots, b_{d-1}$ are letters, we have $pa_0 \cdots a_r =_\eta pa_0 \cdots a_r(b_{[r+1]} \cdots b_{d-1}b_0 \cdots b_{[r]})^d$.*

Proof. We prove the result by induction on the length of the word $a_0 a_1 \cdots a_r$. If the length is 0, the result simply follows from the relation $px =_\eta p$. Suppose by induction that the result holds for a word of length $\leqslant r$, that is

$$pa_0 \cdots a_{r-1} =_\eta pa_0 \cdots a_{r-1}(b_{[r]} \cdots b_{d-1}b_0 \cdots b_{[r-1]})^d \qquad (7)$$

Then we get by (7)

$$pa_0 \cdots a_{r-1}a_r(b_{[r+1]} \cdots b_{d-1}b_0 \cdots b_{[r]})^d$$
$$=_\eta pa_0 \cdots a_{r-1}(b_{[r]} \cdots b_{d-1}b_0 \cdots b_{[r-1]})^d a_r(b_{[r+1]} \cdots b_{d-1}b_0 \cdots b_{[r]})^d$$
$$=_\eta pa_0 \cdots a_{r-1}b_{[r]}(b_{[r+1]} \cdots b_{d-1}b_0 \cdots b_{[r]})^{d-1} \underbrace{b_{[r+1]} \cdots b_{d-1}b_0 \cdots b_{[r-1]}a_r}_{s}$$
$$\underbrace{(b_{[r+1]} \cdots b_{d-1}b_0 \cdots b_{[r]})^{d-1} b_{[r+1]} \cdots b_{d-1}b_0 \cdots b_{[r]}}_{t}$$

Equation (5) allows one to swap s and t and consequently we obtain

$$pa_0 \cdots a_r(b_{[r+1]} \cdots b_{d-1}b_0 \cdots b_{[r]})^d$$
$$=_\eta pa_0 \cdots a_{r-1}b_{[r]}(b_{[r+1]} \cdots b_{d-1}b_0 \cdots b_{[r]})^{d-1} \underbrace{b_{[r+1]} \cdots b_{d-1}b_0 \cdots b_{[r]}}_{t}$$
$$\underbrace{(b_{[r+1]} \cdots b_{d-1}b_0 \cdots b_{[r]})^{d-1} b_{[r+1]} \cdots b_{d-1}b_0 \cdots b_{[r-1]}a_r}_{s}$$
$$=_\eta pa_0 \cdots a_{r-1}(b_{[r]} \cdots b_{d-1}b_0 \cdots b_{[r-1]})^{2d}a_r =_\eta pa_0 \cdots a_{r-1}a_r \quad \text{by (7)},$$

which concludes the induction step. $\qquad\square$

Lemma 4.5. *Let $a_0, a_1, \ldots, a_r$ be letters and let p and $x = b_0 \cdots b_{d-1}$ be two words such that $px =_\eta p$. Setting for each $k \geqslant 0$*

$$z_k = b_0 b_1 \cdots b_{[k-1]}a_{[k]}b_{[k+1]} \cdots b_{d-1}$$

the following relation holds

$$pa_0 \cdots a_r =_\eta px^{d-1}z_0 x^{d-1}z_1 \cdots x^{d-1}z_{[r]}x^{d-1}b_0 \cdots b_{[r]} \qquad (8)$$

Proof. Applying Lemma 4.4 repeatedly yields the formula

$$pa_0 \cdots a_r =_\eta p(b_0 \cdots b_{d-1})^d a_0 (b_1 \cdots b_{d-1} b_0)^d a_1 \cdots$$
$$(b_{[r]} \cdots b_{d-1} b_0 \cdots b_{[r-1]})^d a_r (b_{[r+1]} \cdots b_{d-1} b_0 \cdots b_{[r]})^d \quad (9)$$

It suffices now to observe that the right hand sides of (9) and of (8) are the same word. $\square$

Proposition 4.6. *If $u \sim_d v$, then $u =_\eta v$.*

Proof. Let $u \in L$ and let v be a word such that $u \sim_d v$. We claim that $u =_\eta v$. If $|u| < d$ or $|v| < d$, then $u = v$ and the result is trivial. Thus we may assume that $|u|, |v| \geqslant d$ and by the definition of $\sim_d$, $p_d(u) = p_d(v)$.

Let p and $x = b_0 b_1 \cdots b_{d-1}$ be the words given by Proposition 4.2. Then p is a common prefix of length $< |M|$ of u and v and x is a word of length d such that $px =_\eta p$.

Let $u = pa_0 \cdots a_m$ and $v = pc_0 \cdots c_n$. Since $u \sim_d v$, $|u| \equiv |v|$ mod d and thus $[n] = [m]$. Setting

$$y_k = b_0 b_1 \cdots b_{[k-1]} a_{[k]} b_{[k+1]} \cdots b_{d-1}$$
$$z_k = b_0 b_1 \cdots b_{[k-1]} c_{[k]} b_{[k+1]} \cdots b_{d-1}$$

we get by Lemma 4.5

$$u =_\eta px^{d-1} y_0 x^{d-1} y_1 \cdots x^{d-1} y_{[m]} x^{d-1} b_0 \cdots b_{[m]}$$
$$v =_\eta px^{d-1} z_0 x^{d-1} z_1 \cdots x^{d-1} z_{[n]} x^{d-1} b_0 \cdots b_{[n]}$$

Since L satisfies the equations (5) and (6), one has for each i, j

$$x^{d-1} y_i x^{d-1} y_i =_\eta x^{d-1} y_i \qquad\qquad x^{d-1} z_i x^{d-1} z_i =_\eta x^{d-1} z_i$$
$$x^{d-1} y_i x^{d-1} y_j =_\eta x^{d-1} y_j x^{d-1} y_i \qquad x^{d-1} z_i x^{d-1} z_j =_\eta x^{d-1} z_j x^{d-1} z_i$$

We can now conclude the proof of Proposition 4.6. Since $u \sim_d v$, for each $i \geqslant d$ there is a j such that $j \equiv i$ mod d and $a_i = c_j$. Therefore, for each i there is a j such that $y_i = z_j$. Similarly, for each j there is an i such that $z_j = y_i$. It follows that $u =_\eta v$. $\square$

We are now ready to prove the main result of this section.

Theorem 4.7. *Let L be regular language, let $\eta : A^* \to M$ be its syntactic morphism and let $d = |M|!$. Then the following conditions are equivalent:*

(1) *η satisfies the profinite equations (5) and (6) for all words x, s and t of the same length,*

(2) *L is a finite union of $\sim_d$-classes,*

(3) *$L \in \mathcal{B}_{\mathrm{Reg}}$,*

(4) *$L \in \mathcal{B} \cap \mathrm{Reg}$.*

Proof. Proposition 4.6 proves that (1) implies (2). Proposition 4.3 shows that (2) implies (3), (3) implies (4) is trivial and (4) implies (1) follows from Theorems 3.3 and 3.4.   □

Corollary 4.8. *One can effectively decide whether or not a given a regular language belongs to* $\mathcal{B}$.

Coming back to logic, one could derive the following characterization, in which $=_c$ stands for the set of unary predicates of the form $\{c\}$, for $c \in \mathbb{N}$ and **MOD** stands for the set of modulo predicates, as defined in [4].

Theorem 4.9. *A language belongs to* $\mathcal{B} \cap \mathrm{Reg}$ *if and only if it can be defined by a sentence of* **FO[MOD**, $= c]$.

Acknowledgement. The authors would like to thank Charles Paperman for his useful suggestions.

References

1. Almeida, J.: Residually finite congruences and quasi-regular subsets in uniform algebras. Portugaliæ Mathematica 46, 313–328 (1989)
2. Almeida, J.: Finite semigroups and universal algebra. World Scientific Publishing Co. Inc., River Edge (1994), Translated from the 1992 Portuguese original and revised by the author
3. Barrington, D.A.M., Straubing, H., Thérien, D.: Non-uniform automata over groups. Information and Computation 89, 109–132 (1990)
4. Chaubard, L., Pin, J.-É., Straubing, H.: First order formulas with modular predicates. In: 21st Annual IEEE Symposium on Logic in Computer Science (LICS 2006), pp. 211–220. IEEE (2006)
5. Gehrke, M., Grigorieff, S., Pin, J.-É.: Duality and equational theory of regular languages. In: Aceto, L., Damgård, I., Goldberg, L.A., Halldórsson, M.M., Ingólfsdóttir, A., Walukiewicz, I. (eds.) ICALP 2008, Part II. LNCS, vol. 5126, pp. 246–257. Springer, Heidelberg (2008)
6. Gehrke, M., Grigorieff, S., Pin, J.-É.: A topological approach to recognition. In: Abramsky, S., Gavoille, C., Kirchner, C., Meyer auf der Heide, F., Spirakis, P.G. (eds.) ICALP 2010. Part II. LNCS, vol. 6199, pp. 151–162. Springer, Heidelberg (2010)
7. McKenzie, P., Thomas, M., Vollmer, H.: Extensional uniformity for Boolean circuits. SIAM J. Comput. 39(7), 3186–3206 (2010)
8. Pin, J.-É.: Profinite methods in automata theory. In: Albers, S., Marion, J.-Y. (eds.) 26th International Symposium on Theoretical Aspects of Computer Science (STACS 2009), pp. 31–50. Internationales Begegnungs- und Forschungszentrum für Informatik (IBFI), Schloss Dagstuhl (2009)
9. Pin, J.-É.: Equational descriptions of languages. Int. J. Found. Comput. S. 23, 1227–1240 (2012)
10. Straubing, H.: Constant-depth periodic circuits. Internat. J. Algebra Comput. 1(1), 49–87 (1991)
11. Straubing, H.: Finite automata, formal logic, and circuit complexity. Birkhäuser Boston Inc., Boston (1994)

Computation Width and Deviation Number

Daniel Goč and Kai Salomaa

School of Computing, Queen's University,
Kingston, Ontario K7L 3N6, Canada
{goc,ksalomaa}@cs.queensu.ca

Abstract. The computation width (a.k.a. tree width, a.k.a. leaf size) of a nondeterministic finite automaton (NFA) A counts the number of branches in the computation tree of A on a given input. The deviation number of A on a given input counts the number of nondeterministic paths that branch out from the best accepting computation. Deviation number is a best-case nondeterminism measure closely related to the guessing measure of Goldstine, Kintala and Wotschke (Infrom. Comput. 86, 1990, 179–194). We consider the descriptional complexity of NFAs with similar given deviation number and with computation width.

1 Introduction

Different ways to quantify and measure the amount of nondeterminism in finite automata have been considered in the literature. The degree of ambiguity counts the number of accepting computations and is possibly the most well studied measure [4,10]. Other measures are based on the amount of nondeterminism used in all (accepting as well as non-accepting) computations, and further distinctions arise depending on whether the measure is a best-case or a worst-case measure [4,12].

The *computation width* of an NFA measures the width (i.e., the number of leaves) of the computation tree on a given input. This measure has been previously studied under the name 'tree width' [11] or 'leaf size' [2,7]. On the other hand, the *guessing measure* of an NFA [5] counts the number of bits the automaton needs to make the nondeterministic choices on the "best" accepting path, that is, the path using the least amount of nondeterminism.

The computation width measures the amount of nondeterminism in all possible computations, while guessing is a best case measure that limits the amount of nondeterminism only on a best accepting computation. If the computation width of an NFA A is k then also the guessing of A is at most $k - 1$ but, in general, the guessing of A can be much smaller. In particular, it is possible that an NFA A has finite guessing but the computation width of A is unbounded.

In this paper we study the descriptional complexity trade-offs between NFAs of finite computation width and NFAs where the amount of nondeterminism of only the best accepting computation is bounded. If the minimal NFA for a regular language L has to make a sequence of binary nondeterministic choices where always one of the choices leads to failure (without further nondeterminism),

H. Jürgensen et al. (Eds.): DCFS 2014, LNCS 8614, pp. 150–161, 2014.
© Springer International Publishing Switzerland 2014

then the computation width is equivalent to a best case computation measure. We provide an example where this situation occurs. Our main goal is to provide a construction of the opposite situation where a given limit k on a best case computation measure allows to have a much smaller NFA than the same limit k on the computation width.

As a best case computation measure we consider *deviation number* that counts the number of nondeterministic paths that branch out from a best accepting computation. Deviation number is closely related to the guessing measure [5] and, if all nondeterministic steps of an NFA A have exactly two choices, then the deviation number of A is always exactly the guessing of A. For a descriptional complexity comparison with computation width, we feel that deviation number is a more natural best case measure than guessing. For example, the guessing (as defined in [5]) involved in one transition involving three possible choices is less than the guessing of a computation where we achieve three choices by first making a binary choice between states q_1 and q_2 and then another binary choice in state q_2.

The definition of the guessing measure is natural and it guarantees that the guessing of an NFA A is always the logarithm of the multiplicative branching measure of A [5], however, for our purposes deviation number is more suitable as it directly counts the number of branches the best accepting computation has.

We note that there are known examples of NFAs with finite deviation number where determinization causes a super-polynomial size blow-up. This follows from Theorem 5.4 of [5] by observing that the guessing of an NFA is finite if and only if its deviation number is finite. At first sight the above could seem to yield an example where an NFA with given deviation number is significantly smaller than the minimal equivalent NFA with the same computation width because it is known that determinizing an NFA with finite computation width causes only a polynomial size blow-up [11]. However, the above does not directly imply a size difference between NFAs with finite deviation number and finite computation width, because in the latter result the degree of the polynomial depends on the computation width (and, in fact, the NFAs used in [5] for the super-polynomial size blow-up have the same finite computation width and deviation number). It seems possible that for the language family L_n, $n \geq 1$, used in the proof of Theorem 5.4 of [5] one could construct NFAs with suitably chosen finite deviation number that are smaller than any NFAs of the same computation width, but determining the size difference would require similar, and likely more complicated, estimations than what we use below in Section 4.

In our construction, to obtain worst-case size blow-up from an NFA with given deviation number to an NFA with the same computation width we want to have a regular language L such that any minimal NFA has to make all nondeterministic choices in the beginning.[1] In this case by making the initial nondeterministic choices as a balanced binary tree we obtain an NFA with deviation number ℓ (where ℓ is roughly the logarithm of the number of nondeterministic choices) that is not much larger than the minimal NFA for L while, at least intuitively,

---

[1] We use NFAs that can have multiple initial states.

it seems clear that any NFA with computation width ℓ must be significantly larger. Naturally the intuition is not sufficient for a proof and the goal of our main result is to construct specific languages where we can prove a corresponding lower bound for the size of any NFA having computation width ℓ.

We construct, for all $k \in \mathbb{N}$, languages $L_{k,s}$ over an alphabet of size s such that a minimal NFA for $L_{k,s}$ has $k \cdot s^k$ states and the minimal DFA is of size $k \cdot (2^s - 1)^k$. As the main result we then show that $L_{k,s}$ has a deviation number $k\lceil \log s \rceil$ NFA of size $(k + 1) \cdot s^k - 1$ while any NFA for $L_{k,s}$ with the same computation width of $c \cdot k \log s$ needs $\Omega\left(c \cdot k \log s \, 2^{\frac{s}{c \cdot \log s}}\right)$ states for any $c \geq 1$.

2 Preliminaries

We assume that the reader is familiar with the basics of finite automata [14,15]. Surveys on descriptional complexity include [4,8]. Here we briefly recall some notation and introduce definitions for the computation width and deviation number measures considered in the paper.

The set of strings (or words) over a finite alphabet Σ is Σ^* and ε is the empty string. A *nondeterministic finite automaton* (NFA) is a 5-tuple $A = (Q, \Sigma, \delta, Q_0, F)$ where Q is the finite set of states, Σ is the input alphabet, $\delta : Q \times \Sigma \to 2^Q$ is the transition function, $Q_0 \subseteq Q$ is the set of initial states and $F \subseteq Q$ is the set of final states. The transition function δ is in the usual way extended as a function $Q \times \Sigma^* \to 2^Q$ and the language recognized by A is $L(A) = \{w \in \Sigma^* \mid (\exists q \in Q_0)\delta(q, w) \cap F \neq \emptyset\}$. By the size of the NFA A, size(A), we mean the number of states, that is, the cardinality of Q.

A transition of A is a triple $\mu = (q, a, p)$, $q, p \in Q$, $a \in \Sigma$ such that $p \in \delta(q, a)$. The *branching* of a transition $\mu = (q, a, p)$ is $\beta_A(\mu) = |\delta(q, a)|$. The transition μ is *nondeterministic* if it has branching at least 2, and otherwise μ is deterministic.

A *computation* of A from state s_1 to state s_2 on input $w = a_1 \cdots a_k$, $a_i \in \Sigma$, $i = 1, \ldots, k$, $k \geq 1$, is a sequence of transitions

$$C = (\mu_1, \ldots, \mu_k), \quad \mu_i = (q_i, a_i, q_{i+1}), \ 1 \leq i \leq k, \ s_1 = q_1, s_2 = q_{k+1}. \quad (1)$$

The set of all computations on $w \in \Sigma^*$ from state s_1 to state s_2 is denoted $\text{comp}_A(w, s_1, s_2)$.

The minimal size of a DFA or an NFA recognizing a regular language L is called the deterministic (nondeterministic) state complexity of L and denoted, respectively, $\text{sc}(L)$ and $\text{nsc}(L)$. Note that we allow DFAs to be incomplete and, consequently, the deterministic state complexity of L may differ by one from a definition using complete DFAs.

Theorem 2.1 ([1]). *Let $L \subseteq \Sigma^*$ be a regular language, and suppose there exists a set of pairs $P = \{(x_i, y_i) \mid 1 \leq i \leq n\}$ such that:*

(a) $x_i y_i \in L$ for $1 \leq i \leq n$;
(b) $x_i y_j \notin L$ or $x_j y_i \notin L$ for $1 \leq i, j \leq n$ and $i \neq j$.

Then any NFA accepting L has at least n states.

A set of pairs of strings satisfying the conditions of Theorem 2.1, for a regular language L, is called a *fooling set* for L.

2.1   Computation Width and Deviation Number

The computation width measure counts the total number of paths in computation trees of an NFA. Here we recall some definitions. More details on the computation width measure, a.k.a. tree width or leaf size, can be found in [7,11], however, the definitions used there require an NFA to have only one initial state.

In the following let $A = (Q, \Sigma, \delta, Q_0, F)$ be an NFA and the set of initial states is $Q_0 = \{q_{1,0}, \ldots, q_{k,0}\}$, $k \geq 1$.

For $q \in Q$ and $w \in \Sigma^*$, the q-*computation tree of A on w*, $T_{A,q,w}$, is a finite tree where the nodes are labelled by elements of $Q \times (\Sigma \cup \{\varepsilon, \natural\})$ defined inductively as follows. The tree $T_{A,q,\varepsilon}$ consists of a single node labelled by (q, ε). When $w = au$, $a \in \Sigma$, $u \in \Sigma^*$ and $\delta(q, a) = \emptyset$ we set $T_{A,q,w}$ to be the singleton tree where the only node is labelled by $(q, \natural)$. Then assuming $\delta(q, a) = \{p_1, \ldots, p_m\}$, $m \geq 1$, we define $T_{A,q,w}$ as the tree where the root is labelled by (q, a) and the root has m children where the sub-tree rooted at the ith child is $T_{A,p_i,u}$, $i = 1, \ldots, m$. For our purposes the order of children of a node is not important and we assume that the elements of $\delta(q, a)$ are ordered by a fixed but arbitrary linear order. Note that in $T_{A,q,w}$ every path from the root to a leaf has length at most $|w|$. A path may have length less than w because the corresponding computation of A may become blocked when $\delta(q, a) = \emptyset$.

The *computation forest* of A on word w, $F_{A,w}$ consists of all $q_{0,i}$-computation trees of A on w for initial states $q_{0,i} \in Q_0$, that is,

$$F_{A,w} = \{T_{A,q_{1,0},w}, \ldots, T_{A,q_{k,0},w}\}.$$

The *computation width* of A on $w \in \Sigma^*$, $\mathrm{cw}_A(w)$, is the sum of the numbers of leaves of the trees in the forest $F_{A,w}$. The *computation width* of A is defined as $\mathrm{cw}(A) = \sup\{\mathrm{cw}_A(w) \mid w \in \Sigma^*\}$. We say that A is a *finite computation width* NFA if cw_A is finite.

More generally, one may consider the computation width as a function on input length [12], however, here we mainly concentrate on NFAs with bounded computation width and the distinction between a finite and an unbounded computation width.

The computation width measure counts the number of all computation paths of A on a given input. As a best case nondeterminism measure we consider a measure that counts the least number of nondeterministic branches that deviate, or branch out, from an accepting computation.

If $C = (\mu_1, \ldots, \mu_k) = \mathrm{comp}_A(w, s_1, s_2)$ is a computation of A on string w (from state s_1 to s_2), the *deviation number* of C, $\mathrm{dn}(C)^2$, is defined as

$$\mathrm{dn}(C) = \sum_{i=1}^{k} (\beta(\mu_i) - 1).$$

---

[2] We avoid the name "branching of C" because branching of a nondeterministic computation is defined differently in the literature [5].

For $w \in L(A)$, the *deviation number of A on w*, $\mathrm{dn}_A(w)$ is defined by

$$\mathrm{dn}_A(w) = \min\{\ \mathrm{dn}(C) : (\exists q \in Q_0, p \in F)\ C \in \mathrm{comp}_A(w, q, p)\ \}.$$

The deviation number of A on string w counts the number of nondeterministic paths that branch out from the "best" accepting computation, that is, the computation for which this value is smallest. The deviation number measure is closely related to the *guessing* measure of an NFA A [5], which is another best case measure of nondeterminism. If all nondeterministic transitions of A have branching exactly two, then for any string w the guessing of A on w equals to $\mathrm{dn}_A(w)$.

Again we define the deviation number of A as $\mathrm{dn}(A) = \sup\{\mathrm{dn}_A(w) \mid w \in \Sigma^*\}$. An NFA with finite computation width has always finite deviation number but the converse does not hold, in general.

We will study the descriptional complexity of regular languages by measuring the sizes of optimal finite computation width, or finite deviation number, NFAs. For a regular language L and $m \in \mathbb{N}$, we denote

$$\mathrm{nsc}_{\mathrm{cw} \leq m}(L) = \min\{\mathrm{size}(A) : \mathrm{cw}(A) \leq m,\ L(A) = L\},$$

$$\mathrm{nsc}_{\mathrm{dn} \leq m}(L) = \min\{\mathrm{size}(A) : \mathrm{dn}(A) \leq m,\ L(A) = L\}.$$

Directly from the definitions we get the following lemma.

Lemma 2.1. *For any regular language L and $m \in \mathbb{N}$,*

$$\mathrm{nsc}_{\mathrm{dn} \leq m}(L) \leq \mathrm{nsc}_{\mathrm{cw} \leq m+1}(L).$$

Conversely, since deviation number is a best case measure, for a given NFA A, $\mathrm{dn}(A)$ does not yield any upper bound for $\mathrm{cw}(A)$. Note that the computations of A could always begin with a nondeterministic choice into states q_1 and q_2 where q_1 accepts the remaining input deterministically while q_2 begins a computation involving unbounded nondeterminism.

First we observe that there exist languages for which the inequality of Lemma 2.1 is an equality.

Proposition 2.1. *For every $m \in \mathbb{N}$ there exists a regular language L_m over a three letter alphabet such that*

$$\mathrm{nsc}_{\mathrm{cw} \leq m+1}(L_m) = \mathrm{nsc}_{\mathrm{dn} \leq m}(L_m) < \mathrm{nsc}_{\mathrm{dn} \leq m-1}(L_m).$$

Note that the condition $\mathrm{nsc}_{\mathrm{dn} \leq m}(L_m) < \mathrm{nsc}_{\mathrm{dn} \leq m-1}(L_m)$ guarantees that the minimal NFA for L_m really uses (at least) deviation number m (and computation width m). Without this requirement we could trivially make the inequality of Lemma 2.1 to be an equality (for all m) by choosing L to be a regular language such that the minimal DFA for L is also minimal as an NFA.

The languages L_m constructed in Proposition 2.1 have the property that an accepting computation of a minimal NFA A for L_m, on any input in L_m, has to make a finite sequence of binary choices where always the incorrect choice leads to failure after some deterministic steps. This means that the minimal NFA for L_m has the same finite computation width and deviation number.

In the next section we construct languages for which a given bound on the deviation number allows an NFA to be much smaller than the same bound on computation width, that is, we construct languages L such that, for suitably chosen $m \in \mathbb{N}$, $\mathrm{nsc}_{\mathrm{dn} \leq m}(L)$ is provably much smaller than $\mathrm{nsc}_{\mathrm{cw} \leq m}(L)$.

3   Words Defined by Cyclically Avoiding Letters

Below, we define a language $L_{k,s}$, $k, s \in \mathbb{N}$, consisting of words that 'avoid' a given letter in each position modulo k.

First, we let $E_k : (\Sigma \cup \{\star\})^* \times \{1, 2, \ldots, k\} \to \mathcal{P}(\Sigma)$ be the function defined as follows:

$$E_k(w, i) = \{w_j : w_j \in \Sigma \text{ and } j \equiv i \pmod{k}\}$$

where $w \in (\Sigma \cup \{\star\})^*$, and $i \in \{1, \ldots, k\}$.

Informally, $E_k(w, i)$ is the set of letters 'encountered' at the positions of w that are congruent to i modulo k where $\star$ acts as a 'wildcard' character and is ignored.

We say that a word $w \in (\Sigma \cup \{\star\})^*$ *avoids* the word $x = x_1 x_2 \cdots x_k \in \Sigma^k$ if $\exists i$ such that $x_i \notin E_k(w, i)$. In other words, w and x^ω do not have any coinciding letters when put side by side.

This gives rise to the following definition of a language $L_{k,s} \subset (\Sigma_s \cup \{\star\})^*$, $k, s \in \mathbb{N}$:

$$L_{k,s} = \{w : k \text{ divides } |w| \text{ and } (\exists x \in \Sigma_s^k) \text{ such that } w \text{ avoids } x\}$$

where Σ_s denotes the set $\{a_1, a_2, \ldots, a_s\}$.

Furthermore, we say that a word $w \in (\Sigma_s \cup \{\star\})^*$ is *full with respect to* x if w avoids x but does not avoid any other $y \neq x$ where $|y| = |x|$. We denote the set of all the words full with respect to x by F_x. Formally,

$$F_x = \{w : w \text{ is full with respect to } x\}.$$

We assume an arbitrary but fixed order on the alphabets considered and let f_x denote the lexicographically least word in F_x.

Lemma 3.1. *A word* $w \in (\Sigma \cup \{\star\})^*$ *is full with respect to* $x \in \Sigma^k$ *if and only if* $E_k(w, i) = \Sigma \setminus \{x_i\}$ *for all* $i \in \{1, 2, \ldots, k\}$.

Proof. Suppose that w is full with respect to $x \in \Sigma^k$ and fix $i \in \{1, 2, \ldots, k\}$.

We know that w avoids x and so we have that $E_k(w, i) \subseteq \Sigma \setminus \{x_i\}$.

Now, assume that $E_k(w, i) \subsetneq \Sigma \setminus \{x_i\}$. Then there is another letter $b \neq x_i$ such that $b \notin E_k(w, i)$. Thus, w also avoids the word $y = x_1 \cdots x_{i-1} b x_{i+1} \cdots x_k \neq x$ meaning w is not full with respect to x. This contradicts our assumption.

Therefore, $E_k(w, i) = \Sigma \setminus \{x_i\}$.

To see the other direction, suppose that $E_k(w, i) = \Sigma \setminus \{x_i\}$ for all $i \in \{1, 2, \ldots, k\}$. Then clearly w avoids $x = x_1 x_2 \cdots x_k$. If w avoids some y with

$|y| = |x|$ then for each i we have $y_i \notin E_k(w, i)$. We conclude that for each i, $y_i = x_i$, and so $y = x$. Therefore, w is full with respect to x.   □

Lemma 3.2. *If* $x \in \Sigma^k$ *then* $|f_x| = k \cdot (|\Sigma| - 1)$.

Proof. Let $x = x_1 x_2 \cdots x_k$.

Suppose that $|f_x| < k \cdot (|\Sigma| - 1)$. Then let $m, l \in \mathbb{N}, l < k$ be such that $|f_x| = k \cdot m + l$. This means that $m \geq |E_k(w, k)|$. But $m < |\Sigma| - 1 = |\Sigma \setminus \{x_k\}|$ and so, by Lemma 3.1, w is not full leading to a contradiction. Therefore we conclude that $|f_x| \geq k \cdot (|\Sigma| - 1)$.

On the other hand, the word

$$w = (w_{1,1} w_{1,2} \cdots w_{1,k})(w_{2,1} w_{2,2} \cdots w_{2,k}) \cdots (w_{|\Sigma|-1,1} w_{|\Sigma|-1,2} \cdots w_{|\Sigma|-1,k}),$$

where, for each $i = 1, \ldots, k$, $\{w_{1,i}, w_{2,i}, \ldots, w_{|\Sigma|-1,i}\} = \Sigma - \{x_i\}$, is full with respect to x. Furthermore, by choosing $w_{j,i} < w_{j+1,i}, 1 \leq i \leq k, 1 \leq j < |\Sigma| - 1$, it is clear that w is the lexicographically least word of length $k \cdot (|\Sigma| - 1)$ that can be full with respect to x.   □

3.1 Minimal NFA

Next we determine what is the size of a minimal NFA for the language $L_{k,s}$.

Theorem 3.1. *Let* $k, s \in \mathbb{N}$, *the nondeterministic state complexity of* $L_{k,s}$ *is* $\mathrm{nsc}(L_{k,s}) = k \cdot s^k$.

Proof. To prove a lower bound for the size of a minimal NFA recognizing $L_{k,s}$ we will appeal to the fooling set theorem.

Claim. Let $X = \{(f_x \star^\ell, \star^{k-\ell} f_x) : x \in \Sigma_s^k \text{ and } \ell < k\}$. We claim that X is a fooling set for the language $L_{k,s}$.

> *Proof (Claim).* To verify the claim, we note that $f_x \star^\ell \cdot \star^{k-\ell} f_x \in L_{k,s}$ since $|f_x|$ is a multiple of k by Lemma 3.2 and $f_x \star^k f_x$ avoids x.
>
> Next consider $w = f_x \star^\ell \cdot \star^{k-\ell} f_y$ where $x_i \neq y_i$, for some $1 \leq i \leq k$. Then there is no z that w avoids. To see this, consider $E_k(w, i)$. Since both f_x and f_y are factors of w and aligned modulo k we have that the set of 'encountered' letters $E_k(f_x, i) \subseteq E_k(w, i)$ and $E_k(f_y, i) \subseteq E_k(w, i)$. Applying Lemma 3.1 we see that $\Sigma_s \setminus \{x_i\} \subseteq E_k(w, i)$ and $\Sigma_s \setminus \{y_i\} \subseteq E_k(w, i)$ so we conclude $E_k(w, i) = \Sigma_s$. Therefore, $w \notin L_{k,s}$ as there are no options for z_i.
>
> Lastly, if $w = f_x \star^\ell \cdot \star^m f_y$ where $\ell + m \neq k$ we have $w \notin L_{k,s}$ since the length of w is not a multiple of k.
>
> This concludes the proof of the claim that X is a fooling set for $L_{k,s}$.   □

Next, we compute the cardinality of X: there are s^k choices for x and another k choices for ℓ. This gives a total of at least $k \cdot s^k$ states in a minimal NFA for $L_{k,s}$.

Indeed, an NFA with $k \cdot s^k$ states recognizing $L_{k,s}$ exists. For each $x \in \Sigma_s^k$ we construct an NFA A_x recognizing strings in $L_{k,s}$ avoiding x as follows:

$$A_x = (\{q_1, q_2, \ldots, q_k\}, \Sigma', \delta_x, \{q_1\}, \{q_1\})$$

where $\Sigma' = \Sigma_s \cup \{\star\}$ and

$$\delta_x(q_i, a) = \begin{cases} \{q_j\} & \text{if } a \neq x_i \text{ and } j \equiv i+1 \pmod{k} \\ \emptyset & \text{if } a = x_i. \end{cases}$$

In Figure 1 we give a sketch of the automaton A_x.

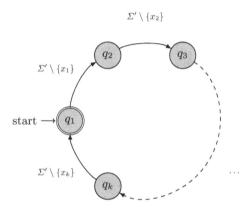

Fig. 1. A sketch of the NFA A_x

Now, we construct the NFA A recognizing $L_{k,s}$ by taking the union of A_x over all $x \in \Sigma_s^k$ in the straightforward way. We have that each A_x has k states and there are s^k words x to choose from, so A has $k \cdot s^k$ states. Note that $w \in L_{k,s}$ iff w avoids some $x \in \Sigma_s^k$ iff $w \in L(A_x)$ for some $x \in \Sigma_s^k$ iff $w \in L(A)$. Thus the minimal NFA recognizing $L_{k,s}$ has at most $k \cdot s^k$ states. This concludes the proof. $\qquad\square$

3.2   Minimal DFA

In this section we use the Myhill-Nerode theorem to give a lower bound for the deterministic state complexity of $L_{k,s}$.

Theorem 3.2. *Let $k, s \in \mathbb{N}$. The (deterministic) state complexity of $L_{k,s}$ is* $\mathrm{sc}(L_{k,s}) = k \cdot (2^s - 1)^k$.

Proof. Let Δ be the power set of Σ_s not including Σ_s itself, namely

$$\Delta = \mathcal{P}(\Sigma_s) \setminus \{\Sigma_s\}.$$

Given an element in Δ^k, $D = D_1 D_2 \cdots D_k$, we let w_D be the lexicographically least word in $(\Sigma_s \cup \{\star\})^{k(s-1)}$ (meaning $|w_D| = k(s-1)$) such that for each $i \in \{1, \ldots, k\}$, we have $E_k(w_D, i) = D_i$. Since $|D_i| \leq (s-1)$ there must exist some word w of length $k(s-1)$ satisfying $E_k(w, i) = D_i$.

Claim. Let $X = \{w_D \cdot \star^\ell : D \in \Delta^k, \ell < k \in \mathbb{N}\}$. We claim that each element in X is in a distinct Myhill-Nerode equivalence class of $L_{k,s}$.

Proof (Claim). First, consider $w_D \cdot \star^\ell, w_E \cdot \star^\ell \in X$ where $D_j \neq E_j$ for some $1 \leq j \leq k$. Without loss of generality, $a \in D_j$ and $a \notin E_j$. Let $x \in \Sigma_s^k$, $x = x_1 x_2 \cdots x_k$ be such that for all $i \in \{1, \ldots, k\}$ $x_i \notin E_i$ and $x_j = a$; such an x exists since $|E_i| \leq (s-1)$. Then let our differentiating suffix be $\star^{k-\ell} f_x$. Since both w_E and f_x avoid x and the length of $w_E \cdot \star^k$ is a multiple of k we have that $w_E \cdot \star^k \cdot f_x$ avoids x as well and so $w_E \star^k f_x \in L_{k,s}$. However, $a = x_j \in D_j$ and so $w_D \cdot f_x$ does not avoid x. Therefore, $w_D \star^k f_x \notin L_{k,s}$ and, consequently, $w_D \cdot \star^\ell$ and $w_E \cdot \star^\ell$ belong to different equivalence classes.

   Second, consider words $w_D \cdot \star^\ell$ and $w_E \cdot \star^{\ell'}$ in X where $\ell \neq \ell'$. Here $\star^{k-\ell}$ acts as a differentiating suffix: $w_D \cdot \star^k \in L_{k,s}$ since it avoids some word x (as shown above) and k divides $\ell + k - \ell$ but k does not divide $\ell' + k - \ell$ so $w_E \cdot \star^{\ell'+k-\ell} \notin L_{k,s}$.

   This concludes the proof of our claim. $\square$

There are $|\Delta|^k = (2^s - 1)^k$ choices for each $D \in \Delta^k$, furthermore there are k choices for ℓ and so $|X| = k \cdot (2^s - 1)^k$. Therefore, $\mathrm{sc}(L_{k,s}) \geq k \cdot (2^s - 1)^k$ as the 'dead state' class is not represented in X (and our definition allows DFAs to be incomplete).

   It is not hard to see that all the Myhill-Nerode equivalence classes of $L_{k,s}$ are captured in X and so we conclude that $\mathrm{sc}(L_{k,s}) = k \cdot (2^s - 1)^k$. $\square$

4   Comparing Computation Width and Deviation Number of NFAs for $L_{k,s}$

Note that the construction of Theorem 3.1 yields an NFA of computation width s^k which is exponential in k and in particular we have that

$$\mathrm{nsc}_{\mathrm{cw} \leq s^k}(L_{k,s}) = k \cdot s^k.$$

This is because at the start of the computation we have to choose which one of the initial states to take and there are s^k choices of these. Likewise the deviation number is s^k for this automaton.

   However, we can construct a not much larger NFA with deviation number that is linear in k (and logarithmic in s.)

Theorem 4.1. *The language $L_{k,s}$ has an NFA with deviation number $k\lceil \log s \rceil + 1$ and at most $(k+1) \cdot s^k - 1$ states.*

Proof. We will construct an NFA B satisfying the above inequality. The bulk of B consists of loops corresponding to automata A_x in the the proof of Theorem 3.1. The rest is a computation tree 'guessing and remembering' which of the loops we select.

First, we have to describe a way of storing our guesses. Let the set $\Gamma \subset \mathcal{P}(\Sigma_s)$ be defined as follows:

1. $\Sigma_s \in \Gamma$
2. if $G \in \Gamma$ where $\Sigma_s \supseteq G = \{i, \ldots, i + \ell\}, \ell \geq 2$ then the sets $G_1 = \{i, \ldots, i + \lceil \frac{\ell}{2} \rceil\}, G_2 = \{i + \lceil \frac{\ell}{2} \rceil + 1, \ldots, i + \ell\}$ are in Γ. We let child(G) denote the set $\{G_1, G_2\}$ of children of G.

Note that Γ has exactly $\lceil \log s \rceil$ 'generations' of children. We further extend the notion of children to elements of Γ^k as follows:

$$H = G_1 \cdots G_{i-1} H_i G_{i+1} \cdots G_k \in \text{child}_i(G) \text{ iff } H_i \in \text{child}(G_i)$$

where $G = G_1 G_2 \cdots G_k$.

Now we construct an automaton recognizing $L_{k,s}$ with deviation number $k \lceil \log s \rceil$. Let $B = (Q_B, \Sigma', \delta, \{p_{\Sigma_s^k, 1}\}, F_B)$ where $\Sigma' = \Sigma_s \cup \{\star\}$ and

$$Q_B = \{p_{G,i} : G \in \Gamma^k, i \in \mathbb{Z}/k\mathbb{Z}\} \cup \{q_{x,i} : x \in \Sigma_s^k, 1 \leq i \leq k\},$$

$p_{G,1} = q_{x,1}$ whenever $G_i = \{x_i\}$ for all $1 \leq i \leq k$, and the state $q_{x,i}$ corresponds to state q_i of the automaton A_x as follows

$$\delta(q, a) = \begin{cases} \delta_x(q_i, a) & \text{if } q = q_{x,i}, \\ \{p_{H,i+1} : H \in \text{child}_i(G) \text{ and } a \notin H_i\} & \text{if } q = p_{G,i} \end{cases}$$

The transitions on the $p_{G,i}$ states make up a binary tree with s^k leaf nodes – each a start state of a unique loop A_x, therefore there are at most $k \cdot s^k + s^k - 1$ states in total as promised. Figure 2 provides an example of such an automaton. In fact, if $s = 2^\ell$ for some $\ell \geq 1$ then this bound is reached. Since this tree is balanced, its depth is $k \lceil \log s \rceil$ and therefore the deviation number of B is $k \lceil \log s \rceil$.   □

Furthermore, if we try to create an NFA of computation width similar to the deviation number above we will see that we need a much larger NFA.

Theorem 4.2. *For $k, s \geq 1$ and a constant $c \geq 1$,*

$$\text{nsc}_{\text{cw} \leq c \cdot k \log s}(L_{k,s}) \in \Omega \left(c \cdot k \log s \, 2^{\frac{s}{c \cdot \log s}} \right).$$

Proof. Let A be an arbitrary NFA for $L_{k,s}$ having computation width ℓ and let n be the size of A. By an easy modification of Lemma 3.3 in [11] (the paper uses a single start state model) we have that:

$$\text{sc}(L_{k,s}) \leq \sum_{i=1}^{\ell} \binom{n}{i} \leq \sum_{i=1}^{\ell} \frac{n^i}{i!} \leq \sum_{i=1}^{\ell} \frac{n^i \ell^{\ell-i}}{\ell!} \leq \sum_{i=1}^{\ell} \frac{n^{i+\ell-i}}{\ell!} \leq \frac{n^\ell}{(\ell-1)!}$$

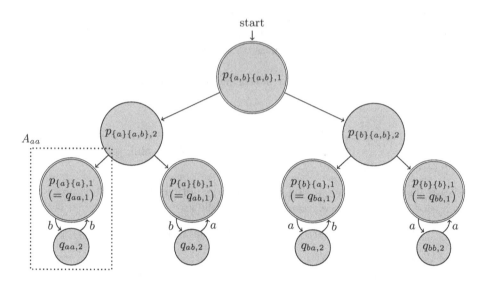

Fig. 2. An example of NFA B where $k = 2$ and $s = 2$ (meaning $\Sigma_s = \{a, b\}$)

But from Theorem 3.2 we have that $\mathrm{sc}(L_{k,s}) \geq k(2^s - 1)^k$ and so if we fix the computation-width to be $\ell = c \cdot k \log s$ we have that

$$n \geq 2^{-3.1} c \cdot k \log(s)(2^s - 1)^{\frac{1}{c \cdot \log s}} \tag{2}$$

Thus, we conclude that $\mathrm{nsc}_{\mathrm{cw} \leq c \cdot k \log s}(L_{k,s}) \in \Omega\left(c \cdot k \log s\, 2^{\frac{s}{c \cdot \log s}}\right).$ □

5   Further Work

The bound presented in Theorem 4.2 seems rather loose; as k grows large the size of the minimal NFA $k \cdot s^k$ grow much faster than $\Omega\left(c \cdot k \log s\, 2^{\frac{s}{c \cdot \log s}}\right)$. This is likely because its proof only uses general properties of bounded computation width automata.

It would be interesting to see if a tighter bound can be achieved exploiting the highly structured nature of the language $L_{k,s}$, perhaps via an argument involving quotient languages.

Acknowledgements. We would like to thank our reviewer for the suggestion of the term 'deviation number' and a minor improvement in the definition of this measure.

References

1. Birget, J.C.: Intersection and union of regular languages and state complexity. Inform. Process. Lett. 43, 85–90 (1992)
2. Björklund, H., Martens, W.: The tractability frontier for NFA minimization. J. Comput. System Sci. 78, 198–210 (2012)
3. Gill, A., Kou, L.T.: Multiple-entry finite automata. J. Comput. System. Sci. 9, 1–19 (1974)
4. Goldstine, J., Kappes, M., Kintala, C.M.R., Leung, H., Malcher, A., Wotschke, D.: Descriptional complexity of machines with limited resources. J. Univ. Comput. Sci. 8, 193–234 (2002)
5. Goldstine, J., Kintala, C.M.R., Wotschke, D.: On measuring nondeterminism in regular languages. Inform. Comput. 86, 179–194 (1990)
6. Holzer, M., Salomaa, K., Yu, S.: On the state complexity of k-entry deterministic finite automata. J. Automata, Languages, and Combinatorics 6, 453–466 (2001)
7. Hromkovič, J., Seibert, S., Karhumäki, J., Klauck, H., Schnitger, G.: Communication complexity method for measuring nondeterminism in finite automata. Inform. Comput. 172, 202–217 (2002)
8. Kutrib, M., Pighizzini, G.: Recent trends in descriptional complexity of formal languages. Bulletin of the EATCS 111, 70–86 (2013)
9. Leung, H.: On finite automata with limited nondeterminism. Acta Inf. 35, 595–624 (1998)
10. Leung, H.: Separating exponentially ambiguous finite automata from polynomially ambiguous finite automata. SIAM J. Comput. 27(4), 1073–1082 (1998)
11. Palioudakis, A., Salomaa, K., Akl, S.G.: State complexity of finite tree width NFAs. Journal of Automata, Languages and Combinatorics 17(2–4), 245–264 (2012)
12. Palioudakis, A., Salomaa, K., Akl, S.G.: Comparisons between measures of nondeterminism on finite automata. In: Jurgensen, H., Reis, R. (eds.) DCFS 2013. LNCS, vol. 8031, pp. 217–228. Springer, Heidelberg (2013)
13. Palioudakis, A., Salomaa, K., Akl, S.G.: Lower bound for converting an NFA with finite nondeterminism in an MDFA. J. Automata, Languages, and Combinatorics (accepted for publication March 2014)
14. Shallit, J.: A Second Course in Formal Languages and Automata Theory. Cambridge University Press (2009)
15. Yu, S.: Regular languages. In: Rozenberg, G., Salomaa, A. (eds.) Handbook of Formal Languages, vol. I, pp. 41–110. Springer (1997)

Boundary Sets of Regular and Context-Free Languages

Markus Holzer and Sebastian Jakobi

Institut für Informatik, Universität Giessen,
Arndtstr. 2, 35392 Giessen, Germany
{holzer,sebastian.jakobi}@informatik.uni-giessen.de

Abstract. We investigate the descriptional and computational complexity of boundary sets of regular and context-free languages. The right (left, respectively) a-boundary set of a language L are those words that belong to L, where the a-predecessor or the a-successor of these words w.r.t. the prefix (suffix, respectively) relation is not in L. For regular languages described by deterministic finite automata (DFAs) we give tight bounds on the number of states for accepting boundary sets. Moreover, the question whether the boundary sets of a regular language is finite is shown to be NL-complete for DFAs, while it turns out to be PSPACE-complete for nondeterministic devices. Boundary sets for context-free languages are not necessarily context free anymore. Here we find a subtle difference of right and left a-boundary sets. While right a-boundary sets of deterministic context-free languages stay deterministic context free, we give an example of a deterministic context-free language the left a-boundary set of which is already non context free. In fact, the finiteness problem for a-boundary sets of context-free languages becomes undecidable.

1 Introduction

In topology, see, e.g., [7], the boundary or frontier of a subset S of a (topological) space is the set of points which can be approached both from S and from the outside of S. Here the term approaching refers to a distance relation induced by the neighborhood property. For instance, the boundary of the half-open set $[0, 1)$ is equal to the two-point set $\{0, 1\}$. This example shows that the elements of the boundary of a set S are *not* necessarily elements of the original set S. Boundaries play an important role by the integration of real or complex valued functions in function theory. We slightly adapt the definition of a boundary or frontier set to the case of formal languages. To this end, we use the prefix- or suffix-relation on the set of words as a suitable neighborhood concept. Due to the left-to-right nature of words it is natural to define two variants of boundary sets, namely left and right versions. Roughly speaking, the right (left, respectively) a-boundary set of a language L are those words that belong to L, where the a-predecessor or the a-successor of these words w.r.t. the prefix (suffix, respectively) relation is *not* in L. By the choice of our definition the elements of a boundary of a formal language L are *always* elements of the original set L.

H. Jürgensen et al. (Eds.): DCFS 2014, LNCS 8614, pp. 162–173, 2014.

Why is it interesting to study boundaries of formal languages? There are two major aspects. First of all, boundaries of formal languages play an important role on the solvability of certain language equations based on almost-equivalence. Two sets are almost-equivalent, if they are equivalent up to a finite number of exceptions. Recently, this form of "equivalence" has attracted a lot of attention in a series of papers on automata with errors, see, e.g., [1,2,4]. Our contribution to language equations with almost-equivalence is just a first step towards a more general theory on this subject. Second, the computation of the boundary set of a language can be seen as an application of a (very complex) combined operation on just one input language. Combined operations, more precisely the descriptional complexity of combined operations, for regular languages were popularized in [8]. In fact, we will show that the left a-boundary set of a language L is equal to

$$\left(L \setminus (a^{-1} \cdot L)\right) \cup a \cdot \left((a^{-1} \cdot L) \setminus L\right);$$

a similar result is also valid for the right a-boundary set of L—here the only non-standard operation is $a^{-1} \cdot L := \{\, u \in \Sigma^* \mid au \in L \,\}$, which refers to the *left quotient* or *left derivative* of $L \subseteq \Sigma^*$ by a letter $a \in \Sigma$.

We define the different boundary sets of languages in Section 2. There we also give an introductory example and explain the connection of boundaries of languages to language equations with almost-equivalence. Here the finiteness of a certain boundary set is a necessary precondition for the solvability of the language equation. Therefore, we later also investigate the decidability status of the finiteness problem for the boundary of languages. But before that, we turn our attention to the descriptional complexity of language boundaries. For regular languages we find the following situation in Section 3: by standard automata constructions on the involved language operations it is easy to see that the boundary set of a regular language is regular, too. Moreover, one may deduce a quartic upper bound of $O(n^4)$ for the a-boundary sets in general. A closer look reveals, that the right a-boundary set obeys a tight linear bound on the number of states for deterministic finite automata (DFAs), while the state complexity of the left a-boundary set of a language accepted by an n-state DFA is at most $n^2 \cdot (n-1)/2 + 2$, and this bound is also tight. The study on boundary sets of regular languages continues in Section 4 by the investigation on the decidability of the finiteness problem for boundaries of regular languages. The question whether the boundary of a regular language is finite is shown to be NL-complete for DFAs, while it turns out to be PSPACE-complete for nondeterministic devices. Finally, our focus is changed to boundary sets of context-free languages. First we show in Section 5 that boundary sets for context-free languages are not necessarily context free anymore. Moreover, we also find a subtle difference of right and left a-boundary sets. While right a-boundary sets of *deterministic* context-free languages stay deterministic context free, we give an example of a deterministic context-free language the left a-boundary set of which is already not context free. In fact, the finiteness problem for a-boundary sets of context-free languages becomes undecidable. This closes our study on boundaries of regular and context-free languages. Nevertheless, some problems remain open.

For instance, the decidability of the finiteness problem for left a-boundary sets of deterministic context-free languages is yet unknown. Due to space constraints most of the proofs are omitted.

2 Preliminaries

We recall some definitions on finite automata as contained in [3]. A *nondeterministic finite automaton* (NFA) is a quintuple $A = (Q, \Sigma, \delta, q_0, F)$, where Q is the finite set of *states*, Σ is the finite set of *input symbols*, $q_0 \in Q$ is the *initial state*, $F \subseteq Q$ is the set of *accepting states*, and $\delta \colon Q \times \Sigma \to 2^Q$ is the *transition function*. The *language accepted* by the finite automaton A is defined as $L(A) = \{\, w \in \Sigma^* \mid \delta(q_0, w) \cap F \neq \emptyset \,\}$, where the transition function is recursively extended to $\delta \colon Q \times \Sigma^* \to 2^Q$. A finite automaton is *deterministic* (DFA) if and only if $|\delta(q, a)| = 1$, for all states $q \in Q$ and letters $a \in \Sigma$. Then we simply write $\delta(q, a) = p$ instead of $\delta(q, a) = \{p\}$, assuming that the transition function is a total mapping $\delta \colon Q \times \Sigma \to Q$.

In this paper we study some aspects on the descriptional and computational complexity of language boundaries. To this end we define for a language $L \subseteq \Sigma^*$ and a letter $a \in \Sigma$, the *left a-boundary of L* as $_a\partial(L) = {_a\partial}^\uparrow(L) \cup {_a\partial}^\downarrow(L)$, where

$$_a\partial^\uparrow(L) = \{\, w \in L \mid aw \notin L \,\} \quad \text{and} \quad {_a\partial}^\downarrow(L) = \{\, aw \in L \mid w \notin L \,\}.$$

Here the former set is referred to as the *upper left a-boundary of L*, while the latter set is called the *lower left a-boundary of L*. Similarly one defines the *right a-boundary of L* as $\partial_a(L) = \partial_a^\uparrow(L) \cup \partial_a^\downarrow(L)$, where $\partial_a^\uparrow(L) = \{\, w \in L \mid wa \notin L \,\}$ is the *upper right a-boundary of the set L* and $\partial_a^\downarrow(L) = \{\, wa \in L \mid w \notin L \,\}$ is the *lower right a-boundary of L*. In order to clarify our notation we give an example.

Example 1. Let $\Sigma = \{a, b\}$. Consider the language $L = \{a, a^2, a^4, b, ab\}$. Then the left a-boundary of L is $_a\partial(L) = \{a, a^2, a^4, ab\}$ because $_a\partial^\uparrow(L) = \{a^2, a^4, ab\}$, and $_a\partial^\downarrow(L) = \{a, a^4\}$. The right a-boundary of L is $\partial_a(L) = \{a, a^2, a^4, b, ab\}$. The computation of the left (right, respectively) a-boundary set can be visualized by a left (right, respectively) language tree—see Figure 1. Taking an edge in a left (right, respectively) language tree corresponds to a left (right, respectively) concatenation of a letter. From the left language tree one can read out the elements of the left a-boundary set $_a\partial(L)$ by simply collecting all words that belong to the language L but where the a-successor or -predecessor does not belong to the set under consideration. Here the a-successor of a node is it's left child in the tree, and a node p is the a-predecessor of a node q, if q is the a-successor of p. A similar construction can be applied for the right a-boundary using the right language tree. □

Boundaries of languages play an important role on the solvability of certain language equations based on almost equivalence. Two languages $R, S \subseteq \Sigma^*$ are said to be *almost equivalent*, for short $R \simeq S$, if their symmetric difference $R \bigtriangleup S := (R \setminus S) \cup (S \setminus R)$ is a finite set. First recall what is known for the

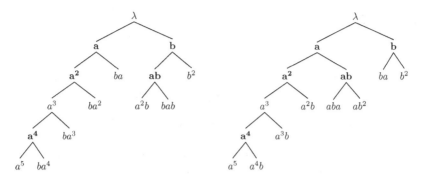

Fig. 1. Left language tree (left) and right language tree (right) of $L = \{a, a^2, a^4, b, ab\}$. Elements that belong to L are typeset in boldface. The left a-boundary set $_a\partial(L)$ can be determined with the help of the left language tree, while the right a-boundary set $\partial_a(L)$ can be read out from the right language tree.

simple language equation[1] $X = a \cdot X + b$, where X is a variable on languages and a and b are letters. Here this equation obeys a unique minimal solution w.r.t. set inclusion, which reads as $X = a^*b$. Note that any other solution of the aforementioned language equation is a superset of a^*b. Now, if one is interested in solutions for the equation $X \simeq a \cdot X + b$, the boundaries of languages come into play. This is seen in the following theorem, which is a reformulation of the symmetric difference property of the left- and right-side of the language equation.

Theorem 2. *Let X be a variable ranging over languages from the alphabet Σ and let $X \simeq a \cdot X + b$, for $a, b \in \Sigma$, be a language equation. Then L is a solution to this language equation if and only if*

1. *the left a-boundary $_a\partial(L)$ of L is finite, and*
2. *the set $L \setminus a\Sigma^*$, i.e., the set of words from L that do not start with letter a, is finite.*   □

Example 3. It is clear that the language equation $X \simeq a \cdot X + b$ does not have a unique solution, since any finite language is already a solution, like, e.g., the language $L = \{a, a^2, a^4, b, ab\}$ from the previous example. Other solutions are, e.g., a^* or $\{a^n \mid n \geq 42\}$, since their left a-boundaries are finite. On the other hand, e.g., ab^* is not a solution, since its left a-boundary is infinite.   □

3   State Complexity of Boundaries of Regular Languages

At first we give an alternative characterization of a-boundary sets in terms of set difference, concatenation, and (left) derivative of languages. The *left derivative*

---

[1] More generally, the language equation $X = S \cdot X + T$, where X is a variable on languages and S and T are sets, has the unique minimal solution $X = S^*T$, if S does not have the empty word property, i.e., if $\lambda \notin S$.

of a language $L \subseteq \Sigma^*$ by a letter $a \in \Sigma$ is defined as $a^{-1} \cdot L = \{ u \in \Sigma^* \mid au \in L \}$. Observe, that $w \in a^{-1} \cdot L$ if and only if $aw \in L$. The characterization reads as follows:

Theorem 4. *Let $L \subseteq \Sigma^*$ and $a \in \Sigma$. Then for the left a-boundary sets we have $_a\partial^\uparrow(L) = L \setminus (a^{-1} \cdot L)$ and $_a\partial^\downarrow(L) = a \cdot ((a^{-1} \cdot L) \setminus L)$.*   □

A similar statement holds for right a-boundary sets. Here left concatenation and left derivatives have to be replaced by right concatenation and right derivatives, respectively. Observe, that there is a certain asymmetry in the characterization of the upper and the lower a-boundary sets. If we would have taken $(a^{-1} \cdot L) \setminus L$ as a definition for the lower left a-boundary set, then the words of this set are not elements of L anymore.

These characterizations and the fact that the family of regular languages is closed under union and all the aforementioned operations, it follows that all a-boundary sets of regular languages are regular, too. Thus, it is worth looking at the descriptional complexity of these sets w.r.t. the number of states of finite automata. By standard automata constructions on the involved language operations one deduces a quadratic upper bound of $O(n^2)$ for DFAs accepting one of the upper or lower a-boundary sets and a quartic upper bound of $O(n^4)$ for the a-boundary sets in general for DFAs. A more closer look will give more specific tight bounds on a-boundary sets accepted by DFAs.

We summarize our results in Table 1. Notice that for unary languages, left and right boundaries are identical. The lower bounds of the boundary operations on non-unary languages are witnessed by binary languages, so we have a complete classification of the state complexity of the different boundary operations with respect to the alphabet size. Next we exemplarily prove the result on the state complexity of the upper left boundary operation on non-unary languages.

Theorem 5. *Let $L \subseteq \Sigma^*$ be a regular language accepted by some n-state DFA, with $n \geq 1$, and let $a \in \Sigma$. Then $n \cdot (n-1)+1$ states are sufficient and necessary in the worst case for a DFA to accept the upper left a-boundary $_a\partial^\uparrow(L)$ of the language L.*

Table 1. State complexity of the different boundary operations on regular languages. The integer n denotes the number of states of the input DFA, and the integer t denotes the number of its final states. All stated bounds are tight bounds. †: the bound for the (general) left boundary is tight for $|\Sigma| \geq 2$ and $n \notin \{3,4\}$—in case $n \in \{3,4\}$ it is tight for $|\Sigma| \geq 3$.

	State complexity of boundary operations		
	$|\Sigma| \geq 2$		$|\Sigma| = 1$
	left	right	(left = right)
upper	$n \cdot (n-1) + 1$	n	n
lower	$n \cdot (n-1) + 2$	$n + \min(t, n-t)$	$n + 1$
general†	$n^2 \cdot (n-1)/2 + 2$	$n + \min(t, n-t)$	$n + 1$

Proof. Let $L = L(A)$ for some DFA $A = (Q, \Sigma, \delta, q_0, F)$, and let $a \in \Sigma$. We use a cross product construction to construct a DFA A' for the upper left a-boundary of L. Let $A' = (Q', \Sigma, \delta', \langle q_0, \delta(q_0, a) \rangle, F')$, where the set of states is $Q' = \{ \langle p, q \rangle \mid p, q \in Q, p \neq q \} \cup \{s\}$, accepting states are $F' = F \times (Q \setminus F)$, and whose transition function is defined as follows for all symbols $b \in \Sigma$:

$$\delta'(\langle p, q \rangle, b) = \begin{cases} \langle \delta(p, b), \delta(q, b) \rangle & \text{if } \delta(p, b) \neq \delta(q, b), \\ s & \text{else,} \end{cases}$$

$$\delta'(s, b) = s.$$

Obviously, A' has $n \cdot (n - 1) + 1$ states. The correctness of this construction is seen as follows. For all words $w \in \Sigma^*$ we have

$$w \in L(A') \iff \delta'(\langle q_0, \delta(q_0, a) \rangle, w) \in F'$$
$$\iff \delta(q_0, w) \in F, \ \delta(q_0, aw) \notin F$$
$$\iff w \in {}_a\partial^\uparrow(L).$$

We now prove this bound to be tight. The case $n = 1$ is easily verified, so let $n \geq 2$, and define the DFA $A = (Q, \{a, b\}, \delta, 1, \{n\})$, with $Q = \{1, 2, \ldots, n\}$, whose transition function δ is defined as follows:

$$\delta(q, a) = \begin{cases} q + 1 & \text{for } 1 \leq q \leq n - 1, \\ 1 & \text{for } q = n, \end{cases} \qquad \delta(q, b) = \begin{cases} q + 1 & \text{for } 2 \leq q \leq n - 1, \\ 1 & \text{for } q \in \{1, n\}. \end{cases}$$

Let $A' = (Q', \{a, b\}, \delta', \langle 1, 2 \rangle, F')$ be the DFA for the boundary language constructed as above. We will show that A' has $n \cdot (n - 1) + 1$ reachable and pairwise inequivalent states. First note that the sink state s is reachable from the initial state by reading b^{n-1}. Now let $p, q \in Q$ with $p < q$. Then the state $\langle p, q \rangle$ is reachable from the initial state $\langle 1, 2 \rangle$ by reading $b^{q-p-1}a^{p-1}$, and state $\langle q, p \rangle$ is reachable from state $\langle q - p, n \rangle$ by reading a^p. To see that no two states are equivalent, first note that the sink state s can not be equivalent to any other state $\langle p, q \rangle$, since the word a^{n-p} is accepted from the latter state. Now let $\langle p, q \rangle$ and $\langle p', q' \rangle$ be two different states of A'. If $p \neq p'$ then a^{n-p} distinguishes the two states. If $p = p'$ then it must be $q \neq q'$, and the states can be distinguished as follows: first we read a^{n-p+1}, which leads to states $\langle 1, r \rangle$ and $\langle 1, r' \rangle$, with $r \neq r'$. These states can then be distinguished by reading b^{n-r+1}, if $r > r'$, and by reading $b^{n-r'+1}$, if $r < r'$, because then one of the states is the sink state, and the other is not. □

4 Decision Problems for a-Boundary Sets

In this section we study the computational complexity of the finiteness problem for a-boundary sets of a language L, which is given by a deterministic or nondeterministic finite automaton. Recall, that the finiteness problem is of particular interest, when considering language equations with the almost equivalence relation. Here it turns out that the finiteness problems related to DFAs are efficiently solvable, while those for NFAs become highly intractable, namely PSPACE-complete. The main result of this section reads as follows:

Theorem 6. *The following decision problems are* NL-*complete: given a DFA A with input alphabet Σ accepting the language $L = L(A)$ and a letter $a \in \Sigma$,*

1. *decide whether $_a\partial^\uparrow(L)$ is finite,*
2. *decide whether $_a\partial^\downarrow(L)$ is finite,*
3. *decide whether $_a\partial(L)$ is finite,*
4. *decide whether $\partial_a^\uparrow(L)$ is finite,*
5. *decide whether $\partial_a^\downarrow(L)$ is finite,*
6. *decide whether $\partial_a(L)$ is finite.*

These problems are PSPACE-*complete, if A is an NFA instead.*   □

The following lemma provides the upper bounds for Theorem 6. The proofs of the lower bounds are omitted due to space constraints.

Lemma 7. *The problems 1-6 from Theorem 6 belong to* NL *if the input automaton A is deterministic, and they belong to* PSPACE, *if A is nondeterministic.*

Proof. Let A be a DFA with input alphabet Σ, and let $a \in \Sigma$. Assume $L = L(A)$. By using the DFA constructions from Section 3, which can be carried out by deterministic logspace-bounded Turing machines, we can reduce the problem of deciding finiteness of the a-boundary of L to the problem of deciding finiteness of DFA-languages. Since this more general problem is NL-complete [6], the NL upper bound follows. If the input automaton A is nondeterministic, a polynomial space-bounded Turing machine can construct (on the fly) the equivalent minimal DFA A', and can run the NL-algorithm on that DFA. This shows the PSPACE upper bound.   □

5   On a-Boundary Sets Beyond Regular Languages

In this section we consider the a-boundaries of context-free languages and their deterministic variants. To this end we have to define deterministic and nondeterministic pushdown automata. A *nondeterministic pushdown automaton* (NPDA) is a 7-tuple $A = (Q, \Sigma, \Gamma, \delta, q_0, \bot, F)$, where Q is the finite set of *states*, Σ is the finite set of *input symbols*, Γ is the finite *stack alphabet*, $q_0 \in Q$ is the *initial state*, $\bot \in \Gamma$ is the *bottom of stack symbol* which initially appears on the pushdown store, $F \subseteq Q$ is the set of *accepting states*, and *transition function* δ maps $Q \times (\Sigma \cup \{\lambda\}) \times \Gamma$ to finite subsets of $Q \times \Gamma^*$. An NPDA A is in *configuration* $c = (q, w, \gamma)$ if A is in state $q \in Q$ with $w \in \Sigma^*$ as remaining input, and $\gamma \in \Gamma^*$ on the pushdown store, the rightmost symbol of γ being the top symbol on the pushdown. We write

$$(q, aw, \gamma Z) \vdash_A (p, w, \gamma\beta),$$

if $(p, \beta) \in \delta(q, a, Z)$, for $a \in \Sigma \cup \{\lambda\}$, $w \in \Sigma^*$, $\gamma, \beta \in \Gamma^*$, and $Z \in \Gamma$. As usual, the reflexive transitive closure of $\vdash_A$ is denoted by $\vdash_A^*$, and the subscript A will be dropped from $\vdash_A$ and $\vdash_A^*$ whenever the meaning remains clear. The *language accepted* by A (*with final state*) is defined by

$$L(A) = \{\, w \in \Sigma^* \mid (q_0, w, \bot) \vdash^* (q, \lambda, \gamma), \text{ for some } q \in F \text{ and } \gamma \in \Gamma^* \,\}.$$

It is well known that NPDAs characterize the family of context-free languages defined by context-free grammars, see, e.g., [3]. Our first goal is to show that context-free languages are not closed under any of the a-boundary operations. The following result relates left and right boundaries *via* the reversal operation on languages.

Theorem 8. *Let $L \subseteq \Sigma^*$. Then $_a\partial^\uparrow(L) = (\partial_a^\uparrow(L^R))^R$ and $_a\partial^\downarrow(L) = (\partial_a^\downarrow(L^R))^R$ where $L^R := \{\, w^R \mid w \in L \,\}$ refers to the reversal of the language L.* □

The next theorem shows how the a-boundary operations can be used to simulate intersection of languages.

Theorem 9. *Let $L \subseteq \Sigma^*$ and $a \notin \Sigma$. Then we have $_a\partial^\uparrow(\Sigma^* \cup a^+L) = \Sigma^* \setminus L$ and $_a\partial^\downarrow(L \cup a^+\Sigma^*) = a \cdot (\Sigma^* \setminus L)$. Moreover, $\partial_a^\uparrow(\Sigma^* \cup La^+) = \Sigma^* \setminus L$ and $\partial_a^\downarrow(L \cup \Sigma^*a^+) = (\Sigma^* \setminus L) \cdot a$.* □

Now we are ready to show that the context-free languages are not closed under any of the a-boundary operations.

Theorem 10. *There is a context-free language $L \subseteq \Sigma^*$ and $a \in \Sigma$, such that neither of the a-boundary sets $_a\partial^\uparrow(L)$, $_a\partial^\downarrow(L)$, nor $_a\partial(L)$ is context free. A similar statement is valid for the right a-boundary sets.*

Proof. It is well known that the family of context-free languages is not closed under complementation [3]. Thus, there is a context-free language $L \subseteq \Sigma^*$, such that $\Sigma^* \setminus L$ is not context free. Let a be a new letter with $a \notin \Sigma$ and assume to the contrary that the family of context-free languages is closed under the a-boundary operations under consideration. Then by Theorem 9 we have $_a\partial^\uparrow(\Sigma^* \cup a^+L) = \Sigma^* \setminus L$ and $a^{-1} \cdot {}_a\partial^\downarrow(L \cup a^+\Sigma^*) = \Sigma^* \setminus L$. Since context-free languages are closed under union, both languages $\Sigma^* \cup a^+L$ and $L \cup a^+\Sigma^*$ on the left of the equations are context free, too. But then also $\Sigma^* \setminus L$ is context free, too, which contradicts our assumption on the the the choice of the language L. For the lower left a-boundary operation we additionally have used the closure of context-free languages under left derivatives.

For the non-closure of the family of context-free languages under the a-boundary operation $_a\partial(\cdot)$ we have to slightly alter our above argumentation. Again, consider the language $\Sigma^* \cup a^+L$ from above. Then by definition

$$_a\partial(\Sigma^* \cup a^+L) = {}_a\partial^\uparrow(\Sigma^* \cup a^+L) \cup {}_a\partial^\downarrow(\Sigma^* \cup a^L L)$$
$$= (\Sigma^* \setminus L) \cup {}_a\partial^\downarrow(\Sigma^* \cup a^+L).$$

It is easy to see that the latter set contains only words that start with the letter a—in fact we even have $_a\partial^\downarrow(\Sigma^* \cup a^+L) \subseteq a \cdot L$. Thus, taking the intersection of $_a\partial(\Sigma^* \cup a^+L)$ with Σ^* results in $\Sigma^* \setminus L$, since $a \notin \Sigma$. Then we can use the same argumentation as above in order to prove the non-closure of the family of context-free languages under the a-boundary operation in general.

Finally, the non-closure results for the right a-boundary operations follow by our investigations on left a-boundary operations above and Theorem 8. □

Close inspection of the previous proof reveals that non-closure under any sort of a-boundary operations also holds already for linear context-free languages.

Corollary 11. *The family of linear context-free languages is not closed under any kind of a-boundary operation considered so far.*   □

For deterministic context-free languages and a-boundary operations, the situation is more involved as we will see in the forthcoming. Before we continue our investigation we have to define deterministic pushdown automata. A pushdown automaton $A = (Q, \Sigma, \Gamma, \delta, q_0, \bot, F)$ is *deterministic* (DPDA), if for all $q \in Q$, $a \in \Sigma$, and $Z \in \Gamma$ it is $|\delta(q, a, Z)| + |\delta(q, \lambda, Z)| \leq 1$. The family of deterministic context-free languages is the class of all languages accepted by DPDAs (with final state), or equivalently by $\mathrm{LR}(k)$ context-free grammars. The class of deterministic context-free languages is strictly contained in the class of context-free languages [3]. The next result shows that right boundaries of deterministic context-free languages are still deterministic context free.

Theorem 12. *Let $L \subseteq \Sigma^*$ be a deterministic context-free language, and $a \in \Sigma$. Then its right boundaries $\partial_a^\uparrow(L)$, $\partial_a^\downarrow(L)$, and $\partial_a(L)$ are deterministic context-free.*

Proof. The basic idea for the proof is the following. Given a DPDA A we construct a DPDA A' that simulates A and uses states of the form $\langle \tau_1, c, \tau_2, q, \tau_3 \rangle$, where the five components have the following meaning: q is the current state of A, letter c is the last read input symbol, τ_1 indicates whether A would have accepted the prefix of the input word to the left of symbol c, τ_2 indicates whether A went through some accepting state some time after reading the input symbol c, and τ_3 indicates whether A will enter an accepting state if the next input symbol is the boundary symbol a. With this information, the set of accepting states can be defined appropriately to yield DPDAs for the different a-boundary languages.

The trickiest part is determining the fifth component of the state of A'. For this we first fix some DFAs related to the DPDA $A = (Q, \Sigma, \Gamma, \delta, q_0, \bot, F)$. Let the state set of A be $Q = \{q_0, q_1, \ldots, q_n\}$. For every state $q_i \in Q$ let

$$L_i = \{\, \gamma \in \Gamma^* \mid (q_i, \lambda, \gamma) \vdash_A^* (q_f, \lambda, \beta), \; q_f \in F, \; \beta \in \Gamma^* \,\}.$$

Informally, the language $L_i \subseteq \Gamma^*$ consists of those pushdown contents that lead A from state q_i to an accepting state, by only using λ-transitions. Notice that L_i is a regular language for every $q_i \in Q$. Therefore, for all $q_i \in Q$ there are DFAs $D_i = (Q_i, \Gamma, \delta_i, s_i, F_i)$ with $L_i = L(D_i)$. Besides the original pushdown symbols, the DPDA A' will also store the states the DFAs D_i are in, when reading the current pushdown content except for the topmost symbol. This information will be used to determine the value of τ_3.

Let $L = L(A)$, $a \in \Sigma$ and assume $+, -, ? \notin Q \cup \Sigma \cup \Gamma$. Now we can construct the DPDA $A' = (Q', \Sigma, \Gamma', \delta', q_0', \bot', F')$ as follows: the set of states is

$$Q' = \{\, \langle \tau_1, c, \tau_2, q, \tau_3 \rangle \mid q \in Q, \; c \in \Sigma \cup \{?\}, \; \tau_1, \tau_3 \in \{+, -, ?\}, \; \tau_2 \in \{+, -\} \,\}$$

and the initial state is $q_0' = \langle ?, ?, +, q_0, ? \rangle$, if $q_0 \in F$, and it is $q_0' = \langle ?, ?, -, q_0, ? \rangle$, if $q_0 \notin F$. The set of final states depends on the type of boundary we want to accept. We define the sets $F'_\downarrow = \{\langle -, a, +, q, \tau_3\rangle \mid q \in Q,\ \tau_3 \in \{+, -\}\}$, and $F'_\uparrow = \{\langle \tau_1, c, +, q, -\rangle \mid \tau_1 \in \{+, -, ?\},\ c \in \Sigma \cup \{?\},\ q \in Q\}$. If we want $L(A') = \partial_a^\downarrow(L)$ then we choose $F' = F'_\downarrow$, if we want $L(A') = \partial_a^\uparrow(L)$, then choose $F' = F'_\uparrow$, and for $L(A') = \partial_a(L)$ we use $F' = F'_\downarrow \cup F'_\uparrow$. The pushdown alphabet of A' is $\Gamma' = Q_0 \times Q_1 \times \cdots \times Q_n \times \Gamma$, and the initial pushdown symbol is $\perp' = (s_0, s_1, \ldots, s_n, \perp)$—recall that Q_i are the state sets of the above defined DFAs D_i, and s_i are the initial states thereof.

The transitions of A' can be grouped in two types. Transitions of the first type directly simulate transitions of A, and lead to states where the fifth component τ_3 is "unknown", i.e., where $\tau_3 = ?$. Transitions of the second type are used to determine the correct value of τ_3.

Assume that for some $q \in Q$, $b \in \Sigma \cup \{\lambda\}$, and $X \in \Gamma$, the DPDA A has the transition

$$\delta(q, b, X) = (\hat{q}, \beta),$$

where $\hat{q} \in Q$, and $\beta = Z_1 Z_2 \ldots Z_k$, with $Z_i \in \Gamma$ for $1 \leq i \leq k$, or $\beta = \lambda$. Then for all $\tau_1 \in \{+, -, ?\}$, $\tau_2, \tau_3 \in \{+, -\}$, $c \in \Sigma \cup \{?\}$, and $X' = (p_0, p_1, \ldots, p_n, X)$, with $p_i \in Q_i$ for $1 \leq i \leq n$, the DPDA A' has the transition

$$\delta'(\langle \tau_1, c, \tau_2, q, \tau_3\rangle, b, X') = (\langle \hat{\tau}_1, \hat{c}, \hat{\tau}_2, \hat{q}, ?\rangle, \beta'),$$

where

$$(\hat{\tau}_1, \hat{c}) = \begin{cases} (\tau_1, c) & \text{if } b = \lambda, \\ (\tau_2, b) & \text{if } b \neq \lambda, \end{cases} \qquad \hat{\tau}_2 = \begin{cases} + & \text{if } \hat{q} \in F, \text{ or if } \tau_2 = + \text{ and } b = \lambda, \\ - & \text{otherwise}, \end{cases}$$

and $\beta' = \lambda$ if $\beta = \lambda$, and otherwise $\beta' = Z'_1 Z'_2 \ldots Z'_k$ with

$$Z'_{i+1} = (\delta_0(p_0, Z_{1\ldots i}), \delta_1(p_1, Z_{1\ldots i}), \ldots, \delta_n(p_n, Z_{1\ldots i}), X), \tag{1}$$

where $Z_{1\ldots i}$ is the prefix $Z_1 Z_2 \ldots Z_i$ of β—the string $Z_{1\ldots 0}$ is the empty word.

Transitions of the second type are defined as follows. For all $\tau_1 \in \{+, -, ?\}$, $\tau_2 \in \{+, -\}$, $c \in \Sigma \cup \{?\}$, $q \in Q$, and $X' = (p_0, p_1, \ldots, p_n, X) \in \Gamma'$ let

$$\delta'(\langle \tau_1, c, \tau_2, q, ?\rangle, \lambda, X') = \begin{cases} (\langle \tau_1, c, \tau_2, q, +\rangle, X') & \text{if } \delta(q, a, X) = (q_i, \gamma), \\ & \text{and } \delta_i(p_i, \gamma) \in F_i, \\ (\langle \tau_1, c, \tau_2, q, -\rangle, X') & \text{otherwise}. \end{cases}$$

This concludes the definition of A'.

Notice that from a state where the last component is $\tau_3 = ?$ the automaton A' always makes a transition of the second type to set τ_3 to $+$ or to $-$, and from a state with $\tau_3 \neq ?$ the automaton always simulates a transition of A, which leads to a state with $\tau_3 = ?$ again. Moreover, one can see that the second component of a state of A' always holds the last input symbol consumed by A—if A has not consumed any input yet, this component is $?$. Therefore, for all $w \in \Sigma^*$, and $\gamma \in \Gamma^*$ there are τ_1, τ_2, and γ' such that

$$(q_0, w, \perp) \vdash_A^\ell (q, \lambda, \gamma) \qquad \Longleftrightarrow \qquad (q'_0, w, \perp') \vdash_{A'}^{2\ell} (\langle \tau_1, c, \tau_2, q, ?\rangle, \lambda, \gamma'),$$

where $c = ?$ if $w = \lambda$, and $c = b$ if $w = vb$, for $v \in \Sigma^*$, and $b \in \Sigma$. Moreover, the pushdown content γ' of A' directly resembles the pushdown content γ of A: by mapping in γ' every pushdown symbol $X' = (p_0, p_1, \ldots, p_n, X) \in \Gamma'$ to its last component $X \in \Gamma$, we obtain γ.

Further, the pushdown of A' always keeps the correct information on the states of the DFAs D_i in the following sense: if $(q'_0, w, \perp') \vdash^*_{A'} (q', \lambda, Z'_1 Z'_2 \ldots Z'_m)$, with $Z'_j = (p_0^{(j)}, p_1^{(j)}, \ldots, p_n^{(j)}, Z_j)$ for $1 \leq j \leq m$, then $p_i^{(j+1)} = \delta_i(s_i, Z_1 Z_2 \ldots Z_j)$, for all i, j with $0 \leq i \leq n$ and $0 \leq j \leq m - 1$. This can be seen by an inductive argument: by definition of $\perp' = (s_0, s_1, \ldots, s_n, \perp)$ the statement holds at the beginning of the computation. Notice that the pushdown is only changed by transitions of the first type, i.e., those which simulate a transition of A. Clearly, if the height of the pushdown is reduced, i.e., the pushdown is replaced by a prefix of itself, the statement still holds. Further, from the definition of δ' one can see that the statement also remains true, if the topmost symbol is replaced by some non-empty pushdown word—see Equation (1).

The correctness of our construction can be concluded from the following claim—due to space constraints the proof of the claim is omitted.

Claim. For all $w \in \Sigma^*$ we have

1. $(q'_0, w, \perp') \vdash^*_{A'} (\langle \tau_1, c, \tau_2, q, \tau_3 \rangle, \lambda, \gamma')$ with $\tau_1 = +$ if and only if $w = vc$ with $v \in L$ and $c \in \Sigma$,
2. $(q'_0, w, \perp') \vdash^*_{A'} (\langle \tau_1, c, \tau_2, q, \tau_3 \rangle, \lambda, \gamma')$ with $\tau_2 = +$ if and only if $w \in L$,
3. $(q'_0, w, \perp') \vdash^*_{A'} (\langle \tau_1, c, \tau_2, q, \tau_3 \rangle, \lambda, \gamma')$ with $\tau_3 = +$ if and only if $wa \in L$. □

Recall that a state $\langle \tau_1, c, \tau_2, q, \tau_3 \rangle$ belongs to $F'_\downarrow$ if and only if $\tau_1 = -$, $c = a$, and $\tau_2 = +$, and it belongs to $F'_\uparrow$ if and only if $\tau_2 = +$, and $\tau_3 = -$. Now the claim implies the following for all words $w \in \Sigma^*$:

$$w \in \partial_a^\downarrow(L) \iff \exists q'_f \in F'_\downarrow, \gamma' \in \Gamma'^* : (q'_0, w, \perp') \vdash^*_{A'} (q'_f, \lambda, \gamma'), \text{ and}$$

$$w \in \partial_a^\uparrow(L) \iff \exists q'_f \in F'_\uparrow, \gamma' \in \Gamma'^* : (q'_0, w, \perp') \vdash^*_{A'} (q'_f, \lambda, \gamma').$$

This concludes the proof of the theorem. □

In contrast to the previous result on right boundaries of deterministic context-free languages, we now show that the class of (deterministic) context-free languages is not closed under taking left boundaries, by presenting a deterministic context-free language, whose left boundaries are not context-free.

Theorem 13. *There exists a deterministic context-free language $L \subseteq \Sigma^*$, such that none of its left boundaries $_a\partial^\uparrow(L)$, $_a\partial^\downarrow(L)$, and $_a\partial(L)$ is context free.* □

We now come back to the problems of deciding whether the boundary of a given language is finite. For context-free languages one can easily prove these problems to be undecidable for both left and right boundaries—even for linear context-free languages. The main ingredient for the proof of the next theorem are the set of valid computations $\mathrm{VALC}(M)$ of a Turing machine M. Roughly speaking, these are histories of accepting Turing machine computations. The complement of $\mathrm{VALC}(M)$ is the set $\mathrm{INVALC}(M)$ of invalid computations of M. For more on valid and invalid computations of Turing machines we refer to [5].

Theorem 14. *The following decision problems are undecidable: given a context-free language $L \subseteq \Sigma^*$, and a symbol $a \in \Sigma$,*

1. *decide whether $_a\partial^\uparrow(L)$ is finite,*
2. *decide whether $_a\partial^\downarrow(L)$ is finite,*
3. *decide whether $_a\partial(L)$ is finite,*

4. *decide whether $\partial_a^\uparrow(L)$ is finite,*
5. *decide whether $\partial_a^\downarrow(L)$ is finite,*
6. *decide whether $\partial_a(L)$ is finite.*

The statement also holds, if L is a linear context-free language.   □

Finally, let us turn to deterministic context-free languages. Since finiteness is decidable for context-free languages in general [3], Theorem 12 readily implies that the finiteness problem for right boundaries of deterministic context-free languages is decidable. Whether the corresponding problem for left boundaries of deterministic context-free languages is also decidable is left as an open problem.

References

1. Badr, A., Geffert, V., Shipman, I.: Hyper-minimizing minimized deterministic finite state automata. RAIRO–Informatique Théorique et Applications Theoretical Informatics and Applications 43(1), 69–94 (2009)
2. Gawrychowski, P., Jeż, A., Maletti, A.: On minimising automata with errors. In: Murlak, F., Sankowski, P. (eds.) MFCS 2011. LNCS, vol. 6907, pp. 327–338. Springer, Heidelberg (2011)
3. Harrison, M.A.: Introduction to Formal Language Theory. Addison-Wesley (1978)
4. Holzer, M., Jakobi, S.: From equivalence to almost-equivalence, and beyond—minimizing automata with errors (extended abstract). In: Yen, H.-C., Ibarra, O.H. (eds.) DLT 2012. LNCS, vol. 7410, pp. 190–201. Springer, Heidelberg (2012)
5. Holzer, M., Kutrib, M.: Descriptional complexity—an introductory survey. In: Martín-Vide, C. (ed.) Scientific Applications of Language Methods, pp. 1–58. World Scientific (2010)
6. Jones, N.: Space-bounded reducibility among combinatorial problems. Journal of Computer and System Sciences 11, 68–85 (1975)
7. Willard, S.: General Topology. Addison-Wesley (1970)
8. Yu, S.: On the state complexity of combined operations. In: Ibarra, O.H., Yen, H.-C. (eds.) CIAA 2006. LNCS, vol. 4094, pp. 11–22. Springer, Heidelberg (2006)

Biclique Coverings, Rectifier Networks and the Cost of ε-Removal

Szabolcs Iván[1,*], Ádám D. Lelkes[2], Judit Nagy-György[1,**], Balázs Szörényi[3,4], and György Turán[2,3,***]

[1] University of Szeged, Hungary
[2] University of Illinois at Chicago
[3] MTA-SZTE Research Group on Artificial Intelligence, Szeged, Hungary
[4] INRIA Lille, SequeL Project, France

Abstract. We relate two complexity notions of bipartite graphs: the *minimal weight biclique covering number* $\mathrm{Cov}(G)$ and the *minimal rectifier network size* $\mathrm{Rect}(G)$ of a bipartite graph G. We show that there exist graphs with $\mathrm{Cov}(G) \geq \mathrm{Rect}(G)^{3/2-\epsilon}$. As a corollary, we establish that there exist nondeterministic finite automata (NFAs) with ε-transitions, having n transitions total such that the smallest equivalent ε-free NFA has $\Omega(n^{3/2-\epsilon})$ transitions. We also formulate a version of previous bounds for the weighted set cover problem and discuss its connections to giving upper bounds for the possible blow-up.

1 Introduction

In the world of descriptive complexity, questions involving the possible blow-up when transforming a description of some mathematical object from a formalism to another is a central topic, with one of the first papers dating back to 1971 [12]. We are primarily interested in the cost of chain rule removals from context-free grammars (CFGs). That is, how large a chain-rule free CFG has to be in the worst case which is equivalent to an input CFG of size n, having chain rules? The obvious upper bound resulting from the standard transformation is $O(n^2)$. The best known lower bound is $\Omega(n^{3/2-\epsilon})$ [2]. The question is interesting since chain rule elimination is the bottleneck part of the transformation to Chomsky Normal Form. Despite the question being well-motivated, we have no knowledge of progress in the last three decades; the gap is still there.

* Corresponding author. szabivan@inf.u-szeged.hu
  This research was supported by the European Union and the State of Hungary, co-financed by the European Social Fund in the framework of TÁMOP 4.2.4.A/2-11-1-2012-0001 National Excellence Program.
** Supported by the European Union and co-funded by the European Social Fund under the project "Telemedicine-focused research activities on the field of Mathematics, Informatics and Medical sciences" of project number "TÁMOP-4.2.2.A-11/1/KONV-2012-0073".
*** Partially supported by NSF grant CCF-0916708.

H. Jürgensen et al. (Eds.): DCFS 2014, LNCS 8614, pp. 174–185, 2014.

The maximal possible blow-up is not known even in the special case of regular languages. When a *regular* language is given (e.g. by a nondeterministic automaton or NFA, possibly having ε-transitions), an equivalent "chain-rule-free" *regular* grammar corresponds to a nondeterministic automaton with no ε-transitions. In order to define the "blow-up", we have to choose a notion for measuring the *size* of an NFA – we say that the size of an NFA is the number of its *transitions*. Regular languages can be represented by a variety of different formalisms, some of which are more concise than the others. For example, transforming a regular expression (RE) to an equivalent NFA can be done within linear bounds, i.e. the cost of this direction is worst-case $\Theta(n)$. From RE to ε-free NFA the worst-case cost is $\Theta(n \log^2 n)$, by the upper bound result of [4] and the matching lower bound of [15]. The lower bound is achieved with a language possessing a linear-size RE as well, thus it is recognized by an NFA of size $O(n)$, hence the cost of the NFA $\rightarrow$ ε-free NFA transformation is $\Omega(n \log^2 n)$. It is also known that from ε-free NFA to RE an exponential blow-up can occur and Kleene's algorithm produces an RE of exponential size from an NFA.

One of the main results of the paper is that the NFA $\rightarrow$ ε-free NFA transformation has worst-case cost $\Omega(n^{3/2-\epsilon})$ for any $\epsilon > 0$. It is interesting that this bound (as well as the upper bound $O(n^2)$) coincides with that of [2] for the seemingly more general problem of chain rule elimination. The methods (as well as the models) are very different but there is also a similarity: for the lower bound of [2], languages consisting of words of length 3 were defined. In our case, we consider languages consisting of words of length 2. Such languages $L \subseteq \Sigma\Delta$ can be viewed as bipartite graphs $G_L = (\Sigma, \Delta, E_L)$ with (a, b) being an edge in the graph iff the word ab belongs to L. When the language is viewed this way, ε-free NFAs recognizing L correspond to biclique coverings [6] of G_L with the size of an NFA corresponding to the weight of the associated biclique covering. Also, NFAs recognizing L correspond to rectifier networks [6] realizing G_L; again, with the size of an NFA corresponding to the size of the associated network.

Hence, proving worst-case lower bounds for the minimum-weight covering of a bipartite graph having a rectifier network of size n, we get as byproduct worst-case lower bounds for the NFA $\rightarrow$ ε-free NFA transformation. Thus the bulk of the paper discusses biclique coverings and rectifier networks. These have also been studied for a long time in various contexts, see Sections 2 and 3.

The paper is organized as follows. In Section 2 we give the notations we use for graphs and automata. In Section 3 we give lower bounds for the possible blow-up between rectifier network size and biclique covering weight. In Section 4 we give upper bounds for this blow-up and consider the biclique covering problem as a weighted set cover problem. An approximation bound for the greedy algorithm given by Lovász for the unweighted case is generalized to the weighted case. We discuss the connection of this bound to possible upper bounds for the blow-up. In Section 5 we relate these graph-theoretic results to automata theory and prove the aforementioned lower bound of $\Omega(n^{3/2-\epsilon})$ for ε-removal.

2 Notations

Graphs, Biclique Coverings and Rectifier Networks. Let $[n]$ stand for the
set $\{1, \ldots, n\}$. For sets A and B, $K_{A,B}$ stands for the complete bipartite graph
$(A, B, A \times B)$. When only the cardinalities a and b of the sets A and B matter,
we write $K_{a,b}$ for $K_{A,B}$. When $G = (A, B, E)$ is a bipartite graph, a *biclique*
of G is a complete bipartite subgraph of G and the *weight of a biclique* is the
number of its vertices. A *biclique covering* of G is a collection $\mathcal{C}$ of its bicliques
such that each edge of G belongs to at least one member of $\mathcal{C}$, the *weight of a
covering* is the sum of the weights of the bicliques present in the covering and
$\mathrm{Cov}(G)$ is the minimum possible weight of a biclique covering of G.

A biclique $K_{a,b}$ has weight $a + b$ while it covers ab edges of G. In our investigations we will frequently use the inverse $\frac{ab}{a+b}$ of the relative cost of covering the
edges by $K_{a,b}$. We introduce the shorthand $\mathrm{H}(a, b)$ to denote the quantity $\frac{ab}{a+b}$.

For a bipartite graph $G = (A, B, E)$, a *rectifier network realizing* G is a directed acyclic graph (DAG) $R = (V, E')$ with A being the set of source nodes of
R and B being the set of sink nodes of R, satisfying the property that $(a, b) \in E$
if and only if b is reachable from a in R. The *size* of a rectifier network is the
number of its edges. The *depth* of a network is the length of its longest path.
We let $\mathrm{Rect}(G)$ stand for the size of the smallest rectifier network realizing G
and $\mathrm{Rect}_k(G)$ for the size of the smallest rectifier network of depth at most k
realizing G. We may assume *w.l.o.g.* that there are no isolated vertices.

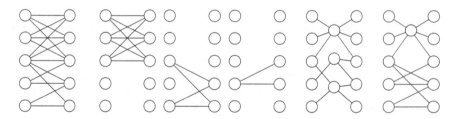

Fig. 1. From left to right: a graph G, three bicliques showing $\mathrm{Cov}(G) \leq 13$, a depth-2
network corresponding to the bicliques having size 13, and another network showing
$\mathrm{Rect}_2(G), \mathrm{Rect}(G) \leq 12$. In the networks, edges are directed from left to right.

There are constructions of graphs for which only large rectifier networks exist
(i.e. having large Rect value), the dates of the results ranging from 1956 till 1996,
e.g. graphs G on n vertices with $\mathrm{Rect}(G)$ being $\Omega(n^{3/2})$ [13], $\Omega(n^{5/3})$ [11,14,16]
and $\Omega(n^{2-\epsilon})$ [8]. Also, it is known that $\mathrm{Rect}(G) \leq \frac{n^2}{\log n}$ [10].

In this paper we are interested in the largest possible gap between Cov and
Rect, thus we seek graph classes having a *small* Rect and a *large* Cov value.

For Cov, a related notion is that of Steiner 2-transitive-closure-spanners [1]
(Steiner-2-TC-Spanners), which is a more general notion for realizing general
graphs. The two notions coincide when we look for spanners of bipartite graphs,

viewed as 2-level layered directed graphs. The authors of [1] show a lower bound for the minimal Steiner-2-TC-Spanner a bipartite graph can have. Applying these results to our problem, we get that there exist graphs with $\mathrm{Rect}(G) = O(n)$ and $\mathrm{Cov}(G) = \Omega(n\,\mathrm{polylog}(n))$ which is exactly the type of result we seek to achieve. We use the asymptotic behaviour operators O, Ω and Θ as well as their "up to a polylogarithmic factor" variants $\tilde{O}$, $\tilde{\Theta}$, e.g. $f(n) = \tilde{O}(g(n))$ is a shorthand for "$f = O(g(n)\log^k g(n))$ for some constant $k \geq 0$".

Automata. A nondeterministic finite automaton, or NFA for short, is a tuple $M = (Q, \Sigma, \delta, q_0, F)$ with Q being an alphabet of states, Σ being the input alphabet, $\delta \subseteq Q \times \Sigma_\varepsilon \times Q$ a transition relation where Σ_ε denotes the set $\Sigma \cup \{\varepsilon\}$, $q_0 \in Q$ being the start state and $F \subseteq Q$ being the set of accepting states. The automaton is ε-free if there is no transition of the form $(p, \varepsilon, q) \in \delta$.

A *run* of the above M is a sequence $(p_1, a_1, r_1) \ldots (p_t, a_t, r_t) \in \delta^*$ such that for each $1 \leq i < t$, $r_i = p_{i+1}$, and $p_1 = q_0$. The run is accepting if $r_t \in F$. The label of the run is the Σ-word $a_0 a_1 \ldots a_t$. The *language* recognized by M is $L(M) = \{w \in \Sigma^* : \text{there is an accepting run of } M \text{ with label } w\}$.

The *size* of an NFA M is the cardinality $|M|$ of its set δ of transitions. It is well-known that for each NFA M there exists an equivalent ε-free automaton M' with $|M'| = \mathcal{O}(|M|^2)$, i.e. ε-elimination can be achieved via a quadratic blow-up.

3   Lower Bounds for the Blow-Up

It is clear that $\mathrm{Rect}_{k+1}(G) \leq \mathrm{Rect}_k(G)$ for each $k \geq 0$, and that there exists some $k \geq 0$ with $\mathrm{Rect}_k(G) = \mathrm{Rect}(G)$ and $\mathrm{Rect}_k(G) = \mathrm{Rect}_{k'}(G)$ for every $k' > k$. Moreover, $\mathrm{Rect}_2(G) \leq \mathrm{Cov}(G) \leq 2 \cdot \mathrm{Rect}_2(G)$: for any collection $\mathcal{C}$ of bicliques one can construct a rectifier network $R = (A \uplus \mathcal{C} \uplus B, E')$ with $(a, K_{A',B'})$ and $(K_{A',B'}, b)$ being an edge iff $a \in A'$ and $b \in B'$, respectively, showing $\mathrm{Rect}_2(G) \leq \mathrm{Cov}(G)$. For $\mathrm{Cov}(G) \leq 2\cdot\mathrm{Rect}_2(G)$, let $R = (A \uplus X \uplus B, E')$ be a depth-2 rectifier network realizing G. Then, edges of E' are directed from A to X, from X to B and also "jump edges" from A directly to B are allowed. First, subdividing each such jump edge and adding the intermediate node to X eliminates jump edges and the resulting network $R' = (A \uplus X' \uplus B, E'')$ still realizes G in depth 2 and due to the subdividing, $|E''| \leq 2 \cdot |E'|$. For a node $x \in X'$, let $A(x)$ be the set of its ancestors (in A) and $B(x)$ be the set of its descendants (in B). Note that if R is minimal, then neither of these sets is empty. Then in G, each member of $B(x)$ is reachable from any member of $A(x)$, hence $K_{A(x),B(x)}$ is a biclique of G and the collection $\mathcal{C} = \{K_{A(x),B(x)} : x \in X'\}$ is a biclique cover of G of size $|E''| \leq 2 \cdot |E'| = 2 \cdot \mathrm{Rect}_2(G)$. Observe that the factor of 2 is tight e.g. in the case of complete matchings.

Since adding or removing isolated nodes to G does not affect either $\mathrm{Cov}(G)$ or $\mathrm{Rect}(G)$, from now on we assume that G has no isolated vertices.

It is also clear that

$$n \leq \mathrm{Rect}(G) \leq \mathrm{Cov}(G) \leq 2|E(G)|$$

where n stands for the number of vertices[1] of G: in any rectifier network the outdegree of each node $a \in A$ is at least one, and the collection $\{K_{\{a\},\{b\}} : (a,b) \in E\}$ of bicliques is a covering of weight $2|E(G)|$. Hence, $\mathrm{Cov}(G) = O(\mathrm{Rect}^2(G))$. However, it is not known whether the quadratic gap is attainable: in the rest of the article we seek an $\alpha > 1$, being as high as possible, such that there exist graphs with arbitrary large $\mathrm{Rect}(G)$ and with $\mathrm{Cov}(G) = \Omega(\mathrm{Rect}^\alpha(G))$.

To this end, we have to construct graph families having *small* Rect and *large* Cov. To show Rect is small (usually it will be $\tilde{O}(n)$ in our candidates) it suffices to give a small realizing network. On the other side, to see that Cov is large, we should have good lower bound methods.

For providing lower bounds, we define the following parameter $\kappa(G)$ of a bipartite graph G: let

$$\kappa(G) = \max\{\mathrm{H}(|A'|,|B'|) : K_{A',B'} \text{ is a biclique of } G\}$$

Observe that by monotonicity, it suffices to take *maximal* bicliques of G into account.

This graph parameter provides lower bounds not only for $\mathrm{Cov}(G)$ but for $\mathrm{Rect}(G)$:

Proposition 1 (See e.g. [6], Lemma 1.10. and Theorem 1.72.). *For any bipartite graph* $G = (A,B,E)$, *it holds that* $\frac{|E|}{\kappa(G)} \leq \mathrm{Cov}(G)$ *and* $\frac{|E|}{\kappa(G)^2} \leq \mathrm{Rect}(G)$.

By a similar argument, we can obtain the following inequality as well:

Proposition 2. *For any bipartite graph* G, *it holds that* $\mathrm{Cov}(G) \leq \mathrm{Rect}(G) \cdot 2\kappa(G)$.

Proof. Claim 1.73. in [6] states the following. Let k be the maximum integer with $K_{k,k}$ being a biclique of G. For any rectifier network $R = (V,E')$ realizing G, call an edge $(u,v) \in E'$ *eligible* iff $|A(u)| \leq k$ and $|B(v)| \leq k$. Then for any edge $(a,b) \in E$ there is a path from a to b in R containing an eligible edge.

In that case $\{K_{A(u),B(v)} : (u,v) \in E' \text{ is eligible}\}$ is a covering of G, consisting of at most $|E'| = \mathrm{Rect}(G)$ bicliques. Each biclique has weight at most $2k$ which in turn is at most $2\kappa(G)$ since $H(a,b) \leq \min\{a,b\}$ holds for any $a,b > 0$. $\square$

It is also worth observing that $k = \Theta(\kappa)$ since $\min\{a,b\} \leq 2H(a,b)$.

Our first result considers the bipartite graph corresponding to the mod 2 inner product function.

Theorem 1. *Let* $d > 0$ *be an even integer and* $G_\perp^d = (A,B,E)$ *be the bipartite graph with* $A = B = \{0,1\}^d$ *and* $(\boldsymbol{u}, \boldsymbol{v}) \in E$ *for the vectors* $\boldsymbol{u}, \boldsymbol{v} \in \{0,1\}^d$ *iff* $\boldsymbol{u} \perp \boldsymbol{v}$ *in* $\mathbb{Z}_2^d$, *i.e. iff* $\sum_{i \in [d]} u_i v_i = 0$ *where sum is taken modulo 2.*

Then $\mathrm{Rect}(G_\perp^d) = \tilde{O}(n)$ *and* $\mathrm{Cov}(G_\perp^d) = \Omega(n^{3/2})$ *where* $n = 2^d$ *is the number of vertices of* $G_\perp^d$.

---

[1] At times n will denote the size of *one* of the two classes of G, introducing a factor of 2 but never causing differences in the growth order.

The proof is broken into two parts. The lower bound follows from the first inequality of Proposition 1 and a special case of Lindsey's lemma [5].

Proposition 3. $\kappa(G_\perp^d) = \frac{\sqrt{n}}{2}$. Thus $\mathrm{Cov}(G_\perp^d) = \Omega(n^{3/2})$.

At the same time, $\mathrm{Rect}(G_\perp^d)$ is small enough. To see this, we show $\mathrm{Rect}(G) = \tilde{O}(n)$ for a specific family of bipartite graphs, which we call *permutation invariant* graphs. A bipartite graph $G = (\{0,1\}^d, \{0,1\}^d, E)$ is permutation invariant if $(\boldsymbol{u}, \boldsymbol{v}) \in E$ implies $(\pi(\boldsymbol{u}), \pi(\boldsymbol{v})) \in E$ for any permutation $\pi : [d] \to [d]$ of the coordinate index set. Here $\pi(u_1, \ldots, u_d)$ is defined to be $(u_{\pi(1)}, \ldots, u_{\pi(d)})$. It is clear that the graphs $G_\perp^d$ are permutation invariant.

For such graphs the following holds (which also state that within this class of graphs, the bound $\mathrm{Cov}(G) = \Omega(\mathrm{Rect}(G)^{3/2})$ is optimal):

Theorem 2. *For permutation invariant graphs* Rect *is* $\tilde{O}(n)$ *and* Cov *is* $\tilde{O}(n^{3/2})$.

Proof. Suppose $G = (A, B, E)$ is permutation invariant with $A = B = \{0,1\}^d$. Let $c : \{0,1\}^d \times \{0,1\}^d \to \{0, \ldots, d\}^{\{0,1\} \times \{0,1\}}$ be the function defined as

$$c((u_1, \ldots, u_d), (v_1, \ldots, v_d))(a, b) = |\{i \in [d] : u_i = a, v_i = b\}|.$$

That is, $c(\boldsymbol{u}, \boldsymbol{v})(a, b)$ is the number of positions i on which $\boldsymbol{u}$ is a and $\boldsymbol{v}$ is b.

Then, G factors through c in the following sense: if $c(\boldsymbol{u}, \boldsymbol{v}) = c(\boldsymbol{u}', \boldsymbol{v}')$, then $(\boldsymbol{u}, \boldsymbol{v}) \in E$ iff $(\boldsymbol{u}', \boldsymbol{v}') \in E$. Indeed, $c(\boldsymbol{u}, \boldsymbol{v}) = c(\boldsymbol{u}', \boldsymbol{v}')$ if and only if there exists a permutation $\pi : [d] \to [d]$ such that $u_i = u'_{\pi(i)}$ and $v_i = v'_{\pi(i)}$ for each $i \in [d]$, yielding $(\boldsymbol{u}, \boldsymbol{v}) \in E$ if and only if $(\boldsymbol{u}', \boldsymbol{v}') \in E$.

Hence there exists a subset C of the finite set $\{0, \ldots, d\}^{\{0,1\} \times \{0,1\}}$ such that $(\boldsymbol{u}, \boldsymbol{v}) \in E$ iff $c(\boldsymbol{u}, \boldsymbol{v}) \in C$.

We define a rectifier network $R = (\{0,1\}^d \times \{0, \ldots, d\}^{\{0,1\} \times \{0,1\}} \times \{0, \ldots, d\})$: the pair $((u_1, \ldots, u_d), f, \ell), ((v_1, \ldots, v_d), f', \ell')$ is an edge of R iff the following conditions hold: $\ell' = \ell + 1$ (so that R is a DAG of depth $d + 1$); for each $i \neq \ell'$, $u_i = v_i$ holds; finally, $f'(u_{\ell'}, v_{\ell'}) = f(u_{\ell'}, v_{\ell'}) + 1$ and for any other $(a, b) \in \{0,1\} \times \{0,1\}$, $f'(a, b) = f(a, b)$.

Then by induction on $\ell' - \ell$ we get that there is a path from $((u_1, \ldots, u_d), f, \ell)$ to $((v_1, \ldots, v_d), f', \ell')$ iff the following conditions hold: $\ell < \ell'$; for each $i \leq \ell$ and $i > \ell'$, $u_i = v_i$; finally, $f'(a, b) = f(a, b) + |\{\ell < i \leq \ell' : u_i = a, v_i = b\}|$.

Now let $R' = (V(R) \uplus \{0,1\}^d, E(R) \uplus E')$ with E' consisting of the edges of the form $(\boldsymbol{v}, f, d) \to \boldsymbol{v}$ with $f \in C$. Then R' realizes G by identifying each $\boldsymbol{u} \in A$ with $(\boldsymbol{u}, \boldsymbol{0}, 0)$ and each $\boldsymbol{v} \in B$ with the element $\boldsymbol{v}$ of this last layer of R' (here $\boldsymbol{0}$ stands for the constant zero function $\boldsymbol{0} : (a, b) \mapsto 0$). Since in R', there are at most $2^d \cdot (\cdot d \cdot \{d+1\}^4) \cdot 2$ edges (each node not belonging to layer d has outdegree 2 in R and in the last step, $2^d \times |C| \leq 2^d \cdot \{d+1\}^4$ edges are added), which is $O(n \log^5 n)$, showing $\mathrm{Rect}(G) = \tilde{O}(n)$.

For $\mathrm{Cov}(G) = \tilde{O}(n^{3/2})$, let X be the set of vertices of R' of the form $(\boldsymbol{u}, f, d/2)$. (That is, nodes of the middle layer of R'.) As before, let $A(x) \subseteq A$, $x \in X$ stand for the set of nodes from which x is reachable in R' and let $B(x) \subseteq B$ stand for the set of those nodes which are reachable from x in R'. Then, since each node of R has indegree at most 2, we have that $|A(x)| \leq 2^{d/2}$. For $|B(x)|$, since each outdegree in R is 2, we get that there are at most $2^{d/2}$ nodes of the form (v, f, d) reachable from x. In E', the outdegree of these nodes is $|C|$ which is at most $(d+1)^4$, hence $|B(x)| \leq 2^{d/2}(d+1)^4 = \tilde{O}(\sqrt{n})$. Thus, the covering $\mathcal{C} = \{K_{A(x),B(x)} : x \in X\}$ has size $\sum_{x \in X}(|A(x)| + |B(x)|)$ which is at most $2^d \cdot (d+1)^4 \cdot (2^{d/2} + 2^{d/2}(d+1)^4) = \tilde{O}(n^{3/2})$. Note that due to the layered structure of R', each $u \to v$ path contains a node belonging to X, so $\mathcal{C}$ is indeed a covering.   □

Thus we have showed that for an arbitrarily $\epsilon > 0$ there are graphs $G = G_\perp^d$ having arbitrarily large $\text{Rect}(G) = \tilde{O}(n)$ and with $\text{Cov}(G) = \Omega(\text{Rect}^{3/2-\epsilon}(G))$. (Observe that any permutation invariant graph with $\kappa = \Theta(\sqrt{n})$ and $\Theta(n^2)$ edges meets this condition.)

As an interesting corollary, we get that $\text{Cov}(G_\perp^d)$ is $\tilde{\Theta}(n^{3/2})$ which is $\tilde{\Theta}(\frac{|E|}{\kappa})$ so in this case the bound of Proposition 1 is optimal up to a log factor.

A general construction for constructing a biclique covering of a graph $G = (A, B, E)$ is the following: starting from a rectifier network $R = (V, E')$ first one chooses a *cut* $E_0 \subseteq E'$ of the edges of R (so that each $a \to b$, $a \in A$, $b \in B$ path contains an edge from E_0), in which case a covering is $\mathcal{C}(E_0) = \{K_{A(x),B(y)} : (x, y) \in E_0\}$. We call coverings of this form *cut-coverings of R*. (In the proof of Theorem 2 we employ a similar construction, choosing a subset X of vertices instead of a subset E_0.) In the following we state without proof that this construction is not optimal, not even up to a polylogarithmic factor, even when R is optimal up to a polylogarithmic factor.

Theorem 3. *Consider the graph $G_\Delta^n = (A, B, E)$ with $A = B = [n]$ and $(i, j) \in E$ iff $d(i, j) \leq \frac{n}{4}$ where $d(i, j)$ is the modulo n distance $\min\{|i - j|, |n + i - j|\}$. (That is, distance on the circle graph C_n.) Then:*

1. *There exists a rectifier network R_n realizing G_Δ^n with $\tilde{O}(n)$ edges.*
2. *Any cut-covering of R_n has size $\Omega(n^2)$.*
3. *At the same time, $\text{Cov}(G_\Delta^n)$ is $O(n^{1+\epsilon})$ for any $\epsilon > 0$ where the O notation hides a constant depending only on ϵ.*

Note that for this graph we have $\kappa = \Theta(n)$ since $K_{[n/4],[n/4]}$ is a biclique. Hence also for this class of graphs, $\frac{|E|}{\kappa} = \Theta(n)$ approximates $\text{Cov} = O(n^{1+\epsilon})$ relatively well. In the next section we show that a closely related formula gives an upper bound for $\text{Cov}(G)$.

4   Upper Bounds for the Blow-Up

In this section we will show that, under certain assumptions, $\text{Cov}(G) = o(\text{Rect}(G)^2)$ or even $\text{Cov}(G) = O(\text{Rect}(G)^{3/2})$ holds. Proposition 2 implies the following result:

Theorem 4. *For any bipartite graph G and $0 < \alpha \leq 1$ with $\text{Cov}(G) \leq \frac{|E|}{\kappa^\alpha}$ we have $\text{Cov}(G) \leq 2\text{Rect}(G)^\beta$ for some $\beta \leq 1 + \frac{1}{1+\alpha} \in [3/2, 2)$. Hence if $\text{Cov}(G) \leq \frac{|E|}{\kappa^\alpha}$ holds for a family of graphs G, then $\text{Cov}(G) = O(\text{Rect}(G)^{2 - \frac{\alpha}{1+\alpha}})$.*

Proof. Let us introduce the following notation: $|E| = n^\delta$ for $n = |V(G)|$, $\text{Rect}(G) = n^r$ and $\text{Cov}(G) = \frac{|E|}{\kappa^\alpha}$, $0 < \alpha \le 1$. We will show that choosing $\beta = \frac{\delta + \alpha \cdot r}{r(1+\alpha)}$ suffices. (Note that since $\delta \le 2$ and $r \ge 1$, β is indeed at most $1 + \frac{1}{1+\alpha}$.)

By $\text{Cov} = \frac{|E|}{\kappa^\alpha}$ we have $\log_n \text{Cov} = \delta - k\alpha$ where $k = \log_n \kappa$. Now assuming for contradiction that $2^{\frac{\alpha}{1+\alpha}} \text{Rect}^\beta < 2\text{Rect}^\beta < \text{Cov}$ we get

$$r\beta = \frac{\delta + \alpha \cdot r}{1 + \alpha} < \delta - k \cdot \alpha - \frac{\alpha}{1+\alpha} \log_n 2.$$

Then direct computation shows that $r < \delta - k(1 + \alpha) - \log_n 2$ which is a contradiction, since by $\text{Rect} \ge \frac{\text{Cov}}{2\kappa}$ we have $r \ge \delta - k(1+\alpha) - \log_n 2$. $\square$

Simple examples show that the assumption of the theorem does not hold for all graphs. A similar argument gives a similar, but somewhat weaker, bound $\text{Cov}(G) = O(\text{Rect}(G)^{2-\varepsilon})$ for some $\varepsilon > 0$ if the condition $\text{Cov}(G) \le \frac{|E|}{\kappa^\alpha}$ is replaced by $\text{Cov}(G) \le \text{polylog} n \max \frac{|E(G')|}{\kappa(G')}$, where the maximum ranges over induced subgraphs G' of G. Thus an affirmative answer to the following open problem would imply $\text{Cov}(G) = O(\text{Rect}(G)^{2-\varepsilon})$ for all bipartite graphs.

Problem 1. *Is it true that for any bipartite graph G on n vertices,*

$$\text{Cov}(G) \le \text{polylog} n \max \frac{|E(G')|}{\kappa(G')}$$

where the maximum ranges over induced subgraphs G' of G?

4.1   The Set Cover Problem

Now we apply the weighted set cover problem to our setting. For a detailed discussion of this problem, and an introduction to approximation methods see [18].

The *weighted set cover* problem is the following: we are given a collection $\mathcal{S} = \{S_1, \ldots, S_t\}$ of subsets of some finite universe A of n elements with $\cup \mathcal{S} = A$, and to each S_i, a cost $c(S_i) > 0$ is associated. The goal is to find a subset $\mathcal{C}$ of $\mathcal{S}$ such that $\cup \mathcal{C} = A$ and the total cost $\sum_{S \in \mathcal{C}} c(S)$ is minimized. The problem is well-known to be NP-complete already for the uniform setting when $c(S_i) = 1$; however, the following greedy algorithm returns a fair enough approximation:

Let $U := A$ and $\mathcal{C} := \emptyset$.
while $U \neq \emptyset$ **do**
   Choose $S \in \mathcal{S}$ such that $\frac{c(S)}{|S \cap U|}$ is the minimum possible value.
   Let $U := U - S$ and $\mathcal{C} := \mathcal{C} \cup \{S\}$.
return $\mathcal{C}$.

The following linear program is the standard relaxation of weighted set cover:

$$\text{minimize} \sum_{i=1}^{t} c(S_i)x_i \qquad \text{subject to} \sum_{i:a \in S_i} x_i \ge 1 \quad \forall a \in A, \quad x_i \ge 0$$

Denote by OPT the optimal solution of the weighted set cover problem. It is well-knownthat the value of the solution returned by the above algorithm is bounded by $\ln n \cdot$ OPT, where $n = |A|$, and even by $\ln n \cdot Z_{LP}^*$, where Z_{LP}^* denotes the value of an optimal solution to the LP relaxation.

Now we define a related combinatorial quantity. For a subset B of A, let $\eta(B)$ stand for the value $\min_{S \in \mathcal{S}} \frac{c(S)}{|S \cap B|}$ which is present inside the loop of the greedy algorithm. Note that this value is positive and finite for any $B \subseteq A$. Also, let η^* stand for $\max_{B \subseteq A} |B| \cdot \eta(B)$. Then we have:

Proposition 4. $\eta^* \leq Z_{LP}^* \leq$ OPT.

Proof. Consider any feasible solution x and a subset B of A. Then,

$$\sum_{i \in [t]} c(S_i)x_i \geq \sum_{i: S_i \cap B \neq \emptyset} c(S_i)x_i = \sum_{i: S_i \cap B \neq \emptyset} \sum_{a \in S_i \cap B} \frac{c(S_i)}{|S_i \cap B|} x_i$$

$$\geq \sum_{i: S_i \cap B \neq \emptyset} \sum_{a \in S_i \cap B} \eta(B)x_i = \eta(B) \sum_{a \in B} \sum_{S_i \ni a} x_i$$

$$\geq \eta(B) \sum_{a \in B} 1 = \eta(B) \cdot |B|.$$

$\square$

On the other hand, this quantity can be used to give an upper bound for OPT as well. The following bound is proven in [9] for the unweighted case.

Proposition 5. *Let* GREEDY *stand for the cost of the solution returned by the greedy algorithm. Then* GREEDY $\leq H_n \eta^*$, *where* H_n *is* $\sum_{i \in [n]} \frac{1}{i} \approx \ln n$.

Proof. Let U_k denote the set of uncovered elements at the beginning of the kth iteration of the loop of the greedy algorithm and let $n_k = |U_k|$. Then

$$\min_{S_i} \frac{c(S_i)}{|S_i \cap U_k|} = \min_{S_i} \frac{c(S_i)}{|S_i \cap U_k|} \frac{|U_k|}{n_k} \leq \max_{B \subseteq A} \min_{S_i} \frac{c(S_i)}{|S_i \cap B|} \frac{|B|}{n_k}$$

$$= \max_{B \subseteq A} \frac{|B|}{n_k} \min_{S_i} \frac{c(S_i)}{|S_i \cap B|} = \max_{B \subseteq A} \frac{|B|}{n_k} \eta(B) = \frac{1}{n_k} \eta^*.$$

Thus, the covering of the elements covered in the kth iteration costs at most $\frac{1}{n_k} \eta^*$ for each such element. Since $n_1 = |A| = n$ and at each iteration, n_k is strictly decreasing, the total cost is at most $\sum_{i \in [n]} \frac{1}{i} \eta^* = H_n \eta^*$. $\square$

Thus we have the following chain of inequalities:

$$\eta^* \leq Z_{LP}^* \leq \text{OPT} \leq \text{GREEDY} \leq H_n \cdot \eta^*.$$

4.2   Application to Biclique Coverings

Determining $\text{Cov}(G)$ for $G = (A, B, E)$ can be viewed as a set cover problem: the universe is E, the allowed sets are bicliques of G, and the cost of a biclique $K_{A', B'}$ is $|A'| + |B'|$.

In this problem, η is the following: given a subset E' of E (that is, a *subgraph* $G' = (A', B', E')$ of G), $\eta(E')$ is defined as

$$\min_{K_{A'', B''} \subseteq E} \frac{|A''| + |B''|}{|E' \cap (A'' \times B'')|}$$

and η^* is the maximal possible value of $|E'|/\eta(E')$. By Proposition 5 we have that $\mathrm{Cov}(G) \le \eta^* \cdot H_n$.

Observe that for the biclique $K_{A_0, B_0} = \arg\min_{K_{A', B'} \subseteq E} \frac{|A'| + |B'|}{|E' \cap (A' \times B')|}$ we have $A_0 \subseteq A'$ and $B_0 \subseteq B'$. Indeed, otherwise $K_{A_0 \cap A', B_0 \cap B'}$ would be a better biclique. Thus, the minimizer biclique K_{A_0, B_0} is a biclique of the subgraph of G *induced* by $A_0 \cup B_0$. It is also clear that for induced subgraphs G' we have $\frac{1}{\kappa(G')} = \eta(E')$ and $|E'|/\eta(E') = \frac{|E(G')|}{\kappa(G')}$.

Hence, there are two cases: either η^* takes its value on some *induced* subgraph of G up to a polylogarithmic factor, in which case the bound $\mathrm{Cov}(G) \ge \max \frac{|E(G')|}{\kappa(G')}$ is essentially optimal up to a polylog factor, or not, in which case there are graphs having much larger η^* than $\frac{|E|}{\kappa}$. In the first case, the remarks following Theorem 4 imply a subquadratic upper bound for the blow-up.

Problem 2. *Determine the gap possible between η^* and $\max \frac{|E(G')|}{\kappa(G')}$, where the maximum is taken over all induced subgraphs.*

5   Application: The Cost of ε-Removal

Let A and B be disjoint alphabets (nonempty finite sets) and $L \subseteq AB$ a (finite) language consisting of two-letter words. Then L can also be viewed as a bipartite graph $G_L = (A, B, L)$ where the notation for L is slightly abused (i.e. (a, b) is an edge iff the word ab belongs to the language). Without loss of generality we may assume that for each $a \in A$ ($b \in B$, resp.) there exists a $b \in B$ ($a \in A$, resp.) such that ab is in L.

Proposition 6. *There is some NFA M recognizing L with $|M| = O(\mathrm{Rect}(G_L))$.*

Proof. Let $R = (V, E)$ be a rectifier network for G_L with $|E| = \mathrm{Rect}(G_L)$. Then the automaton $M = (V \uplus \{q_0, q_f\}, A \cup B, \delta, q_0, \{q_f\})$ with

$$\delta = \{(q_0, a, a) : a \in A\} \cup \{(b, b, q_f) : b \in B\} \cup \{(p, \varepsilon, q) : p \to q \in E\}$$

recognizes L with $|M| = |E| + |A| + |B| = O(\mathrm{Rect}(G_L))$.   □

Proposition 7. *For any ε-free NFA M recognizing L, $\mathrm{Cov}(G_L) \le |M|$. Moreover, there exists a ε-free NFA M recognizing L with $|M| = \mathrm{Cov}(G_L)$.*

Proof. Let $M = (Q, A \cup B, \delta, q_0, F)$ be an ε-free NFA recognizing L of minimal size. Since L is prefix-free and M is minimal, $F = \{q_f\}$ is a singleton set. Also, M is *trim*, i.e. for each state $p \in Q$ there exist words x, y with $p \in q_0 x$ and

$q_f \in py$. Since every word in L has the same length 2, to each state p there is an integer $0 \leq n_p \leq 2$ such that whenever $p \in q_0 x$ for some word x, then $|x| = n_p$. Otherwise if $p \in q_0 x_1$ and $p \in q_0 x_2$ for words x_1, x_2 of different length, then $x_1 y$ and $x_2 y$ are members of L of different length for any word y with $q_f \in py$, a contradiction. Also, it is clear that $n_p = 0$ only for $p = q_0$ and $n_p = 2$ only for $p = q_f$. Thus if X stands for $Q - \{q_0, q_f\}$, we get that M is a layered automaton with transitions of the form (q_0, a, p) for $a \in A$ and $p \in X$ and (p, b, q_f) for $b \in B$ and $p \in X$. Hence, letting $A(p)$ to stand for the set $\{a \in A : (q_0, a, p) \in \delta\}$ and $B(p)$ stand for the set $\{b \in B : (p, b, q_f) \in \delta\}$ we get that there is an associated biclique covering $\mathcal{C}_M$ of L to M consisting of the bicliques $K_{A(p),B(p)}$, $p \in X$ that has the same size as M.

Observe that the transformation is invertible in the sense that to each biclique covering $\mathcal{C}$ such an automaton of the same size can be constructed, showing the second part of the claim.   □

Since by Theorem 1 there exist graphs with arbitrary large $\mathrm{Rect}(G)$ and $\mathrm{Cov}(G) = \Omega(\mathrm{Rect}(G)^{3/2-\epsilon})$, we have the following as byproduct:

Theorem 5. *For any $\epsilon > 0$ and for arbitrarily large n there exist languages (consisting of two-letter words only) which are recognizable by NFAs of size n but are only recognizable by ε-free NFAs of size $\Omega(n^{3/2-\epsilon})$.*

6   Conclusion, Future Directions

We proved a lower bound for the blow-up when transforming NFAs to ε-free NFAs. We showed that the cost of ε-removal from NFAs is worst-case $\Omega(n^{3/2-\epsilon})$, improving the previous bound $\Omega(n \log^2 n)$. The largest possible gap is between $\Omega(n^{3/2-\epsilon})$ and $O(n^2)$, just like in the case of going from CFGs to chain rule free CFGs. Narrowing these gaps seem to be nontrivial open problems.

We used a graph-theoretic approach by translating the problem into finding large blow-ups between two complexity measures for bipartite graphs: the rectifier network size Rect and the minimal weight biclique covering Cov. We proved that there are graphs with arbitrarily large Rect value n such that Cov $= \Omega(n^{3/2-\epsilon})$ for any $\epsilon > 0$. We gave partial results for determining the largest possible blow-up between these quantities. These include a sufficient condition for a subquadratic upper bound, and the sharpness of a combinatorial bound for the minimal weight biclique covering (obtained by proving a bound for the general weighted set covering problem). We also formulated two open problems about related combinatorial bounds, which appear to be of interest in themselves. Solving these problems may also be useful for determining the largest possible blow-up. The relationship between Rect and Cov can be viewed as a size-depth trade-off problem for depth-2 and unrestricted depth circuits computing sets of Boolean disjunctions [17]. As far as we know, there are many other related open problems, such as establishing a bounded-depth hierarchy.

Note Added in Proof. We learned from Stasys Jukna about the following references, which supersede some results of the paper and overlap with others: [3]

(Theorem 1) and [7] (Problem 7.6). In particular, Theorem 1 of the first paper proves a quadratic lower bound (up to a logarithmic factor) for the worst-case blow-up of ε-removal.

We are very grateful to Stasys Jukna for his comments.

References

1. Berman, P., Bhattacharyya, A., Grigorescu, E., Raskhodnikova, S., Woodruff, D.P., Yaroslavtsev, G.: Steiner transitive-closure spanners of low-dimensional posets. Combinatorica, 1–24 (2014)
2. Blum, N.: More on the power of chain rules in context-free grammars. Theoretical Computer Science 27(3), 287–295 (1983), Special Issue Ninth International Colloquium on Automata, Languages and Programming (ICALP), Aarhus, Summer (1982)
3. Hromkovic, J., Schnitger, G.: Comparing the size of NFAs with and without ε-transitions. Theor. Comput. Sci. 380(1-2), 100–114 (2007)
4. Hromkovic, J., Seibert, S., Wilke, T.: Translating regular expressions into small epsilon-free nondeterministic finite automata. In: Reischuk, R., Morvan, M. (eds.) STACS 1997. LNCS, vol. 1200, pp. 55–66. Springer, Heidelberg (1997)
5. Jukna, S.: Boolean Function Complexity - Advances and Frontiers. Algorithms and combinatorics, vol. 27. Springer (2012)
6. Jukna, S.: Computational complexity of graphs. In: Dehmer, M., Emmert-Streib, F. (eds.) Advances in Network Complexity, pp. 99–153. Wiley (2013)
7. Jukna, S., Sergeev, I.: Complexity of linear Boolean operators. Foundations and Trends in Theoretical Computer Science 9(1), 1–123 (2013)
8. Kollár, J., Rónyai, L., Szabó, T.: Norm-graphs and bipartite Turán numbers. Combinatorica 16(3), 399–406 (1996)
9. Lovász, L.: A kombinatorika minimax tételeiről. Matematikai Lapok 26, 209–264 (1975)
10. Lupanov, O.B.: On rectifier and switching-and-rectifier schemes. Dokl. Akad. Nauk SSSR 111, 1171–1174 (1956)
11. Mehlhorn, K.: Some remarks on Boolean sums. In: Becvar, J. (ed.) MFCS 1979. LNCS, vol. 74, pp. 375–380. Springer, Heidelberg (1979)
12. Meyer, A.R., Fischer, M.J.: Economy of description by automata, grammars, and formal systems. In: SWAT (FOCS), pp. 188–191. IEEE Computer Society (1971)
13. Nechiporuk, E.I.: On a Boolean matrix. Systems Theory Res. 21, 236–239 (1971)
14. Pippenger, N., Valiant, L.G.: Shifting graphs and their applications. J. ACM 23(3), 423–432 (1976)
15. Schnitger, G.: Regular expressions and nFAs without e-transitions. In: Durand, B., Thomas, W. (eds.) STACS 2006. LNCS, vol. 3884, pp. 432–443. Springer, Heidelberg (2006)
16. Wegener, I.: A new lower bound on the monotone network complexity of Boolean sums. Acta Informatica 13(2), 109–114 (1980)
17. Wegener, I.: The Complexity of Boolean Functions. John Wiley & Sons, Inc., New York (1987)
18. Williamson, D.P., Shmoys, D.B.: The Design of Approximation Algorithms, 1st edn. Cambridge University Press, New York (2011)

Small Universal Non-deterministic Petri Nets with Inhibitor Arcs

Sergiu Ivanov, Elisabeth Pelz, and Sergey Verlan

Laboratoire d'Algorithmique, Complexité et Logique, Université Paris Est
61, av. du gén. de Gaulle, 94010 Créteil, France
{sergiu.ivanov,pelz,verlan}@u-pec.fr

Abstract. This paper investigates the universality problem for Petri nets with inhibitor arcs. Four descriptional complexity parameters are considered: the number of places, transitions, inhibitor arcs, and the maximal degree of a transition. Each of these parameters is aimed to be minimized, a special attention being given to the number of places. Four constructions are presented having the following values of parameters (listed in the above order): $(5, 877, 1022, 729)$, $(5, 1024, 1316, 379)$, $(4, 668, 778, 555)$, and $(4, 780, 1002, 299)$. The decrease of the number of places with respect to previous work is primarily due to the consideration of non-deterministic computations in Petri nets. Using equivalencies between models our results can be translated to multiset rewriting with forbidding conditions, or to P systems with inhibitors.

1 Introduction

The research of small universal computing devices is an amazing and continuous research topic since many decades. It started by A. Turing proposal of an universal (Turing) machine [17] capable of simulating the computation of any other (Turing) machine. This universal machine takes as input a description of the machine to simulate, the contents of its input tape, and computes the result of its execution on the given input.

More generally, the universality problem for a class of computing devices (or functions) $\mathfrak{C}$ consists in finding a fixed element $\mathcal{M}$ of $\mathfrak{C}$ able to simulate the computation of any element $\mathcal{M}'$ of $\mathfrak{C}$ using an appropriate fixed encoding. More precisely, if $\mathcal{M}'$ computes y on an input x (we will write this as $\mathcal{M}'(x) = y$), then $\mathcal{M}'(x) = f(\mathcal{M}(g(\mathcal{M}'), h(x)))$, where h and f are the encoding and decoding functions, respectively, and g is the function retrieving the number of $\mathcal{M}'$ in some fixed enumeration of $\mathfrak{C}$. Although general recursive functions may be used for encoding and decoding, we would prefer to see the computing device, and not the encoders and decoders, do most of the work. Hence we prefer computationally very simple encoding and decoding functions [18], like the typically used $f(x) = \log_2(x)$ and $h(x) = 2^x$.

In what follows, we will keep to the terminology considered by Korec [6] and call the element $\mathcal{M}$ (weakly) universal for $\mathfrak{C}$. We shall call $\mathcal{M}$ strongly universal (for $\mathfrak{C}$) if the encoding and decoding functions are identities.

H. Jürgensen et al. (Eds.): DCFS 2014, LNCS 8614, pp. 186–197, 2014.

Some authors [7,6] implicitly consider only the strong notion of universality as the encoding and decoding functions can perform quite complicated transformations, which are not necessarily doable in the original devices. For example, Minsky's proof of (weak) universality of register machines with two counters [11] makes use of exponential (resp. logarithmic) encoding (resp. decoding) functions, while it is known that such functions cannot be computed directly (without encoding) on these machines [2,16]. We refer to [6] for a detailed discussion of different variants of the universality and to [8] for a survey on this topic. Generally, the class of all partially recursive functions is considered as $\mathfrak{C}$, but it is possible to have a narrower class, e.g. the class of all primitive recursive functions, which is known to admit a universal generally recursive function [7]. We remark that in the case of devices not working with integers directly, some natural coding of integers should be used in order to consider the above notions.

Small universal devices have mostly theoretical importance as they demonstrate the minimal ingredients needed to achieve a complex (universal) computation. Their construction is a long-standing and fascinating challenge involving a lot of interconnections between different models, constructions, and encodings.

In [1] a small universal maximally parallel multiset rewriting system is constructed. Due to equivalences between Petri nets and multiset rewriting systems, this result can be seen as a universal Petri net working with max step semantics. For the traditional class of Petri nets there were no known universality constructions for a long time. Recently, Zaitsev has investigated the universality of Petri nets with inhibitor arcs and priorities [20] and has constructed a small universal net with 14 places and 29 transitions (for nets without priorities the same author obtained a universal net with 500 places and 500 transitions [19]). We remark that inhibitor arcs and priorities are equivalent extensions for Petri nets in terms of computational power, so using both concepts together is not necessary for universality constructions.

In [4], a series of small strongly and weakly universal Petri nets was constructed and compared to Zaitsev' results. The cited paper only focuses on Petri nets with deterministic evolution, however. The present article considers a natural question of whether allowing non-deterministic evolution may be exploited to further minimize certain parameters. We show some trade-offs: trying to reduce the number of places of a universal Petri net seems to imply an increase in the other parameters, e.g., the maximal transition degree. We also show how linear programming could be used to minimize this degree while keeping the number of places small.

Quite importantly, due to the equivalence between Petri nets, multiset rewriting, and asynchronous P systems [13], the results from this paper can be immediately translated to corresponding universality statements for these models.

2 Preliminaries

In this section we only recall some basic notions and notations; see [15] for further details. An alphabet is a finite non-empty set of symbols. Given an alphabet V, we designate by V^* the set of all strings over V, including the empty string,

λ. For each $x \in V^*$ and $a \in V$, $|x|_a$ denotes the number of occurrences of the symbol a in x. A finite multiset over V is a mapping $X : V \longrightarrow \mathbb{N}$, where $\mathbb{N}$ denotes the set of non-negative integers. $X(a)$ is said to be the multiplicity of a in X.

2.1   Register Machines

A deterministic *register machine* is defined as a 5-tuple $M = (Q, R, q_0, q_f, P)$, where Q is a set of states, $R = \{R_1, \ldots, R_k\}$ is the set of registers, $q_0 \in Q$ is the initial state, $q_f \in Q$ is the final state and P is a set of instructions (called also rules) of the following form:

1. *(Increment)* $(p, RiP, q) \in P$, $p, q \in Q, p \neq q, R_i \in R$ (being in state p, increment register R_i and go to state q).
2. *(Decrement)* $(p, RiM, q) \in P$, $p, q \in Q, p \neq q, R_i \in R$ (being in state p, decrement register R_i and go to state q).
3. *(Zero check)* $(p, Ri, q, s) \in P$, $p, q, s \in Q, R_i \in R$ (being in state p, go to q if register R_i is not zero or to s otherwise).
4. *(Zero test and decrement)* $(p, RiZM, q, s) \in P$, $p, q, s \in Q, R_i \in R$ (being in state p, decrement register R_i and go to q if successful or to s otherwise).
5. *(Stop)* $(q_f, STOP)$ (may be associated only to the final state q_f).

Note that a $RiZM$ instruction can be used to simulate both a RiM instruction and a Ri instruction.

A configuration of a register machine is given by $(q, n_1, \ldots, n_k)$, where $q \in Q$ and $n_i \in \mathbb{N}, 1 \leq i \leq k$, describe the current state of the machine as well as the contents of all registers. A transition of the register machine consists in updating/checking the value of a register according to an instruction of one of the types above and in changing the current state to another one. We say that the machine stops if it reaches the state q_f. We say that M computes a value $y \in \mathbb{N}$ on the *input* $x_1, \ldots, x_n$, $x_i \in \mathbb{N}$, $1 \leq i \leq n \leq k$, if, starting from the initial configuration $(q_0, x_1, \ldots, x_n, 0, \ldots, 0)$, it reaches the final configuration $(q_f, y, 0, \ldots, 0)$.

It is well-known that register machines compute all partial recursive functions and only them [11]. Therefore, every register machine M with n registers can be associated with the function it computes: an m-ary partial recursive function Φ_M^m, where $m \leq n$. Let $\Phi_0, \Phi_1, \Phi_2, \ldots$, be a fixed enumeration of the set of unary partial recursive functions. Then, a register machine M is said to be *strongly universal* [6] if there exists a recursive function g such that $\Phi_x(y) = \Phi_M^2(g(x), y)$ holds for all $x, y \in \mathbb{N}$. A register machine M is said to be *(weakly) universal* if there exist recursive functions f, g, h such that $\Phi_x(y) = f(\Phi_M^2(g(x), h(y)))$ holds for all $x, y \in \mathbb{N}$. We remark that here the meaning of the term weakly universal is different from the Turing machines case, where it is commonly used to denote a universal machine working on a tape that has an infinite initial configuration with one constant word repeated to the right, of and another to the left of the input [9].

2.2   Petri Nets

A *Place-Transition-net* or for short, *PT-net, with inhibitor arcs* is a construct
$N = (P, T, W, M_0)$ where P is a finite set of *places*, T is a finite set of *transitions*,
with $P \cap T = \emptyset$, $W : (P \times T) \cup (T \times P) \to \mathbb{N} \cup \{-1\}$ is the *weight function* and
M_0 is a multiset over P called the *initial marking*.

PT-nets are usually represented by diagrams where places are drawn as circles,
transitions are drawn as squares annotated with their location, and a directed
arc (x, y) is added between x and y if $W(x, y) \geq 1$. These arcs are then annotated
with their weight if this one is 2 or more. Arcs having the weight -1 are called
inhibitor arcs and are drawn such that the arcs end with a small circle on the
side of the transition.

The *degree* of a transition t is defined as the sum of the weights of the incoming
and outgoing arcs involved with it plus the number of inhibitor arcs:

$$degree(t) = \sum_{p \in P} |W(p, t)| + |W(t, p)|.$$

Note that the degree is not the number of valuated arcs adjacent to the transition,
but rather the number of single arcs they represent.

Given a PT-net N, the *pre-* and *post-multiset* of a transition t are respectively
the multiset $pre_{N(t)}$ and the multiset $post_{N(t)}$ such that, for all $p \in P$, for which
$W(p, t) \geq 0$, $pre_{N(t)}(p) = W(p, t)$ and $post_{N(t)}(p) = W(t, p)$. A state of N,
which is called a *marking*, is a multiset M over P; in particular, for every $p \in P$,
$M(p)$ represents the number of *tokens* present inside place p. A transition t is
enabled at a marking M if the multiset $pre_{N(t)}$ is contained in the multiset
M and all inhibitor places p (such that $W(p, t) = -1$) are empty. An enabled
transition t at marking M can *fire* and produce a new marking M' such that
$M' = M - pre_{N(t)} + post_{N(t)}$ (i.e., for every place $p \in P$, the firing transition
t consumes $pre_{N(t)}(p)$ tokens and produces $post_{N(t)}(p)$ tokens). We denote this
as $M \xrightarrow{t} M'$.

For the purposes of this paper, we have to define which kind of PT-nets can
execute computations (e.g. compute partially recursive functions). In such a net
some distinguished places $i_1, \ldots, i_k$, $k > 0$ from P are called *input* places (which
are normally different from the places marked in M_0 containing the control
tokens) and one other, $i_0 \in P$, is called the *output* place. The computation of
the net N on the input vector $(n_1, \ldots, n_k)$ starts with the initial marking M_0'
such that $M_0'(i_j) = n_j$ and $M_0'(x) = M_0(x)$, for all $x \neq i_j$, $1 \leq j \leq k$. This net
will evolve by firing transitions until deadlock in some marking M_f, i.e. in M_f
no transition is enabled. Thus we have $M_0' \xrightarrow{*} M_f$ and there are no M_f' and
$t \in T$ such that $M_f \xrightarrow{t} M_f'$. The result of the computation of N on the vector
$(n_1, \ldots, n_k)$, denoted by $\Phi_N^k(n_1, \ldots, n_k)$, is defined as $M_f(i_0)$, i.e. the number
of tokens in place (i_0) in the final state. Since in the general case Petri nets are
non-deterministic, the function Φ_N^k could compute a set of numbers.

If, for any reachable marking M, there is at most one transition t and one
marking M' such that $M \xrightarrow{t} M'$, the Petri net is called *deterministic*. This

corresponds to labeled deterministic Petri nets in which all transitions are labeled with the same symbol [14]. Otherwise the Petri net is called non-deterministic, and it is this kind of nets that we will focus on in this paper.

The *size* of a Petri net is the vector (p, t, h, d) where p is the number of places, t is the number of transitions, h is the number of inhibitor arcs, and d is the maximal degree of a transition. These parameters of the Petri net provide the fundamental information about its structure and can be further used to reason about its other features (e.g., the average number of inhibitor arcs per transition). Moreover, each of these parameters has a direct equivalent in the multiset rewriting interpretation of Petri nets as the cardinality of the alphabet, number of rules, inhibitors and maximal rule size. Remember that, when counting the degree of a transition, we take into account the weights of the arcs it involves, even though the degree is often considered to be the sum of the number of input and output places of the transition.

3 Universal Register Machines with Few Registers

A rather well-known result on the computational power of register machines is that there exists a strongly universal machine with 3 registers and a weakly universal machine with 2 registers only [10]. Since it seems natural to represent registers as places [4], we have special interest in machines with a small number of registers, and, for the purposes of this paper, we have actually constructed a strongly universal 3-register machine U_3 and a weakly universal 2-register machine U_2 following the ideas proposed in [10].

Roughly, the construction is based on exponential coding of the configuration of an arbitrary register machine and simulating increments and decrements as multiplications and divisions. Two registers are enough for these purposes, and a third one is required for exponentiation of the input and computing of the logarithm to retrieve the output. To construct the weakly universal U_2, we simulate the 20-state 8-register weakly universal register machine U_{20} described in [6]. The 2-register machine obtained in this way has 112 decrement and 165 increment instructions: 278 states all in all (including a final state). Simulating the 22-state 8-register machine U_{22} described in the same paper, and further adding exponentiation and logarithm, allows building the strongly universal 3-register machine U_3 with 147 decrement and 217 increment instructions: 365 states in total.

Clearly, U_2 and U_3 simulate the corresponding machines from [6] with exponential slowdown. However, the machines by Korec simulate partial recursive functions with exponential slowdown, too (cf. Theorem 4.2 in [6]). This means that the slowdown of the machines U_2 and U_3 with respect to the (indirectly) simulated partial recursive functions is doubly exponential.

Both register machines use the register R_0 to store the exponentially-coded values of the simulated register machine, and the register R_1 to keep the intermediate results. The input of U_2 should thus be provided in coded form in R_0. The register machine U_3, on the other hand, reads its input from and produces its output in the third register, R_2.

An important remark with regard to the strong universality of U_3 is due here: since we use one register for input, we are only able to directly simulate unary partial recursive functions. Nevertheless, Section 9 of [6] describes a way to construct register machines simulating n-ary partial recursive functions; the machines use a coding to store the values of the n arguments in one of the working registers. This approach can be pretty naturally adapted to the register machine U_2 to obtain strongly universal register machines with n input registers, read by successive decrements at the start of the computation, and which only have two working registers. Such register machines can be translated into Petri nets by the same techniques as the ones we will show for U_2 and U_3 in the coming sections.

Section 2 of [5] describes the construction of a universal three-register machine with 130 states. However, in that paper compound instructions are assigned to single states (e.g., any increment of a register by m is treated as a single instruction). Writing out the corresponding construction in terms of elementary register machine commands as we use in U_2 and U_3, would yield more than 450 instructions.

4 Non-deterministic Simulation of Register Machines by Petri Nets

In this section we will show that a register machine with n registers can be simulated by a non-deterministic Petri net with only $n + 2$ places. We recall that in the non-deterministic semantics, we only consider those branches of computation which halt. The basic idea is representing a state number of a register machine in unary encoding as the number of tokens in a (single) place of the Petri net, and then using another place to assure that the transformations of state numbers happen correctly.

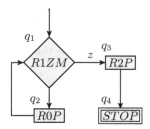

Fig. 1. The toy register machine S

Consider the following register machine $S = (Q, R, q_1, q_4, P)$, where $Q = \{q_1, q_2, q_3, q_4\}$, $R = \{R_0, R_1, R_2\}$ and the set of instructions P is defined as $P = \{(q_1, R_1 M, q_2, q_3), (q_2, R0P, q_1), (q_3, R2P, q_4), (q_4, STOP)\}$. This machine is depicted on Figure 1 using a standard flow-chart notation.

This machine adds the contents of the register r_1 to the register r_0, eventually sets r_1 to zero, and increments r_2 once. The corresponding non-deterministic Petri net N is shown in Figure 2. This Petri net uses a place to store the value of each register, and two more places to store state numbers and validate state transitions. Note that we do not need to represent the final state q_4 in the Petri net. It suffices that N carries out the operation associated with q_3 and halts.

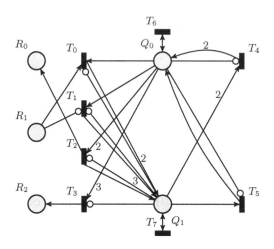

Fig. 2. The non-deterministic Petri net N simulating S

The Petri net N starts with one token in the state place Q_0, which corresponds to the number of the initial state q_1 of S. The simulation of an arc in the graph of the register machine S is carried out in two phases. In the first phase, the current state is read from the place Q_0, the corresponding registers are checked and/or modified, and the number of the next state is put into the state place Q_1. The first phase of the simulation is carried out by the transitions T_0 through T_3. The goal of the second phase is checking that state places are always read completely, i.e., that every transition which fires consumes all the tokens from Q_0 or Q_1. The second phase corresponds to the firing of one of the transitions T_4 or T_5.

Transitions T_0 and T_1 simulate state q_1 of S in the following way. If R_1 is nonempty, the transition T_0 can fire, consuming one token from the place R_1 and putting two tokens into Q_1, which corresponds to moving into state q_2. If, however, R_1 is empty, transition T_1 fires and places three tokens into Q_1, which corresponds to moving into state q_3. Similarly, transition T_2 simulates the behavior of S in state q_2, and transition T_3 corresponds to the state q_3. Note that T_3 only empties Q_0 and adds nothing to Q_1, moreover, Q_1 should be empty in order for T_3 to fire. Therefore, after T_3 fires, no more transitions can fire and N halts.

The goal of transitions T_4 and T_5 is moving the number of the next state to simulate from Q_1 to Q_0. At the same time, the two transitions verify that the correct instruction has just been carried out. Indeed, suppose that Q_0 contains two tokens and R_1 is empty. Then both transition T_0 and T_2 can fire. However, if T_0 fires, it will place two tokens into Q_1, and one token will still be left in Q_0. In this case transitions T_0 through T_5 will be blocked, and the only enabled transitions T_6 and T_7 will make the net loop forever. Similarly, if transition T_5 fires when Q_1 contains more than one token, N will never halt. Therefore, in a halting computation of N, either the place Q_0 or Q_1 contains tokens, but not both at the same time, which assures the correct simulation of the register machine S.

We remark that the number of places in the above simulation does not depend on the number of states of the simulated machine. Hence, it should be rather clear that the same reasoning can be repeated for the 3-register universal machine U_3 to obtain a 5-place strongly universal non-deterministic Petri net N_1. It will have 511 transitions simulating the activity of U_3, 364 more transitions moving the code of the next state from Q_1 to Q_0, and two loops; 877 transitions all in all. It will also have 1022 inhibitor arcs and the maximal transition degree will be 729. In an analogous fashion we can build a 4-place weakly universal Petri net N_2 with 668 transitions of maximal degree 555 and with 778 inhibitor arcs.

5 Decreasing the Transition Degree

In this section we will show that it is possible to almost halve the maximal transition degree of the transitions in the strongly universal non-deterministic Petri net N_1 at the cost of a slight increase in the number of transitions.

First of all, remark that the maximal transition degree in the Petri net N_1 is largely determined by the transitions moving tokens from Q_1 to Q_0: since U_3 contains 365 states including a final one, N_1 must have a transition moving 364 tokens from Q_1 to Q_0, which, together with the inhibitor arc coming from Q_0, results in degree 729. Remember that we try to be precise from the graph-theoretic point of view and count the weights of the arcs as well, while in many Petri-net-related works the degree of such transitions would be considered 3, as they have only one input and two output places.

On the other hand, the transitions actually simulating the activity of U_3 need not be this large: it is possible to code states in such a way that these transitions have an input degree which is much smaller than the output degree or vice versa. The idea therefore is as follows: mirror the transitions which simulate the activity of U_3, so that the simulation happens both when tokens are moved from Q_0 to Q_1 *and* when tokens are moved from Q_1 to Q_0. In the case of our toy register machine S, this will result in the Petri net shown in Figure 3. Observe that the transition T_i' simulates the same instruction of S as T_i does.

In the case of this net, maximal transition degree is determined solely by the way in which the states are represented. Namely, consider a transition T which corresponds to the move from state q_i to state q_j in the simulated register machine. The way in which states are encoded will be given by the function $c : Q \rightarrow \mathbb{N}$, where Q is the set of states. Then, the degree of T is $c(q_i) +$

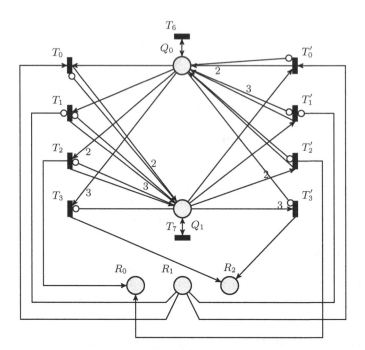

Fig. 3. A mirrored non-deterministic Petri net simulating S

$c(q_j) + 2$, where we add 2 because all transitions read and/or modify a register and are inhibited by either Q_0 or Q_1. We can now define the mapping c via a minimization problem, keeping in mind the goal to minimize the worst transition degree.

We start by remarking that c takes values in the interval 1 through $|Q| - 1$ (we do not need to code the final state). In the optimization problem, we use the following family of variables:

$$c_{i,i'} = \begin{cases} 1, & \text{if } c(q_i) = i', \\ 0, & \text{otherwise,} \end{cases}$$

with the following normalization conditions:

$$\forall q_i \in Q : \sum_{1 \leq i' \leq |Q|-1} c_{i,i'} = 1 \text{ and } \forall 1 \leq i' \leq |Q| - 1 : \sum_{q_i \in Q} c_{i,i'} = 1,$$

which require that every state have exactly one code and that each code be picked exactly once.

The family of variables $c_{i,i'}$ can be used to express the cost $c(q_i) + c(q_j)$ for a transition from state q_i to q_j in the following way:

$$c(q_i) + c(q_j) = \sum_{1 \leq i' \leq |Q|-1} i' \cdot c_{i,i'} + \sum_{1 \leq j' \leq |Q|-1} j' \cdot c_{j,j'}.$$

For convenience, we will also define the set of pairs of states between which there exists an arc in the graph of the register machine:

$$B = \{(q_i, q_j) \in Q \times Q \mid (q_i, RkP, q_j) \in P,$$
$$\text{or } (q_i, RkP, q_j, s) \in P, \text{ or } (q_i, RkP, s, q_j) \in P\}.$$

We can now write the linear programming problem optimizing the cost of the "worst" transition:

Minimize C

Subject to $\displaystyle\sum_{1 \le i' \le |Q|-1} i' \cdot c_{i,i'} + \sum_{1 \le j' \le |Q|-1} j' \cdot c_{j,j'} \le C, (q_i, q_j) \in B,$

$\forall q_i \in Q : \displaystyle\sum_{1 \le i' \le |Q|-1} c_{i,i'} = 1,$

$\forall 1 \le i' \le |Q| - 1 : \displaystyle\sum_{q_i \in Q} c_{i,i'} = 1.$

To attack the instances of this linear programming for U_3 and U_2, we used the Gurobi Optimizer [12]. The problem itself was formulated in the AMPL model description language [3].

Remark that the number of variables in this linear programming problem is $|Q|^2 + 1$, which results in rather large linear programming models in the case of the universal 3- and 2-register machines. U_3, for example, has 364 non-final states, which means 132 496 variables. Furthermore, the first line in the definition of the linear programming problem introduces 511 constraints, each constraint being a linear combination of 729 variables. Finally, the last two lines introduce 728 more constraints involving 364 variables each. This amounts to 1239 constraints involving either 364 or 729 variables.

Due to limited computing resources, we were not able to find the optimal solutions for U_3 and U_2 until now. Nevertheless, we managed to obtain some reasonably good solutions which allow reducing the maximal transition degree to 379 in the case of U_3 and to 277 for U_2. Thus, we obtained a 5-place strongly universal non-deterministic Petri net N_3 with 1024 transitions of maximal degree 379 and 1316 inhibitor arcs, and a 4-place weakly universal non-deterministic Petri net N_4 with 780 transitions of maximal degree 299 and 1002 inhibitor arcs.

6   Main Results

To summarize the results concerned with strong universality, we formulate the following statement.

Theorem 1. *There exist strongly universal non-deterministic Petri nets of sizes* $(5, 877, 1022, 729)$ *and* $(5, 1024, 1316, 379)$.

Similarly we can state the following with respect to weakly universal non-deterministic Petri nets:

Theorem 2. *There exist weakly universal non-deterministic Petri nets of sizes* $(4, 668, 778, 555)$ *and* $(4, 780, 1002, 299)$.

7   Conclusion

In this paper we constructed two strongly universal and two weakly universal non-deterministic Petri nets. We have shown that dropping the restriction of a deterministic evolution allows a dramatic minimization of the number of places, but produces an important increase in the values of the other parameters.

We remark that, while the strong universality results obtained in this paper implicitly suppose that corresponding Petri nets have a single input (thus computing unary functions), it is possible to generalize them to an n-ary input using the ideas from [6]. Corresponding nets will have $n+2$ places and a linear increase in the number of inhibitors and maximal transition degree.

While our main goal was to minimize the number of places, we did also show that a trade-off still existed between the maximal transition degree on the one side and the number of transitions and inhibitor arcs on the other. It could be interesting to look for other similar trade-offs.

It might be possible to achieve strong universality with 4 places, because the input/output register of U_3 is only modified during the initial and final phases of execution of the machine, when input is read and output is produced [10]. We conjecture that 3 places or less are insufficient to achieve strong universality, however. On the other hand, it may be possible to reduce the values of some other parameters, while keeping the number of states at 5.

In Section 3, we have cited a universal 3-register machine from [5] and said that, if only elementary increments and decrements with zero check are allowed as instructions, this machine would have more instruction than U_3. However, compound instructions do map naturally on Petri net transitions: multiple increments can be carried out in one step by an arc with multiplicity greater than 1. Therefore, the register machine from [5] could be simulated directly, thus reducing the number of transitions, and of inhibitor arcs, but also increasing the maximal transition degree. Constructing the corresponding Petri net might be a worthwhile task that could potentially lead to a better maximal degree measure.

The universality results shown in this paper indirectly rely on exponential coding, which imposes operations of considerable time complexity. A different approach could be used to reduce simulation time, but this would almost certainly result in an increase in the size of the Petri nets.

Finally, we would like to stress that the results we give in this paper can be straightforwardly translated to the domain of P systems [13] and, more generally, multiset rewriting [1].

References

1. Alhazov, A., Verlan, S.: Minimization strategies for maximally parallel multiset rewriting systems. Theoretical Computer Science 412(17), 1581–1591 (2011)
2. Barzdin, I.M.: Ob odnom klasse machin Turinga (machiny Minskogo), russian. Algebra i Logika 1, 42–51 (1963)
3. Fourer, R., Gay, D.M., Kernighan, B.W.: AMPL: A Modeling Language for Mathematical Programming, 2nd edn. Duxbury Press, Brooks/Cole Publishing Company (2002)

4. Ivanov, S., Pelz, E., Verlan, S.: Small universal Petri nets with inhibitor arcs. In: Computability in Europe (2014)
5. Koiran, P., Moore, C.: Closed-form analytic maps in one and two dimensions can simulate universal turing machines. Theor. Comput. Sci. 210(1), 217–223 (1999)
6. Korec, I.: Small universal register machines. Theoretical Computer Science 168(2), 267–301 (1996)
7. Malcev, A.I.: Algorithms and Recursive Functions. Wolters-Noordhoff Pub. Co., Groningen (1970)
8. Margenstern, M.: Frontier between decidability and undecidability: A survey. Theoretical Computer Science 231(2), 217–251 (2000)
9. Margenstern, M.: An algorithm for buiding inrinsically universal automata in hyperbolic spaces. In: Arabnia, H.R., Murgin, M. (eds.) FCS, pp. 3–9. CSREA Press (2006)
10. Minsky, M.: Size and structure of universal Turing machines using tag systems. In: Recursive Function Theory: Proceedings, Symposium in Pure Mathematics, Provelence, vol. 5, pp. 229–238 (1962)
11. Minsky, M.: Computations: Finite and Infinite Machines. Prentice Hall, Englewood Cliffts (1967)
12. Gurobi Optimization, Inc. Gurobi optimizer reference manual (2014)
13. Păun, G.: Membrane Computing. An Introduction. Springer, Berlin (2002)
14. Pelz, E.: Closure properties of deterministic Petri nets. In: Brandenburg, F.J., Wirsing, M., Vidal-Naquet, G. (eds.) STACS 1987. LNCS, vol. 247, pp. 371–382. Springer, Heidelberg (1987)
15. Rozenberg, G., Salomaa, A. (eds.): Handbook of Formal Languages, vol. 1–3. Springer (1997)
16. Schroeppel, R.: A two counter machine cannot calculate 2N. In: AI Memos. MIT AI Lab (1972)
17. Turing, A.M.: On computable numbers, with an application to the Entscheidungsproblem. Proceedings of the London Mathematical Society 42(2), 230–265 (1936)
18. Woods, D., Neary, T.: The complexity of small universal Turing machines: A survey. Theor. Comput. Sci. 410(4-5), 443–450 (2009)
19. Zaitsev, D.A.: Universal Petri net. Cybernetics and Systems Analysis 48(4), 498–511 (2012)
20. Zaitsev, D.A.: A small universal Petri net. EPTCS 128, 190–202 (2013); In Proceedings of Machines, Computations and Universality (MCU 2013), arXiv:1309.1043

A Full and Linear Index of a Tree for Tree Patterns

Jan Janoušek, Bořivoj Melichar, Radomír Polách, Martin Poliak,
and Jan Trávníček

Department of Theoretical Computer Science,
Faculty of Information Technology,
Czech Technical University in Prague,
Thákurova 9, 160 00 Prague 6, Czech Republic
{Jan.Janousek,Borivoj.Melichar,Radomir.Polach,Martin.Poliak,
Jan.Travnicek}@fit.cvut.cz

Abstract. A new and simple method of indexing a tree for tree patterns is presented. A tree pattern is a tree whose leaves can be labelled by a special symbol S, which serves as a placeholder for any subtree. Given a subject tree T with n nodes, the tree is preprocessed and an index, which consists of a standard string compact suffix automaton and a subtree jump table, is constructed. The number of distinct tree patterns which match the tree is $\mathcal{O}(2^n)$, and the size of the index is $\mathcal{O}(n)$. The searching phase uses the index, reads an input tree pattern P of size m and computes the list of positions of all occurrences of the pattern P in the tree T. For an input tree pattern P in linear prefix notation $pref(P) = P_1 S P_2 S \ldots S P_k$, $k \geq 1$, the searching is performed in time $\mathcal{O}(m + \sum_{i=1}^{k} |occ(P_i)|))$, where $occ(P_i)$ is the set of all occurrences of P_i in $pref(T)$.

1 Introduction

Indexing a data subject preprocesses the subject and constructs an index that allows to answer queries related to the content of the subject efficiently. For example, occurrences of input patterns in the subject can be located repeatedly and quickly, in time typically not depending on the size of the subject.

The theory of text indexing, which is a result of Stringology research [9, 10, 11], is very well-researched and uses many sophisticated data structures: suffix tree and suffix array are most widely used structures for text indexing, providing efficient solutions for a wide range of applications. The Directed Acyclic Word Graph [5], also known as suffix (or factor) automaton [8], is another elegant structure. The compacted and minimized version of both suffix trees and suffix automata is represented by compact suffix automaton [12]. Another text indexing structure, called position heap, was proposed recently [13]. Generally, the number of substrings in a text is quadratic to the size of the text, but the size of the text index structure for substrings is typically linear to the size of the text. By

H. Jürgensen et al. (Eds.): DCFS 2014, LNCS 8614, pp. 198–209, 2014.

means of the suffix tree or the compact suffix automaton, the list of positions of all occurrences of an input string pattern y of size m can be computed in time $\mathcal{O}(m + |occ(y)|)$, where $occ(y)$ is the set of all occurrences of y in the text [9].

Subtree matching and tree pattern matching are often declared to be analogous to problems of string pattern matching [2, 14]. There is simple but a key property of linear notations of trees: the linear notation of a subtree is a substring of the linear notation of the tree [17]. A tree pattern is a tree whose leaves can be labelled by a special symbol S, which serves as a placeholder for any subtree. Given a tree with n nodes, the number of distinct tree patterns which match the tree is $\mathcal{O}(2^n)$. A tree pattern in a linear notation corresponds to a substring of the linear notation of the tree, where the symbols S are replaced with linear notations of subtrees. Therefore, tree pattern matching is analogous to the problem of matching string patterns with specific gaps.

The problem of matching string patterns with gaps has been explored in many methods. The methods differ in the kinds of considered gaps, in the achieved complexity and in the fact whether the method is based on indexing, where the subject text is preprocessed, or it is based on the principle that a string pattern is preprocessed and the subject text is read as the input of the searching phase. A method of indexing a text for string patterns with gaps is described in [3], where an index is constructed for matching with wildcards. A wildcard matches any single character, which cannot be used for tree patterns in a linear notation. In [16], an index is constructed for matching with variable length gaps. Unfortunately, the searching time depends on the gaps sizes, which is not time efficient for matching tree patterns, where the gaps can be of any size.

Tree pattern pushdown automaton represents a full index of a tree for tree patterns but its size is not linear to the size of the subject tree [17]. Also, a finite tree automaton accepting all tree patterns that match the tree can be constructed but its size is exponential to the size of the subject tree [6, 7, 15].

In [4], a matching algorithm, where the patterns are preprocessed, for variable length gap matching problem was proposed. This solution is incompatible with the tree pattern matching problem because of different interpretation of gaps. Moreover, this solution is not indexing but matching, when the pattern is preprocessed. By analogy for trees, many tree pattern matching methods, which preprocess tree patterns and not the subject tree, have been proposed [6, 14].

In this paper, a new and simple method of indexing a tree for tree patterns is presented. Given a subject tree T with n nodes, the tree is preprocessed and an index, which consists of a standard string compact suffix automaton and a subtree jump table, is constructed. We note that any convenient text index can possibly be used instead of the compact suffix automaton, which has been chosen here because of its good space and time complexities. Despite the fact that the number of distinct tree patterns which match the tree is $\mathcal{O}(2^n)$, the size of the index presented in this paper is $\mathcal{O}(n)$.

The searching phase uses the index, reads an input tree pattern P of size m and computes the list of positions of all occurrences of the pattern P in the tree T. For an input tree pattern P in linear prefix notation $pref(P) = P_1 S P_2 S \ldots S P_k,$

$k \geq 1$, the searching is performed in time $\mathcal{O}(m + \sum_{i=1}^{k} |occ(P_i)|)$, where $occ(P_i)$ is the set of all occurrences of P_i in $pref(T)$. We are not aware of any other known method of full and linear indexing a tree for tree patterns with these time and space complexities.

The rest of the paper is organised as follows. Basic definitions are given in section 2. The third section describes our method of indexing a tree for tree patterns. The fourth section deals with the searching phase. The complexities are formally described in the fifth section. The last section is the conclusion.

2   Basic Notions

A *ranked alphabet* is a finite nonempty set of symbols each of which has a unique nonnegative *arity* (or *rank*). Given a ranked alphabet $\mathcal{A}$, the arity of a symbol $a \in \mathcal{A}$ is denoted $Arity(a)$. The set of symbols of arity p is denoted by $\mathcal{A}_p$. Elements of arity $0, 1, 2, \ldots, p$ are respectively called nullary (constants), unary, binary, ..., p-ary symbols. We assume that $\mathcal{A}$ contains at least one constant. In the examples we use numbers at the end of the identifiers for a short declaration of symbols with arity. For instance, $a2$ is a short declaration of a binary symbol a. We use $|\mathcal{A}|$ notation for the size of set $\mathcal{A}$.

Because of lack of space we do not provide definitions of basic graph theory notions, such as graph, directed graph, and others (see [1]). Based on concepts from graph theory, a tree over an alphabet $\mathcal{A}$ can be defined as follows:

A *rooted and directed tree* T is an acyclic connected directed graph $T = (N, R)$ with a special node $r \in N$, called the *root*, such that

(1) r has in-degree 0,
(2) all other nodes of T have in-degree 1,
(3) there is just one path from the root r to every $f \in N$, where $f \neq r$.

A node g is a *direct descendant* of node f if a pair $(f, g) \in R$.

A *labelled, (rooted, directed) tree* is a tree having the following property:

(4) every node $f \in N$ is labelled by a symbol $a \in \mathcal{A}$, where $\mathcal{A}$ is an alphabet.

A *ranked, (labelled, rooted, directed) tree* is a tree labelled by symbols from a ranked alphabet and out-degree of a node f labelled by symbol $a \in \mathcal{A}$ equals to $Arity(a)$. Nodes labelled by nullary symbols (constants) are called *leaves*.

An *ordered, (ranked, labelled, rooted, directed) tree* is a tree where direct descendants $a_{f1}, a_{f2}, \ldots, a_{fn}$ of a node a_f having an $Arity(a_f) = n$ are ordered.

A *subtree* of tree $T = (N, R)$ rooted at node $f \in N$ is a tree $T_f = (N_f, R_f)$, such that f is the root of T_f and N_f, R_f is the greatest possible subset of N, R, respectively.

If g is not a root node, then there exists $(f, g) \in R$. A *right sibling* of node g is a node h that is the smallest node greater than g that satisfies $(f, h) \in R$.

The *prefix notation* $pref(T)$ of tree T is defined as follows:

1. $pref(a) = a_0$ if a is a leaf,

2. $pref(T) = a_n\ pref(b_1)\ pref(b_2)\ldots pref(b_n)$, where a_n is the root of tree T and $b_1, b_2, \ldots b_n$ are subtrees of T rooted at the respective direct descendants of a_n.

Let $w = a_1 a_2 \ldots a_m$, $m \geq 1$, be a string over a ranked alphabet $\mathcal{A}$. Then, the *arity checksum* $ac(w) = arity(a_1) + arity(a_2) + \ldots + arity(a_m) - m + 1 = \sum_{i=1}^{m} arity(a_i) - m + 1$. Let $pref(T)$ and w be a tree T in prefix notation and a substring of $pref(T)$, respectively. Then, w is the prefix notation of a subtree of T, if and only if $ac(w) = 0$, and $ac(w_1) \geq 1$ for each proper prefix w_1 of w (ie. $w = w_1 x$, $x \neq \varepsilon$) [17].

Example 1. Consider a ranked alphabet $\mathcal{A} = \{a4, a0, b0\}$. Consider an ordered, ranked, labelled, rooted, and directed tree $T_1 = (\{a4_1, a4_2, a4_3, a0_4, b0_5, a0_6, a0_7, a0_8, b0_9, a0_{10}, a0_{11}, a0_{12}, b0_{13}\}, R_1)$ over an alphabet $\mathcal{A}$, where R_1 is a set of the following ordered pairs:

$$R_1 = \{(a4_1, a4_2), (a4_1, a0_{11}), (a4_1, a0_{12}), (a4_1, b0_{13}), (a4_2, a4_3), (a4_2, a0_8),$$
$$(a4_2, b0_9), (a4_2, a0_{10}), (a4_3, a0_4), (a4_3, b0_5), (a4_3, a0_6), (a4_3, a0_7)\}.$$

Prefix notation of tree T_1 is $pref(T_1) = a4_1 a4_2 a4_3 a0_4\, b0_5 a0_6 a0_7\, a0_8 b0_9 a0_{10}\, a0_{11}\, a0_{12} b0_{13}$. Tree T_1 is illustrated in Figure 1.   □

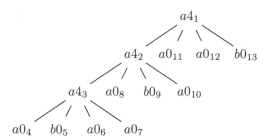

Fig. 1. Tree T_1 from Example 1

To define a *tree pattern*, we use a special nullary symbol $S \notin \mathcal{A}$, $Arity(S) = 0$, which serves as a placeholder for any subtree. A tree pattern is defined as a labelled ordered tree over an alphabet $\mathcal{A} \cup \{S\}$. In this paper we assume that the tree pattern has at least one node labelled by a symbol from $\mathcal{A}$.

A tree pattern P with $k \geq 0$ occurrences of symbol S *matches* a subject tree T at node n if there exist subtrees $T_1, T_2, \ldots, T_k$ (not necessarily the same) of tree T such that tree P', obtained from tree pattern P by substituting subtree T_i for the i-th occurrence of symbol S in P, $i = 1, 2, \ldots, k$, is equal to the subtree T_s of tree T rooted at node n. Tree T_s is the *matched subtree* of tree T.

Let a tree pattern P match a subject tree T at node n and let m be the number of nodes in the matched subtree T_s. Let i be the index of node n in $pref(T) = a_1 a_2 \ldots a_i a_{i+1} \ldots a_{i+m-1} a_{i+m} \ldots$. An *occurrence* of tree pattern P in subject tree T is a pair $(i, i + m)$. The pair $(i, i + m)$ is also an *occurrence* of substring $pref(T_s)$ in string $pref(T)$.

A *compact suffix automaton* [9] for a text is a finite automaton that accepts all suffixes of the text. By means of the compact suffix automaton, the list of positions of all occurrences of an input string pattern y of size m can be computed in time $\mathcal{O}(m + |occ(y)|)$, where $occ(y)$ is the set of all occurrences of y in the text.

Example 2. Consider tree T_1 from Example 1.

Consider subtree P' over alphabet $\mathcal{A}$, $P' = (\{a4_1, a0_2, b0_3, a0_4, a0_5\}, R_{P'})$, $pref(P') = a4\ b0\ a0\ a0\ a0$ and $R_{P'} = \{(a4_1, a0_2), (a4_1, b0_3), (a4_1, a0_4), ((a4_1, b0_5)\}$.

Consider tree pattern P'' over an alphabet $\mathcal{A} \cup \{S\}$, $P'' = (\{a4_1, S_2, a0_3, S_4, S_5\}, R_{P''})$. Tree pattern P'' in prefix notation is $pref(P'') = a4\ S\ a0\ S\ S$ and $R_{P''} = \{(a4_1, S_2), (a4_1, a0_3), (a4_1, S_4), (a4_1, S_5)\}$.

Tree patterns P' and P'' are illustrated in Figure 2. Tree pattern P' has one occurrence in tree T_1 – it matches T_1 at node 3. Tree pattern P'' has two occurrences in tree T_1 – it matches T_1 at nodes $a4_1$ and $a4_2$.

Compact suffix automaton $M(pref(T_1))$ is illustrated in Figure 3.   □

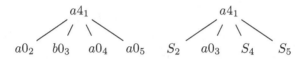

Fig. 2. Subtree P' (left) and tree pattern P'' (right) from Example 2

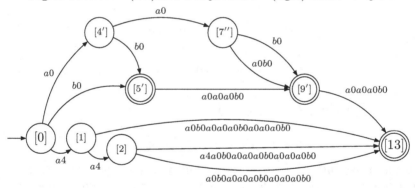

Fig. 3. Transition diagram of compact suffix automaton $M(pref(T_1))$ for tree T_1 from Example 3. The long edge labels can be represented by pairs of beginning and ending indices into $pref(T_1)$, see [9].

3   Indexing a Tree for Tree Patterns

The section deals with the preprocessing phase, in which an index of a subject tree T is constructed. The index consists of two parts:

- A compact suffix automaton [9] for $pref(T)$, by which occurrences of all substrings of $pref(T)$ can be located. We note that not all substrings of $pref(T)$ are subtrees in the prefix notation.

- A *subtree jump table*, a linear-size structure needed for finding positions of ends of subtrees represented by special symbols S.

Definition 1. Let T and $pref(T) = a_1a_2\ldots a_n$, $n \geq 1$, be a tree and its prefix notation, respectively. A *subtree jump table* $SJT(T)$ is defined as a mapping from set $\{1..n\}$ into set $\{2..n+1\}$. If $a_ia_{i+1}\ldots a_{j-1}$ is the prefix notation of a subtree of tree T, then $SJT(T)[i] = j$, $1 \leq i < j \leq n+1$.

Algorithm 1. Construction of subtree jump table

Name: ConstructSubtreeJumpTable
Input: Tree T in prefix notation $pref(T)$, index of current node
   $rootIndex$, reference to an empty subtree jump table $SJT(T)$
Output: index $exitIndex$, subtree jump table $SJT(T)$
1 $index = rootIndex + 1$;
2 **for** $i = 1$ **to** $Arity(pref(T)[rootIndex])$ **do**
3     $index = $ ConstructSubtreeJumpTable$(pref(T), index, SJT(T))$;
4 $SJT(T)[rootIndex] = index$;
5 **return** $index$;

Lemma 1. Given $pref(T)$ and $rootIndex$ equal to 1, Algorithm 1 constructs subtree jump table $SJT(T)$.   □

Example 3. Consider tree T_1 over $\mathcal{A}$ from Example 1, $pref(T_1) = a4_1a4_2a4_3a0_4$ $b0_5a0_6a0_7\,a0_8b0_9a0_{10}\,a0_{11}a0_{12}b0_{13}$. Compact suffix automaton $M(pref(T_1))$ [9] is illustrated in Figure 3. Subtree jump table $SJT(T_1)$, constructed by Alg 1, is in Table 1.   □

Table 1. Subtree jump table for tree T_1 from Example 3

i	1	2	3	4	5	6	7	8	9	10	11	12	13
$SJT[i]$	14	11	8	5	6	7	8	9	10	11	12	13	14

Furthermore, the array $Rev_{\mathcal{S}}^n$ serves as a working data structure for the main matching algorithm (Alg. 5) during the searching phase and its initial value, denoted $Rev_{\{\}}^n$, is to be set once and constructed during the preprocessing phase.

Definition 2. Let $\mathcal{S} = \{(first_1, last_1), \ldots (first_k, last_k)\}$ be a set of pairs of positive integers such that $last_i \neq last_j$ if $i \neq j$, $1 \leq i \leq k$, $1 \leq j \leq k$. Array $Rev_{\mathcal{S}}^n$ is an array of integers such that $Rev_{\mathcal{S}}^n[last_h] = first_h$ for all $1 \leq h \leq k$. For all other values $1 \leq v \leq n$, $Rev_{\mathcal{S}}^n[v] = -1$.

Example 4. Array $Rev_{\{(1,11),(2,8),(3,5)\}}^{13}$, which represents occurrences of tree pattern P'' from Example 2 in tree T_1 from Example 1, is illustrated in Table 2.   □

Table 2. Array $Rev_{\{(1,11),(2,8),(3,5)\}}^{13}$ from Example 4

i	1	2	3	4	5	6	7	8	9	10	11	12	13
$Rev_{\{(1,11),(2,8),(3,5)\}}^{13}[i]$	-1	-1	-1	-1	3	-1	-1	2	-1	-1	1	-1	-1

4   Computing Positions of All Occurrences of an Input Tree Pattern

The section describes the searching phase using the index. Algorithm 5 (Match-Pattern) computes the list of all occurrences of an input pattern P in the tree T. This main algorithm uses three algorithms, which are presented before Alg. 5:

- Algorithm 2 (VerifyArityChecksum) [17] computes the arity checksum of the input $pref(P)$ so that it would be verified that $pref(P)$ is a valid prefix notation of a tree pattern.
- Algorithm 3 (FindOccurences) is a substring matching algorithm from [9]. This algorithm is reproduced in listing 3, with some technical details omitted for the sake of brevity and clarity.
- Algorithm 4 (MergeOccurrences) is an algorithm that merges two sets of occurrences in linear time. A linear time merging algorithm would be simple if the sets of occurrences $(first, last)$ were in the form of lists sorted by index $first$ or by index $last$. Such a principle is used in related work [4]. Unfortunately, we did not manage to gain a sorted list of occurrences from the compact suffix automaton in a linear time. Therefore, we have avoided such sorting completely, reaching the linear time with a special merge operation, which uses the working data structure Rev introduced in the previous section.

Definition 3. Let $pref(P) = P_1SP_2S \ldots SP_k$ be the prefix notation of a tree pattern P over an alphabet $A \cup \{S\}$, where no substring P_i, $1 \le i \le k$, contains any symbol S. The substring P_i is called a *subpattern* of P at index i.

Example 5. Consider $pref(P'') = a4Sa0SS$, the prefix notation of tree pattern P'' from Example 2. Tree pattern P'' has four subpatterns, $pref(P'') = P_1SP_2SP_3SP_4$, where $P_1 = a4$, $P_2 = a0$, $P_3 = \varepsilon$ and $P_4 = \varepsilon$.   □

Definition 4. Let $pref(T) = a_1a_2...a_n$ be the prefix notation of a tree T. Let $pref(P) = P_1SP_2S...P_k$ be the prefix notation of a tree pattern P. An *occurrence* of subpattern P_i in $pref(T)$ is a pair $(first, last)$, where:

- if $P_i = \varepsilon$, $1 < first = last \le n + 1$,
- if $P_i \ne \varepsilon$, $1 \le first < last \le n + 1$ and $a_{first}a_{first+1} \ldots a_{last-1} = P_i$.

The set of all occurrences of subpattern P_i in $pref(T)$ is denoted by $occ^T(P_i)$. If tree T is obvious from the context, the set can be denoted by $occ(P_i)$.

Example 6. Consider subpattern $P_2 = a0$ of tree pattern P'' from Example 2. Subpattern P_2 has seven occurrences in tree T: $occ^T(P_2) = \{ (4,5), (6,7), (7,8), (8,9), (10,11), (11,12), (12,13) \}$.   □

Definition 5. Let $pref(P) = P_1SP_2S \ldots SP_k$ be the prefix notation of a tree pattern P. Then any string $P_1SP_2S \ldots SP_{k'}$, $k' \le k$, or $P_1SP_2S \ldots SP_{k''}S$, $k'' < k$, is called a *tree pattern prefix* of tree pattern P, abbreviated $TPP(P)$.

Example 7. Consider tree pattern P'', $pref(P'') = a4Sa0SS$, from Example 2. Then $\{a4, a4S, a4Sa0, a4Sa0S, a4Sa0SS\}$ is a set of tree pattern prefixes of tree pattern P''. □

Definition 6. Let P be a tree pattern and T be a tree. An *occurrence of tree pattern prefix* $TPP(P) = P_1SP_2S \ldots SP_k$ in tree T is a pair $(first, last)$, where $(first, last_1)$ is an occurrence of subpattern P_1 in $pref(T)$, pair $(SJT(T)[last_1], last_2)$ is an occurrence of subpattern P_2 in $pref(T)$, ..., and pair $(SJT(T)[last_{k-1}], last)$ is an occurrence of subpattern P_k in $pref(T)$. The set of all occurrences of a tree pattern prefix $TPP(P)$ in $pref(T)$ is denoted by $occ^T(TPP(P))$. If tree T is obvious from the context, the set can be denoted by $occ(TPP(P))$.

Example 8. Consider tree pattern prefix $TPP_1(P'') = a4S$ of tree pattern P'' from Example 2. Consider tree T_1 from Example 1. Then $occ^{T_1}(TPP_1(P'')) = \{(1, 11), (2, 8), (3, 5)\}$. □

Lemma 2. Let $(first, last)$ be an occurrence of a tree pattern prefix $pref(P)$ in a tree T, $pref(P) = P_1SP_2 \ldots SP_k$, $pref(T) = a_1a_2 \ldots a_{first}a_{first+1} \cdots a_{last-1}a_{last} \ldots a_n$. Then pattern P matches tree T at node a_{first}. Node a_{last-1} is the rightmost leaf of the subtree rooted at node a_{first}. □

We note that an *occurrence of tree pattern* P in tree T is an occurrence of tree pattern prefix $pref(P)$ in $pref(T)$.

Example 9. Consider tree pattern P'' from Example 2. Tree pattern P'' has two occurrences in tree T_1: $occ^{T_1}(P'') = \{(1, 14), (2, 11)\}$. □

Lemma 3. Let T be a tree and P be a tree pattern. Let pairs $(first_A, last_A)$ and $(first_B, last_B)$, $first_A \neq first_B$, be occurrences of tree pattern prefix $TPP(P) = P_1SP_2S \ldots$ in tree T. If $TPP(P) \neq pref(P)$, then $last_A \neq last_B$. □

Algorithm 2. Verification with the use of arity checksum [17]

Name: VerifyArityChecksum
Input: String over a ranked alphabet $str = a_1a_2 \ldots a_n$, $n \geq 1$.
Output: Decision whether $str = pref(T)$ for a tree T.

```
1  Set ac(T) := 1;
2  for i := 1 to n do
3      ac(str) := ac(str) + Arity(a_i) - 1;
4      if i < n and ac(str) = 0 then
5          return false;
6  return ac(str) = 0 ? true : false;
```

Algorithm 3. Finding Occurrence of Subpatterns [9]

Name: FindOccurrences
Input: Compact suffix automaton $M(pref(T))$, subpattern P_i of tree
pattern P
Output: $occ^T(P_i)$

1 Let q be the state of the M reached after processing P_i from input;
2 Find all paths from state q that lead to a final state of M;
3 For each path of length $length$ from state q to a final state, there is an
occurrence $(n - length - |P_i|, n - length)$ of subpattern P_i;

Algorithm 4. Merging Occurrences

Name: MergeOccurrences
Input: A set $prevOcc = occ^T(TPP(P))$, a set $subOcc = occ^T(P_k)$, an
array $Rev_{\{\}}^{|pref(T)|}$
Output: A set $mergedOcc = occ^T(TPP(P)P_k)$

1 $mergedOcc := \{\}$;
2 **foreach** $(first, last)$ in $prevOcc$ **do** $Rev_{prevOcc}^{|pref(T)|}[last] := first$;
3 **foreach** $(first', last')$ in $subOcc$ **do**
4   **if** $Rev_{prevOcc}^{|pref(T)|}[first'] \neq -1$ **then**
        $mergedOcc := mergedOcc \cup \{(Rev_{prevOcc}^{|pref(T)|}[first'], last')\}$;
5 **foreach** $(first, last)$ in $prevOcc$ **do** $Rev_{\{\}}^{|pref(T)|}[last] := -1$;
6 **return** $mergedOcc$;

Theorem 1. Let $TPP'(P) = P_1 S P_2 S \ldots S P_{k-1} S P_k$ be a tree pattern prefix
of a tree pattern P. Let $prevOcc = occ^T(TPP(P))$ be a set of occurrences of
a tree pattern prefix $TPP(P) = P_1 S P_2 S \ldots S P_{k-1} S$; let $subOcc = occ^T(P_k)$
be a set of occurrences of a subpattern P_k. Given $prevOcc$ and $subOcc$ on
input, Algorithm 4 (MergeOccurrences) computes occurrences $mergedOcc = occ^T(TPP'(P))$ of tree pattern prefix $TPP'(P)$.   □

Example 10. Consider the prefix notation $pref(P'') = a4Sa0SS$ of tree pattern
P'', illustrated in Figure 2. Tree pattern P'' can be rewritten as $pref(P'') = P_1 S P_2 S P_3 S P_4$, where $P_1 = a4$, $P_2 = a0$ and $P_3 = P_4 = \varepsilon$.

We consider the run of Algorithm 5 (MatchPattern) using tree pattern P'',
compact suffix automaton $M(pref(T_1))$ and subtree jump table SJT(T_1):

Algorithm 2 (VerifyArityChecksum) returns true for tree pattern P'' because
P'' is a valid tree pattern (if you replaced S symbols with $a0$ symbols in the
prefix notation of the pattern, you would get a prefix notation of a tree).

At $i = 1$, after Algorithm 3 (FindOccurrences) is executed, $prevOcc = \{(1, 2), (2, 3), (3, 4)\}$. Using subtree jump table $SJT(T_1)$, $prevOcc$ is then rewritten to
$prevOcc = \{(1, 11), (2, 8), (3, 5)\}$.

Algorithm 5. Searching for occurrences of a tree pattern

Name: MatchPattern
Input: Tree pattern $pref(P) = P_1SP_2S\ldots P_k$, compact suffix automaton
$M(pref(T))$, subtree jump table $SJT(T)$, Array $Rev_{\{\}}^{|pref(T)|}$
Output: List of occurrences of tree pattern P

1 **if** not VerifyArityChecksum(P) **then**
2   **return** $ERROR - invalid\ pattern$;
3 $prevOcc := \{\}$;
4 **for** $i := 1$ **to** k **do**
5   **if** $P_i \neq \varepsilon$ **then**
6          $occ :=$ FindOccurrences(M,P_i);
7   **if** $i = 1$ **then** $prevOcc := occ$;
8   **else** $prevOcc :=$ MergeOccurrences($prevOcc$,occ,$Rev_{\{\}}^{|pref(T)|}$);
9   **if** $i \neq k$ **then foreach** occurrence $(first, last)$ in $prevOcc$ **do**
10          $(first, last) := (first, SJT(T)[last])$;
11 **return** $prevOcc$;

At $i = 2$, after Algorithm 3 is executed, $occ = \{((4,5),(6,7),(7,8),(8,9),$
$(10,11),(11,12),(12,13)\}$. Using Algorithm 4 (MergeOccurences), $prevOcc$ is
rewritten to $prevOcc = \{(1,12),(2,9)\}$. Using $SJT(T_1)$, $prevOcc$ is then rewritten to $prevOcc = \{(1,13),(2,10)\}$.

At $i = 3$, algorithm uses $SJT(T_1)$ to rewrite $prevOcc$ to $prevOcc = \{(1,14),$
$(2,11)\}$.

At $i = 4$, $prevOcc$ is not modified because subpattern P_4 is the empty string
and the algorithm returns set of occurrences $\{(1,14),(2,11)\}$.

Algorithm 5 has found two occurrences of tree pattern P'': the first one starting at position 1 (ending at position 14) and the second one at position 2 (ending at position 11) in $pref(T_1)$.

$\square$

Theorem 2. Algorithm 5 (MatchPattern) finds all occurrences $occ^T(P)$ of tree
pattern $P = P_1SP_2S\ldots SP_k$ in tree T. $\square$

5 Time and Space Complexities

Lemma 4. Algorithm 1 (ConstructSubtreeJumpTable) runs in $\mathcal{O}(n)$ time, where
n is the number of nodes of the subject tree T. Size of subtree jump table is n.

Proof. The algorithm is based on a depth-first search traversal of the subject
tree, where at each node only a constant amount work is performed (line 7).
Thus, its running time is bound by the number of nodes n. Counting assignment
operations, the running time is at worst $7n$. $\square$

Theorem 3. Construction of index takes time $\mathcal{O}(n)$ time and produces index
of $\mathcal{O}(n)$ size.

Proof. The creation of compact suffix automaton of size $\mathcal{O}(n)$ [9] and the creation of an array of integers of size n require $\mathcal{O}(n)$ time. Algorithm 1 that creates the subtree jump table is proved to be linear in time and space in Lemma 4. The array Rev is created in time $\mathcal{O}(n)$. $\square$

Lemma 5. Algorithm 4 (MergeOccurrences) runs in $\mathcal{O}(|prevOcc| + |occ|)$ time, where $|prevOcc| + |occ|$ is the number of occurrences in both input sets.

Proof. The algorithm uses array Rev of size n prepared during the indexing phase. This array is used for the fast lookup. The algorithm runs in three loops whose lengths are determined by $|prevOcc| + |occ|$ and at each iteration in each loop, the amount of work is constant. Thus, the total running time holds. Counting assignment operations, the running time is at most $1 + 2|prevOcc| + min(|occ|, |prevOcc|)$. $\square$

Theorem 4. Let $pref(P) = P_1 S P_2 S \ldots S P_k$ of length m be the prefix notation of a tree pattern P. Algorithm 5 (MatchPattern) runs in $\mathcal{O}(m + \sum_{i=1}^{k} |occ'(P_i)|)$ time, where $occ'(P_i) = occ(P_i)$ if $P_i \neq \varepsilon$; otherwise, $occ'(P_i) = occ'(P_{i-1})$.

Proof. Verification of the arity checksum for the pattern takes $\mathcal{O}(m)$ time. Finding the occurrences of subpattern $P_i \neq \varepsilon$ takes time $\mathcal{O}(|P_i| + |occ(P_i)|)$. Summing over all subpatterns yields total time $\mathcal{O}(m + \sum_{i=1, P_i \neq \varepsilon}^{k} |occ(P_i)|)$.

The merging time will be the sum of running times of all calls of Algorithm 4 with input size $\mathcal{O}(|occ(P_i)|)$, $P_i \neq \varepsilon$. Algorithm 4 outputs a list whose size is less than or equal to the minimum of the sizes of the two provided lists of occurrences. Thus, remembering that merging is not performed for $P_i = \varepsilon$, it must hold that the running time of all calls of Algorithm 4 will be less than or equal to $\mathcal{O}(\sum_{i=1, P_i \neq \varepsilon}^{k} (2 * |occ(P_i)|)) = \mathcal{O}(\sum_{i=1}^{k} |occ'(P_i)|)$. $\square$

6   Conclusion

A new and simple method of a full and linear index of a tree for tree patterns has been presented. The presented algorithms can be modified also for unranked trees. The modification is simple and is based on the use of the prefix bar linear notation of the tree [17] instead of the prefix notation.

Acknowledgments. This research has been partially supported by the Czech Science Foundation (GAČR) as project No. GA-13-03253S and by Technology Agency of the Czech Republic (TAČR)) as project No. TA03010964 in α programme.

References

[1] Aho, A.V., Ullman, J.D.: The theory of parsing, translation, and compiling. Prentice-Hall Englewood Cliffs, N.J (1972)

[2] Bille, P.: Pattern Matching in Trees and Strings. PhD thesis.FIT University of Copenhagen, Copenhagen (2008)

[3] Bille, P., Gørtz, I.L., Vildhøj, H.W., Vind, S.: String indexing for patterns with wildcards. In: Fomin, F.V., Kaski, P. (eds.) SWAT 2012. LNCS, vol. 7357, pp. 283–294. Springer, Heidelberg (2012)

[4] Bille, P., Li Gørtz, I., Vildhøj, H.W., Wind, D.K.: String matching with variable length gaps. In: Chavez, E., Lonardi, S. (eds.) SPIRE 2010. LNCS, vol. 6393, pp. 385–394. Springer, Heidelberg (2010)

[5] Blumer, A., Blumer, J., Haussler, D., Ehrenfeucht, A., Chen, M.T., Seiferas, J.I.: The smallest automaton recognizing the subwords of a text. Theor. Comput. Sci. 40, 31–55 (1985)

[6] Cleophas, L.: Tree Algorithms. Two Taxonomies and a Toolkit. PhD thesis.Technische Universiteit Eindhoven, Eindhoven (2008)

[7] Comon, H., Dauchet, M., Gilleron, R., Löding, C., Jacquemard, F., Lugiez, D., Tison, S., Tommasi, M.: Tree automata techniques and applications (release November 18, 2008), http://www.grappa.univ-lille3.fr/tata

[8] Crochemore, M.: Transducers and repetitions. Theor. Comput. Sci. 45(1), 63–86 (1986)

[9] Crochemore, M., Hancart, C., Lecroq, T.: Algorithms on strings. Cambridge Univ. Pr. (2007)

[10] Crochemore, M., Rytter, W.: Text Algorithms. Oxford University Press (1994)

[11] Crochemore, M., Rytter, W.: Jewels of stringology. World Scientific (2002)

[12] Crochemore, M., Vérin, R.: Direct construction of compact directed acyclic word graphs. In: Hein, J., Apostolico, A. (eds.) CPM 1997. LNCS, vol. 1264, pp. 116–129. Springer, Heidelberg (1997)

[13] Ehrenfeucht, A., McConnell, R.M., Osheim, N., Woo, S.-W.: Position heaps: A simple and dynamic text indexing data structure. J. Discrete Algorithms 9(1), 100–121 (2011)

[14] Hoffmann, C.M., O'Donnell, M.J.: Pattern matching in trees. J. ACM 29(1), 68–95 (1982)

[15] Janoušek, J.: Tree indexing by deteministic automata. Dagstuhl reports 3(5), 6 (2013)

[16] Lewenstein, M.: Indexing with gaps. In: Grossi, R., Sebastiani, F., Silvestri, F. (eds.) SPIRE 2011. LNCS, vol. 7024, pp. 135–143. Springer, Heidelberg (2011)

[17] Melichar, B., Janoušek, J., Flouri, T.: Arbology: Trees and pushdown automata. Kybernetika 48(3), 402–428 (2012)

Prefix-Free Languages: Right Quotient and Reversal

Jozef Jirásek[1,*], Galina Jirásková[2,**], Monika Krausová[1], Peter Mlynárčik[2], and Juraj Šebej[1,***]

[1] Institute of Computer Science, Faculty of Science, P.J. Šafárik University,
Jesenná 5, 040 01 Košice, Slovakia
{mon.krausova,juraj.sebej}@gmail.com, jozef.jirasek@upjs.sk
[2] Mathematical Institute, Slovak Academy of Sciences,
Grešákova 6, 040 01 Košice, Slovakia
jiraskov@saske.sk, mlynarcik1972@gmail.com

Abstract. We investigate the right quotient and the reversal operations on the class of prefix-free languages. We get the tight bounds $n - 1$ and $2^{n-2} + 1$ on the state complexity of right quotient and reversal, respectively. To prove the tightness of the bound for reversal, we use a ternary alphabet. Moreover, we prove that this bound cannot be met by any binary language. In the binary case, we get a lower bound $2^{n-2} - 7$ infinitely often. Our calculations show that this lower bound cannot be exceeded.

1 Introduction

A language is prefix-free if it does not contain two distinct strings one of which is a prefix of the other. Prefix-free languages are used in coding theory. In prefix codes, like variable-length Huffman codes or country calling codes, there is no codeword that is a proper prefix of any other codeword. With such a code, a receiver can identify each codeword without any special marker between words.

Motivated by prefix codes, the class of prefix-free regular languages has been recently investigated. It is known that every minimal deterministic automaton recognizing a prefix-free regular language must have exactly one final state, from which all transitions go to a dead state. Using this property, tight bounds on the state complexity of basic operations such as union, intersection, concatenation, star, and reversal have been obtained in [4] and strengthened in [8,10]. The nondeterministic state complexity of basic regular operations has been investigated in [5,8], while the complexity of combined operations on prefix-free regular languages has been studied in [6].

In [7] it has been shown that the tight bound on the state complexity of cyclic shift on prefix-free languages is $(2n-3)^{n-2}$. To prove the tightness of this bound,

---

* Research supported by grants APVV-0035-10 and VEGA 1/0479/12.
** Research supported by grant APVV-0035-10.
*** Research supported by the Slovak Grant Agency for Science under contract VEGA 1/0479/12.

H. Jürgensen et al. (Eds.): DCFS 2014, LNCS 8614, pp. 210–221, 2014.

the authors used a quaternary alphabet, and they proved that this bound cannot be met by any ternary languages. On the other hand, they showed that lower bounds in the binary and ternary cases are still exponential.

In this paper, we investigate the right quotient and the reversal operations on the class of prefix-free languages. In the case of right quotient, we get an upper bound $n-1$ on its state complexity, and we prove that it is tight for an alphabet with at least two symbols. Recall that in the general case of regular languages, the tight bound is n [20].

In the second part of the paper, we study the reversal operation defined as $L^R = \{w^R \mid w \in L\}$, where w^R stands for the string w written backwards. The operation preserves regularity as shown already by Rabin and Scott in 1959 [14]: A nondeterministic finite automaton for the reverse of a regular language can be obtained from an automaton recognizing the given language by swapping the role of initial and final states, and by reversing the transitions. This gives the upper bound 2^n on the state complexity of reversal. Its tightness in the ternary case has been pointed out already by Mirkin [13], who noticed that a ternary Lupanov's witness automaton for determinization [12] is a reverse of a deterministic automaton. The binary witness languages meeting the upper bound 2^n have been presented in [9,11].

In the case of prefix-free languages, the upper bound on the state complexity of reversal is $2^{n-2}+1$, and in the first part of Section 4, we present a simple proof of its tightness in the ternary case. Then, we show that this upper bound cannot be met by any binary language. In the case of binary prefix-free languages, we get a lower bound $2^{n-2}-7$ whenever $n \not\equiv 2 \pmod 3$, and $2^{n-3}-6$ if $n \equiv 2 \pmod 3$.

We also did some calculations. While for some small values of n our lower bounds can be exceeded, starting with $n = 9$, we were not be able to find any language exceeding our lower bounds $2^{n-2}-7$ in the case of $n \not\equiv 2 \pmod 3$. We strongly conjecture that this is also an upper bound. However, we think that it is almost impossible to prove this conjecture.

2   Preliminaries

In this section, we recall some basic definitions and preliminary results. For further details and all unexplained notions, the reader may refer to [17,19].

Let Σ be a finite alphabet and Σ^* the set of all strings over the alphabet Σ including the empty string ε. A language is any subset of Σ^*. The cardinality of a finite set A is denoted by $|A|$, and its power-set by 2^A.

A *nondeterministic finite automaton* (NFA) is a quintuple $A = (Q, \Sigma, \delta, I, F)$, where Q is a finite set of states, Σ is a finite alphabet, $\delta \colon Q \times \Sigma \to 2^Q$ is the transition function which is extended to the domain $2^Q \times \Sigma^*$ in the natural way, $I \subseteq Q$ is the set of initial states, and $F \subseteq Q$ is the set of final states. The *language accepted by* A is the set $L(A) = \{w \in \Sigma^* \mid \delta(I, w) \cap F \neq \emptyset\}$.

An NFA A is *deterministic* (DFA) (and complete) if $|I| = 1$ and $|\delta(q, a)| = 1$ for each q in Q and each a in Σ. In such a case, we write $\delta(q, a) = q'$ instead of $\delta(q, a) = \{q'\}$. A non-final state q is a *dead state* if $\delta(q, a) = q$ for each a in Σ.

The *state complexity* of a regular language L, $\mathrm{sc}(L)$, is the number of states in the minimal DFA for L. It is well known that a DFA is minimal if all its states are reachable from its initial state, and no two of its states are equivalent.

Every NFA $A = (Q, \Sigma, \delta, I, F)$ can be converted to an equivalent DFA $A' = (2^Q, \Sigma, \cdot, I, F')$, where $R \cdot a = \delta(R, a)$ and $F' = \{R \in 2^Q \mid R \cap F \neq \emptyset\}$ [14]. The DFA A' is called the *subset automaton* of the NFA A. The subset automaton need not be minimal since some of its states may be unreachable or equivalent.

The *reverse* w^R *of a string* w is defined as $\varepsilon^R = \varepsilon$ and if $w = va$ for a string v in Σ^* and a symbol a in Σ, then $w^R = av^R$. The *reverse of a language* L is the language $L^R = \{w^R \mid w \in L\}$. The *reverse of a DFA* $A = (Q, \Sigma, \delta, s, F)$ is the NFA A^R obtained from A by reversing all the transitions and by swapping the role of initial and final states, that is, $A^R = (Q, \Sigma, \delta^R, F, \{s\})$, where $\delta^R(q, a) = \{p \in Q \mid \delta(p, a) = q\}$. The reverse of a DFA A recognizes the language $L(A)^R$. For the sake of completeness, we give a short proof of the following fact [1,2,13].

Lemma 1 ([1,2,13]). *Let A be a DFA, in which all states are reachable. Then all the states of the subset automaton of the NFA A^R are pairwise distinguishable.*

Proof. Let q be a state of the NFA A^R. Since q is reachable in A, there is a string w_q that is accepted by A^R from the state q. Moreover, the string w_q cannot be accepted by A^R from any other state because otherwise the automaton A would not be deterministic. Now, two distinct subsets of the subset automaton of the NFA A^R differ in a state q, and the string w_q distinguishes the two subsets. □

If $w = uv$ for some strings u and v, then u is a *prefix* of w. If, moreover, the string v is non-empty, then u is a proper prefix of w.

A language is *prefix-free* if it does not contain two distinct strings one of which is a prefix of the other. It is known that a language L is prefix-free if and only if the minimal DFA for L contains exactly one final state that goes to the dead state on every input symbol [4].

3   Right Quotient on Prefix-Free Languages

Recall that the right quotient of a language L by a string w is the language $L/w = \{x \mid xw \in L\}$, and the right quotient of a language L by a language K is the language $L/K = \bigcup_{w \in K} L/w$.

If a language L is accepted by an n-state DFA $A = (Q, \Sigma, \delta, s, F)$, then the language L/K is accepted by a DFA that is exactly the same as the DFA A except for the set of final states that consists of all the states q of A such that there exists a string w in K with $\delta(q, w) \in F$ [20]. Thus $\mathrm{sc}(L/K) \leq n$. The tightness of this upper bound is shown by taking the language $K = \{\varepsilon\}$ with $\mathrm{sc}(K) = 2$ in [20].

We first show that the tightness can be shown by taking a language K with $\mathrm{sc}(K) = m$ for every m with $m \geq 2$.

Proposition 1. *Let $m, n \geq 2$. There exists binary regular languages K and L with $\mathrm{sc}(K) = m$ and $\mathrm{sc}(L) = n$ such that $\mathrm{sc}(L/K) = n$.*

Proof. Let $K = \{\varepsilon, b^{m-2}\}$ and $L = \{a^{n-2}\}$. Then $\mathrm{sc}(K) = m$, $\mathrm{sc}(L) = n$ and $L/K = L$. □

Now, we consider the right quotient operation on prefix-free languages. Our aim is to show that the tight bound in the prefix-free case is $n - 1$. We start with the upper bound.

Lemma 2. *Let* $m, n \geq 3$. *Let* K *and* L *be prefix-free languages with* $\mathrm{sc}(K) = m$ *and* $\mathrm{sc}(L) = n$. *Then* $\mathrm{sc}(L/K) \leq n - 1$.

Proof. Since L is prefix-free, the minimal DFA for L has a unique final state f that goes to the dead state on every symbol. Since $m \geq 3$ and K is prefix-free, the empty string is not in K. Therefore in the DFA for L/K, the state f is non-final, and it is equivalent to the dead state. This gives the upper bound. □

Our next lemma gives the matching lower bound in the binary case.

Lemma 3. *Let* $m, n \geq 3$. *There exist binary prefix-free languages* K *and* L *with* $\mathrm{sc}(K) = m$ *and* $\mathrm{sc}(L) = n$ *such that* $\mathrm{sc}(L/K) = n - 1$.

Proof. Let $K = \{a, b^{m-2}\}$ and $L = \{a^{n-2}\}$. Then the languages K and L are prefix-free with $\mathrm{sc}(K) = m$ and $\mathrm{sc}(L) = n$. Next we have $L/K = \{a^{n-3}\}$, and so $\mathrm{sc}(L/K) = n - 1$. □

Taking into account also the small values of m and n, we get the following result.

Theorem 1 (Right Quotient on Prefix-Free Languages; $|\Sigma| \geq 2$). *The state complexity of right quotient on prefix-free languages over an alphabet with at least two symbols is given by the function*

$$f(m, n) = \begin{cases} 1, & \text{if } m = 1 \text{ or } n = 1, \\ n, & \text{if } m = 2 \text{ and } n >= 2, \\ n - 1, & \text{otherwise.} \end{cases}$$

Proof. If $m = 1$ or $n = 1$, then K or L must be empty, and the language L/K is empty as well. If $m = 2$, then $K = \{\varepsilon\}$, and therefore $L/K = L$. The tight bound $n - 1$ in the case of $m, n \geq 3$ is given by Lemma 2 and Lemma 3. □

The situation is different in the case of a unary alphabet. Recall that the only unary prefix-free language with state complexity n, where $n \geq 2$, is $\{a^{n-2}\}$.

Theorem 2 (Right Quotient on Prefix-Free Languages; $|\Sigma| = 1$). *The state complexity of right quotient on unary prefix-free languages is given by the function*

$$f(m, n) = \begin{cases} 1, & \text{if } m = 1 \text{ or } m > n, \\ n - m + 2, & \text{otherwise.} \end{cases}$$

Proof. If $m = 1$ or $m > n$, then $L/K = \emptyset$. Otherwise, we have $K = \{a^{m-2}\}$ and $L = \{a^{n-2}\}$. Thus $L/K = \{a^{n-m}\}$ and $\mathrm{sc}(L/K) = n - m + 2$. □

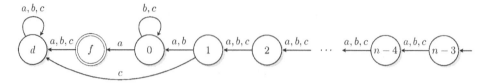

Fig. 1. A DFA of a ternary prefix-free language meeting the bound $2^{n-2}+1$ for reversal

4   Reversal on Prefix-Free Languages

Recall that a minimal DFA A accepting a prefix-free language L has exactly one final state f that goes to the dead state on every symbol of the input alphabet. To construct an NFA for the language L^R from the DFA A, omit the dead state, make the state f initial, make the initial state of A final, and reverse all the transitions. In the resulting NFA, no transition goes to the state f, and therefore, in the corresponding subset automaton, no subset containing f is reachable, except for $\{f\}$. This gives an upper bound $2^{n-2}+1$ on the state complexity of L^R [4].

To meet this upper bound it is enough to take a language $L_{n-2}c$, where L_{n-2} is the language over $\{a, b\}$ accepted by an $(n-2)$-state DFA meeting the upper bound 2^{n-2} [9,11,16]. However, having a ternary alphabet, we can provide a witness language with a very simple proof of reachability of $2^{n-2}+1$ subsets.

Lemma 4. *Let $n \geq 3$. There exists a ternary prefix-free regular language L with $\mathrm{sc}(L) = n$ such that $\mathrm{sc}(L^R) = 2^{n-2}+1$.*

Proof. Let L be the ternary prefix-free language accepted by the DFA shown in Fig. 1. Construct an NFA for the language L^R as shown in Fig. 2, and consider the corresponding subset automaton. The initial state of the subset automaton is $\{f\}$, and it goes by a^{i+1} to the singleton set $\{i\}$ with $0 \leq i \leq n-3$, and by b to the empty set. Let $1 \leq k \leq n-3$ and assume that each subset of $\{0, 1, \ldots, n-3\}$ of size k is reachable. Let $S = \{i_0, i_1, \ldots, i_k\}$ be a subset of $\{0, 1, \ldots, n-3\}$ with $0 \leq i_0 < i_1 < \cdots < i_k \leq n-3$. Then the set $\{0, i_2 - i_1, i_3 - i_1 \ldots, i_k - i_1\}$ is reachable by the induction hypothesis, and it goes to S by the string $bc^{i_1-i_0-1}a^{i_0}$. This proves the reachability of $2^{n-2}+1$ subsets. By Lemma 1, all these subsets are pairwise distinguishable, which completes the proof.   □

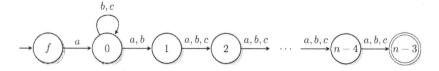

Fig. 2. The reverse of the DFA from Fig. 1

4.1   Binary Case

The aim of this subsection is to show that the upper bound $2^{n-2} + 1$ on the state complexity of reversal cannot be met by any binary prefix-free language. On the other hand, we show that the bound $2^{n-2} - 7$ can be met by a binary prefix-free language infinitely many often. Let us start with the upper bound.

Lemma 5. *Let $n \geq 5$. Let L be a binary prefix-free language with $sc(L) = n$. Then $sc(L^R) < 2^{n-2} + 1$.*

Proof. Let A be a minimal DFA over $\{a, b\}$ for a prefix-free language L with the state set $\{1, 2, \ldots, n-2, f, d\}$, of which f is a unique final state and d is the dead state. Since automaton A is deterministic and complete, each of its states has exactly two out-transitions; one on symbol a and the other on symbol b. It means that in the corresponding reversed automaton A^R, each state has just two in-transitions, one on a and the other on b. This is a very important fact that we will use in the next considerations.

Since the language L is prefix-free, the DFA A has just one final state f and one dead state, and moreover, both transitions from the final state go into the dead state. Therefore, in the reversed automaton A^R, the dead state is unreachable, and we can omit it. The initial state of A^R is f and no in-transitions go to f. It follows that in the subset automaton of the NFA A^R, no subset containing state f is reachable, except for the initial set $\{f\}$. Let δ^R be the transition function of the NFA A^R.

(1) First, assume that there is a state q that is reached from f by both a and b, that is, $q \in \delta^R(f, a) \cap \delta^R(f, b)$. It follows that no other transitions can go to state q. Therefore, just two sets containing the state q are reachable in the subset automaton, namely, $\delta^R(f, a)$ and $\delta^R(f, b)$. So there are at least $2^{n-3} - 2$ unreachable sets. Since $n \geq 5$, at least two sets are unreachable in this case.

(2) Now, assume that no state of A^R is reached from f by both a and b. Without loss of generality, we may assume that $\delta^R(f, a)$ is non-empty.

(2.1) If we have $\delta^R(f, a) = \{1, 2, \ldots, n-2\}$, then no other transition on a may go to each of the states in $\{1, 2, \ldots, n-2\}$. On the other hand, each of these states must have an in-transition on b from one of these states. Therefore, in the subset automaton, only the sets $\{f\}$, $\{1, 2, \ldots, n-2\}$, and $\emptyset$ are reachable.

(2.2) Next, assume that $\delta^R(f, a) \neq \{1, 2, \ldots, n-2\}$, and, without loss of generality, let 1 be a state in $\delta^R(f, a)$. Since 1 must also have an in-transition on b, there must be a state i in $\{1, 2, \ldots, n-2\}$ and such that i goes to 1 by b.

(2.2.1) If i goes also to some other state j by b, that is $\delta^R(i, b) \supseteq \{1, j\}$, then every reachable subset containing 1 either contains also j, or is equal to $\delta^R(f, a)$. This gives at most $2^{n-4} + 1$ subsets containing state 1, and therefore, at least $2^{n-4} - 1$ subsets are unreachable. Since $n \geq 5$, at least one set is unreachable.

(2.2.2) Let i go only to 1 by b, that is $\delta^R(i, b) = \{1\}$. To meet the upper bound, all two-element subsets $\{1, 2\}, \{1, 3\}, \ldots, \{1, n-2\}$ must be reached.

(2.2.2.1) First, assume that none of them is reached from $\{f\}$ by a. Then, the set $\{1, 2\}$ must be reached by b from a set containing states i and j_2, where j_2 is a state in $\{1, 2, \ldots, n-2\}$ such that j_2 goes only to 2 by b in A^R. Thus, to reach all the above mentioned two-element sets, there must be pairwise distinct states $j_2, j_3, \ldots, j_{n-2}$, all different from i, and such that $\delta^R(j_\ell, b) = \{\ell\}$. However, then the set $\{1, 2, \ldots, n-2\}$ goes to itself by b, and since b is a permutation on $\{1, 2, \ldots, n-2\}$, the set $\{1, 2, \ldots, n-2\}$ cannot be reached by b from any other set. Since it could be reached by a only from $\{f\}$, and we assumed that it is not, this set is unreachable in this case.

(2.2.2.2) Now assume that one of these two-element subsets is reached from f by a, and without loss of generality, let it be the set $\{1, 2\}$. Then we must be able to reached all subsets $\{1, 3\}, \{1, 4\}, \ldots, \{1, n-2\}$, as well as the set $\{1, 2, 3\}$. To do this, similarly as above, there must be pairwise distinct states $j_3, j_4, \ldots, j_{n-2}$ with $\delta(j_\ell, b) = \{\ell\}$. Then, to reach $\{1, 2, 3\}$, there must be another state j_2 with $\delta(j_2, b) = \{2\}$. In the same way as above, we show that the set $\{1, 2, \ldots, n-2\}$ is unreachable.

Hence we have shown that in each case, some sets are unreachable, and the lemma follows. □

To get a lower bound, we could use a similar method as described in [3]. We could encode symbols a, b, c by $00, 01, 11$, respectively, and construct a $2n$-state prefix-free DFA over $\{0, 1\}$ for the language $L_{n-2} c$, where L_{n-2} is an $(n-2)$-state Šebej's binary automaton [9] meeting the upper bound 2^{n-2} for reversal. The corresponding subset automaton for reversal will have at least $2^{n-2}+1$ reachable states. This would result in a lower bound $2^{n/2}$ on the state complexity of reversal of binary prefix-free languages.

However, in some cases we can do much better. In the next part of this subsection, we show that the bound $2^{n-2} - 7$ for reversal can be met by a binary prefix-free language infinitely many often.

First, let us recall the well-known Chinese Remainder Theorem [18, p. 130]: If $m_1, m_2, \ldots, m_n$ are pairwise relatively prime and greater than 1, and $a_1, a_2, \ldots, a_n$ are integers, then there is a solution x to the following simultaneous congruences:

$$x \equiv a_1 \pmod{m_1},$$
$$x \equiv a_2 \pmod{m_2},$$
$$\cdots$$
$$x \equiv a_n \pmod{m_n}.$$

Now, consider the n-state NFA show in Fig. 3, in which the transitions are defined as follows:

- on a, there are two cycles (q_0, q_1, q_2) and $(0, 1, \ldots, n-4)$;
- on b, there is a cycle $(q_2, 0)$, state q_0 goes to $\{q_0, q_1\}$, each state i with $1 \le i \le n-5$ goes to $\{i\}$, and states q_1 and $n-4$ go to the empty set.

Let us prove the following observation.

Lemma 6. *Let $n \ge 5$. If $n \not\equiv 0 \pmod{3}$, then the subset automaton of the NFA shown in Fig. 3 has $2^n - 8$ reachable subsets.*

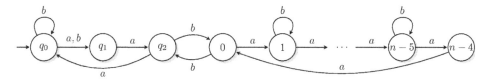

Fig. 3. A binary n-state NFA with $2^n - 8$ reachable subsets in the corresponding subset automaton

Proof. Since $n \not\equiv 0 \pmod{3}$, the numbers 3 and $n-3$ are relatively prime, and the Chinese Remainder Theorem guarantees that there always exists an integer $x(i, j)$ such that

$$x(i, j) \equiv i \pmod{3}, \tag{1}$$
$$x(i, j) \equiv j \pmod{(n-3)}. \tag{2}$$

We will use this result several times in our proof.

The proof will have four parts, in which we show that the following subsets are reachable in the subset automaton:

(a) all the subsets of $\{q_0, q_1, q_2\}$;
(b) all the subsets $\{q_0, q_1\} \cup S$ with $S \subseteq \{0, 1, \ldots, n-5\}$;
(c) all the subsets $R \cup S$ with $R \subseteq \{q_0, q_1, q_2\}$ and $S \subseteq \{0, 1, \ldots, n-5\}$;
(d) all the subsets $R \cup S$ with $R \subseteq \{q_0, q_1, q_2\}$ and $S \subsetneq \{0, 1, \ldots, n-4\}$.

Thus, only eight subsets $R \cup \{0, 1, \ldots, n-4\}$ with $R \subseteq \{q_0, q_1, q_2\}$ will be unreachable, and the lemma will follow.

(a) The initial subset is $\{q_0\}$. From this set, we can reach subsets $\{q_1\}$ and $\{q_2\}$ by a and aa, respectively. By b, the set $\{q_0\}$ goes to $\{q_0, q_1\}$, and we again use the strings a and aa to reach $\{q_1, q_2\}$ and $\{q_0, q_2\}$. The empty set is reached from $\{q_1\}$ by b, and the set $\{q_0, q_1, q_2\}$ is reached from $\{q_0, q_2\}$ by bb.

(b) We prove the reachability of subsets $\{q_0, q_1\} \cup S$ with $S \subseteq \{0, 1, \ldots, n-5\}$ by induction on the size of S. The basis, with $|S| = 0$, is proved in case (a). Let $S = \{i_1, i_2, \ldots, i_k\}$, where $1 \leq k \leq n-4$ and $0 \leq i_1 < i_2 < \cdots < i_k \leq n-5$, be a set of size k. Then the set $\{q_0, q_1\} \cup \{i_2 - i_1, \ldots, i_k - i_1\}$ is reachable by the induction hypothesis, and since we have

$$\{q_0, q_1\} \cup \{i_2 - i_1, \ldots, i_k - i_1\} \xrightarrow{a^{x(2,0)}} \{q_0, q_2\} \cup \{i_2 - i_1, \ldots, i_k - i_1\} \xrightarrow{b}$$
$$\{q_0, q_1\} \cup \{0, i_2 - i_1, \ldots, i_k - i_1\} \xrightarrow{a^{x(0,i_1)}} \{q_0, q_1\} \cup \{i_1, i_2, \ldots, i_k\},$$

the set $\{q_0, q_1\} \cup S$ is reachable; remind that $x(i, j)$ is a solution of (1) and (2). This completes the second part.

(c) Each set $\{q_0, q_1\} \cup S$ goes to the set $\{q_1, q_2\} \cup S$ by the string $a^{x(1,0)}$, and it goes to $\{q_0, q_2\} \cup S$ by $a^{x(2,0)}$. Next, we have

$$\{q_0, q_2\} \cup S \xrightarrow{bb} \{q_0, q_1, q_2\} \cup S, \text{ and}$$
$$\{q_1, q_2\} \cup S \xrightarrow{bb} \{q_2\} \cup S \xrightarrow{a^{x(1,0)}} \{q_0\} \cup S \xrightarrow{a^{x(1,0)}} \{q_1\} \cup S \xrightarrow{bb} S,$$

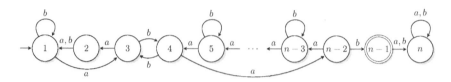

Fig. 4. The prefix-free DFA meeting the bound $2^{n-2} - 7$ for reversal; $n \not\equiv 2 \pmod 3$

which completes the third part.

(d) Let $R \subseteq \{q_0, q_1, q_2\}$ and $S \subsetneq \{0, 1, \ldots, n-4\}$. If $n-4 \notin S$, then $R \cup S$ is reachable as shown in (a)-(c). Let $n-4 \in S$. Since S is a proper subset of $\{0, 1, \ldots, n-4\}$, there is a state i with $i \notin S$. Let $S' = \{s + (n-4-i) \mid s \in S\}$. Then $n-4 \notin S'$, so the set $R \cup S'$ is reachable as shown in (a)-(c). The set $R \cup S'$ goes to the set $R \cup S$ by the string $a^{x(0,i+1)}$. This proves the fourth part.

Now let $S = \{0, 1, \ldots, n-4\}$. Then no set $R \cup S$ can be reached from any set by symbol b since no b goes to the state $n-4$. However, the symbol a is a permutation on the state set $\{q_0, q_1, q_2\} \cup \{0, 1, \ldots, n-4\}$, and in the subset automaton we have three cycles

$$\{q_0, q_1, q_2\} \cup S \xrightarrow{a} \{q_0, q_1, q_2\} \cup S,$$

$$\{q_0, q_1\} \cup S \xrightarrow{a} \{q_1, q_2\} \cup S \xrightarrow{a} \{q_0, q_2\} \cup S \xrightarrow{a} \{q_0, q_1\} \cup S, \text{ and}$$

$$\{q_0\} \cup S \xrightarrow{a} \{q_1\} \cup S \xrightarrow{a} \{q_2\} \cup S \xrightarrow{a} \{q_0\} \cup S.$$

No subset on these cycles can be reached from any other subset on b since all these subsets contain $n-4$. Since a is a permutation symbol, the sets on the cycles cannot be reached on a from any set outside the cycles. Therefore, the eight sets $R \cup S$ with $R \subseteq \{q_0, q_1, q_2\}$ and $S = \{0, 1, \ldots, n-4\}$ are unreachable. This concludes the proof. □

Using the above result, we can get a lower bound $2^{n-2} - 7$ on the state complexity of the reversal on binary prefix-free languages infinitely many times.

Lemma 7. *Let $n \geq 5$ and $n \not\equiv 2 \pmod 3$. There exists a binary prefix-free language L with $\mathrm{sc}(L) = n$ such that $\mathrm{sc}(L^R) = 2^{n-2} - 7$.*

Proof. Consider the binary prefix-free language accepted by the DFA A shown in Fig. 4. In the subset automaton of the NFA A^R, the initial subset is $\{n-1\}$, and it goes to $\{1\}$ by *baba*. Since we have $(n-2) \not\equiv 0 \pmod 3$, by Lemma 6, from the set $\{1\}$ we can reach $2^{n-2} - 8$ subsets of $\{1, 2, \ldots, n-2\}$ in the subset automaton of the NFA A^R. This gives $2^{n-2} - 7$ reachable subsets. By Lemma 1, all these subsets are pairwise distinguishable, and the lemma follows. □

In the next lemma, we consider the remaining cases of n.

Lemma 8. *Let $n \geq 5$ and $n \equiv 2 \pmod 3$. There exists a binary prefix-free language L with $\mathrm{sc}(L) = n$ such that $\mathrm{sc}(L^R) = 2^{n-3} - 6$.*

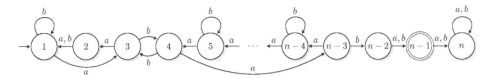

Fig. 5. The prefix-free DFA meeting the bound $2^{n-3} - 6$ for reversal; $n \equiv 2 \pmod 3$

Proof. This time, we consider the binary prefix-free language accepted by the DFA shown in Fig. 5. Since we have $(n - 3) \not\equiv 0 \pmod 3$, the lower bound $2^{n-3} - 6$ again follows from Lemma 6. $\qquad\square$

Let us summarize the results of this subsection in the following theorem.

Theorem 3 (Reversal on Binary Prefix-Free Languages). *Let $n \geq 5$. Let L be a binary prefix-free language with $\mathrm{sc}(L) = n$. Then $\mathrm{sc}(L^R) < 2^{n-2} + 1$. The bound $2^{n-2} - 7$ can be met whenever $n \not\equiv 2 \pmod 3$, and the bound $2^{n-3} - 6$ can be met whenever $n \equiv 2 \pmod 3$.* $\qquad\square$

Hence, although the upper bound $2^{n-2} - 1$ cannot be met by binary prefix-free languages, the lower bound is smaller just by eight if $n \not\equiv 2 \pmod 3$, and it is still exponential in $2^{n-O(1)}$ in the remaining cases.

4.2 Calculations

We did some calculations concerning the reversal operation on binary prefix-free languages. Our result are summarized in Table 1.

Table 1. Calculations: Reversal on binary prefix-free languages

n	DFA A	DFA B	max	
5	8	8		8
6	$9 = 2^4 - 7$	14		14
7	$25 = 2^5 - 7$	27		27
8	21	59		59
9	$121 = 2^7 - 7$	121		121
10	$249 = 2^8 - 7$	246	$\geq$	249
11	307	500	$\geq$	500
12	$1017 = 2^{10} - 7$	1007	$\geq$	1017
13	$2041 = 2^{11} - 7$	2026	$\geq$	2041
14	3207	4067	$\geq$	4081
15	$8185 = 2^{13} - 7$	8153	$\geq$	8185
16	$16377 = 2^{14} - 7$	16333	$\geq$	16377
17	29071	32700	$\geq$	32700

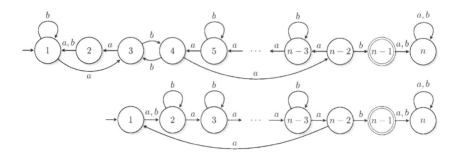

Fig. 6. The DFA A given by Lemma 7 (top), and the DFA B (bottom)

Notice that while for small values of n with $n \not\equiv 2 \pmod 3$, our lower bound $2^{n-2} - 7$ given by Lemma 7 can be exceeded, starting with $n = 9$ we were not be able to find a binary prefix-free language exceeding this lower bound. We strongly conjecture that this is also an upper bound, although, to prove this seems to be almost impossible.

For the values of n with $n \equiv 2 \pmod 3$, the binary language recognized by the DFA B shown in Fig. 6 exceeds our lower bound $2^{n-3} - 6$, however, we are not able to get the state complexity of its reversal for an arbitrary n. Next, notice that even the language recognized by the DFA A in Fig. 6 exceeds our lower bound $2^{n-3} - 6$ that we have obtained in Lemma 8 by using a modification of the DFA A.

5   Conclusions

We investigated the right quotient and the reversal operations on the class of prefix-free languages. We have obtained the upper bound $n - 1$ on the state complexity of right quotient, and we proved that it is tight in the binary case.

In the case of reversal on prefix-free languages, we showed that the upper bound $2^{n-2} + 1$ is tight in the ternary case. Moreover, we proved that this bound cannot be met by any binary prefix-free language. In the binary case, we obtained the lower bound $2^{n-2} - 7$ whenever $n \not\equiv 2 \pmod 3$. Our calculations show that this could also be an upper bound, however, to prove this seems to be almost impossible.

References

1. Brzozowski, J.A.: Canonical regular expressions and minimal state graphs for definite events. In: Fox, J. (ed.) Proceedings of the Symposium on Mathematical Theory of Automata, New York, NY, April 24-26. MRI Symposia Series, vol. 12, pp. 529–561. Polytechnic Press of the Polytechnic Institute of Brooklyn, Brooklyn (1963)
2. Champarnaud, J.-M., Khorsi, A., Paranthoën, T.: Split and join for minimizing: Brzozowski's algorithm, http://jmc.feydakins.org/ps/c09psc02.ps

3. Cmorik, R., Jirásková, G.: Basic operations on binary suffix-free languages. In: Kotásek, Z., Bouda, J., Černá, I., Sekanina, L., Vojnar, T., Antoš, D. (eds.) MEMICS 2011. LNCS, vol. 7119, pp. 94–102. Springer, Heidelberg (2012)
4. Han, Y.-S., Salomaa, K., Wood, D.: Operational state complexity of prefix-free regular languages. In: Automata, Formal Languages, and Related Topics, pp. 99–115. Institute of Informatics, University of Szeged (2009)
5. Han, Y.-S., Salomaa, K., Wood, D.: Nondeterministic state complexity of basic operations for prefix-free regular languages. Fund. Inform. 90, 93–106 (2009)
6. Han, Y.-S., Salomaa, K., Yu, S.: State complexity of combined operations for prefix-free regular languages. In: Dediu, A.H., Ionescu, A.M., Martín-Vide, C. (eds.) LATA 2009. LNCS, vol. 5457, pp. 398–409. Springer, Heidelberg (2009)
7. Jirásek, J., Jirásková, G.: Cyclic shift on prefix-free languages. In: Bulatov, A.A., Shur, A.M. (eds.) CSR 2013. LNCS, vol. 7913, pp. 246–257. Springer, Heidelberg (2013)
8. Jirásková, G., Krausová, M.: Complexity in prefix-free regular languages. In: McQuillan, I., Pighizzini, G., Trost, B. (eds.) Proc. 12th DCFS, pp. 236–244.
9. Jirásková, G., Šebej, J.: Reversal of binary regular languages. Theoret. Comput. Sci. 449, 85–92 (2012)
10. Krausová, M.: Prefix-free regular languages: Closure properties, difference, and left quotient. In: Kotásek, Z., Bouda, J., Černá, I., Sekanina, L., Vojnar, T., Antoš, D. (eds.) MEMICS 2011. LNCS, vol. 7119, pp. 114–122. Springer, Heidelberg (2012)
11. Leiss, E.: Succinct representation of regular languages by Boolean automata. Theoret. Comput. Sci. 13, 323–330 (1981)
12. Lupanov, U.I.: A comparison of two types of finite automata. Problemy Kibernetiki 9, 321–326 (1963) (in Russian)
13. Mirkin, B.G.: On dual automata. Kibernetika (Kiev) 2, 7–10 (1966) (in Russian); English translation: Cybernetics 2, 6–9 (1966)
14. Rabin, M., Scott, D.: Finite automata and their decision problems. IBM Res. Develop. 3, 114–129 (1959)
15. Salomaa, A., Wood, D., Yu, S.: On the state complexity of reversals of regular languages. Theoret. Comput. Sci. 320, 315–329 (2004)
16. Šebej, J.: Reversal of regular languages and state complexity. In: Pardubská, D. (ed.) Proc. 10th ITAT, pp. 47–54. Šafárik University, Košice (2010)
17. Sipser, M.: Introduction to the theory of computation. PWS Publishing Company, Boston (1997)
18. Yan, S.Y.: Number theory for computing, 2nd edn. Springer (2002)
19. Yu, S.: Regular languages. In: Rozenberg, G., Salomaa, A. (eds.) Handbook of Formal Languages, vol. I, ch. 2, pp. 41–110. Springer, Heidelberg (1997)
20. Yu, S., Zhuang, Q., Salomaa, K.: The state complexity of some basic operations on regular languages. Theoret. Comput. Sci. 125, 315–328 (1994)

Complement on Prefix-Free, Suffix-Free, and Non-Returning NFA Languages

Galina Jiráskovᇠand Peter Mlynárčik

Mathematical Institute, Slovak Academy of Sciences
Grešákova 6, 040 01 Košice, Slovakia
jiraskov@saske.sk, mlynarcik1972@gmail.com

Abstract. We prove that the tight bound on the nondeterministic state complexity of complementation on prefix-free and suffix-free languages is 2^{n-1}. To prove tightness, we use a ternary alphabet, and we show that this bound cannot be met by any binary prefix-free language. On non-returning languages, the upper bound is $2^{n-1} + 1$, and it is tight already in the binary case. We also study the unary case in all three classes.

1 Introduction

The complement of a formal language L over an alphabet Σ is the language $L^c = \Sigma^* \setminus L$, where Σ^* is the set of all strings over an alphabet Σ. The complementation is an easy operation on regular languages represented by deterministic finite automata (DFAs) since to get a DFA for the complement of a regular language, it is enough to interchange the final and non-final states in a DFA for this language.

On the other hand, complementation on regular languages represented by nondeterministic finite automata (NFAs) is an expensive task. First, we have to apply the subset construction to a given NFA, and only after that, we may interchange the final and non-final states. This gives an upper bound 2^n.

Sakoda and Sipser [17] gave an example of languages over a growing alphabet size meeting this upper bound on the nondeterministic state complexity of complementation. Birget claimed the result for a three-letter alphabet [3], but later corrected this to a four-letter alphabet. Ellul [7] gave binary $O(n)$-state witness languages. Holzer and Kutrib [12] proved the lower bound 2^{n-2} for a binary n-state NFA language. Finally, a binary n-state NFA language meeting the upper bound 2^n was described by Jirásková in [15]. In the unary case, the complexity of complementation is known to be in $e^{\Theta(\sqrt{n \ln n})}$ [12, 14].

In this paper, we investigate the complementation operation on prefix-free, suffix-free, and non-returning languages. A language is prefix-free if it does not contain two distinct strings one of which is a prefix of the other. The suffix-free languages are defined in a similar way. We call a language non-returning if a minimal NFA for this language does not have any transitions going to the initial state.

---

* Research supported by grant APVV-0035-10.

H. Jürgensen et al. (Eds.): DCFS 2014, LNCS 8614, pp. 222–233, 2014.

Prefix-free languages are used in coding theory. In prefix codes, like variable-length Huffman codes or country calling codes, there is no codeword that is a proper prefix of any other codeword. With such a code, a receiver can identify each codeword without any special marker between words.

The complexity of basic regular operations on prefix-free and suffix-free languages, both in the deterministic and nondeterministic cases, was studied by Han *at al.* in [8–11]. For the nondeterministic state complexity of complementation, they obtained an upper bound $2^{n-1} + 1$ in both classes, and lower bounds 2^{n-1} and $2^{n-1} - 1$ for prefix-free and suffix-free languages, respectively. The question of a tight bound remained open. In this paper, we solve this open question, and prove that in both classes, the tight bound is 2^{n-1}. To prove tightness, we use a ternary alphabet, and in the case of prefix-free languages, we show that this bound cannot be met by any binary language.

Eom *et al.* in [6] investigated also the class of so called non-returning regular languages, the minimal DFA for which has no transitions going to the initial state. It is known that every suffix-free language is non-returning, but the converse does not hold. Here we study the complementation on so called non-returning NFA languages, defined as languages represented by a minimal non-returning NFA. We show that the upper bound on the complexity of complementation in this class is $2^{n-1} + 1$, and we prove that it is tight already in the binary case.

We also study the unary case, and prove that the nondeterministic state complexity of complementation is in $\Theta(\sqrt{n})$ in the class of prefix-free or suffix-free languages, and it is in $2^{\Theta(\sqrt{n \log n})}$ in the class of non-returning NFA languages.

To prove the minimality of nondeterministic finite automata, we use a fooling set lower-bound technique [1, 3, 5, 13].

Definition 1. *A set of pairs of strings* $\{(x_1, y_1), (x_2, y_2), \ldots, (x_n, y_n)\}$ *is called a fooling set for a language* L *if for all* i, j *in* $\{1, 2, \ldots, n\}$,

(F1) $x_i y_i \in L$, *and*

(F2) *if* $i \neq j$, *then* $x_i y_j \notin L$ *or* $x_j y_i \notin L$.

Lemma 1 ([3, 5, 13]). *Let* $\mathcal{F}$ *be a fooling set for a language* L. *Then every NFA (with multiple initial states) for the language* L *has at least* $|\mathcal{F}|$ *states.* □

2 Complement on Prefix-Free Languages

Let us start with complementation on prefix-free languages. The following two observations are easy to prove.

Proposition 1 ([9]). *Let* $n \geq 2$ *and* $A = (Q, \Sigma, \delta, s, F)$ *be a minimal n-state DFA for a language* L. *Then* L *is prefix-free if and only if* A *has a dead state* q_d *and exactly one final state* q_f *such that* $\delta(q_f, a) = q_d$ *for each* a *in* Σ. □

Proposition 2 ([10]). *Let* $N = (Q, \Sigma, \delta, s, F)$ *be a minimal NFA for a prefix-free language. Then* N *has exactly one final state* q_f, *and* $\delta(q_f, a) = \emptyset$ *for each* a *in* Σ. □

Han *et al.* in [10] obtained an upper bound $2^{n-1} + 1$ and a lower bound 2^{n-1} on the nondeterministic complexity of complementation on prefix-free languages. Our first result shows that the upper bound can be decreased by one. Recall that the nondeterministic state complexity of a regular language L, $\mathrm{nsc}(L)$, is defined as the smallest number of states in any NFA recognizing the language L.

Lemma 2. *Let $n \geq 3$. Let L be a prefix-free regular language with $\mathrm{nsc}(L) = n$. Then $\mathrm{nsc}(L^c) \leq 2^{n-1}$.*

Proof. Let N be an n-state NFA for a prefix-free language L. Construct the subset automaton of the NFA N and minimize it. Then, all the final states are equivalent, and they go to the dead state on each input. Thus L is accepted by a DFA $A = (Q, \Sigma, \delta, s, \{q_f\})$ with at most $2^{n-1} + 1$ states, with a dead state q_d which goes to itself on each symbol, and one final state q_f which goes to the dead state on each symbol, thus $\delta(q_d, a) = q_d$ and $\delta(q_f, a) = q_d$ for each a in Σ.

To get a DFA for the language L^c, we interchange the final and non-final states in the DFA A, thus L^c is accepted by the $(2^{n-1} + 1)$-state DFA $A^c = (Q, \Sigma, \delta, s, Q \setminus \{q_f\})$. We show that using nondeterminism, we can save one state, that is, we describe a 2^{n-1}-state NFA for the language L^c.

Construct a 2^{n-1}-state NFA N^c for L^c from the DFA A^c by omitting state q_d, and by replacing each transition (q, a, q_d) by two transitions (q, a, q_f) and (q, a, s). Formally, construct an NFA $N^c = (Q \setminus \{q_d\}, \Sigma, \delta', s, Q \setminus \{q_f, q_d\})$, where

$$\delta'(q, a) = \begin{cases} \{\delta(q, a)\}, & \text{if } \delta(q, a) \neq q_d, \\ \{q_f, s\}, & \text{if } \delta(q, a) = q_d. \end{cases}$$

Let us show that $L(N^c) = L(A^c)$.

Let $w = a_1 a_2 \cdots a_k$ be a string in $L(A^c)$, and let $s, q_1, q_2, \ldots, q_k$ be the computation of the DFA A^c on the string w. If $q_k \neq q_d$, then each q_i is different from q_d since q_d goes to itself on each symbol. It follows that $s, q_1, q_2, \ldots, q_k$ is also a computation of the NFA N^c on the string w. Now assume that $q_k = q_d$. Then there exists an ℓ such that the states $q_\ell, q_{\ell+1}, \ldots, q_k$ are equal to q_d, and the states $s, q_1, \ldots, q_{\ell-1}$ are not equal to q_d. If $\ell = k$, then $\delta(q_{k-1}, a_k) = q_d$, so $s \in \delta'(q_{k-1}, a_k)$. It follows that $s, q_1, q_2, \ldots, q_{k-1}, s$ is an accepting computation of N^c on w. If $\ell < k$, then we have $q_\ell = q_{\ell+1} = \cdots = q_k = q_d$, and therefore the string w is accepted in N^c through the accepting computation $s, q_1, \ldots, q_{\ell-1}, q_f, q_f, \ldots, q_f, s$ since we have $\delta'(q_{\ell-1}, a_\ell) = \{q_f, s\}$, and $\delta'(q_f, a) = \{q_f, s\}$ for each a in Σ.

Now assume that a string $w = a_1 a_2 \cdots a_k$ is rejected by the DFA A^c. Let $s = q_0, q_1, q_2, \ldots, q_k$ be the rejecting computation of the DFA A^c on the string w. Since the only non-final state of the DFA A^c is q_f, we must have $q_k = q_f$. It follows that each state q_i is different from q_d, and therefore in the NFA N^c, we have $\delta'(q_{i-1}, a_i) = \{\delta(q_{i-1}, a_i)\}$. This means that $s = q_0, q_1, q_2, \ldots, q_k$ is a unique computation of N^c on w. Since this computation is rejecting, the string w is rejected by the NFA N^c. $\qquad\square$

To prove tightness, we use the same languages as in [10]. We provide a simple alternative proof, in which we use a fooling-set lower bound technique.

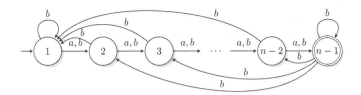

Fig. 1. An NFA of a binary regular language K with $\mathrm{nsc}(K^c) = 2^{n-1}$

Lemma 3. *Let $n \geq 3$. There exists a ternary prefix-free language such that* $\mathrm{nsc}(L) = n$ *and* $\mathrm{nsc}(L^c) \geq 2^{n-1}$.

Proof. Let K be the language accepted by the NFA over $\{a, b\}$ shown in Fig. 1 with $n - 1$ states. Set $L = K \cdot c$. Then L is a prefix-free language recognized by an n-state NFA in Fig. 2. As shown in [15, Theorem 5], there exists a fooling set $\mathcal{F} = \{(x_S, y_S) \mid S \subseteq \{1, 2, \ldots, n - 1\}\}$ of size 2^{n-1} for the language K^c. Then the set of pairs of strings $\mathcal{F}' = \{(x_S, y_S \cdot c) \mid S \subseteq \{1, 2, \ldots, n - 1\}\}$ is a fooling set of size 2^{n-1} for the language L^c. Hence, by Lemma 1, every NFA for the language L^c requires at least 2^{n-1} states.   □

We summarize the results given in Lemma 2 and Lemma 3 in the following theorem which provides the tight bound on the nondeterministic state complexity of complementation on prefix-free languages. This solves an open problem from [10].

Theorem 1 (Complement on Prefix-Free Languages, $|\Sigma| \geq 3$). *Let $n \geq 3$. Let L be a prefix-free regular language over an alphabet Σ with $\mathrm{nsc}(L) = n$. Then $\mathrm{nsc}(L^c) \leq 2^{n-1}$, and the bound is tight if $|\Sigma| \geq 3$.*   □

Notice that the bound 2^{n-1} is tight for an alphabet with at least three symbols. In Section 6, we prove that this bound cannot be met by any binary prefix-free language.

3   Complement on Suffix-Free Languages

In this section, we study the complementation operation on suffix-free languages. We first recall some definitions and known facts.

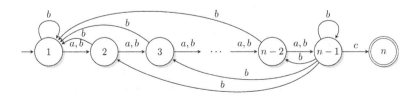

Fig. 2. An NFA of a ternary prefix-free language L with $\mathrm{nsc}(L^c) = 2^{n-1}$

An automaton $A = (Q, \Sigma, \delta, s, F)$ is *non-returning* if the initial state s has no in-transitions, that is, for each state q and each symbol a, we have $s \notin \delta(q, a)$.

Proposition 3 ([8, 11]). *Every minimal DFA (NFA) for a non-empty suffix-free language is non-returning.*

Proposition 4. *Let $A = (Q, \Sigma, \delta, s, F)$ be a minimal DFA for a non-empty suffix-free regular language. Then A has a dead state d. Moreover, for each symbol a in Σ, there is a state q_a with $q_a \neq d$ such that $\delta(q_a, a) = d$.*

Proof. Let $a \in \Sigma$. Consider the string a^m with $m \geq |Q|$. We must have $\delta(s, a^m) = d$, where d is a dead state, because otherwise, the DFA A would accept strings $a^m w$ and $a^\ell w$ with $\ell < m$, which would be a contradiction with suffix-freeness of $L(A)$. Since $s \neq d$, there is a state q_a with $q_a \neq d$ such that $\delta(q_a, a) = d$.   □

Han and Salomaa in [11] have obtained an upper bound $2^{n-1} + 1$ on the nondeterministic state complexity of complementation on suffix-free languages. Our next result shows that this upper bound can be again decreased by one.

Lemma 4. *Let $n \geq 3$. Let L be a suffix-free regular language with $\mathrm{nsc}(L) = n$. Then $\mathrm{nsc}(L^c) \leq 2^{n-1}$.*

Proof. Let N be a non-returning n-state NFA for a suffix-free language L. The subset automaton $A = (Q, \Sigma, \delta, s, F)$ of the NFA N has at most $1 + 2^{n-1}$ reachable states since the only reachable subset that contains the initial state of N is the initial state of the subset automaton. The initial state of the subset automaton is non-final since L does not contain the empty string.

After interchanging the final and non-final states, we get a DFA $A^c = (Q, \Sigma, \delta, s, Q \setminus F)$ for L^c of $1 + 2^{n-1}$ states. The initial state of A^c is final and has no in-transitions. The state d is final as well, and it accepts every string.

Construct a 2^{n-1}-state NFA N^c from the DFA A^c as follows. Let Q_d be the set of states of A^c different from d and such that they have a transition to the state d, that is, $Q_d = \{q \in Q \setminus \{d\} \mid$ there is an a in Σ such that $\delta(q, a) = d\}$; remind that by Proposition 4, for each symbol a, there is a state q_a in Q_d that goes to d by a. Replace each transition (q, a, d) by transitions (q, a, p) for each p in Q_d, and moreover add the transition (q, a, s). Then, remove the state d. Formally, let $N^c = (Q \setminus \{d\}, \Sigma, \delta', s, (Q \setminus \{d\}) \setminus F)$, where

$$\delta'(q, a) = \begin{cases} \{\delta(q, a)\}, & \text{if } \delta(q, a) \neq d, \\ \{s\} \cup Q_d, & \text{if } \delta(q, a) = d. \end{cases}$$

In a similar way as in the case of prefix-free languages, it can be shown that $L(N^c) = L(A^c)$.   □

As for a lower bound, Han and Salomaa in [11] claimed that there exists a ternary suffix-free language meeting the bound $2^{n-1} - 1$. In the next lemma, we increase this lower bound by one.

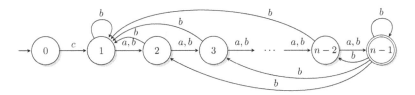

Fig. 3. An NFA of a ternary suffix-free language L with $\mathrm{sc}(L^c) = 2^{n-1}$

Lemma 5. *Let* $n \geq 3$. *There exists a ternary suffix-free language such that* $\mathrm{nsc}(L) = n$ *and* $\mathrm{nsc}(L^c) \geq 2^{n-1}$.

Proof. Let K be the language accepted by the NFA over $\{a, b\}$ shown in Fig. 1 with $n-1$ states. Set $L = c \cdot K$. Then L is a suffix-free language recognized by an n-state NFA shown in Fig 3. As shown in [15, Theorem 5], there exists a fooling set $\mathcal{F} = \{(x_S, y_S) \mid S \subseteq \{1, 2, \ldots, n-1\}\}$ of size 2^{n-1} for the language K^c. Then the set of pairs of strings $\mathcal{F}' = \{(c \cdot x_S, y_S) \mid S \subseteq \{1, 2, \ldots, n-1\}\}$ is a fooling set of size 2^{n-1} for the language L^c.   □

We can summarize the results of this section in the following theorem which provides the tight bound on the nondeterministic state complexity of complementation on suffix-free languages over an alphabet with at least three symbols. Whether or not this bound can be met by binary languages remains open.

Theorem 2 (Complement on Suffix-Free Languages, $|\Sigma| \geq 3$). *Let* $n \geq 3$. *Let* L *be a suffix-free regular language over an alphabet* Σ *with* $\mathrm{nsc}(L) = n$. *Then* $\mathrm{nsc}(L^c) \leq 2^{n-1}$, *and the bound is tight if* $|\Sigma| \geq 3$.   □

4   Complement on Non-Returning Languages

In this section, we consider languages that are recognized by non-returning NFAs. We call a regular language *non-returning* if it is accepted by a minimal non-returning NFA. Notice that every suffix-free language is non-returning, but the converse does not hold.

The state complexity of basic regular operations on languages represented by non-returning DFAs has been investigated by Eom *et al* in [6].

Here we study the nondeterministic state complexity of complementation on non-returning NFA languages. Our next theorem shows that in this case, the tight bound is $2^{n-1} + 1$. Moreover, this bound is tight already in the binary case.

Theorem 3 (Complement on Non-Returning Languages, $|\Sigma| \geq 2$). *Let* $n \geq 3$. *Let* L *be a non-returning language over an alphabet* Σ *with* $\mathrm{nsc}(L) = n$. *Then* $\mathrm{nsc}(L^c) \leq 2^{n-1} + 1$, *and the bound is tight if* $|\Sigma| \geq 2$.

Proof. Let $N = (Q, \Sigma, \delta, s, F)$ be an n-state non-returning NFA for a language L. In the subset automaton of the NFA N, no subset containing the initial state s is reachable, except for the initial subset $\{s\}$. Therefore, the subset automaton has at most $2^{n-1} + 1$ reachable subsets. After interchanging the final and non-final states in the subset automaton, we get a $(2^{n-1} + 1)$-state DFA for L^c. This gives the upper bound.

To prove tightness, consider a non-returning language $L = b \cdot K$, where K is the language accepted by the NFA shown in Fig. 1. The n-state NFA N for the language L is shown in Fig. 4.

Let $\mathcal{F} = \{(x_S, y_S) \mid S \subseteq \{1, 2, \ldots, n-1\}\}$ be the fooling set for the language K^c described in [15, Theorem 5]; notice that x_S is a string, by which the initial state 1 of the NFA in Fig. 1 goes to the set S. Let us show that the set

$$\mathcal{F}' = \{(\varepsilon, b^{n-2}), (a, b^n)\} \cup \{(bx_S, y_S) \mid S \subseteq \{1, 2, \ldots, n-1\} \text{ and } S \neq \emptyset\}$$

is a fooling set for the language L^c.

(F1) The strings b^{n-2} and ab^n are rejected by N, so they are in L^c. Each string $x_S y_S$ is in K^c, which means that the string $bx_S y_S$ is in L^c.

(F2) If S and T are distinct and non-empty subset of $\{1, 2, \ldots, n-1\}$, then at least one of the strings $x_S y_T$ and $x_T y_S$ is in K, so at least one of $bx_S y_T$ and $bx_T y_S$ is in L, so it is not in L^c. Let S be a non-empty set of $\{1, 2, \ldots, n-1\}$. The initial state 0 goes to the set S by $b \cdot x_S$. Since S is non-empty, both strings b^{n-2} and b^n are accepted from S since they are accepted from each state in $\{1, 2, \ldots, n-1\}$. It follows that the NFA N accepts the strings $bx_S \cdot b^n$ and $bx_S \cdot b^{n-2}$, so these strings are not in L^c. Finally, the string $\varepsilon \cdot b^n$ is accepted by the NFA N, so it is not in L^c.

Hence $\mathcal{F}'$ is a fooling set for the language L^c of size $2^{n-1} + 1$. By Lemma 1, every NFA for the language L^c requires at least $2^{n-1} + 1$.   □

5   Unary Alphabet

In this section, we consider the complementation operation on unary prefix-free, suffix-free, and non-returning languages. Our aim is to show that while in the case of prefix-free and suffix-free unary languages, the nondeterministic state complexity of complementation is in $\Theta(\sqrt{n})$, in the case of non-retuning unary languages, it is in $2^{\Theta(\sqrt{n \log n})}$. Let us start with the following observation.

Lemma 6. *Let $n \geq 3$ and $L = \{a\}^* \setminus \{a^n\}$. Then $\sqrt{n/3} \leq \mathrm{nsc}(L) \leq 6\sqrt{n}$.*

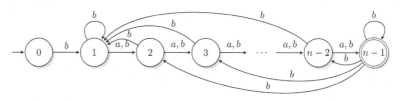

Fig. 4. An NFA of a binary non-returning language L with $\mathrm{sc}(L^c) = 2^{n-1} + 1$

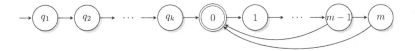

Fig. 5. An NFA A that does not accept a^n and accepts all the longer strings; $m = \lfloor \sqrt{n} \rfloor$, $k = n - (m^2 - m - 1) \le 3\sqrt{n}$

Proof. First consider a lower bound, and let us show that every NFA for L requires at least $\sqrt{n/3}$ states. Assume for a contradiction that there is an NFA N for L with less than $\sqrt{n/3}$ states. Then the tail in the Chrobak normal form of N is of size less that $3 \cdot (\sqrt{n/3})^2$ [4, 19], thus less than n. Since a^n must be rejected, each cycle in the Chrobak normal form must contain a rejecting state. It follows that infinitely many strings are rejected, which is a contradiction.

Now let us prove the upper bound. Let $m = \lfloor \sqrt{n} \rfloor$, and consider relatively prime numbers m and $m+1$. It is known that the maximal integer that cannot be expressed as $xm + y(m+1)$ for non-negative integers x and y is $(m-1)m-1 = m^2 - m - 1$ [21]. Let $k = n - (m^2 - m - 1)$. Then $0 < k \le 3\sqrt{n}$. Next, the NFA A shown in Fig. 5 and consisting of a path of length k and two overlapping cycles of lengths m and $m+1$ does not accept a^n, and accepts all strings a^i with $i \ge n+1$.

It remains to accept the shorter strings. To this aim let $p_1, p_2, \ldots, p_\ell$ be the first ℓ primes such that $p_1 p_2 \cdots p_\ell > n$. Then $\ell \le \lceil \log n \rceil$. Thus $p_1 + p_2 + \cdots + p_\ell = \Theta(\ell^2 \ln \ell) \le \sqrt{n}$ [2]. Consider an NFA B consisting of an initial state s that is connected to ℓ cycles of lengths $p_1, p_2, \ldots, p_\ell$. Let the states in the $j-th$ cycle be $0, 1, \ldots, p_j - 1$, where s is connected to state 1. The state $n \bmod p_j$ is non-final, and all the other states are final. Then this NFA does not accept a^n, but accepts all strings a^i with $i \le n - 1$ since we have $(i \bmod p_1, i \bmod p_2, \ldots, i \bmod p_\ell) \ne (n \bmod p_1, n \bmod p_2, \ldots, n \bmod p_\ell)$. The NFA B for $n = 24$ is shown in Fig. 6.

Now we get the resulting NFA for the language L of at most $6\sqrt{n}$ states as the union of NFAs A and B. □

Using the above result, we get that the nondeterministic state complexity of complementation on unary prefix-free or suffix-free languages is in $\Theta(\sqrt{n})$.

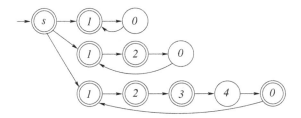

Fig. 6. The NFA B; $n = 24$

Theorem 4 (Complement on Unary Prefix- and Suffix-Free Languages). *Let L be a unary prefix-free or suffix-free regular language with* $\mathrm{nsc}(L) = n$. *Then* $\mathrm{nsc}(L^c) = \Theta(\sqrt{n})$.

Proof. The only prefix-free or suffix-free unary language with $\mathrm{nsc}(L) = n$ is the singleton language $\{a^{n-1}\}$. Its complement is $\{a\}^* \setminus \{a^{n-1}\}$, and the theorem follows from Lemma 6.   □

Now, we turn our attention to unary non-returning NFA languages.

In the NFA-to-DFA conversion of unary languages, a crucial role is played by the function $F(n) = \max\{\mathrm{lcm}(x_1, \ldots, x_k) \mid x_1 + \cdots + x_k = n\}$. It is known that $F(n) \in e^{\Theta(\sqrt{n \ln n})}$ and that $O(F(n))$ states suffice to simulate an n-state NFA by a DFA [4]. This means that $O(F(n))$ states are sufficient for an NFA to accept the complement of a unary NFA language. Moreover, in [12] a unary n-state NFA language is described such that every NFA accepting its complement needs at least $F(n-1)$ states. In [14], using a fooling set method, the lower bound $F(n-1)+1$ is proved for a non-returning language. For the sake of completeness, we recall this proof here.

Lemma 7. *Let $n \geq 3$. There exits a unary n-state non-returning NFA N such that every NFA for the complement of $L(N)$ requires at least $F(n-1)+1$ states.*

Proof. Let $i_1, i_2, \ldots, i_k$ be the integers, for which the maximum in the definition of $F(n-1)$ is attained. Consider an n-state NFA N shown in Fig. 7. The NFA N consists of the initial state s and k disjoint cycles of lengths $i_1, i_2, \ldots, i_k$. The initial and rejecting state s is nondeterministically connected to the rejecting states $q_{1,0}, q_{2,0}, \ldots, q_{k,0}$. All the remaining states are accepting.

Denote $m = F(n-1) = \mathrm{lcm}(i_1, i_2, \ldots, i_k)$. Consider the set of $m+1$ pairs of strings $\mathcal{F} = \{(\varepsilon, \varepsilon)\} \cup \{(a^i, a^{m+1-i}) \mid 1 \leq i \leq m\}$, and let us show that $\mathcal{F}$ is a fooling set for the language $L(N)^c$.

(F1) The strings ε and aa^m are not accepted by N since the initial state s is rejecting, and every computation on aa^m ends up in a rejecting state $q_{j,0}$ because each i_j divides m. Hence $\varepsilon\varepsilon$ and $a^i a^{m+1-i}$ with $1 \leq i \leq m$ are in $L(N)^c$.

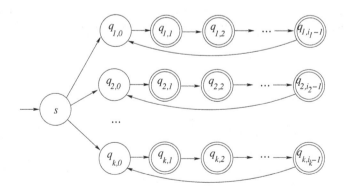

Fig. 7. A unary non-returning NFA N meeting the bound $F(n-1)+1$ for complement

(F2) If $1 \leq \ell < m = \text{lcm}(i_1, i_2, \ldots, i_k)$, then some i_j does not divide ℓ. This means that the computation on aa^ℓ beginning with states s and $q_{j,0}$ ends up in an accepting state in $\{q_{j,1}, q_{j,2}, \ldots, q_{j,i_j-1}\}$. It follows that the strings $\varepsilon a^m, \varepsilon a^{m-1}, \ldots, \varepsilon a^2$ and $a^m \varepsilon$, as well as the strings $a^i a^{m+1-j} = aa^{m-(j-i)}$, where $1 \leq i < j \leq m$, are accepted by N, and therefore they are not in $L(N)^c$.

Thus $\mathcal{F}$ is a fooling set for the language $L(N)^c$, and the lemma follows.   □

Hence we get the following result.

Theorem 5 (Complement on Unary Non-Returning NFA Languages).
Let L be a unary non-returning NFA language with $\text{nsc}(L) = n$. Then $\text{nsc}(L^c)$ is in $2^{O(\sqrt{n \log n})}$. The bound $2^{\Omega(\sqrt{n \log n})}$ can be met infinitely many times.

Proof. Every unary n-state NFA can be simulated by a $[2n^2 + n + F(n)]$-state DFA [4, 19]. After interchanging the final and non-final states, we get a DFA for the complement with the same number of states. Since $2n^2 + n + F(n)$ is in $2^{O(\sqrt{n \log n})}$, this gives the upper bound. For the lower bound, consider the language L accepted by the n-state NFA N shown in Fig. 7.

First, we show that the language L is a non-returning NFA language. We denoted $m = F(n-1)$, thus $n < m$. Let N' be an n'-state NFA for the language $L(N)$ with $n' \leq n$. Then N' must accept all strings a^i with $2 \leq i \leq m$ since all these strings are in $L(N)$. Assume for a contradiction that N' is not non-returning. Then, the initial state of N' is in a cycle of length ℓ with $1 \leq \ell \leq n' < m$. But then N' accepts the string $a^{m+1} = a^\ell \cdot a^{m+1-\ell}$ which is a contradiction since a^{m+1} is not in $L(N)$.

Now, let $k = \min\{\ell \mid F(\ell) = F(n-1)\}$. Let us show that $k \leq \text{nsc}(L) \leq k+1$. Recall that $m = F(n-1)$, thus $m = F(k)$. Let $F(k) = \text{lcm}(x_1, \ldots, x_r)$. Then L is accepted by a $(k+1)$-state NFA consisting of an initial state that is nondeterministically connected to r disjoint cycles of lengths $x_1, \ldots, x_r$.

Next, assume for a contradiction that L is accepted by an n'-state NFA N' with $n' < k$. Then in the Chrobak normal form of the NFA N', the number of states in cycles is at most n'. It follows that L is accepted by a DFA A, the loop of which is of length at most $F(n') < m$. Then there is an integer $\hat{t}$ such that the computation of the DFA A on the string $aa^{\hat{t}m}$ ends in the loop. However, all the strings $aa^{\hat{t}m} \cdot a^i$ with $1 \leq i \leq m-1$ must be accepted since they are in L. It follows that all the states in the loop of the DFA A must be final. But then the DFA A accepts a co-finite language, which is a contradiction since the language L is not co-finite. Since $F(n-1) + 1$ is in $2^{\Omega(\sqrt{n \log n})}$, the theorem follows.   □

6 Binary Alphabet

In this section, we study the complementation operation on binary prefix-free and suffix-free languages. We prove that the nondeterministic state complexity of complementation in this case is still exponential in $2^{\Omega(\sqrt{n \log n})}$. In the case of prefix-free binary languages, we prove that the upper bound 2^{n-1} given by Lemma 2 cannot be met. Whether or not this bound can be met by binary suffix-free languages remains open. Let us start with lower bounds.

Lemma 8. *There exists a binary prefix-free (suffix-free) n-state NFA N such that every NFA for the complement of $L(N)$ requires at least $F(n-2)+1$ states.*

Proof. Let L be the unary language accepted by an $(n-1)$-state NFA given by Lemma 7. Let $\mathcal{F} = \{(x_i, y_i) \mid i = 1, 2, \ldots, F(n-2)+1\}$ be the fooling set for L^c given in the proof of Lemma 7.

In the prefix-free case, we take an n-state NFA for the binary prefix-free language Lb. Then the set $\{(x_i, y_i\,b) \mid (x_i, y_i) \in \mathcal{F}\}$ is the fooling set for $(Lb)^c$ of size $F(n-2)+1$, and the lemma follows.

In the suffix-free case, we take an n-state NFA for the language bL. This time, the fooling set for $(bL)^c$ is $\{(b\,x_i, y_i) \mid (x_i, y_i) \in \mathcal{F}\}$.   □

The next lemma provides an upper bound on the nondeterministic state complexity of complementation on binary prefix-free languages.

Lemma 9. *Let $n \geq 12$. Let L be a binary prefix-free language with $\mathrm{nsc}(L) = n$. Then $\mathrm{nsc}(L^c) \leq 2^{n-1} - 2^{n-3} + 1$.*

Proof. Let N be a minimal NFA for L. Let $\{1, 2, \ldots, n\}$ be the state set of N. Let n be the final state of N. Without loss of generality, the state n is reached from the state $n-1$ on a in N.

If there is no transition (i, a, j) with $i, j \in \{1, 2, \ldots, n-1\}$, then the automaton on states $\{1, 2, \ldots, n-1\}$ is unary. It follows that in the subset automaton of N, at most $O(F(n-1)) < 2^{n-1} - 2^{n-3}$ subsets of $\{1, 2, \ldots, n-1\}$ can be reached, and the lemma follows in this case.

Now consider a transition (i, a, j) with $i, j \in \{1, 2, \ldots, n-1\}$. Let us show that no subset of $\{1, 2, \ldots, n-1\}$ containing states i and $n-1$ may be reachable. Assume for contradiction, that a set $S \cup \{i, n-1\}$ is reached from the initial state of the subset automaton by a string u. Since N is minimal, the final state n is reached from the state j by a non-empty string v. However, the set $S \cup \{i, n-1\}$ goes to a final set $S' \cup \{j, n\}$ by a, and then to a final set $S'' \cup \{n\}$ by v. It follows that the subset automaton accepts the strings ua and uav, which is a contradiction with the prefix-freeness of the accepted language. Thus at least 2^{n-3} subsets of $\{1, 2, \ldots, n-1\}$ are unreachable. Therefore, the subset automaton has at most $2^{n-1} - 2^{n-3} + 1$ states. After exchanging the accepting and the rejecting states we get a DFA of the same size for the complement of $L(N)$, and the lemma follows.   □

Now we summarize the results given by Lemma 9 and Lemma 8 in the following theorem; recall that $F(n) = \max\{\mathrm{lcm}(x_1, \ldots, x_k) \mid x_1 + \cdots + x_k = n\}$, and that $F(n)$ is in $2^{\Theta(\sqrt{n \log n})}$.

Theorem 6 (Complement on Binary Prefix-Free Languages). *Let L be a binary prefix-free language with $\mathrm{nsc}(L) = n$. Then $\mathrm{nsc}(L^c) \leq 2^{n-1} - 2^{n-3} + 1$. The lower bound $F(n-2)+1$ can be met for infinitely many n.*   □

References

1. Aho, A.V., Ullman, J.D., Yannakakis, M.: On notions of information transfer in VLSI circuits. In: Johnson, D.S., et al. (eds.) STOC 1983, pp. 133–139. ACM (1983)
2. Bach, E., Shallit, J.: 2.7 in Algorithmic Number Theory. Efficient Algorithms, vol. 1. MIT Press, Cambridge (1996)
3. Birget, J.C.: Partial orders on words, minimal elements of regular languages, and state complexity. Theoret. Comput. Sci. 119, 267–291 (1993), ERRATUM: Partial orders on words, minimal elements of regular languages, and state complexity (2002), http://clam.rutgers.edu/~birget/papers.html
4. Chrobak, M.: Finite automata and unary languages. Theoret. Comput. Sci. 47, 149–158 (1986)
5. Glaister, I., Shallit, J.: A lower bound technique for the size of nondeterministic finite automata. Inform. Process. Lett. 59, 75–77 (1996)
6. Eom, H.-S., Han, Y.-S., Jirásková, G.: State complexity of basic operations on non-returning regular languages. In: Jurgensen, H., Reis, R. (eds.) DCFS 2013. LNCS, vol. 8031, pp. 54–65. Springer, Heidelberg (2013)
7. Ellul, K.: Descriptional complexity measures of regular languages. Master's thesis. University of Waterloo (2002)
8. Han, Y.-S., Salomaa, K.: State complexity of basic operations on suffix-free regular languages. Theoret. Comput. Sci. 410, 2537–2548 (2009)
9. Han, Y.-S., Salomaa, K., Wood, D.: Operational state complexity of prefix-free regular languages. In: Automata, Formal Languages, and Related Topics, pp. 99–115. Institute of Informatics, University of Szeged (2009)
10. Han, Y.-S., Salomaa, K., Wood, D.: Nondeterministic state complexity of basic operations for prefix-free regular languages. Fundam. Inform. 90, 93–106 (2009)
11. Han, Y.-S., Salomaa, K.: Nondeterministic state complexity for suffix-free regular languages. In: McQuillan, I., Pighizzini, G. (eds.) DCFS 2010. EPTCS, vol. 31, pp. 189–196 (2010)
12. Holzer, M., Kutrib, M.: Nondeterministic descriptional complexity of regular languages. Int. J. Found. Comput. Sci. 14, 1087–1102 (2003)
13. Hromkovič, J.: Communication complexity and parallel computing. Springer (1997)
14. Jirásková, G.: State complexity of some operations on regular languages. In: Csuhaj-Varjú, E., et al. (eds.) DCFS 2003. MTA SZTAKI, pp. 114–125. Hungarian Academy of Sciences, Budapest (2003)
15. Jirásková, G.: State complexity of some operations on binary regular languages. Theoret. Comput. Sci. 330, 287–298 (2005)
16. Maslov, A.N.: Estimates of the number of states of finite automata. Soviet Math. Doklady 11, 1373–1375 (1970)
17. Sakoda, W.J., Sipser, M.: Nondeterminism and the size of two-way finite automata. In: Proc. 10th Annual ACM Symposium on Theory of Computing, pp. 275–286 (1978)
18. Sipser, M.: Introduction to the theory of computation. PWS Publishing Company, Boston (1997)
19. To, A.W.: Unary finite automata vs. arithmetic progressions. Inform. Process. Lett. 109, 1010–1014 (2009)
20. Yu, S.: Regular languages. In: Rozenberg, G., Salomaa, A. (eds.) Handbook of Formal Languages, vol. I, ch. 2, pp. 41–110. Springer, Heidelberg (1997)
21. Yu, S., Zhuang, Q., Salomaa, K.: The state complexity of some basic operations on regular languages. Theoret. Comput. Sci. 125, 315–328 (1994)

On the State Complexity of Closures and Interiors of Regular Languages with Subwords

Prateek Karandikar[1,2,*] and Philippe Schnoebelen[2,**]

[1] Chennai Mathematical Institute
[2] LSV, ENS Cachan, CNRS

Abstract. We study the state complexity of the set of subwords and superwords of regular languages, and provide new lower bounds in the case of languages over a two-letter alphabet. We also consider the dual interior sets, for which the nondeterministic state complexity has a doubly-exponential upper bound. We prove a matching doubly-exponential lower bound for downward interiors in the case of an unbounded alphabet.

1 Introduction

Quoting from [1], *"State complexity problems are a fundamental part of automata theory that has a long history. [. . .] However, many very basic questions, which perhaps should have been solved in the sixties and seventies, have not been considered or solved."*

In this paper, we are concerned with *(scattered) subwords* and the associated operations on regular languages: computing closures and interiors (see definitions in Section 2). Our motivations come from automatic verification of channel systems, see, e.g., [2,3]. Other applications exist in data processing or bioinformatics [4]. Closures and interiors wrt subwords and superwords are very basic operations, and the above quote certainly applies to them.

It has been known since [5] that $\downarrow L$ and $\uparrow L$, the downward closure and, respectively, the upward closure, of a language $L \subseteq \Sigma^*$, are regular for any L.

In [6], Gruber et al. explicitly raised the issue of the state complexity of downward and upward closures of regular languages (less explicit precursors exist, e.g. [7]). Given a n-state automaton A, constructing an automaton A' for $\downarrow L(A)$ or for $\uparrow L(A)$ can be done by simply adding extra transitions to A. However, when A is a DFA, the resulting A' is in general not deterministic, and determinization of A' may entail an exponential blowup in general. Gruber et al. proved a $2^{\Omega(\sqrt{n}\log n)}$ lower bound on the number of states of any DFA for $\downarrow L(A)$ or $\uparrow L(A)$, to be compared with the 2^n upper bound that comes from the simple closure+determinization algorithm.

Okhotin improved on these results by showing a $2^{\frac{n}{2}-2}$ and a $2^{n-2}+1$ lower bound for $\downarrow L(A)$ and, respectively, $\uparrow L(A)$ (again for an unbounded alphabet).

---

* Partially funded by Tata Consultancy Services.
** Supported by Grant ANR-11-BS02-001.

H. Jürgensen et al. (Eds.): DCFS 2014, LNCS 8614, pp. 234–245, 2014.

The second bound is known to be tight [8,9]. However, all these lower bounds assume an unbounded alphabet.

Okhotin also considered the case of languages over a *fixed alphabet* with $|\Sigma| = 3$ letters, in which case he demonstrated an exponential $2^{\sqrt{2n+30}-6}$ and $\frac{1}{5}4^{\sqrt{n/2}}n^{-\frac{3}{4}}$ lower bound for $\downarrow L(A)$ and, respectively, $\uparrow L(A)$ [8]. The construction and the proof are quite involved, and they leave open the case where $|\Sigma| = 2$ (the 1-letter case is trivial). It turns out that, in the 2-letter case, Héam had already proved a $\Omega(r^{\sqrt{n}})$ lower bound for $\uparrow L(A)$, here with $r = (\frac{1+\sqrt{5}}{2})^{\frac{1}{\sqrt{2}}}$ [10], so that the main remaining question is whether $\downarrow L(A)$ may require an exponential number of states even when $|\Sigma| = 2$.

Dual to closures are *interiors*. The *upward interior* and *downward interior* of a language L, denoted $\Uparrow L$ and $\Downarrow L$, are the largest upward-closed and, resp., downward-closed, sets included in L. Building closures and interiors are essential operations when reasoning with subwords, e.g., when model-checking lossy channel systems [11]. The state complexity of interiors has not yet been considered in the literature. When working with DFAs, computing interiors reduces to computing closures, thanks to duality. However, when working with NFAs, the simple complement+closure+complement algorithm only yields a quite large 2^{2^n} upper-bound on the number of states of an NFA for $\Uparrow L(A)$ or $\Downarrow L(A)$ —it actually yields DFAs— and one would like to improve on this, or to prove a matching lower bound.

Our contribution. Regarding closures, we prove in Section 3 an exponential lower bound on $\downarrow L(A)$ in the case of a two-letter alphabet, answering the open question raised above. We also give some new proofs for known results, usually relying on simpler examples demonstrating hard cases. For example, we prove a tighter 2^{n-1} lower bound for $\downarrow L(A)$ when the alphabet is unbounded.

Regarding interiors on NFAs, we show in Section 4 a doubly-exponential lower bound for downward interiors when the alphabet is not bounded. In the case of upward interiors, or in the case of fixed alphabets, we are left with an exponential gap between lower bounds and upper bounds. A partial result is a doubly-exponential lower bound for a restricted version of these problems. Table 1 shows a summary of the known results.

Finally, we analyze in Section 5 the computational complexity of deciding whether $L(A)$ is upward or downward-closed for a DFA or a NFA A.

2   Basic Notions and Results

Fix a finite alphabet $\Sigma = \{a, b, \ldots\}$. We say that a ℓ-letter word $x = a_1 a_2 \cdots a_\ell$ is a subword of y, written $x \sqsubseteq y$, when $y = y_0 a_1 y_1 \cdots y_{\ell-1} a_\ell y_\ell$ for some factors $y_0, \ldots, y_\ell \in \Sigma^*$, i.e., when there are positions $p_1 < p_2 < \cdots < p_\ell$ s.t. $x[i] = y[p_i]$ for all $1 \leqslant i \leqslant \ell = |x|$. For a language $L \subseteq \Sigma^*$, its downward closure is $\downarrow L \stackrel{\text{def}}{=} \{x \in \Sigma^* \mid \exists y \in L : x \sqsubseteq y\}$. Symmetrically, we consider an upward closure operation and we let $\uparrow L \stackrel{\text{def}}{=} \{x \in \Sigma^* \mid \exists y \in L : y \sqsubseteq x\}$. For singletons, we may write $\uparrow x$ and $\downarrow x$ for $\uparrow\{x\}$ and $\downarrow\{x\}$, e.g., $\downarrow a b b = \{\epsilon, a, b, a b, b b, a b b\}$. Closures

Table 1. A summary of the results. Each cell shows (a bound on) the maximum number of states that can result when the operation is applied to an automaton with n states and the output is minimized.

Operation	NFA	DFA
Upward closure	n	$2^{\Theta(n)}$, and $2^{\Omega(n^{1/2})}$ for $|\Sigma| = 2$
Downward closure	n	$2^{\Theta(n)}$, and $2^{\Omega(n^{1/3})}$ for $|\Sigma| = 2$
Upward interior	$\leqslant 2^{2^n}$, $\Omega(2^n)$	same as downward closure
Downward interior	$2^{2^{\Theta(n)}}$	same as upward closure

distribute over union, that is, $\uparrow L = \bigcup_{x \in L} \uparrow x$ and $\downarrow L = \bigcup_{x \in L} \downarrow x$. A language L is *downward-closed* (or *upward-closed*) if $L = \downarrow L$ (respectively, if $L = \uparrow L$). Note that L is downward-closed if, and only if, $\Sigma^* \smallsetminus L$ is upward-closed.

Upward-closed languages are also called *shuffle ideals* since they satisfy $L = L \shuffle \Sigma^*$. They correspond exactly to level $\frac{1}{2}$ of Straubing's hierarchy [12].

Since, by Higman's Lemma, any L has only finitely many minimal elements wrt the subword ordering, one deduces that $\uparrow L$ is regular for any L.

Effective construction of a finite-state automaton for $\downarrow L$ or $\uparrow L$ is easy when L is regular (see Section 3), is possible when L is context-free [13,14], and is not possible in general since this would allow deciding the emptiness of L.

The *upward interior* of L is $\circlearrowright L \overset{\text{def}}{=} \{x \in \Sigma^* \mid \uparrow x \subseteq L\}$. Its *downward interior* is $\circlearrowleft L \overset{\text{def}}{=} \{x \in \Sigma^* \mid \downarrow x \subseteq L\}$. Alternative characterizations are possible, e.g., by noting that $\circlearrowright L$ (respectively, $\circlearrowleft L$) is the largest upward-closed (respectively, downward-closed) language contained in L, or by using the following dualities:

$$\circlearrowleft L = \Sigma^* \smallsetminus \uparrow(\Sigma^* \smallsetminus L), \qquad \circlearrowright L = \Sigma^* \smallsetminus \downarrow(\Sigma^* \smallsetminus L). \qquad (1)$$

If L is regular, one may compute automata for the interiors of L by combining complementations and closures as in Eq. (1).

When considering a finite automaton $A = (\Sigma, Q, \delta, I, F)$, we usually write n for $|Q|$ (the number of states), m for $|\delta|$ (the number of transitions, seeing $\delta \subseteq Q \times \Sigma \times Q$ as a table), and k for $|\Sigma|$ (the size of the alphabet). For a regular language L, $n_D(L)$ and $n_N(L)$ denote the minimum number of states of a DFA (resp., a NFA) that accepts L.

We now illustrate a well-known technique for proving lower bounds on $n_N(L)$:

Lemma 2.1 (Extended fooling set technique, [15]). *Let L be a regular language. Suppose there exists a set of pairs of words $S = \{(x_i, y_i)\}_{1 \leqslant i \leqslant n}$ such that for all i, j, $x_i y_i \in L$ and at least one of $x_i y_j$ and $x_j y_i$ is not in L. Then $n_N(L) \geqslant n$.*

Lemma 2.2 (An application of the fooling set technique). *Fix Σ and define the following two languages:*

$$U \overset{\text{def}}{=} \{x \mid \forall a \in \Sigma : \exists i : x[i] = a\}, \qquad V \overset{\text{def}}{=} \{x \mid \forall i \neq j : x[i] \neq x[j]\}. \qquad (2)$$

Then $n_N(U) \geqslant 2^{|\Sigma|}$ and $n_N(V) \geqslant 2^{|\Sigma|}$.

Proof. The same proof applies to U and V: note that U has all words where every letter in Σ appears at least once, while V has all words where no letter appears twice.

With any $\Gamma \subseteq \Sigma$, we associate two words x_Γ and y_Γ, where x_Γ (respectively, y_Γ) has exactly one occurrence of each letter from Γ (respectively, each letter not in Γ). Then $x_\Gamma y_\Gamma$ is in U and V, while for any $\Delta \neq \Gamma$ one of $x_\Gamma y_\Delta$ and $x_\Delta y_\Gamma$ is not in U (and one is not in V). One concludes with Lemma 2.1.   □

3   State Complexity of Closures

3.1   Nondeterministic Automata

For a regular language L, an NFA for the upward or downward closure of L is obtained by simply adding transitions to an NFA for L, without increasing the number of states. More precisely, given an NFA A for L, an NFA for $\uparrow L$ is obtained by adding to A self-loops $q \xrightarrow{a} q$ for every state q of A and every letter $a \in \Sigma$. Similarly, an NFA for $\downarrow L$ is obtained by adding to A epsilon transitions $p \xrightarrow{\epsilon} q$ for every transition $p \to q$ of A (on any letter).

3.2   Deterministic Automata

Since every DFA is an NFA, Section 3.1 along with the powerset construction shows that if a language has an n-state DFA, then both its upward and downward closures have DFAs with at most 2^n states. An exponential blowup is also necessary as we now illustrate.

Let $\Sigma = \{a_1, \dots, a_k\}$ and define $L_1 \overset{\text{def}}{=} \{a\,a \mid a \in \Sigma\}$, i.e., L_1 contains all words consisting of two identical letters. The minimal DFA for L_1 has $n = k + 2$ and $m = 2k$, see Fig. 1.

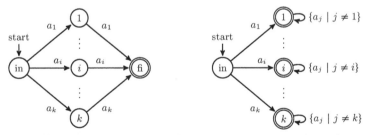

Fig. 1. DFAs for $L_1 = \bigcup_{a \in \Sigma} a\,a$ (left) and $L_2 = \bigcup_{a \in \Sigma} a \cdot (\Sigma \setminus a)^*$ (right)

Now $\uparrow L_1 = \{x \in \Sigma^{\geq 2} \mid \exists j > i : x[i] = x[j]\} = \bigcup_{a \in \Sigma} \Sigma^* \cdot a \cdot \Sigma^* \cdot a \cdot \Sigma^*$, i.e., $\uparrow L_1$ has all words where *some letter reappears*, i.e., $\uparrow L_1$ is the complement of V from Lemma 2.2. A DFA for $\uparrow L_1$ has to record all letters previously read: the minimal (complete) DFA has $2^k + 1$ states. Hence $2^{n-2} + 1$ states are sometimes required for the minimal DFA recognizing the upward closure of an n-state DFA.

Further define $L_2 \stackrel{\text{def}}{=} \{x \in \Sigma^+ \mid \forall i > 1 : x[i] \neq x[1]\} = \bigcup_{a \in \Sigma} a \cdot (\Sigma \smallsetminus \{a\})^*$, i.e., L_2 has words where *the first letter does not reappear*. The minimal DFA for L_2 has $n = k + 1$ and $m = k^2$, see Fig. 1. Now $\downarrow L_2 = \{x \mid \exists a \in \Sigma : \forall i > 1 : x[i] \neq a\} = \epsilon + \Sigma \cdot \bigcup_{a \in \Sigma} (\Sigma \smallsetminus \{a\})^*$, i.e., $\downarrow L_2$ has all words x such that the first suffix $x[2, \dots, \ell]$ does not use all letters. Equivalently $x \in \downarrow L_2$ iff $x \in L_2$ or x does not use all letters, i.e., $\downarrow L_2$ is the union of L_2 and the complement of U from Lemma 2.2. The minimal DFA for $\downarrow L_2$ just records all letters previously encountered *except the first*, hence has exactly 2^k states. Thus 2^{n-1} states may be required for a DFA recognizing the downward closure of an n-state DFA.

The above simple examples use a linear-sized alphabet to establish the lower bounds. This raises the question of whether exponential lower bounds still apply in the case of a fixed alphabet. The 1-letter case is degenerate since then both $n_D(\uparrow L)$ and $n_D(\downarrow L)$ are $\leqslant n_D(L)$. In the 3-letter case, exponential lower bounds are shown in [8].

In the 2-letter case, an exponential lower bound for upward closure is shown with the following witness: For $n > 0$, let $L_n = \{a^i b a^{2j} b a^i \mid i + j + 1 = n\}$. Then $n_D(L_n) = (n+1)^2$, while $n_D(\uparrow L_n) \geqslant \frac{1}{7}(\frac{1+\sqrt{5}}{2})^n$ for $n \geqslant 4$ [10, Prop. 5.11]. However, the downward closure of these languages does not demonstrate a state blowup, in fact $n_D(\downarrow L_n) = n^2 + 3n - 1$ for $n \geqslant 2$.

We now show an exponential lower bound for downward closures in the case of a two-letter alphabet. Interestingly, the same languages can also serve as hard case for upward closure (but it gives weaker bounds than in [10]).

Theorem 3.1. *The state complexity of computing downward closure for DFAs is in $2^{\Omega(n^{1/3})}$. The same result holds for upward closure.*

We now prove the theorem. Fix a positive integer n. Let

$$S = \{n, n+1, \dots, 2n\},$$

and define morphisms $c, h : S^* \to \{a, b\}^*$ with, for any $i \in S$:

$$c(i) \stackrel{\text{def}}{=} a^i b^{3n-i}, \qquad\qquad h(i) \stackrel{\text{def}}{=} c(i) c(i).$$

Note that $c(i)$ always has length $3n$, begins with at least n a's, and ends with at least n b's. If we now let

$$L \stackrel{\text{def}}{=} \{c(i)^n \mid i \in S\},$$

L is a finite language of $n+1$ words, each of length $3n^2$ so that clearly $n_D(L)$ is in $3n^3 + O(n^2)$. (In fact, $n_D(L) = 3n^3 + 1$.) In the rest of this section we show that both $n_D(\uparrow L)$ and $n_D(\downarrow L)$ are in $2^{\Omega(n)}$.

Lemma 3.2. *For $i, j \in S$, the longest prefix of $c(i)^\omega$ that embeds in $h(j) = c(j) c(j)$ is $c(i)$ if $i \neq j$ and $c(i) c(i)$ if $i = j$.*

Proof (Sketch). The case $i = j$ is clear. Fig. 2 displays the leftmost embedding of $c(i)^\omega$ in $h(j)$ in a case where $i > j$. The remaining case, $i < j$, is similar.   □

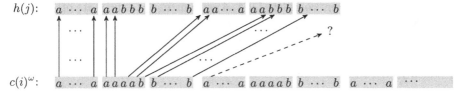

Fig. 2. Case "$i > j$" in Lemma 3.2: here $i = n + 4$ and $j = n + 2$ for $n = 5$

For each $i \in S$, let the morphisms $\eta_i, \theta_i : S^* \to (\mathbb{N}, +)$ be defined by

$$\eta_i(j) \overset{\text{def}}{=} \begin{cases} 1 & \text{if } i \neq j, \\ 2 & \text{if } i = j, \end{cases} \qquad \theta_i(j) \overset{\text{def}}{=} \begin{cases} 2 & \text{if } i \neq j, \\ 1 & \text{if } i = j. \end{cases}$$

Thus for $\sigma = p_1 p_2 \cdots p_s \in S^*$, $\eta_i(\sigma)$ is s plus the number of occurrences of i in σ, while $\theta_i(\sigma)$ is $2s$ minus the number of these occurrences of i.

Lemma 3.3. *Let $\sigma \in S^*$. The smallest ℓ such that $c(\sigma)$ embeds in $c(i)^\ell$ is $\theta_i(\sigma)$.*

Proof. We write $\sigma = p_1 p_2 \cdots p_s$ and prove the result by induction on s. The $s = 0$ case is trivial. The $s = 1$ case follows from Lemma 3.2, since for any p_1 and i, $c(p_1) \sqsubseteq c(i)$ iff $p_1 = i$, and $c(p_1) \sqsubseteq h(i) = c(i)^2$ always.

Assume now $s > 1$, write $\sigma = \sigma' p_s$ and let $\ell' = \theta_i(\sigma')$. By the induction hypothesis, $c(\sigma') \not\sqsubseteq c(i)^{\ell'-1}$ and $c(\sigma') \sqsubseteq c(i)^{\ell'} = c(i)^{\ell'-1} a^i b^{3n-i}$. Write now $c(i)^{\ell'} = w\,v$ where w is the shortest prefix of $c(i)^{\ell'}$ with $c(\sigma') \sqsubseteq w$. Since $c(\sigma')$ ends with a b that only embeds in the $a^i b^{3n-i}$ suffix of $c(i)^{\ell'}$, v is necessarily b^r for some r. So for all $z \in \{a, b\}^*$, $c(p_s) \sqsubseteq z$ if and only if $c(p_s) \sqsubseteq v z$. We have $c(p_s) \sqsubseteq c(i)^{\theta_i(p_s)}$ and $c(p_s) \not\sqsubseteq v c(i)^{\theta_i(p_s)-1}$. Noting that $\sigma = \sigma' p_s$, we get $c(\sigma) \sqsubseteq c(i)^{\theta_i(\sigma)}$ and $c(\sigma) \not\sqsubseteq c(i)^{\theta_i(\sigma)-1}$. □

We now derive a lower bound on the number of states in the minimal complete DFA for $\downarrow L$. For every subset X of S of size $n/2$ (assume n is even), let $w_X \in \{a, b\}^*$ be defined as follows: let the elements of X be $p_1 < p_2 < \cdots < p_{n/2}$ and let

$$w_X \overset{\text{def}}{=} c(p_1 p_2 \cdots p_{n/2}).$$

Note that $\theta_i(p_1 p_2 \cdots p_{n/2}) = n$ if $i \notin X$ and $\theta_i(p_1 p_2 \cdots p_{n/2}) = n - 1$ if $i \in X$.

Lemma 3.4. *Let X and Y be subsets of S of size $n/2$ with $X \neq Y$. There exists a word $v \in \{a, b\}^*$ such that $w_X v \in \downarrow L$ and $w_Y v \notin \downarrow L$.*

Proof. Let $i \in X \setminus Y$. Let $v = c(i)$. Then

- By Lemma 3.3, $w_X \sqsubseteq c(i)^{n-1}$, and so $w_X v \sqsubseteq c(i)^n$. So $w_X v \in \downarrow L$.
- By Lemma 3.3, the smallest ℓ such that $w_Y v \sqsubseteq c(i)^\ell$ is $n + 1$. Similarly, for $j \neq i$, the smallest ℓ such that $w_Y v \sqsubseteq c(j)^\ell$ is at least $n - 1 + 2 = n + 1$ (the w_Y contributes at least $n - 1$ and the v contributes 2). So $w_Y v \notin \downarrow L$. □

This shows that for any complete DFA A recognizing $\downarrow L$, the state of A reached from the start state by every word in $\{w_X \mid X \subseteq S, |X| = n/2\}$ is distinct. Thus A has at least $\binom{n+1}{n/2}$ states, which is $\approx \frac{2^{n+3/2}}{\sqrt{\pi n}}$.

For $n_D(\uparrow L)$, the reasoning is similar:

Lemma 3.5. *Let $\sigma \in S^*$. For all $i \in S$, the longest prefix of $c(i)^\omega$ that embeds in $h(\sigma)$ is $c(i)^{\eta_i(\sigma)}$.*

Proof. By induction on the length of σ and applying Lemma 3.2.   □

For every subset X of S of size $n/2$ (assume n is even), let $w'_X \in \{a, b\}^*$ be defined as follows: let the elements of X be $p_1 < p_2 < \cdots < p_{n/2}$ and let

$$w'_X \stackrel{\text{def}}{=} h(p_1 p_2 \cdots p_{n/2}) = c(p_1 p_1 p_2 p_2 \cdots p_{n/2} p_{n/2}) .$$

Lemma 3.6. *Let X and Y be subsets of S of size $n/2$ with $X \neq Y$. There exists a word $v \in \{a, b\}^*$ such that $w'_X v \in \uparrow L$ and $w'_Y v \notin \uparrow L$.*

Proof. Let $i \in X \setminus Y$. Let $v = c(i)^{n-(n/2+1)} = c(i)^{n/2-1}$.

- By Lemma 3.5, $c(i)^{n/2+1} \sqsubseteq w'_X$, thus $c(i)^n \sqsubseteq w'_X v$, hence $w'_X v \in \uparrow L$.
- By Lemma 3.5, the longest prefix of $c(i)^n$ that embeds in $w'_Y v$ is at most $c(i)^\ell$ where $\ell = n/2 + n/2 - 1 = n - 1$. The longest prefix of $c(j)^n$ that embeds in $w'_Y v$ for $j \neq i$ is at most $c(j)^\ell$ where

$$\ell = \frac{n}{2} + 1 + \left\lceil \frac{n/2 - 1}{2} \right\rceil \leqslant n - 1$$

Therefore $c(j)^n \not\sqsubseteq w'_Y v$ when $j = i$ and also when $j \neq i$. Thus $w'_Y v \notin \uparrow L$.   □

With Lemma 3.6 we reason exactly as we did for $n_D(\downarrow L)$ after Lemma 3.4 and conclude that $n_D(\uparrow L) \geqslant \binom{n+1}{n/2}$ here too.

4   State Complexity of Interiors

Recall Eq. (1) that expresses interiors with closures and complements. Since complementation of DFAs does not increase the number of states, the bounds on interiors are the same as the bounds on closures in the case of DFAs.

For NFAs, Eq. (1) provides an obvious 2^{2^n} upper bound on the NFA state complexity of both the upward and the downward interior, simply by combining the powerset construction for complementation and the results of Section 3.1. (Alternatively, it is possible to design a "powerset-like construction" that directly builds a DFA for the interior, upward or downward, of a language recognized by a DFA: this returns the same DFA as with the complement+closure+complement procedure.) Note that both procedures yield DFAs for the interiors while we are looking for better bounds on their NFA state complexity.

Proposition 4.1. *The NFA state complexity of the downward interior is in* $2^{2^{\Theta(n)}}$ *(assuming an unbounded alphabet).*

Proof. Let ℓ be a positive integer, and let $\Sigma = \{0,1\}^{\ell}$, so that $k = |\Sigma| = 2^{\ell}$. Let

$$L \overset{\text{def}}{=} \Sigma^* \smallsetminus \{a\,a \mid a \in \Sigma\} = \{w \mid |w| \neq 2\} \cup \{a\,b \mid a,b \in \Sigma, a \neq b\}\,.$$

Two letters in Σ, viewed as ℓ-bit sequences, are distinct if and only if they differ in at least one bit. An NFA can check this by guessing the position in which they differ and checking that the letters indeed differ in this position. Fig. 3 shows an NFA for $\{a\,b \mid a \neq b\}$ with $2\ell + 2$ states.

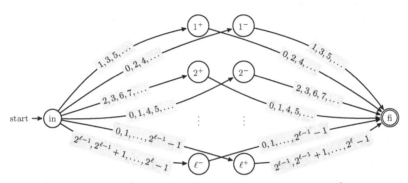

Fig. 3. DFA for $\{a\,b \mid a,b \in \Sigma, a \neq b\}$ with $2\ell + 2$ states and $\ell 2^{\ell}$ transitions

Since now $\{w \mid |w| \neq 2\}$ is recognized by an NFA with 4 states, L is recognized by an NFA with $n = 2\ell + 6$ states.

Finally, $\mathcal{O}L$ consists of all words where every letter is distinct (equivalently, no letter appears more than once), a language called V in Eq. (2). We conclude with Lemma 2.2 showing $n_N(V) \geqslant 2^{|\Sigma|} = 2^{2^{\ell}} = 2^{2^{n/2-3}}$.   □

Proposition 4.2. *The NFA state complexity of the upward interior is in* $\Omega(2^n)$ *(assuming an unbounded alphabet).*

Proof. For Σ a k-letter alphabet we consider $L_3 \overset{\text{def}}{=} \Sigma^* \smallsetminus L_2$ with the same L_2 used earlier, see Fig. 1 in Section 3.2. Thus L_3 contains all words where the first letter reappears. (It also contains the empty word). By complementing the DFA for L_2, one sees that a minimal DFA for L_3 has $n = k + 2$ states.

We noted in Section 3.2 that $\downarrow L_2 = L_2 \cup (\Sigma^* \smallsetminus U)$, where U is the language of all words where each letter from Σ occurs at least once. Hence $\mathcal{O}L_3 = \Sigma^* \smallsetminus \downarrow L_2 = (\Sigma^* \smallsetminus L_2) \cap U = L_3 \cap U$.

Observe now that for any $a \in \Sigma$ and $w \in \Sigma^*$, $a\,w \in L_3 \cap U$ iff $w \in U$. Thus any NFA for $L_3 \cap U$ can be transformed into an NFA for U by simply changing the initial states, and so a state lower bound for U implies the same lower bound for $L_3 \cap U$. With Lemma 2.2 we get $n_N(\mathcal{O}L_3) = n_N(L_3 \cap U) \geqslant n_N(U) \geqslant 2^k = 2^{n-2}$, witnessing the required exponential lower bound.   □

The above results leave us with an exponential gap between lower and upper bounds for $n_N(\mathcal{O}L)$ —and even for $n_D(\mathcal{O}L)$— when L is given by a NFA. We have not been able to close this gap and we do not yet feel able to formulate a conjecture on whether an exponential $2^{n^{O(1)}}$ bound exists or not. Trying to find hard cases by exhaustive or heuristic search is difficult because the search space is huge even for small n, and for most languages the upward interior is trivial. For NFAs with $n = 3$ states and with $|\Sigma| = k = 3$ letters, a worst case example is $L = \left((a+b)(a+b+c)^*(a+b) + (b+c)(a+b+c)^*(a+c)\right)^*$. Here $n_N(L) = 3$ and $n_N(\mathcal{O}L) = 10$, which is well below the 2^{2^n} upper bound.

In the rest of this section, we establish a doubly exponential lower bound for a more general construction called *restricted interior*.

Let Σ be an alphabet and let $X \subseteq \Sigma$. For words u, v, we write $u \sqsubseteq_X v$ if u is obtained from v by deleting some (occurrences of) letters in X, necessarily keeping letters in $\Sigma \smallsetminus X$ intact. For example, $abba \sqsubseteq_{\{b,c\}} abcbcbcac$, but $bba \not\sqsubseteq_{\{b,c\}} abcbcbcac$. Closures and interiors are defined as one would expect:

$$\downarrow_X L \overset{\text{def}}{=} \{w \mid \exists v \in L : w \sqsubseteq_X v\}, \quad \mathcal{O}_X L \overset{\text{def}}{=} \{w \mid \downarrow_X\{w\} \subseteq L\} = \Sigma^* \smallsetminus \uparrow_X(\Sigma^* \smallsetminus L),$$

$$\uparrow_X L \overset{\text{def}}{=} \{w \mid \exists v \in L : v \sqsubseteq_X w\}, \quad \mathcal{O}_X L \overset{\text{def}}{=} \{w \mid \uparrow_X\{w\} \subseteq L\} = \Sigma^* \smallsetminus \downarrow_X(\Sigma^* \smallsetminus L).$$

Theorem 4.3. *The NFA state complexity of the restricted upward interior is $\leqslant 2^{2^n}$ and in $2^{2^{\Omega(\sqrt{n})}}$. The lower bound holds with a 3-letter alphabet.*

As with $\downarrow L$, one can obtain an NFA for $\downarrow_X L$ from an NFA for L by simply adding transitions, without adding new states. Hence the upper bound is clear in Theorem 4.3, and we only need to prove the lower bound.

Fix $n \in \mathbb{N}$. Let $\Sigma = \{0, 1, \#\}$, and $\Sigma_{01} = \{0, 1\}$. Define the following languages:

- N is the set of all words over Σ in which the sum of the number of 0s and the number of 1s is divisible by n;
- $B = ((\epsilon + \#)(0 + 1)^n)^*$. Note that $B \subseteq N$;
- H_2 is the set of all words over Σ with exactly two occurrences of $\#$.

Let $L \subseteq \Sigma^*$ consists of all the following words:

- words in $(N \smallsetminus H_2) \cup (N \smallsetminus B)$;
- words in $H_2 \cap B$ such that the factors of length n immediately following the two occurrences of $\#$ are distinct.

Both $N \smallsetminus H_2$ and $N \smallsetminus B$ are recognized by NFAs with $O(n)$ states. The second summand of L is recognized by an NFA with $O(n^2)$ states, as the n-length factors immediately following the two occurrences of $\#$ being unequal can be checked by guessing a position at which they differ. So L is recognized by an NFA with $O(n^2)$ states. Note that $L \subseteq N$.

Consider $\mathcal{O}_{\{\#\}}L$. This is the set of all words in L such that no matter how we insert occurrences of $\#$, the resulting word remains in L.

Let $\Gamma = \{0, 1\}^n$, considered as an alphabet. Define the homomorphism $h : \Gamma^* \to \Sigma^*$, as $h(x) = x$ for all x. As in Lemma 2.2, let $V \subseteq \Gamma^*$ consist of all words over Γ in which no letter appears twice, and define $V' \subseteq \Sigma^*$ as $h(V)$. Note that $V' \subseteq N \cap \Sigma_{01}^*$.

Lemma 4.4. $V' = (\mho_{\{\#\}}L) \cap \Sigma_{01}^*$.

Proof. ($\subseteq$:) Let $w \in V'$ and v be such that $w \sqsubseteq_{\{\#\}} v$. Since $V' \subseteq N = \uparrow_{\{\#\}}(N)$, we know that $v \in N$. If $v \notin H_2 \cap B$, then $v \in L$. Otherwise, if $v \in H_2 \cap B$, then by the definition of V, it is easy to see that $v \in L$.

($\supseteq$:) Conversely, let $w \in (\mho_{\{\#\}}L) \cap \Sigma_{01}^*$. In particular, $w \in L$, and so $w \in N$. $|w|$ is divisible by n, and so $w = h(x)$ for some $x \in \Gamma^*$. By inserting two copies of $\#$ at suitable positions in w, and using the fact that the resulting word belongs to L, one concludes that $x \in V$, and so $w \in V'$. $\square$

Lemma 4.5. $2^{2^n} \leqslant n_N(V) \leqslant n_N(V') \leqslant n_N(\mho_{\{\#\}}L)$.

Proof. Lemma 2.2 gives $2^{2^n} \leqslant n_N(V)$. Then $V = h^{-1}(V')$ gives $n_N(V) \leqslant n_N(V')$ since it is easy to transform an NFA for V' into an NFA for $h^{-1}(V')$ (for any morphism h, in fact) with the same number of states. Finally, $V' = (\mho_{\{\#\}}L) \cap \Sigma_{01}^*$ gives $n_N(V') \leqslant n_N(\mho_{\{\#\}}L)$ since an NFA for V' can be obtained from an NFA for $\mho_{\{\#\}}L$ by deleting all transitions labelled by $\#$. $\square$

Since $n_N(L)$ is in $O(n^2)$, Lemma 4.5 concludes the proof of Theorem 4.3.

5 Complexity of Decision Problems on Subwords

In automata-based procedures for logic and verification, the state complexity of automata constructions is not always the best measure of computational complexity. In this section we give elementary proofs showing that the problem of deciding whether $L(A)$ is upward-closed, or downward-closed, is unsurprisingly PSPACE-complete for NFAs, and NL-complete for DFAs. (For upward-closedness, this is already shown in [10], and quadratic-time algorithms that decide upward-closedness of $L(A)$ for a DFA A already appear in [16,12].)

Proposition 5.1. *Deciding whether $L(A)$ is upward-closed or downward-closed is PSPACE-complete when A is a NFA, even in the 2-letter alphabet case.*

Proof (Sketch). A PSPACE algorithm simply tests for inclusion between two automata, A and its closure. PSPACE-hardness can be shown by adapting the proof for hardness of universality. Let R be a length-preserving semi-Thue system and x, x' two strings of same length. It is PSPACE-hard to say whether $x \xrightarrow{*}_R x'$, even for a fixed R over a 2-letter alphabet Σ. We reduce (the negation of) this question to our problem.

Fix x and x' of length $n > 1$: a word $x_1 x_2 \cdots x_m$ of length $n \times m$ encodes a derivation if $x_1 = x$, $x_m = x'$, and $x_i \to_R x_{i+1}$ for all $i = 1, \ldots, m-1$. The language L of words that do *not* encode a derivation from x to x' is regular and recognized by a NFA with $O(n)$ states. Now, there is a derivation $x \xrightarrow{*}_R x'$ iff $L \neq \Sigma^*$. Since L contains all words of length not divisible by $n > 1$, it is upward-closed, or downward-closed, iff $L = \Sigma^*$, iff $\neg(x \xrightarrow{*}_R x')$. $\square$

Proposition 5.2. *Deciding whether $L(A)$ is upward-closed or downward-closed is NL-complete when A is a DFA, even in the 2-letter alphabet case.*

Proof. Since L is downward-closed if, and only if, $\Sigma^* \setminus L$ is upward-closed, and since one easily builds a DFA for the complement of $L(A)$, it is sufficient to prove the result for upward-closedness.

We rely on the following easy lemma: L is upward-closed iff for all $u, v \in \Sigma^*$, $u\,v \in L$ implies $u\,a\,v \in L$ for all $a \in \Sigma$. Therefore, $L(A)$ is not upward-closed —for $A = (\Sigma, Q, \delta, \{i\}, F)$— iff there are states $p, q \in Q$, a letter a, and words u, v such that $\delta(i, u) = p$, $\delta(p, a) = q$, $\delta(p, v) \in F$ and $\delta(q, v) \notin F$. If such words exist, in particular one can take u and v of length $< n = |Q|$ and respectively $< n^2$. Hence testing (the negation of) upward-closedness can be done in nondeterministic logarithmic space by guessing u, a, and v within the above length bounds, finding p and q by running u and then a from i, then running v from both p and q.

For hardness, one may reduce from vacuity of DFAs, a well-known NL-hard problem that is essentially equivalent to GAP, the Graph Accessibility Problem. Note that for any DFA A (in fact any NFA) the following holds:

$$L(A) = \varnothing \quad \text{iff} \quad L(A) \cap \Sigma^{<n} = \varnothing \quad \text{iff} \quad L(A) \cap \Sigma^{<n} \text{ is upward-closed,}$$

where n is the number of states of A. This provides the required reduction since, given a FSA A, one easily builds a FSA for $L(A) \cap \Sigma^{<n}$.   □

6   Concluding Remarks

For words ordered by the (scattered) subword relation, we considered the state complexity of computing closures and interiors, both upward and downward, of regular languages given by finite-state automata. These operations are essential when reasoning with subwords, e.g., in symbolic model checking for lossy channel systems, see [11, Section 6]. We completed the known results on closures by demonstrating an exponential lower bound on downward closures even in the case of a two-letter alphabet.

The state complexity of interiors is a new problem that we introduced in this paper and for which we only have partial results: we show that the doubly-exponential upper bound for interiors of NFAs is matched by a doubly-exponential lower bound in the case of downward interiors when the alphabet is unbounded. For upward interiors of NFAs, or for fixed alphabets, there remains an exponential gap between the existing upper and lower bounds.

These results contribute to a more general research agenda: what are the right data structures and algorithms for reasoning with subwords and superwords? The algorithmics of subwords has mainly been developed in string matching and combinatorics [4,17] but other applications exist that require handling sets of strings rather than individual strings, e.g., model-checking and constraint solving. For these applications, there are many different ways of representing closed sets and automata-based representation are not always the preferred option, see, e.g., the SREs used for downward-closed languages in [2]. The existing trade-offs between all the available options are not yet well understood and certainly deserve scrutiny.

Acknowledgments. We thank S. Schmitz and the anonymous reviewers for their helpful comments.

References

1. Sheng, Y.: State complexity: Recent results and open problems. Fundamenta Informaticae 64(1–4), 471–480 (2005)
2. Abdulla, P.A., Collomb-Annichini, A., Bouajjani, A., Jonsson, B.: Using forward reachability analysis for verification of lossy channel systems. Formal Methods in System Design 25(1), 39–65 (2004)
3. Haase, Ch., Schmitz, S., Schnoebelen, Ph.: The power of priority channel systems. In: D'Argenio, P.R., Melgratti, H. (eds.) CONCUR 2013. LNCS, vol. 8052, pp. 319–333. Springer, Heidelberg (2013)
4. Baeza-Yates, R.A.: Searching subsequences. Theoretical Computer Science 78(2), 363–376 (1991)
5. Haines, L.H.: On free monoids partially ordered by embedding. Journal of Combinatorial Theory 6(1), 94–98 (1969)
6. Gruber, H., Holzer, M., Kutrib, M.: More on the size of Higman-Haines sets: Effective constructions. Fundamenta Informaticae 91(1), 105–121 (2009)
7. Birget, J.-C.: Partial orders on words, minimal elements of regular languages and state complexity. Theoretical Computer Science 119(2), 267–291 (1993)
8. Okhotin, A.: On the state complexity of scattered substrings and superstrings. Fundamenta Informaticae 99(3), 325–338 (2010)
9. Brzozowski, J.A., Jirásková, G., Li, B.: Quotient complexity of ideal languages. Theoretical Computer Science 470, 36–52 (2013)
10. Héam, P.-C.: On shuffle ideals. RAIRO Theoretical Informatics and Applications 36(4), 359–384 (2002)
11. Bertrand, N., Schnoebelen, Ph.: Computable fixpoints in well-structured symbolic model checking. Formal Methods in System Design 43(2), 233–267 (2013)
12. Pin, J.-É., Weil, P.: Polynomial closure and unambiguous product. Theory of Computing Systems 30(4), 383–422 (1997)
13. van Leeuwen, J.: Effective constructions in well-partially-ordered free monoids. Discrete Mathematics 21(3), 237–252 (1978)
14. Courcelle, B.: On constructing obstruction sets of words. EATCS Bulletin 44, 178–185 (1991)
15. Gruber, H., Holzer, M.: Finding lower bounds for nondeterministic state complexity is hard. In: Ibarra, O.H., Dang, Z. (eds.) DLT 2006. LNCS, vol. 4036, pp. 363–374. Springer, Heidelberg (2006)
16. Arfi, M.: Polynomial operations on rational languages. In: Brandenburg, F.J., Wirsing, M., Vidal-Naquet, G. (eds.) STACS 1987. LNCS, vol. 247, pp. 198–206. Springer, Heidelberg (1987)
17. Elzinga, C.H., Rahmann, S., Wang, H.: Algorithms for subsequence combinatorics. Theoretical Computer Science 409(3), 394–404 (2008)

State Complexity of Regular Tree Languages for Tree Pattern Matching

Sang-Ki Ko, Ha-Rim Lee, and Yo-Sub Han

Department of Computer Science, Yonsei University,
50, Yonsei-Ro, Seodaemun-Gu, Seoul 120-749, Korea
{narame7,hrlee,emmous}@cs.yonsei.ac.kr

Abstract. We study the state complexity of regular tree languages for tree matching problem. Given a tree t and a set of pattern trees L, we can decide whether or not there exists a subtree occurrence of trees in L from the tree t by considering the new language L' which accepts all trees containing trees in L as subtrees. We consider the case when we are given a set of pattern trees as a regular tree language and investigate the state complexity. Based on the sequential and parallel tree concatenation, we define three types of tree languages for deciding the existence of different types of subtree occurrences. We also study the deterministic top-down state complexity of path-closed languages for the same problem.

Keywords: tree automata, state complexity, tree pattern matching, regular tree languages.

1 Introduction

State complexity is one of the most interesting topics in automata and formal language theory [6,7,18,19]. The state complexity of finite automata has been studied since the 60's [8,10,11]. Maslov [9] initiated the problem of finding the operational state complexity and Yu et al. [19] investigated the state complexity for basic operations.

Recently, the state complexity problem has been extended to regular tree languages. Regular tree languages and tree automata theory provide a formal framework for XML schema languages such as XML DTD, XML Schema, and Relax NG [12]. XML schema languages can process a set of XML documents by specifying the structural properties. Piao and Salomaa [14,15] considered the state complexity between different models of unranked tree automata. They also investigated the state complexity of concatenation [17] and star [16] for regular tree languages. Two of the authors studied the state complexity of subtree-free regular tree languages, which are a proper subclass of regular tree languages [3].

Since a regular tree language is a set of trees, it is suitable for representing a set of structural documents such as XML documents, web documents, or RNA secondary structures. This implies that a regular tree language can be used as a theoretical toolbox for processing of the structured documents. When it comes to the string case, many researchers often use regular languages to process a

H. Jürgensen et al. (Eds.): DCFS 2014, LNCS 8614, pp. 246–257, 2014.

set of strings efficiently. Consider the case that we have a set of strings which is a regular language L. Now we want to find any occurrence of strings in L from a text T. The most common way is to construct an FA A that accepts a regular language $\Sigma^* L$ [2]. Then, we read T using A and check whether or not A reaches a final state. When A reaches a final state, we find that there is an occurrence of a matching string of L in T. We extend this approach to the tree matching problem [5]. First, we formally define the *tree matching problem* to be the problem of finding subtree occurrences of a tree in L from a set of trees T. Since a tree can be processed in a bottom-up or a top-down fashion, we need to consider different types of tree languages for the tree matching problem.

Here we consider three types of tree substructures called a *subtree*, a *topmost subtree* and an *internal subtree*. Given a tree language L, we construct three types of tree languages recognizing trees which contain the trees in L as subtrees, topmost subtrees and internal subtrees. Note that these tree languages can be used for the tree matching problem as we have used $\Sigma^* L$ for the string pattern matching problem. In particular, we tackle the deterministic state complexity of regular tree languages and path-closed languages. Interestingly, the tree language consisting of trees that have a subtree belonging to a path-closed language language need not be path-closed and therefore cannot recognized by deterministic top-down tree automata (DTTAs).

We give basic notations and definitions in Section 2. We define the three types of tree languages for tree matching in Section 3. We present the results on the state complexity of regular tree languages and path-closed languages in Section 4 and Section 5. In Section 6, we conclude the paper.

2   Preliminaries

We briefly recall definitions and properties of finite tree automata and regular tree languages. We refer the reader to the books [1,4] for more details on tree automata. A ranked alphabet Σ is a finite set of characters and we denote the set of elements of rank m by $\Sigma_m \subseteq \Sigma$ for $m \geq 0$. The set F_Σ consists of Σ-labeled trees, where a node labeled by $\sigma \in \Sigma_m$ always has m children. We use F_Σ to denote a set of trees over Σ that is the smallest set S satisfying the following condition: if $m \geq 0, \sigma \in \Sigma_m$ and $t_1, \ldots, t_m \in S$, then $\sigma(t_1, \ldots, t_m) \in S$. Let $t(u \leftarrow s)$ be the tree obtained from a tree t by replacing the subtree at a node u of t with a tree s. The notation is extended for a set U of nodes of t and $S \subseteq F_\Sigma : t(U \leftarrow S)$ is the set of trees obtained from t by replacing the subtree at each node of U by some tree in S.

A *nondeterministic bottom-up tree automaton* (NBTA) is specified by a tuple $A = (\Sigma, Q, Q_f, g)$, where Σ is a ranked alphabet, Q is a finite set of states, $Q_f \subseteq Q$ is a set of final states and g associates each $\sigma \in \Sigma_m$ to a mapping $\sigma_g : Q^m \longrightarrow 2^Q$, where $m \geq 0$. For each tree $t = \sigma(t_1, \ldots, t_m) \in F_\Sigma$, we define inductively the set $t_g \subseteq Q$ by setting $q \in t_g$ if and only if there exist $q_i \in (t_i)_g$, for $1 \leq i \leq m$, such that $q \in \sigma_g(q_1, \ldots, q_m)$. Intuitively, t_g consists of the states of Q that A may reach by reading t. Thus, the tree language accepted by A is defined as follows: $L(A) = \{t \in F_\Sigma \mid t_g \cap Q_f \neq \emptyset\}$. The automaton A is a *deterministic bottom-up tree automaton* (DBTA) if, for each $\sigma \in \Sigma_m$, where $m \geq 0$, σ_g is a partial function $Q^m \longrightarrow Q$.

A *nondeterministic top-down tree automaton* (NTTA) is specified by a tuple $A = (\Sigma, Q, Q_0, g)$, where Σ is a ranked alphabet, Q is a finite set of states, $Q_0 \subseteq Q$ is a set of initial states, and g associates each $\sigma \in \Sigma_m, m \geq 0$, a mapping $\sigma_g : Q \longrightarrow 2^{Q^m}$. As a convention, we denote the m-tuples $q_1, \ldots, q_m$ by $[q_1, \ldots, q_m]$. A top-down tree automaton A is *deterministic* if Q_0 is a singleton set and for all $q \in Q, \sigma \in \Sigma_m$, and $m \geq 1$, σ_g is a partial function $Q \longrightarrow Q^m$.

The nondeterministic (bottom-up or top-down) and deterministic bottom-up tree automata accept the family of *regular tree languages* whereas the deterministic top-down tree automata accept a proper subfamily of regular tree languages—*path-closed languages* [1,4].

3   Tree Languages for Tree Pattern Matching

Pattern matching is the problem of finding occurrences of a pattern in a text. Given an FA A for the pattern L over Σ, we can solve the problem by building a new FA for the language $\Sigma^* L$. Then, we run the new FA with the text and report the occurrence when the FA reaches a final state [2]. For tree pattern matching problem, we consider the case when we are given a set of pattern trees as a tree automaton. Note that a tree can be processed in a bottom-up way with a bottom-up TA or a top-down way with a top-down TA. Therefore, we consider three types of tree languages that can be used for tree pattern matching problem.

First we introduce our definitions for different tree substructures. We provide graphical examples for the definitions in Fig. 1.

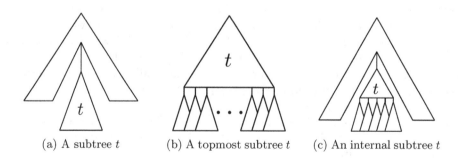

(a) A subtree t (b) A topmost subtree t (c) An internal subtree t

Fig. 1. We define three types of subtrees called a subtree, a topmost subtree and an internal subtree. These figures depict the examples.

Definition 1. *A subtree of a tree t is a tree consisting of a node in t and all of its descendants in t.*

If a tree t_1 is a subtree of t_2, then we call t_2 is a supertree of t_1. Given a tree t and a regular tree language L, we first compute a new regular tree language L' that accepts all possible supertrees of trees in L. Then, we decide whether or not a given tree t occurs as a subtree of a tree in L by deciding $t \in L'$. Similarly, we define the *topmost subtree* and the *internal subtree* as follows:

Definition 2. *A topmost subtree of a tree t is a tree consisting of a set of nodes in t including the root node such that from any node in the set, there exists a path to the root node through the nodes in the set.*

Definition 3. *An internal subtree of a tree t can be defined as a topmost subtree of a subtree of t.*

Recall that we build a new FA that accepts $\Sigma^* L$, which is a concatenation of a universal language Σ^* and a given language L, for matching a language L of string patterns. For tree pattern matching problem, we need to consider how to define the concatenation of trees properly. Recently, Piao and Salomaa [17] studied the state complexity of the concatenation of regular tree languages. They defined the sequential σ-concatenation and parallel σ-concatenation where the substitutions can occur at σ-labeled leaves.

We consider a more generalized operation that allows substitution to occur at all leaves regardless of labels. We denote the set of leaves of a tree t by $\mathtt{leaf}(t)$. Then, for $T_1 \subseteq F_\Sigma$ and $t_2 \in F_\Sigma$, we define the *sequential concatenation* of T_1 and t_2 to be $T_1 \cdot^s t_2 = \{t_2(u \leftarrow t_1) \mid u \in \mathtt{leaf}(t_2), t_1 \in T_1\}$. In other words, $T_1 \cdot^s t_2$ is a set of trees obtained from t_2 by replacing a leaf with a tree in T_1. We extend the sequential concatenation operation to the tree languages $T_1, T_2 \subseteq F_\Sigma$ as follows:

$$T_1 \cdot^s T_2 = \bigcup_{t_2 \in T_2} T_1 \cdot^s t_2.$$

The *parallel concatenation* of T_1 and t_2 is

$$T_1 \cdot^p t_2 = \{t_2(\mathtt{leaf}(t_2) \leftarrow t_1) \mid t_1 \in T_1\}.$$

Thus, $T_1 \cdot^p t_2$ is a set of trees obtained from t_2 by replacing all leaves with a tree in T_1. We can also extend the parallel concatenation to tree languages. Note that Definition 2 can be presented more nicely using the parallel concatenation operation. A tree t_2 is a topmost subtree of t_1 if $t_1 \in F_\Sigma \cdot^p t_2$.

Relying on the sequential and parallel tree concatenations, we construct three types of tree languages from a regular tree language L for the tree pattern matching problem. See Fig. 2. Given a tree language L,

(i) $L \cdot^s F_\Sigma$ is a set of trees where each element contains a subtree occurrence of a tree in L,

(ii) $F_\Sigma \cdot^p L$ is a set of trees where each element contains a topmost subtree occurrence of a tree in L, and

(iii) $F_\Sigma \cdot^p L \cdot^s F_\Sigma$ contains trees having an internal subtree occurrence of a tree of L.

Notice that a leaf node of a tree can be replaced with any other nodes for the topmost subtree occurrence and the internal subtree occurrence.

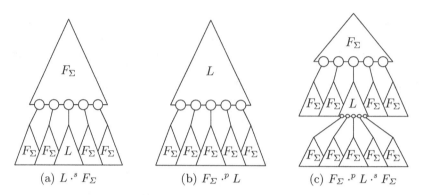

Fig. 2. Three types of tree languages for tree pattern matching problem

4   State Complexity of DBTAs

First we study the state complexity of $F_\Sigma \cdot^p L$ which can be used for finding subtree occurrences of a tree in L.

Lemma 1. *Given a DBTA* $A = (\Sigma, Q, Q_F, g)$ *with* n *states for a regular tree language* L, 2^{n-k} *states are sufficient for recognizing* $F_\Sigma \cdot^p L$ *if* $|\{\sigma_g \mid \sigma \in \Sigma_0\}| = k$.

Proof. Without loss of generality, we assume $Q_F \cap \{\sigma_g \mid \sigma \in \Sigma_0\} = \emptyset$ because otherwise $F_\Sigma \cdot^p L(A) = F_\Sigma$. We present an upper bound construction of a DBTA B for $F_\Sigma \cdot^p L(A)$. Namely, $L(B) = F_\Sigma \cdot^p L(A)$. We define $B = (\Sigma, Q', Q'_F, g')$, where

$$Q' = \{X \cup \{\sigma_g \mid \sigma \in \Sigma_0\} \mid X \in 2^{Q \setminus \{\sigma_g \mid \sigma \in \Sigma_0\}}\}, Q'_F = \{q \in Q' \mid q \cap Q_F \neq \emptyset\},$$

and the transitions of g' are defined as follows:

For $\tau \in \Sigma_0$, we define

$$\tau_{g'} = \{\sigma_g \mid \sigma \in \Sigma_0\}.$$

For $\tau \in \Sigma_m, m \geq 1$, and $P_1, P_2, \ldots, P_m \in Q'$, we define

$$\tau_{g'}(P_1, P_2, \ldots, P_m) = \tau_g(P_1, P_2, \ldots, P_m) \cup \{\sigma_g \mid \sigma \in \Sigma_0\}.$$

Now we explain how B recognizes the tree language $F_\Sigma \cdot^p L$. Note that we define every target state of g' to be the union of the set of states reachable by g and the set of states reachable by reading leaf nodes. Since every target state of g' is not empty, a new DBTA B is complete although A may not be complete. This implies that a state of B contains at least the states in $\{\sigma_g \mid \sigma \in \Sigma_0\}$ that are the set of states by reading leaf nodes in A. After reading any tree in F_Σ, the state of B contains $\{\sigma_g \mid \sigma \in \Sigma_0\}$, and thus can simulate the trees in $L(A)$.   □

The upper bound in Lemma 1 is reachable when a DBTA accepts a set of unary trees. If a DBTA accepts a set of unary trees, then we can regard the DBTA as a DFA with multiple initial states. Since the upper bound reaches the maximum when $k = 1$, we consider the state complexity of catenation of L and Σ^*. Let L be a regular language whose state complexity is n. Then, the state complexity of $\Sigma^* L$ is 2^{n-1} [19] which is the same as the bound in Lemma 1. Furthermore, we show that the upper bound is tight for any $1 \le k \le n$.

Choose $\Sigma = \Sigma_0 \cup \Sigma_1$, where $\Sigma_0 = \{\sigma_1, \sigma_2, \ldots, \sigma_k\}$ and $\Sigma_1 = \{a, b\}$. We define a DBTA $C_1 = (\Sigma, Q_{C_1}, Q_{C_1, F}, g_{C_1})$, where $Q_{C_1} = \{0, 1, \ldots, n - 1\}, Q_{C_1, F} = \{n - 1\}$ and the transition function g_{C_1} is defined by setting:

- $(\sigma_i)_{g_{C_1}} = i - 1 \ (1 \le i \le k)$,
- $a_{g_{C_1}}(i) = i + 1 \mod n$,
- $b_{g_{C_1}}(i) = i \ (0 \le i < k)$,
- $b_{g_{C_1}}(i) = i + 1 \mod n \ (k \le i < n)$.

Based on the construction of the proof of Lemma 1, we construct a DBTA $D_1 = (\Sigma, Q_{D_1}, Q_{D_1, F}, g_{D_1})$ recognizing $F_\Sigma \cdot^p L(C_1)$, where $Q_{D_1} = \{P \mid \{0, 1, \ldots, k - 1\} \subseteq P, P \subseteq Q_{C_1}\}, Q_{D_1, F} = \{P \mid P \in Q_{D_1}, P \cap Q_{C_1, F} \ne \emptyset\}$, and the transition function g_{D_1} is defined as follows:

- $(\sigma_i)_{g_{D_1}} = \{0, 1, \ldots, k - 1\} \ (0 \le i \le k)$,
- $a_{g_{D_1}}(P) = a_{g_{C_1}}(P) \cup \{0, 1, \ldots, k - 1\}$,
- $b_{g_{D_1}}(P) = b_{g_{C_1}}(P) \cup \{0, 1, \ldots, k - 1\}$.

Notice that $L(D_1) = F_\Sigma \cdot^p L(C_1)$ by Lemma 1. In the following lemma, we establish that D_1 is a minimal DBTA by showing that all states of D_1 are reachable and pairwise inequivalent.

Lemma 2. *All states of D_1 are reachable and pairwise inequivalent.*

Proof. First, we prove the reachability of all states of D_1. Note that each state of D_1 is a set of states in C_1. By the construction, the size of a state P in Q_{D_1} satisfies $k \le |P| \le n$ since $\{0, 1, \ldots, k-1\} \subseteq P$. Using induction on $|P|$, we show that all states of D_1 are reachable. For the basis, we have a state $\{0, 1, \ldots, k-1\}$ of size k that is reachable by reading a leaf node. Assuming that all states P are reachable for $|P| \le x$, we will show that any state P' is reachable when $|P'| = x + 1$. Let $P' = \{0, 1, \ldots, k-1, q_k, q_{k+1}, \ldots, q_x\}$ be a state of size $x + 1$. The state P' is reachable from a state $\{0, 1, \ldots, k-1, q_{k+1} - q_k + k - 1, \ldots, q_x - q_k + k - 1\}$ by reading a sequence of unary symbols $ab^{q_k - k}$. Therefore, all states are reachable by induction.

Next we prove that all states of D_1 are pairwise inequivalent. Pick any two distinct states P_1 and P_2. Assume $p \in P_1 \backslash P_2$. (The other possibility is completely symmetric.) After reading a sequence of unary symbols a^{n-p-1}, a final state is reached from state P_1 whereas P_2 reaches a non-final state. Therefore, all states of D_1 are pairwise inequivalent.   □

Since we have shown that there exists a corresponding lower bound for the upper bound, the bound is tight.

Theorem 1. *Given a DBTA A with n states for a regular tree language L, 2^{n-k} states are necessary and sufficient in the worst-case for the minimal DBTA of $F_\Sigma \cdot^p L$ if $|\{\sigma_g \mid \sigma \in \Sigma_0\}| = k$.*

Now we consider $L \cdot^s F_\Sigma$—a tree language consists of all trees that have trees in L as subtrees. In other words, for any tree t in L, we have all possible supertrees of t in L'. Given a regular tree language L, it is known that $L \cdot^s F_\Sigma$ is also a regular tree language [17]. We study the state complexity of $L \cdot^s F_\Sigma$.

Lemma 3. *Given a DBTA $A = (\Sigma, Q, Q_F, g)$ with n states for a regular tree language L, $n + 1$ states are sufficient for recognizing $L \cdot^s F_\Sigma$.*

Proof. We construct a new DBTA $B = (\Sigma, Q', Q'_F, g')$ for $L \cdot^s F_\Sigma$, where $Q' = Q \cup \{q_{new}\}$, $Q'_F = Q_F$, and the transition function g' is defined as follows:
For $\tau \in \Sigma_0$, we define

$$\tau_{g'} = \begin{cases} \tau_g & \text{if } \tau_g \text{ is defined,} \\ q_{new} & \text{otherwise.} \end{cases}$$

For $\tau \in \Sigma_m, m \geq 1, q_1, q_2, \ldots, q_m \in Q'$, and $q_f \in Q'_F$, we define

$$\tau_{g'}(q_1, q_2, \ldots, q_m) = \begin{cases} \tau_g(q_1, q_2, \ldots, q_m) & \text{if } \tau_g(q_1, q_2, \ldots, q_m) \text{ is defined and} \\ & \{q_1, q_2, \ldots, q_m\} \cap Q_f = \emptyset, \\ q_f & \text{if } \{q_1, q_2, \ldots, q_m\} \cap Q_f \neq \emptyset, \\ q_{new} & \text{otherwise.} \end{cases}$$

Now we explain how B accepts a set of all trees that are supertrees of trees in L. We define the transition function g' to be complete by setting the target state of the undefined transition as the new state q_{new}. Then, B moves to q_{new} by reading trees in $\overline{L}$ while moving to one of its final states by reading trees in L. Assume that B accepts a tree in L and arrives at the final state q_f. After then, B stays in q_f by reading any sequence of states including the final state q_f.   □

We cannot reach the upper bound $n+1$ with any DFA in this case since the state complexity of $L\Sigma^*$ is n which is the same as that of L, even for the incomplete DFAs. Thus, we show that there exists a lower bound DBTA of $n + 1$ states for accepting $L \cdot^s F_\Sigma$ where the state complexity of L is n to prove the tightness of the upper bound.

Choose $\Sigma = \Sigma_0 \cup \Sigma_1 \cup \Sigma_2$, where $\Sigma_0 = \{c\}$, $\Sigma_1 = \{a\}$ and $\Sigma_2 = \{b\}$. We define a DBTA $C_2 = (\Sigma, Q_{C_2}, Q_{C_2,F}, g_{C_2})$, where $Q_{C_2} = \{0, 1, \ldots, n-1\}$, $Q_{C_2,F} = \{n-1\}$, and the transition function g_{C_2} is defined by setting:

- $c_{g_{C_2}} = 0$,
- $a_{g_{C_2}}(i) = b_{g_{C_2}}(i, i) = i + 1 \mod n$.

All transitions of g_{C_2} not listed above are undefined. Based on the construction of the proof of Lemma 3, we construct a DBTA $D_2 = (\Sigma, Q_{D_2}, Q_{D_2,F}, g_{D_2})$ recognizing $L(C_2) \cdot^s F_\Sigma$, where $Q_{D_2} = Q_{C_2} \cup \{n\}$, $Q_{D_2,F} = Q_{C_2,F}$ and the transition function g_{D_2} is defined as follows:

$$- c_{gD_2} = 0,$$
$$- a_{gD_2}(i) = b_{gD_2}(i,i) = i+1 \ (0 \le i \le n-2),$$
$$- a_{gD_2}(n-1) = b_{gD_2}(n-1,i) = b_{gD_2}(i,n-1) = n-1 \ (0 \le i \le n-1),$$
$$- a_{gD_2}(n) = b_{gD_2}(i,j) = n \ (i \ne j, i \ne n-1, j \ne n-1).$$

Notice that $L(D_2) = L(C_2) \cdot^s F_\Sigma$ by Lemma 3. In the following lemma, we establish that D_2 is a minimal DBTA by showing that all states in Q_{D_2} are reachable and pairwise inequivalent.

Lemma 4. *All states of D_2 are reachable and pairwise inequivalent.*

From two lemmas, we establish the following theorem.

Theorem 2. *Given a DBTA A with n states for a regular tree languages L, $n+1$ states are necessary and sufficient in the worst-case for the minimal DBTA of $L \cdot^s F_\Sigma$.*

We lastly consider the state complexity of $F_\Sigma \cdot^p L \cdot^s F_\Sigma$. Note that the sequential catenation of trees is not associative whereas the parallel catenation of trees is associative. That means that there exist trees t_1, t_2 and t_3 such that $(t_1 \cdot^s t_2) \cdot^s t_3$ and $t_1 \cdot^s (t_2 \cdot^s t_3)$ do not coincide. This also applies to the catenation of tree languages and thus, leads to $(L_1 \cdot^s L_2) \cdot^s L_3 \ne L_1 \cdot^s (L_2 \cdot^s L_3)$ for some regular tree languages L_1, L_2, and L_3. However, we consider a special tree language F_Σ for L_1 and L_3 that makes $(F_\Sigma \cdot^s L_2) \cdot^s F_\Sigma = F_\Sigma \cdot^s (L_2 \cdot^s F_\Sigma)$. Thus, we simply denote the language by $F_\Sigma \cdot^p L \cdot^s F_\Sigma$ instead of $(F_\Sigma \cdot^s L_2) \cdot^s F_\Sigma$ or $F_\Sigma \cdot^s (L_2 \cdot^s F_\Sigma)$.

Now we tackle the state complexity of $F_\Sigma \cdot^p L \cdot^s F_\Sigma$.

Lemma 5. *Given a DBTA $A = (\Sigma, Q, Q_F, g)$ with n states for a regular tree language L, $2^{n-t-k}+1$ states are sufficient for recognizing $F_\Sigma \cdot^p L \cdot^s F_\Sigma$ if $|Q_F| = t$ and $|\{\sigma_g \mid \sigma \in \Sigma_0\}| = k$.*

Proof. Without loss of generality, we assume that $Q_F \cap \{\sigma_g \mid \sigma \in \Sigma_0\} = \emptyset$ because otherwise $F_\Sigma \cdot^p L(A) \cdot^s F_\Sigma = F_\Sigma$. We give an upper bound construction of DBTA B that recognizes $F_\Sigma \cdot^p L(A) \cdot^s F_\Sigma$. Namely, $L(B) = F_\Sigma \cdot^p L(A) \cdot^s F_\Sigma$. We define $B = (\Sigma, Q', Q'_F, g')$, where

$$Q' = \{X \cup \{\sigma_g \mid \sigma \in \Sigma_0\} \mid X \in 2^{Q \setminus (Q_F \cup \{\sigma_g \mid \sigma \in \Sigma_0\})}\} \cup \{Q_F\}, Q'_F = \{Q_F\},$$

and the transitions of g' are defined as follows:
For $\tau \in \Sigma_0$, we define $\tau_{g'} = \{\sigma_g \mid \sigma \in \Sigma_0\}$.
For $\tau \in \Sigma_m, m \ge 1$, and $P_1, P_2, \ldots, P_m \in Q'$, we define

$$\tau_{g'}(P_1, P_2, \ldots, P_m) = \begin{cases} \tau_g(P_1, P_2, \ldots, P_m) \cup \{\sigma_g \mid \sigma \in \Sigma_0\} \text{ if } \bigcup_{i=1}^{m} P_i \cap Q_F = \emptyset, \\ Q_F \qquad\qquad\qquad\qquad\qquad \text{otherwise.} \end{cases}$$

Here we do not explain how B accepts $F_\Sigma \cdot^p L \cdot^s F_\Sigma$ because the construction can be explained as a simple combination of two constructions given in Lemma 1 and Lemma 3. $\qquad\square$

We give a lower bound example that reaches the upper bound $2^{n-t-k}+1$.

Choose $\Sigma = \Sigma_0 \cup \Sigma_1$, where $\Sigma_0 = \{\sigma_1, \sigma_2, \ldots, \sigma_k\}$ and $\Sigma_1 = \{a, b, c\}$. We define a DBTA $C_3 = (\Sigma, Q_{C_3}, Q_{C_3,F}, g_{C_3})$, where $Q_{C_3} = \{0, 1, \ldots, n-1\}, Q_{C_3,F} = \{n-t, n-t+1, \ldots, n-1\}$ and the transition function g_{C_3} is defined by setting:

- $(\sigma_i)_{g_{C_3}} = i - 1 \ (1 \le i \le k)$,
- $a_{g_{C_3}}(i) = i + 1 \mod n$,
- $b_{g_{C_3}}(i) = i \ (0 \le i \le k)$,
- $b_{g_{C_3}}(i) = i + 1 \mod n \ (k \le i < n)$,
- $c_{g_{C_3}}(i) = i + 1 \mod n$ if $i \ne n - t - 1, c_{g_{C_3}}(n-t-1) = 0$.

Based on the construction in the proof of Lemma 5, we construct a DBTA $D_3 = (\Sigma, Q_{D_3}, Q_{D_3,F}, g_{D_3})$ recognizing $F_\Sigma \cdot^p L(C_3) \cdot^s F_\Sigma$, where $Q_{D_3} = \{P \mid \{0, 1, \ldots, k-1\} \subseteq P, \ P \subseteq Q_{C_3} \setminus Q_{C_3,F}\}, Q_{D_3,F} = \{Q_{C_3,F}\}$, and the transition function g_{D_3} is defined as follows:

- $(\sigma_i)_{g_{D_3}} = \{0, 1, \ldots, k-1\}$,
- $a_{g_{D_3}}(P) = P' \cup \{0, 1, \ldots, k-1\}$ if $a_{g_{C_3}}(P) \cap Q_{C_3,F} = \emptyset$,
- $a_{g_{D_3}}(P) = \{Q_{C_3,F}\}$ if $a_{g_{C_3}}(P) \cap Q_{C_3,F} \ne \emptyset$,
- $b_{g_{D_3}}(P) = P' \cup \{0, 1, \ldots, k-1\}$ if $b_{g_{C_3}}(P) \cap Q_{C_3,F} = \emptyset$,
- $b_{g_{D_3}}(P) = \{Q_{C_3,F}\}$ if $b_{g_{C_3}}(P) \cap Q_{C_3,F} \ne \emptyset$,
- $c_{g_{D_3}}(P) = P' \cup \{0, 1, \ldots, k-1\}$ if $c_{g_{C_3}}(P) \cap Q_{C_3,F} = \emptyset$,
- $a_{g_{D_3}}(\{Q_{C_3,F}\}) = b_{g_{D_3}}(\{Q_{C_3,F}\}) = c_{g_{D_3}}(\{Q_{C_3,F}\}) = \{Q_{C_3,F}\}$.

Notice that $L(D_3) = F_\Sigma \cdot^p L(C_3) \cdot^s F_\Sigma$ by Lemma 5. In the following lemma, we establish that D_3 is a minimal DBTA by showing that all states in Q_{D_3} are reachable and pairwise inequivalent.

Lemma 6. *All states of D_3 are reachable and pairwise inequivalent.*

Proof. We prove the reachability of all non-final states of D_3 using induction on the size of P. Note that any non-final state $P \in Q_{D_3}$ satisfies $k \le |P| \le m - t$ because $Q_{C_3,F} \cap P = \emptyset$ and $\{\sigma_c \mid \sigma \in \Sigma_0\} \subseteq P$ by the construction. A state $\{0, 1, \ldots, k-1\}$ of size k is reachable by reading a leaf node. Assume that all states P is reachable for $|P| \le x$. Then, we show that any state P' of size $x+1$ is reachable. Let $P' = \{0, 1, \ldots, k-1, q_k, q_{k+1}, \ldots, q_x\}$ be a state of size $x+1$. Then, the state P' is reached from a state $\{0, 1, \ldots, k-1, q_{k+1}-q_k+k-1, \ldots, q_x-q_{k+1}+ k-1\}$ after reading a sequence of unary symbols ab^{q_k-k}. From the induction, it is easy to verify that all states except $Q_{C_3,F}$ are reachable. Furthermore, the only final state $Q_{C_3,F}$ is reachable from a non-final state $\{0, 1, \ldots, n-t-1\}$ by reading a unary symbol a.

Next we prove that all states of D_3 are pairwise inequivalent. Pick any two distinct states P_1 and P_2. Assume $p \in P_1 \setminus P_2$. (The other possibility is symmetric.) From P_1, a final state is reached by reading a sequence of unary symbols $c^{n-t-1-p}a$ whereas P_2 does not reach the final state. Therefore, any two states in Q_{D_3} are pairwise inequivalent. $\qquad \square$

Theorem 3. *Given a DBTA $A = (\Sigma, Q, Q_F, g)$ with n states for a regular tree language L, $2^{n-t-k}+1$ states are necessary and sufficient in the worst-case for the minimal DBTA of $F_\Sigma \cdot^p L \cdot^s F_\Sigma$ if $|Q_F| = t$ and $|\{\sigma_g \mid \sigma \in \Sigma_0\}| = k$.*

5   State Complexity of DTTAs

It is well known that every NBTA can be converted into an equivalent NTTA [1,4]. However, it does not mean that there always exists a DTTA for every regular tree language. This implies that a class of regular tree languages accepted by DTTAs is a proper subclass of regular tree languages accepted by NBTAs and NTTAs. It is known that DTTAs recognize exactly the class of *path-closed languages* which is a proper subclass of regular tree languages [1,4].

We study the state complexity of path-closed languages for tree matching. Following the previous results, we consider three types of tree languages $F_\Sigma \cdot^p L$, $L \cdot^s F_\Sigma$, and $F_\Sigma \cdot^p L \cdot^s F_\Sigma$, where L is a tree language. However, two tree languages $L \cdot^s F_\Sigma$ and $F_\Sigma \cdot^p L \cdot^s F_\Sigma$ appear to be not path-closed languages. Nivat and Podelski [13] argued that path-closed languages can be characterized by a property called the subtree exchange property as follows:

Corollary 1 (Nivat and Podelski [13]). *A regular tree language L is path-closed if and only if, for every $t \in L$ and every node $u \in t$, if $t(u \leftarrow a(t_1, \ldots, t_m)) \in L$ and $t(u \leftarrow a(s_1, \ldots, s_m)) \in L$, then $t(u \leftarrow a(t_1, \ldots, s_i, \ldots, t_m)) \in L$ for each $i = 1, \ldots, m$.*

Using the subtree exchange property, we prove that given a tree language L, $L \cdot^s F_\Sigma$ and $F_\Sigma \cdot^p L \cdot^s F_\Sigma$ are not path-closed languages.

Lemma 7. *There exists a path-closed language L such that $L \cdot^s F_\Sigma$ is not a path-closed language.*

Proof. Let $\Sigma = \Sigma_2 \cup \Sigma_0$, where $\Sigma_2 = \{b\}$, and $\Sigma_0 = \{a, c\}$. A singleton language L contains a single-node tree c, namely $L = \{c\}$. It is straightforward to verify that F_Σ contains every binary tree where leaf nodes are labeled by a or c, and non-leaf nodes are labeled by b. Then, $L \cdot^s F_\Sigma$ is a set of binary trees where every tree contains at least one leaf labeled by c. Therefore, $b(a, c) \in L \cdot^s F_\Sigma$, $b(c, a) \in L \cdot^s F_\Sigma$, and $b(a, a) \notin L \cdot^s F_\Sigma$ hold. However, if $L \cdot^s F_\Sigma$ is path-closed, $b(a, a)$ should exist in $L \cdot^s F_\Sigma$ by the subtree exchange property. This implies that $L \cdot^s F_\Sigma$ is not a path-closed language.   □

Lemma 8. *There exists a path-closed language L such that $F_\Sigma \cdot^p L \cdot^s F_\Sigma$ is not a path-closed language.*

Proof. Let $\Sigma = \Sigma_2 \cup \Sigma_0$, where $\Sigma_2 = \{a, b\}$, and $\Sigma_0 = \{c\}$. A singleton language L contains a tree $a(c, c)$, namely $L = \{a(c, c)\}$. It is easy to verify that F_Σ contains every binary tree where all leaf nodes are labeled by c and non-leaf nodes are labeled by a or b.

Then, $F_\Sigma \cdot^p L \cdot^s F_\Sigma$ is a set of binary trees where every tree contains at least one non-leaf node labeled by a. Therefore, $b(a(c, c), c) \in F_\Sigma \cdot^p L \cdot^s F_\Sigma, b(c, a(c, c)) \in F_\Sigma \cdot^p L \cdot^s F_\Sigma$, and $b(c, c) \notin F_\Sigma \cdot^p L \cdot^s F_\Sigma$. However, due to the subtree exchange property, $b(c, c)$ should be in $F_\Sigma \cdot^p L \cdot^s F_\Sigma$ if the language $F_\Sigma \cdot^p L \cdot^s F_\Sigma$ is path-closed. This means that $F_\Sigma \cdot^p L \cdot^s F_\Sigma$ is not a path-closed language.   □

We define the *deterministic top-down state complexity* of a path-closed language L to be the number of states that are necessary and sufficient in the worst-case for the minimal DTTA recognizing L.

Theorem 4. *Given a DTTA $A = (\Sigma, Q, Q_0, g)$ with n states for a path-closed language L, n states are necessary and sufficient in the worst-case for the minimal DTTA of $F_\Sigma \cdot^p L$.*

Proof. We construct a new DTTA $B = (\Sigma, Q', Q_0', g')$ for $F_\Sigma \cdot^p L$, where $Q' = Q$, $Q_0' = Q_0$, and the transition function g' is defined as follows:

For $\tau \in \Sigma_m, m \geq 0$ and $q \in Q'$, we define

$$
\tau_{g'}(q) = \begin{cases} \tau_g(q) & \text{if } \sigma_g(q) \neq \lambda \text{ for any } \sigma \in \Sigma_0, \\ \underbrace{[q, q, \ldots, q]}_{m \text{ times}} & \text{otherwise.} \end{cases}
$$

Now we explain how B simulates $F_\Sigma \cdot^p L$ with n states. Since trees in $F_\Sigma \cdot^p L$ have the same topmost parts with trees in L and leaves can be substituted with any tree in F_Σ, B simulates from the same initial state with A. Let us assume that a state $q \in Q'$ may end the top-down computation with generating a leaf node since $\sigma_g(q) = \lambda$. Once B arrives at q, the new transition function g' continues the computation by reading a non-leaf label of rank m and generating a sequence $[q, q, \ldots, q]$ of states whose length is m. This makes a new DTTA B to generate any subtree in F_Σ at the point where the computation may end with generating leaves and, thus, recognize the language $F_\Sigma \cdot^p L$.

It is easy to verify that n states are necessary to recognize $F_\Sigma \cdot^p L$. Consider a path-closed language of unary trees whose state complexity correspond to that of regular string languages. Since the state complexity of $L\Sigma^*$ is n if the state complexity of L is n, this case can be a lower bound for the path-closed language $F_\Sigma \cdot^p L$.   □

6   Conclusions

We have considered three tree languages $F_\Sigma \cdot^p L$, $L \cdot^s F_\Sigma$, and $F_\Sigma \cdot^p L \cdot^s F_\Sigma$ motivated from the tree pattern matching problem and have established the state complexity of these languages described by DBTAs and DTTAs. We have also shown that $L \cdot^s F_\Sigma$ and $F_\Sigma \cdot^p L \cdot^s F_\Sigma$ are not recognizable by DTTAs even when L is a path-closed language since they are not necessarily path-closed languages. In addition, we have demonstrated that $L \cdot^s F_\Sigma$ and $F_\Sigma \cdot^p L \cdot^s F_\Sigma$ need not be path-closed and therefore cannot recognized by DTTAs. In future, we aim to investigate the descriptional complexity of unranked tree automata, which are a more generalized model than tree automata over ranked alphabet, for recognizing $L \cdot^s F_\Sigma$ and $F_\Sigma \cdot^p L \cdot^s F_\Sigma$.

References

1. Comon, H., Dauchet, M., Jacquemard, F., Lugiez, D., Tison, S., Tommasi, M.: Tree Automata Techniques and Applications (2007), Electronic book available at http://www.tata.gforge.inria.fr

2. Crochemore, M., Hancart, C.: Automata for Matching Patterns. In: Handbook of formal languages, vol. 2, pp. 399–462 (1997)
3. Eom, H.-S., Han, Y.-S., Ko, S.-K.: State complexity of subtree-free regular tree languages. In: Jurgensen, H., Reis, R. (eds.) DCFS 2013. LNCS, vol. 8031, pp. 66–77. Springer, Heidelberg (2013)
4. Gécseg, F., Steinby, M.: Tree languages. In: Rozenberg, G., Salomaa, A. (eds.) Handbook of Formal Languages. Beyond Words, vol. 3, pp. 1–68. Springer-Verlag New York, Inc., (1997)
5. Hoffmann, C.M., O'Donnell, M.J.: Pattern matching in trees. Journal of the ACM 29(1), 68–95 (1982)
6. Holzer, M., Kutrib, M.: Descriptional and computational complexity of finite automata – a survey. Information and Computation 209, 456–470 (2011)
7. Kutrib, M., Pighizzini, G.: Recent trends in descriptional complexity of formal languages. Bulletin of the EATCS 111, 70–86 (2013)
8. Lupanov, O.: A comparison of two types of finite sources. Problemy Kibernetiki 9, 328–335 (1963)
9. Maslov, A.: Estimates of the number of states of finite automata. Soviet Mathematics Doklady 11, 1373–1375 (1970)
10. Meyer, A., Fisher, M.: Economy of description by automata, grammars and formal systems. In: Proceedings of the 12th Annual Symposium on Switching and Automata Theory, pp. 188–191 (1971)
11. Moore, F.: On the bounds for state-set size in the proofs of equivalence between deterministic, nondeterministic and two-way finite automata. IEEE Transactions on Computers C-20, 1211–1214 (1971)
12. Neven, F.: Automata theory for XML researchers. ACM SIGMOD Record 31(3), 39–46 (2002)
13. Nivat, M., Podelski, A.: Minimal ascending and descending tree automata. SIAM Journal on Computing 26(1), 39–58 (1997)
14. Piao, X., Salomaa, K.: State trade-offs in unranked tree automata. In: Holzer, M., Pighizzini, G., kutrib, M. (eds.) DCFS 2011. LNCS, vol. 6808, pp. 261–274. Springer, Heidelberg (2011)
15. Piao, X., Salomaa, K.: Transformations between different models of unranked bottom-up tree automata. Fundamenta Informaticae 109(4), 405–424 (2011)
16. Piao, X., Salomaa, K.: State complexity of Kleene-star operations on trees. In: Dinneen, M.J., Khoussainov, B., Nies, A. (eds.) WTCS 2012 (Calude Festschrift). LNCS, vol. 7160, pp. 388–402. Springer, Heidelberg (2012)
17. Piao, X., Salomaa, K.: State complexity of the concatenation of regular tree languages. Theoretical Computer Science 429, 273–281 (2012)
18. Shallit, J.: A Second Course in Formal Languages and Automata Theory, 1st edn. Cambridge University Press, New York (2008)
19. Yu, S., Zhuang, Q., Salomaa, K.: The state complexities of some basic operations on regular languages. Theoretical Computer Science 125(2), 315–328 (1994)

On the Complexity of L-reachability

Balagopal Komarath*, Jayalal Sarma, and K.S. Sunil

Department of Computer Science & Engineering, Indian Institute of Technology
Madras, Chennai – 36, India
{baluks,jayalal,sunil}@cse.iitm.ac.in

Abstract. We initiate a complexity theoretic study of the language
based reachability problem (L-reach) : Fix a language L. Given a graph
whose edges are labelled with alphabet symbols and two special vertices
s and t, test if there is path P from s to t in the graph such that the concatenation of the symbols seen from s to t in the path P forms a string in
the language L. We study variants of this problem with different graph
classes and different language classes and obtain complexity theoretic
characterisations for all of them. Our main results are the following:

- Restricting the language using formal language theory we show that
  the complexity of L-reachability increases with the power of the formal language class. We show that there is a regular language for
  which the L-reachability problem is NLOG-complete even for undirected graphs. In the case of linear languages, the complexity of
  L-reach does not go beyond the complexity of L itself. Further, there
  is a deterministic context-free language L for which L−DAGREACH is
  LogCFL-complete.
- We use L-reachability as a lens to study structural complexity. In
  this direction we show that there is a language A in LOG for which
  A–DAGREACH is NP-complete. Using this we show that P vs NP
  question is equivalent to P vs DAGREACH^{-1}(P) question. This leads
  to the intriguing possibility that by proving DAGREACH^{-1}(P) is contained in some subclass of P, we can prove an upward translation of
  separation of complexity classes. Note that we do not know a way
  to upward translate the separation of complexity classes.

1 Introduction

Reachability problems in mathematical structures are a well-studied problem
in space complexity. An important example is the graph reachability problem
where given a graph G and two special vertices s and t, is there a path from
s to t in the graph G. This problem exactly captures the space complexity of
problems solvable in nondeterministic logarithmic space. Various restrictions of
the problem have been studied - reachability in undirected graphs characterises
deterministic log space [9], reachability in constant width graphs characterises
NC^1 [3], reachability in planar constant width graphs characterises ACC^0 [5] and
the version in upward planar constant width graphs characterises AC^0 [4].

---

* The author was supported by the TCS Ph.D. Fellowship.

H. Jürgensen et al. (Eds.): DCFS 2014, LNCS 8614, pp. 258–269, 2014.

A natural extension of the problem using formal language theory is the L-reachability problem : Fix a language L defined over a finite alphabet Σ. Given a graph whose edges are labelled by alphabet symbols and two special vertices s and t, test if there is path from s to t in the graph such that the concatenation of the symbols seen from s to t forms a string in the language L. Indeed, if L is Σ^*, then the string on any path from s to t will be in the language. Hence the problem reduces to the graph reachability problem.

Although L-reachability problem has not been studied from a space complexity theory perspective, a lot is known about its complexity. An immediate observation is that, the L-reachability problem is at least as hard as the membership problem of L. Indeed, given a string x, to check for membership in L it suffices to test L-reachability in a simple path of length $|x|$ where the edges are labelled by the symbols in x in that sequence. The literature on the problem is spread over two main themes. One is on restricting the language from the formal language perspective, and the other is by restricting the family of graphs in terms of structure.

An important special case of the problem that was studied, is when the language is restricted to be a context-free language(CFLs). This is called the CFL-reachability problem. A primary motivation to study this problem is their application in various practical situations like inter-procedural slicing and inter-procedural data flow analysis [8,10,11]. These are used in code optimisation, vectorisation and parallelization phases of compiler design where one should have information about reaching definitions, available expressions, live variables, etc. associated with the program elements. The goal of inter-procedural analysis is to perform static examination of above properties of a program that consists of multiple procedures. Once a program is represented by its program dependence graph [10], the slicing problem is simply the CFL-reachability problem.

Our Results: The results in this paper are in two flavours.
Results Based on Chomsky Hierarchy and Graph Classes: Firstly we study restrictions of L-reachability problem when L is restricted using formal language hierarchy and the graph is restricted to various natural graph classes. Our results on this front are listed in the table below (for the sake of completeness, we include the some known results too).

Language Class	Tree-Reach	DAG-Reach	UReach/Reach
Regular	LOG-complete[15]	NLOG-complete[15]	NLOG-complete (Theorem 1/[15])
Linear	NLOG-complete (Theorem 2)	NLOG-complete (Theorem 2)	NLOG-complete (Theorem 2)
Context-free	LogCFL-complete (Prop. 2)	LogCFL-complete (Prop. 2)	P-complete (Theorem 4/[14])
Context-sensitive	PSPACE-complete (Prop. 3)	PSPACE-complete (Prop. 3)	Undecidable ([2])

Results on the Structural Complexity Front: Now we take a complexity theoretic view, we will view L-reach as an operator on languages. It is shown in Barrett et. al. [2] that even for languages in LOG, the languages L–REACH and L–UREACH are undecidable. Therefore in this section, we consider only DAGs. Note that for any L, the language L–DAGREACH is decidable.

It is natural to ask whether increasing the complexity of L increases the complexity of L–DAGREACH. More concretely , does $A \leq_m^{LOG} B \implies$ A–DAGREACH $\leq_m^{LOG}$ B–DAGREACH? The following theorem, along with the fact that there exists a language L that is LOGCFL-complete for which L–DAGREACH remains LOGCFL-complete shows that such a result is highly unlikely.

Theorem 6: *There exists a language* $A \in$ LOG *for which* A–DAGREACH *is* NP-*complete.*

For any complexity class C, we consider the class of languages defined as DAGREACH^{-1}(C) = {L : L–DAGREACH $\in$ C}. We have the following theorems for natural choices of C. Note that for any class C, we have DAGREACH^{-1}(C) $\subseteq$ C.

Theorem : *We show the following structural theorems:*

1. *(Theorem 7)* DAGREACH^{-1}(PSPACE) = PSPACE, DAGREACH^{-1}(NP) = NP.
2. *(Theorem 8)* P $\neq$ DAGREACH^{-1}(P) $\iff$ P $\neq$ NP.
3. *(Theorem 9)* DAGREACH^{-1}(NLOG) $\neq$ NLOG $\iff$ NP $\neq$ NLOG.

The above theorem shows that separating DAGREACH^{-1}(P) from P would separate P from NP. This gives us an upward translation of lower bounds on complexity classes provided we can prove that DAGREACH^{-1}(P) is contained in some subclass of P. Hence the question whether we can identify some "natural" complexity class containing DAGREACH^{-1}(P) becomes very interesting. It is clear that DAGREACH^{-1}(P) contains LogCFL-complete problems but is highly unlikely to contain some problems in LOG. If DAGREACH^{-1}(P) contains some P-complete problem, then proving that DAGREACH^{-1}(P) is contained in some subclass of P would be very hard. In this connection, we show the following:

Theorem 10: *If* L *is* P-*complete under syntactic read-once log space reductions, then* L–DAGREACH *is* NP-*complete.*

If we are able to extend the above theorem to all types of reductions, then it implies that DAGREACH^{-1}(P) is unlikely to contain P-complete problems. In other words, the above theorem could be interpreted as evidence (albeit very weak evidence) that DAGREACH^{-1}(P) may indeed be contained in some subclass of P.

We also remark that Theorem 1 holds with NLOG instead of P. However, since we know that DAGREACH^{-1}(NLOG) contains NLOG-complete (under logspace reductions) languages, item (3) of Theorem 1 is not as promising (as item (2)).

2   Preliminaries

In this section, we define language restricted reachability problems and make some observations on their complexity. The definitions for standard complexity classes and their complete problems that we are using in this paper can be found in standard complexity theory textbooks [1]. We use LOG and NLOG to stand for the complexity classes deterministic logspace and nondeterministic logspace respectively.

Definition 1. *For any* $L \in \Sigma^*$, *we consider graphs G where each edge in G is labelled by an element from Σ. For any path in G we define the yield of the path as the string formed by concatenating the symbols found in the path in that order. Then we define the language* L–REACH *as the set of all (G, s, t) such that there exists a path from s to t in G with yield in* L.

By restricting the graph in Definition 1, we obtain similar definitions for L–DAGREACH (DAGs), L–UREACH (Undirected Graphs) and L–TREEREACH (Orientations of Undirected Trees).

Observation 1. *Any language L is reducible to L–TREEREACH via projection reductions.*

Clearly, the above observation holds for any reachability variant based on the graph. This is because L–TREEREACH is a restriction of the other reachability variants. In fact the following observation shows that L–TREEREACH is not much harder than L.

Observation 2. *For any language L, the language L–TREEREACH is logspace reducible to* L.

Observation 2 holds because in logspace we can find the unique path (and hence its yield) from s to t and run the algorithm for L on the yield.

Next we define classes of languages based on language restricted reachability.

Definition 2. *For any class of languages C, we define the set of languages C–REACH as the class of all languages L–REACH where L is in C.*

Again, by restricting graphs in Definition 2, we obtain similar definitions for C–DAGREACH, C–UREACH and C–TREEREACH.

For a class of languages C and a complexity class D, we say that C–REACH *is complete for* D if the following conditions are satisfied.

– For all $L \in C$, the language L–REACH is in D.
– There exists a language L in C such that the language L–REACH is hard for D.

Definition 3. *For any complexity class C, we define* REACH$^{-1}(C)$ *as the set of all languages L such that L–REACH is in C.*

Again, by restricting graphs in Definition 3, we obtain similar definitions for DAGREACH^{-1}(C), UREACH^{-1}(C) and TREEREACH^{-1}(C).

Note that by Observation 1, for any class C the relations REACH^{-1}(C) $\subseteq$ DAGREACH^{-1}(C) $\subseteq$ TREEREACH^{-1}(C) $\subseteq$ C holds. In this paper, we will be mainly studying DAGREACH^{-1}(C) for many interesting complexity classes C.

Our motivation in studying DAGREACH^{-1}(C) is that it seems that it may be helpful in proving upward translation of separation of complexity classes. Note that we already know, by a standard padding argument, how to translate separations of complexity classes downwards. For example, we know that NEXP $\neq$ EXP $\implies$ P $\neq$ NP. The central question that we address is the following - For a class C, what is the complexity of DAGREACH^{-1}(C)? Clearly DAGREACH^{-1}(C) is contained in C. But for many natural complexity classes C, D and E, if we can show that if DAGREACH^{-1}(C) is contained in some subclass D of C, then separating C and D is equivalent to separating C from some complexity class E that contains C.

We now state a known result with its proof idea which will be used later in the paper.

Proposition 1 ([11]). CFL–REACH *is in* P.

Proof. (Sketch) The proof is a dynamic programming algorithm. The algorithm maintains for each pair of vertices u and v a table entry $Y[u, v]$ such that $Y[u, v]$ is the set of all non-terminals V in the grammar such that there is a path from u to v with yield that can be derived from V. The algorithm can be modified to output the derivation for x where $x \in$ L is the yield of a path from s to t. Note that this implies that for all "yes" instances there exists a string with length of the derivation at most polynomial in the size of the graph. $\square$

We use REG, CFL and CSL to stand for well-known formal language classes of regular, context-free and context-sensitive languages respectively[7]. The formal language class LIN, called the set of all linear languages, is the set of all languages with a context-free grammar where the right-hand side of each production consists of at most one non-terminal. The class LIN can also be characterised as CFLs that can be decided by 1-turn PDAs.

Sudborough [13] studied the class of all languages logspace reducible to a CFL. This class is called LOGCFL. Sudborough [13] also showed that LOGCFL can be characterised as the set of all languages accepted by an AuxPDA(poly). An AuxPDA(poly) is an NTM with a read-only input tape and a logspace read-write work tape. It also has a pushdown stack available for auxiliary storage. The machine is only allowed to run for a polynomial number (in the input length) of steps. It is also known that the language NBC(D_2) (Nondeterministic block choice Dyck$_2$) complete for the class LOGCFL. The language NBC(D_2) consists of all strings of the form $x_1[x_2 \# x_3][x_4 \# x_5] \ldots [x_k \# x_{k+1}]$ where each x_i is a string of two types of parenthesis. The string between "[" and "]" is called a block and the # separates choices in a block. The string is in the language iff there is a choice of x_i from each block such that the final string (after all choices have been made) is in D_2.

3   Formal Language Class Restricted Reachability

We know that REG–REACH is in NLOG[15]. The algorithm works by constructing the product automata of the input graph and the DFA for the regular language. The problem then reduces to the reachability problem on the product automata. One problem with this approach is that even if the input graph is an undirected graph, the product automata will be a directed graph. We know that reachability in directed graphs is harder than reachability in undirected graphs. The following theorem shows that for regular language restricted directed and undirected reachability are equivalent.

Theorem 1. *If* L *is the regular language* L((ab)*) *then* L–UREACH *is* NLOG-*complete.*

Proof. To show that L–UREACH is NLOG-hard, we give a logspace reduction from REACH. Given an instance (G, s, t) of REACH we construct an instance (G', s, t) of L–UREACH where G' is a labelled undirected graph where each edge is labelled either a or b. The vertex set of G' is given by $V(G') = V(G) \cup \{m_{uv} : (u, v) \in E(G)\}$. For each edge $(u, v) \in E(G)$, we add two undirected edges $\{u, m_{uv}\}$ labelled a and $\{m_{uv}, v\}$ labelled b to $E(G')$. It is easy to see that any directed path from s to t corresponds to a path from s to t in G' labelled by a string in L and vice versa.   □

So we know that REG–REACH is NLOG-complete and CFL–REACH is at least as hard as LOGCFL. So it is interesting to consider the complexity of LIN–REACH. We know that REG ⊆ LIN ⊆ CFL in the formal language theory setting. The following theorem shows that LIN–REACH is equivalent to REG–REACH.

Theorem 2. LIN–TREEREACH,   LIN–DAGREACH,   LIN–UREACH   *and* LIN–REACH *are all* NLOG-*complete.*

Proof. There is an NLOG-complete language in LIN [12]. The hardness follows from this fact. Now we show that all these problems are in NLOG. The Dynamic Programming algorithm for CFL–REACH runs in poly-time and produces a polynomial length derivation for the output string (string yielded by the path). For any language in LIN, a polynomial length derivation can only produce a polynomial length string (and hence polynomial length path). Let us say that the length of the path is bounded by n^k where n is the size of the graph and k is a constant. Then our algorithm will search for a path of length at most n^k by nondeterministically guessing the next vertex at each step and simultaneously parsing the string at each step (Using a 1-turn PDA.). This can be implemented by a 1-turn AuxPDA that runs in time n^k and takes $\log(n)$ space. By a result due to Sudborough [12], this class is exactly the same as NLOG.   □

The following theorem shows that for solving reachability for DCFLs (which are nondeterministic), some nondeterminism is unavoidable.

Theorem 3. DCFL–DAGREACH *is* LogCFL-*complete.*

Proof. Let L ∈ DCFL. We will describe an AuxPDA(poly) that decides the language L–DAGREACH. The machine starts with the source vertex s as the current vertex. At each step it nondeterministically moves to an out-neighbour of the current vertex. When the machine takes the edge (u, v) it executes one step of the DPDA for L, using the stack and finite control, with the label on (u, v) as the current input symbol. The machine accepts iff it reaches t and the DPDA accepts.

For hardness, we reduce NBC(D_2) to DCFL–DAGREACH. The reduction results in a series-parallel graph as shown in Figure 1. In the figure, a dashed arrow represents a simple path labelled by the given string. Note that the language D_2 is a DCFL.   □

$$x_1[x_2\#x_3][x_4\#x_5]$$

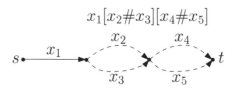

Fig. 1. Reducing NBC(D_2) to DCFL–DAGREACH

Proposition 2. CFL–TREEREACH *and* CFL–DAGREACH *are* LOGCFL*-complete.*

Proof. Sudborough [12] defines a context-free language that is complete for the class LOGCFL. This shows the hardness. To show membership in LOGCFL consider an AuxPDA(poly) that starts with s as the current vertex and at each step guesses the next vertex while simultaneously using the stack to simulate the parsing of the CFL. This machine accepts iff the current vertex is t at some point and the PDA is in an accepting state at the same time. It is easy to see that this AuxPDA(poly) decides these languages.

We now give a simplified presentation of a known result that says that CFL–REACH is P-complete. Also observe that

Theorem 4 ([14]). *If* D_2 *is the CFL (ε-free Dyck$_2$) given by the grammar*

$$S \rightarrow (S) \mid [S] \mid SS \mid (\,) \mid [\,]$$

then D_2–REACH *is* P*-complete.*

Proof. This theorem has been proved by [14] using a different terminology. Here we give a simplified presentation of the proof using our terminology for the hardness of this language. We show the hardness for P by reducing MCVP (Monotone Circuit Value Problem where fan-out and fan-in of each gate is at most 2) to D_2–REACH. We may assume wlog that each gate in the input circuit has fan-out at most 2. The reduction works by replacing each gate by a gadget as shown in Figure 2. Each gadget in the construction has an input vertex and an output vertex. The gadgets for input gates are straightforward. For an AND gate we add 3 new vertices and connect them to the gadgets for two gates feeding input to the AND gate. Suppose that the left input to the AND gate comes from the 2nd (1st) output wire of the left input gate. Then the first and second edges are labelled by "[" ("(" resp.) and "]" (")" resp.) respectively.

We use proof by induction on the level of the output gate of the circuit to prove the correctness of this reduction. The inductive hypothesis is that there is a valid path from the input vertex to the output vertex of a gadget iff the output of the gate is 1 and any path that enters a gadget through its input gate and leaves it from some vertex other than its output vertex will be invalid. This holds trivially for gadgets for input gates. Now any valid path from the input vertex to the output vertex of the AND gadget must consist of valid subpaths within the gadgets for the gates feeding input to this AND gate. The only exception is when some path leaves this gadget for the AND gate from some vertex other than its output vertex. Note that by the inductive hypothesis such a path can only leave from vertex w or z of the gadget. But the vertex w (z) has out-degree at most 2 and the other edge will be labelled by a closing bracket that does not match the type of bracket on the edge (u, v). This mismatch invalidates the path. A similar argument holds for OR gates. This closes the induction.   □

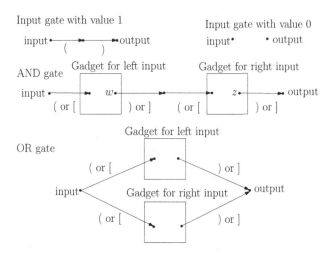

Fig. 2. Reducing MCVP to D_2–REACH

Now we prove a theorem similar in spirit to Theorem 1 for CFLs. The proof uses the same idea to make the undirected version as hard as the directed one.

Theorem 5. *If* DD_2 *is the CFL given by the grammar*

$$S \rightarrow (aSb) \mid [cSd] \mid SS \mid (a\ b) \mid [c\ d]$$

then DD_2–UREACH *is P-complete.*

Proof. CFL–UREACH is in P by [14]. We prove hardness by reducing from D_2–REACH. The reduction works by replacing each edge of the D_2–REACH instance by an undirected path of length two. If for two vertices a, b, the directed edge from a to b is labelled "(" (respectively ")","[" and "]") then replace it by an undirected path of length two with yield "(a"(respectively "b)","[c" and "d]") when read from vertex a to vertex b. The correctness of the reduction is easy to see.   □

We state the following proposition, which follows from Theorem 7.

Proposition 3. CSL–TREEREACH *and* CSL–DAGREACH *are* PSPACE-*complete.*

4   Complexity Class Restricted Reachability

Now we consider the complexity of L–REACH and its variants when L is chosen from complexity classes. We only consider L–DAGREACH as even for languages decidable in LOG as Barrett et. al. [2] has shown that even for languages in LOG, the languages L–REACH and L–UREACH are undecidable. But note that for any decidable L, the language L–DAGREACH is decidable. So we restrict our study only to L–DAGREACH in this section.

We have seen that moving up in the Chomsky hierarchy increases the complexity of reachability. It is natural to ask whether such an observation also holds wrt complexity classes. i.e., increasing the complexity of L increases the complexity of L–DAGREACH. More concretely , does A $\leq_m^{\mathsf{LOG}}$ B imply A–DAGREACH $\leq_m^{\mathsf{LOG}}$ B–DAGREACH. The following theorem shows that this is very unlikely.

Theorem 6. *There is an* A $\in$ LOG *for which* A–DAGREACH *is* NP-*complete.*

Proof. The language A can be thought of as encoding satisfying assignments for some satisfiable formula. Each string in A consists of 3 parts. The first part encodes n, the number of variables. The second part consists of n bits which can be thought of as the assignment to n variables. The next part consists of blocks of the form (id, b) where id uniquely identifies one of the variables $x_1, \ldots, x_n$ and b is 0 or 1. The string is in the language A iff each block is consistent with the assignment stated in the second part of the string. Clearly A $\in$ LOG. The language A–DAGREACH is in NP where the non-deterministic Turing machine guesses the path and then verifies whether the yield of the path is in A. To show NP-hardness, we reduce 3SAT to A–DAGREACH.

The proof is depicted in Figure 3. An edge labelled (x_i, b) is actually a simple path labelled by a string encoding (x_i, b). We use $enc(x)$ to stand for the standard binary encoding of the positive integer x. Note that the graph also contains three parts corresponding to the three parts of a generic string in A. Suppose the input formula is satisfiable. Then we can construct a valid path in the graph from a satisfying assignment for the formula as follows. Take the edge labelled by the value of x_i when we reach the ith block in the second part. When we reach the ith block in the third part, look at the ith clause in the formula. Suppose that

x_i ($\overline{x_i}$) is a satisfied literal in this clause. Then we take the path labelled $(x_i, 1)$ $((\overline{x_i}, 0))$in this block. For the other direction, take any valid path in the graph. Note that we can cross the ith block in the third part of the graph only by taking a path labelled by (x_i, b) where assigning $x_i = b$ satisfies the ith clause. So a valid path exactly corresponds to a satisfying assignment for the formula.   □

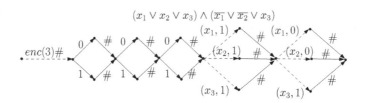

Fig. 3. Reducing 3SAT to A–DagReach

We are now going to see how the above result can be used for translating separations of complexity classes upwards 8. For any complexity class C, we consider the class of languages defined as $\text{DagReach}^{-1}(C) = \{L : L\text{–DagReach} \in C\}$. We have the following theorems for natural choices of C. Note that for any class C, we have $\text{DagReach}^{-1}(C) \subseteq C$.

Theorem 7. $\text{DagReach}^{-1}(\text{PSPACE}) = \text{PSPACE}$ *and* $\text{DagReach}^{-1}(\text{NP}) = \text{NP}$.

Proof. Let $L \in \text{PSPACE}$, then given an instance of L–DagReach we enumerate all paths from s to t and run the PSPACE algorithm for L on the yield. This is a PSPACE algorithm for L–DagReach. Similarly if $L \in \text{NP}$, then a path from s to t along with the certificate for the yield on that path is a poly-time verifiable certificate for the L–DagReach problem.   □

Theorem 8. $P \neq \text{DagReach}^{-1}(P) \iff P \neq \text{NP}$.

Proof. Suppose $P \neq \text{DagReach}^{-1}(P)$ and let $L \in P \setminus \text{DagReach}^{-1}(P)$. Now L–DagReach is in NP by the previous theorem. By the choice of L we also have L–DagReach is not in P.

For the other direction: suppose $\text{DagReach}^{-1}(P) = P$. We know that there is a language $L \in P$ for which L–DagReach is NP-complete. Hence, $P = \text{NP}$.   □

The above theorem shows that separating $\text{DagReach}^{-1}(P)$ from P would separate P from NP. This gives us an upward translation of lower bounds on complexity classes provided we can prove that $\text{DagReach}^{-1}(P)$ is contained in some subclass of P. The interesting question is whether we can identify some "natural" complexity class containing $\text{DagReach}^{-1}(P)$.

By using similar arguments, we also have

Theorem 9. $\text{DagReach}^{-1}(\text{NLOG}) \neq \text{NLOG} \iff \text{NP} \neq \text{NLOG}$.

However $\mathrm{DAGREACH}^{-1}(\mathsf{NLOG})$ contains NLOG-complete languages (See Theorem 2). So proving that $\mathrm{DAGREACH}^{-1}(\mathsf{NLOG})$ is separate from NLOG could be very hard.

The following theorem can be viewed as evidence that $\mathrm{DAGREACH}^{-1}(\mathsf{P})$ could be separate from P. A language L is syntactic read-once logspace (This notion was considered by Hartmanis et. al. in [6]) reducible to another language L' iff there is a logspace reduction from L to L' and in the configuration graph for this reduction all paths from the start configuration to the accepting configuration reads each input variable at most once. It shows that if we restrict our attention to syntactic read-once logspace reductions, then $\mathrm{DAGREACH}^{-1}(\mathsf{L})$ for a P-complete problem L is NP-complete. Note that many natural P-complete problems such as CVP (Circuit Value Problem) remains P-complete even under syntactic read-once logspace reductions.

Theorem 10. *If* L *is* P-*complete under syntactic read-once logspace reductions, then* L–$\mathrm{DAGREACH}$ *is* NP-*complete.*

Proof. Let $\mathsf{V} \in \mathsf{NP}$ via a poly time verifier N. Let W be the witness language for V. i.e., $\mathsf{W} = \{(x, w) : N(x, w) = 1 \text{ and } |w| = |x|^k \text{ for some } k\}$. Since L is P-complete W is read-once log space reducible to L via M. We reduce V to L–$\mathrm{DAGREACH}$. Let x be our input. Take the configuration graph G of M on length $|x| + |x|^k$ inputs (after fixing the value of x) and label each edge by the symbol output by the machine M in that step. This graph H is considered as an input to the language L–$\mathrm{DAGREACH}$. First we prove that $H \in \mathsf{L}$–$\mathrm{DAGREACH}$ implies that $x \in \mathsf{V}$. Consider a path from s to t in H labelled by a string in L. This path corresponds to a witness string for x. Therefore there exists a string w for which (x, w) in W which implies $x \in \mathsf{V}$. For the other direction let $x \in \mathsf{V}$. Therefore there exists a string w such that $(x, w) \in \mathsf{W}$. Now take the path in G that corresponds to this w. The yield of this path is a member of the language L since M outputs this yield when given (x, w) as input. □

5 Discussion and Open Problems

The main result of our work is the observation that if we can prove that the class $\mathrm{DAGREACH}^{-1}(\mathsf{P})$ is contained in some complexity class that is a subclass of P, then we can translate separation of complexity classes upwards. We propose the following open problem.

Open Problem 1: Prove that $\mathrm{DAGREACH}^{-1}(\mathsf{P}) \subseteq \mathsf{NC}$.

It would be interesting to study the behaviour of $\mathrm{DAGREACH}^{-1}(.)$ operator on complexity classes below NLOG. Let AC^0 be the class of all languages computable by poly-sized, constant depth uniform Boolean circuits. Can we say anything about the set of languages $\mathrm{DAGREACH}^{-1}(\mathsf{AC}^0)$. The only languages L for which we know that L–$\mathrm{DAGREACH}$ is in AC^0 are finite languages. Intuitively, any language L for which $\mathrm{DAGREACH}^{-1}(\mathsf{L})$ is in AC^0 should somehow strictly reduce the complexity of the usual reachability problem (Recall that $\mathsf{DAGREACH}$

is NLOG-complete and we know that NLOG $\neq$ AC^0). This leads us to our second open problem.

Open Problem 2: Prove that L–DAGREACH $\in$ AC^0 $\implies$ L is finite.

References

1. Arora, S., Barak, B.: Computational Complexity: A Modern Approach. Cambridge University Press (2009)
2. Barrett, C.L., Jacob, R., Marathe, M.V.: Formal-language-constrained path problems. SIAM Journal of Computing 30(3), 809–837 (2000)
3. Mix Barrington, D.A.: Bounded-width polynomial-size branching programs recognize exactly those languages in NC^1. Journal of Computer and System Sciences 38(1), 150–164 (1989)
4. Mix Barrington, D.A., Lu, C.-J., Miltersen, P.B., Skyum, S.: Searching constant width mazes captures the AC^0 hierarchy. In: Meinel, C., Morvan, M., Krob, D. (eds.) STACS 1998. LNCS, vol. 1373, pp. 73–83. Springer, Heidelberg (1998)
5. Hansen, K.A.: Constant width planar computation characterizes ACC^0. In: Diekert, V., Habib, M. (eds.) STACS 2004. LNCS, vol. 2996, pp. 44–55. Springer, Heidelberg (2004)
6. Hartmanis, J., Immerman, N., Mahaney, S.R.: One-way log-tape reductions. In: Proceedings of 19th Annual Symposium on Foundations of Computer Science, pp. 65–72 (1978)
7. Hopcroft, J.E., Motwani, R., Ullman, J.D.: Introduction to automata theory, languages, and computation - international edition, 2nd edn. Addison-Wesley (2003)
8. Horwitz, S., Reps, T.W., Binkley, D.: Interprocedural slicing using dependence graphs. ACM Transactions on Programming Languages and Systems 12(1), 26–60 (1990)
9. Reingold, O.: Undirected connectivity in log-space. Journal of the ACM 55(4) (2008)
10. Reps, T.W.: On the sequential nature of interprocedural program-analysis problems. Acta Informatica 33(8), 739–757 (1996)
11. Reps, T.W.: Program analysis via graph reachability. Information & Software Technology 40(11-12), 701–726 (1998)
12. Sudborough, I.H.: A note on tape-bounded complexity classes and linear context-free languages. Journal of the ACM 22(4), 499–500 (1975)
13. Sudborough, I.H.: On the tape complexity of deterministic context-free languages. Journal of the ACM 25(3), 405–414 (1978)
14. Ullman, J.D., van Gelder, A.: Parallel complexity of logical query programs. Algorithmica 3, 5–42 (1988)
15. Yannakakis, M.: Graph-theoretic methods in database theory. In: Proceedings of the 9th ACM Symposium on Principles of Database Systems, pp. 230–242 (1990)

Positive and Negative Proofs for Circuits and Branching Programs

Olga Dorzweiler, Thomas Flamm, Andreas Krebs, and Michael Ludwig

WSI - University of Tübingen, Germany, Sand 13, 72076 Tübingen, Germany
{dorzweiler,flamm,krebs,ludwigm}@informatik.uni-tuebingen.de

Abstract. We extend the # operator in a natural way and derive a new type of counting complexity. While $\#\mathcal{C}$ classes (where $\mathcal{C}$ is some circuit-based class like $\mathbf{NC^1}$) only count proofs for acceptance of some input in circuits, one can also count proofs for rejection. The here proposed ZAP-$\mathcal{C}$ complexity classes implement this idea. We show that ZAP-$\mathcal{C}$ lies between $\#\mathcal{C}$ and GAP-$\mathcal{C}$. In particular we consider ZAP-$\mathbf{NC^1}$ and polynomial size branching programs of bounded and unbounded width. We find connections to planar branching programs since the duality of positive and negative proofs can be found again in the duality of graphs and their co-graphs. This links to possible applications of our contribution, like closure properties of complexity classes.

1 Introduction

Besides Turing machines, circuits are a well studied model of computation for the study of low level complexity classes. Measures of complexity in circuits include depth and the number of gates which are roughly speaking an analogue to time and space complexity in Turing machines. When regarding parallels between Turing machines and circuits, a natural question is, what the counterpart to non-determinism in circuits is. A non-deterministic Turing machine can have more than one accepting computation on some input. In fact, the counterpart to the presence of multiple accepting computations is the presence of proof trees in circuits. A proof tree is a sub-tree of the tree unfolding of a circuit, which is a witness for acceptance of some input word. When looking at the circuit-based characterization of, say $\mathbf{NP}$, one can observe that the number of accepting computations and the number of proof trees coincide [Ven92].

How can we calculate the number of proof trees in a circuit? It can be verified easily that if we move to an arithmetic interpretation of the circuit, it computes the number of proof trees [VT89]. I.e. we interpret AND as $\times$ and OR as $+$. Since we cannot treat negation in this setting directly, we assume w.l.o.g. the circuit to be monotone. If we want to address functions counting proof trees in circuits (or equivalently arithmetic circuits) of some complexity bound, say $\mathbf{NC^1}$, we write $\#\mathbf{NC^1}$.

It is an open question whether $\#\mathcal{C}$ functions are closed under subtraction. Here, the case we are most interested in is $\mathcal{C} = \mathbf{NC^1}$. This motivated another type of counting complexity: GAP. Where $\#\mathcal{C}$ functions range over non-negative

H. Jürgensen et al. (Eds.): DCFS 2014, LNCS 8614, pp. 270–281, 2014.

integers, GAP-$\mathcal{C}$ functions range over integers. GAP-$\mathcal{C}$ functions are realized by arithmetic circuits with gates of types $\{+, -, \times\}$. By [FFK94, All04] we know that GAP-$\mathcal{C} = \#\mathcal{C} - \#\mathcal{C}$, what motivated its naming. That means that each GAP function can be computed by an arithmetic circuit only having a single subtraction gate.

Boolean circuits can also compute arithmetic functions. Such a circuit has as many output gates as necessary to display the result integer in binary representation. Hence one can ask e.g. if $\mathbf{NC^1} = \#\mathbf{NC^1}$ or even $\mathbf{NC^1} = \text{GAP-}\mathbf{NC^1}$. By Jung [Jun85] we know, that those classes lie extremely close, but it is still unknown if they coincide.

At this point our contribution comes into play. We propose a new type of counting complexity which fits in between $\#\mathcal{C}$ and GAP-$\mathcal{C}$ very naturally. The starting point for our definition is the observation that we can extend the notion of a proof tree. A (now called positive) proof tree is a witness for a word being accepted. If a word is rejected, there are also witnesses: negative proof trees. To our knowledge, negative proof trees haven't been considered before even though the duality of positive and negative proofs is appealing. We call[1] our new counting complexity classes ZAP-$\mathcal{C}$. ZAP-$\mathcal{C}$ functions are of the form $\Sigma^* \rightarrow \mathbb{Z} \setminus \{0\}$. The image is positive iff there are positive proof trees and negative in the case of the existence of negative proof trees.

Providing the base of the ZAP definition we show an arithmetic interpretation of circuits which then calculate the corresponding ZAP-$\mathcal{C}$ function. By the nature of ZAP, we are not restricted to monotone circuits any more in contrast to the $\#$ case. The second interesting result is that in the case of $\mathcal{C} \in \{\mathbf{NC^i}, \mathbf{AC^i} | i \geq 0\}$ the ZAP-$\mathcal{C}$ functions can be written as differences of $\#\mathcal{C}$ functions with the restriction that the result must not be 0. This uses the fact that each circuit can be transformed in a way that each input has exactly either one negative or one positive proof tree. Those two results place ZAP-$\mathcal{C}$ right between $\#\mathcal{C}$ and GAP-$\mathcal{C}$. So ZAP might give us new possibilities to examine the differences between $\#\mathcal{C}$ and GAP-$\mathcal{C}$.

ZAP circuits are the one major topic of this work. The other one is (nondeterministic) ZAP branching programs (BP). Starting point is the celebrated work of Barrington which showed that bounded width polynomial size branching programs (BWBP) are equally powerful as $\mathbf{NC^1}$ circuits. In the case of BPs we are also interested in counting. In BPs a witness for acceptance is a path from source to target. By [CMTV98] we have that the task of counting paths can be expressed as matrix multiplication which is possible in $\#\mathbf{NC^1}$. However we do not know if counting proof trees in $\mathbf{NC^1}$ circuits is possible in $\#\mathbf{BWBP}$ hence $\#\mathbf{NC^1} \stackrel{?}{=} \#\mathbf{BWBP}$ is an open question.

We extend the ZAP idea to BPs. To do so, we need an analogon to the negative proof trees known from circuits. We found this in the notion of *cuts*. A cut in our sense is a partition of the BP's nodes in two, so that source and target are separated and no undesired edge goes between the two parts. From this it is clear that given a BP and some input there is a path iff there is no cut. In the

---

[1] The naming is motivated by the set of integers $\mathbb{Z}$ and GAP.

case of circuits, positive and negative proofs are dual by negating the circuit. This in a way is inherited by BPs. If we have a planar BP, the counter part to negation is moving to the dual graph. The number of paths becomes the number of cuts and vice versa, so we have switched the sign of the function. We show how ZAP BPs are related to ZAP-$\mathbf{NC^1}$. We have a construction to simulate ZAP-$\mathbf{NC^1}$ functions with BPs. The BPs generated that way are planar but not bounded. This raises questions concerning boundedness and planarity in BPs and ZAP-$\mathbf{NC^1}$ whose answers could give insights in the $\#\mathbf{NC^1} \overset{?}{=} \#\mathbf{BWBP}$ problem.

The paper is structured as follows: We begin with a *Preliminaries* section providing all definitions necessary concerning computational complexity, circuits, arithmetic circuits and BPs. The following section, *Results*, summarizes and states our results. In the subsequent sections we give proofs for the theorems. In the end we give an overview and outlook. Two proofs can be found in the *Appendix* to comply with the length constraint.

We thank the anonymous referees for their helpful comments.

2   Preliminaries

In this paper words are always built from the alphabet $\Sigma = \{0, 1\}$. By $\mathbb{Z}$ we denote integers and by $\mathbb{N}$ the non-negative integers. We assume the reader to be acquainted with Turing Machines and elementary complexity classes like $\mathbf{L}$, $\mathbf{NL}$, $\mathbf{P}$ and $\mathbf{NP}$. See e.g. [Pap94] for basics in computational complexity.

A Turing machine M accepts a word w, if there is a computation on the input resulting in an accepting state. For non-deterministic machines, there can be more than one. We denote the number of accepting computations by $\#acc_M(w)$. The set of functions $\#acc_M \colon \Sigma^* \to \mathbb{N}$ for Turing machines M in the bound of some complexity class, say $\mathbf{NP}$, is denoted by $\#\mathbf{P}$. Similarly we get $\#\mathbf{L}$ from $\mathbf{NL}$.

A circuit C is a connected acyclic graph with a designated output node and n input nodes, where n in the in-degree of the circuit. All other nodes are assigned Boolean functions like AND, OR and NEGATION. There are also other gates, e.g. threshold gates or modulo gates, but those are not in the scope of this work.

A word is accepted by a circuit if the computation results to 1. Since a circuit can only treat words of some constant length, we need to speak about circuit families: $(C_n)_{n \in \mathbb{N}}$. If such a family is computable in the limits of some complexity class, we speak of a uniform circuit family, e.g. DLOGTIME-uniformity. If we bound circuit families in the number of gates, depth or fan-in, we get circuit based complexity classes like $\mathbf{AC^0}$, $\mathbf{ACC^0}$, $\mathbf{TC^0}$ and $\mathbf{NC^1}$. The latter one corresponds to circuits of logarithmic depth, polynomial size and a limit of the fan-in of gates of two. Check e.g. [Vol99] for basics in circuit complexity.

Circuits made up of AND and OR gates can be transformed in a *negation normal form*. We duplicate the input nodes, so that each one has a negated twin. By repeated application of DeMorgan's law, we get a monotone circuit computationally equivalent to the original one, i.e. without negation gates (only negated input gates may be present).

If we take a monotone circuit and replace AND by $\times$ and OR by $+$, then we have arithmetized the circuit, i.e. a circuit which operates over the semi-ring $(\mathbb{N}, +, \times, 0, 1)$. It now computes a function $\Sigma^n \to \mathbb{N}$. The set of functions $\Sigma^* \to \mathbb{N}$ generated by the arithmetic interpretation circuit families in some complexity class $\mathcal{C}$ is denoted as $\#\mathcal{C}$. This way we get e.g. $\#\mathbf{AC^0}$ and $\#\mathbf{NC^1}$. Due to [Ven92] we know that this is consistent with counting complexity as it is defined for Turing machines: If we take the monotone circuit characterization of $\mathbf{NP}$, then counting accepting computations is the same as arithmetizing the corresponding circuits. It is noteworthy that this is not the only way to arithmetize. In [Bei93], Beigel surveys different possibilities.

Also, we know by [VT89] that arithmetic circuits count proofs trees, which can be seen as levels of acceptance in circuits. Unfortunately this tells us nothing about the efficiency of a circuit since the absence of many proof trees does not imply the absence of redundancy in the circuit structure [Fla12]. Given a circuit C, we can unfold it by iteratively treating gates with fan-out greater than one. If g is such a gate with fan-out $k > 1$, let A be the sub-circuit of C, whose output gate is the output of g. By adding $k - 1$ more duplicates of A, we can build an equivalent circuit, where g and its duplicates only have fan-out one. Finally we get an equivalent circuit which is a tree. The number of sub-trees whose root is the circuit's output gate and which result to 1 is the number of proof trees.

In $\#$ classes the closure under (modified) subtraction is unknown which motivated another type of arithmetic class. If we consider arithmetic circuits over $\mathbb{Z}$ with gates of types in $\{+, -, \times\}$, we get GAP complexity classes [FFK94]. A function f is in GAP-$\mathcal{C}$ iff there are $g, h \in \#\mathcal{C}$ so that $f = g - h$.

If we want to compare $\#$ and GAP classes, we do this on non-negative output values. We can even compare those counting classes with Boolean classes. In this case the Boolean circuit has to have several outputs computing the bits of the binary representation of the resulting integer. For a survey on arithmetic circuit complexity, see the survey of Allender [All04].

Another model of computation are non-deterministic branching programs (BP). Note that counting in deterministic BPs is trivial. BPs are directed acyclic layered edge-labeled multi-graphs. A graph is layered if the set of vertices $V = V_1 \,\dot{\cup}\, V_2 \,\dot{\cup}\, \cdots \,\dot{\cup}\, V_k$ so that edges only exist between adjacent layers V_i and V_{i+1}. Further we set $V_1 = \{s\}$ and $V_k = \{t\}$. A BP gets input words from Σ^n. Labels are elements from the set $\{x_1, \ldots, x_n, \overline{x}_1, \ldots, \overline{x}_n\}$. A word w is accepted by some BP if there is a path from s to t so that the labels are *consistent* with the input. I.e. if $w_i = 1$ then no label $\overline{x}_i$ must be read and if $w_i = 0$ vice versa. A label is *inconsistent* with the input if it is not consistent. Since BPs are a computation model which receives words of constant length, we need to consider families of BPs as well. The set of languages which are accepted by some family of polynomial-size BPs is written in bold face: $\mathbf{BP}$. By [Raz91] and [RA97], we know that $\mathbf{BP} = \mathbf{NL}/poly = \mathbf{UL}/poly$. Thus $\mathbf{BP}$ is closed under complement.

To restrict some BP's computational power, one can make constraints. A major one is the restriction of the width of a BP, i.e. the size of the largest layer. We call polynomial-size BPs of bounded (constant) width BWBP. An important result is that $\mathbf{BWBP} = \mathbf{NC^1}$ [Bar89]. As a by-product, we know, that width 5 is always sufficient in the bounded width case. Another restriction is planarity which gives us - in combination with boundedness - the classes $\mathbf{PBP}$ and $\mathbf{PBWBP}$. By [Han08] and [BLMS98] we know that planar bounded-width BPs are related to $\mathbf{AC^0}$ and $\mathbf{ACC^0}$.

Counting classes have already been defined for Turing machines and circuits. By counting the number of paths from s to t being consistent with the input word, we get BP counting classes: $\#\mathbf{BP}$, $\#\mathbf{BWBP}$, $\#\mathbf{PBP}$ and $\#\mathbf{PBWBP}$. We know that $\#\mathbf{BWBP} \subseteq \#\mathbf{NC^1}$ [CMTV98] but we do not know if the inclusion is strict. But then again, we know that $\mathrm{GAP}\text{-}\mathbf{NC^1}$ functions are differences of $\#\mathbf{BWBP}$ functions, so $\mathrm{GAP}\text{-}\mathbf{NC^1} = \mathrm{GAP}\text{-}\mathbf{BWBP}$ [CMTV98].

3   Results

In this section, we define ZAP counting complexity and show how it embeds in the context of $\#\mathcal{C}$ and $\mathrm{GAP}\text{-}\mathcal{C}$. We will also have a look at $\mathrm{ZAP}\text{-}\mathbf{NC^1}$ in particular and BPs. The justification of our results is given in later sections.

In the context of $\#\mathcal{C}$ we spoke of proof trees. We now call these proof trees *positive*, since we consider the case that a word is accepted by a circuit. But now imagine a circuit C rejects some input. Let C' be the negation[2] of C. If a word is rejected, we have, as many proofs for rejection in C as for acceptance in C'. So, analogously we define negative proof trees which are sub-trees of the unfolded circuit which show that the input is rejected. Obviously, there are positive proof trees iff there are no negative ones.

Given a family of circuits C, by $\#acc_C^+ \colon \Sigma^* \to \mathbb{N}$ we denote the function which gives us the number of positive proof trees for some input. $\#acc_C^-$ is the function for the negative proof trees. Further let

$$\#acc_C := \#acc_C^+ - \#acc_C^-,$$

i.e. we subtract the number of negative proof trees from the number of positive proof trees, which gives the number of negative proof trees a negative sign.

The notion of negative proof trees allows us to define the ZAP operator which gives us new counting complexity classes.

Definition 1. *Let $\mathcal{C}$ be some circuit-based complexity class. Then $\mathrm{ZAP}\text{-}\mathcal{C}$ is the set of all functions counting proof trees (positive and negative) in circuits in the limits of $\mathcal{C}$, i.e. all functions of the form $\#acc_C$ for some $C \in \mathcal{C}$.*

$\mathrm{ZAP}\text{-}\mathcal{C}$ functions are always of the form $\Sigma^* \to \mathbb{Z} \setminus \{0\}$. In the $\#\mathcal{C}$ case it turned out that counting proof trees is equivalent to the computation arithmetic circuits perform. But this only holds within circuits in negation normal form because negation gates are not directly treatable. The absence of the notion of negative proof trees limits one to monotone circuits. In the ZAP case we can also arithmetize - and this not only in monotone circuits.

---

[2] Insert a negation gate after the output node and make this negation gate the new output.

Theorem 1. *For a circuit family C, the function $\#acc_C \in$ ZAP-C can be calculated by an arithmetic circuit based on C resulting from the following interpretation of the gates:*

- NEGATION *inverts the sign of its input: $x \mapsto -x$.*
- *An* AND *gate with inputs $x_1, \ldots, x_k$ gets assigned the function*

$$x_1, \ldots, x_k \mapsto \prod_{i=1}^{k} \max\{0, x_i\} + \sum_{i=1}^{k} \min\{0, x_i\}$$

- *An* OR *gate with inputs $x_1, \ldots, x_k$ gets assigned the function*

$$x_1, \ldots, x_k \mapsto \sum_{i=1}^{k} \max\{0, x_i\} - \prod_{i=1}^{k} \max\{0, -x_i\}$$

- *Input gates now hold values in $\{-1, 1\}$; in particular the Boolean 0 becomes -1 and 1 stays the same.*

As one can see, if an AND gate receives only true/positive inputs, then the values are multiplied. If there are negative inputs, those are added. The case OR gates is symmetric.

This lemma gives us an analogue to the $\#C$ correspondence between positive proof trees and arithmetic circuits. The next theorem provides an analogy to the fact that GAP-C functions are differences of $\#C$ functions.

Theorem 2. *Let C be in $\{\mathbf{AC}^i, \mathbf{NC}^i | i \geq 0\}$. A function f is in ZAP-C iff there are functions g and h in $\#C$, so that $g(w) = 0 \Leftrightarrow h(w) \neq 0$ for all inputs w and $f = g - h$.*

Theorem 1 shows ZAP-C to be a generalization of $\#C$ and theorem 2 a restriction of GAP-C. That means ZAP-C lies between $\#C$ and GAP-C and suggests that we have a natural definition at hand.

Next we consider $\mathbf{NC}^1$ and BPs. First we need to define what a ZAP BP is; especially we need a counterpart to negative proof trees in circuits. This part of the paper is based on [Dor13]. Positive proof trees coincide with paths from s to t in BPs. If some input is rejected, there are negative proof trees in the circuit and no path in the BP from s to t being consistent with the input word. If there is no path, then source and target are separated, i.e. there is a cut. So, we discovered that counting cuts gives us exactly what we need. Formally, a cut is a partition of the vertices V in two sets V_s and V_t, so that $s \in V_s$, $t \in V_t$ and all edges between those sets go from V_s to V_t and are inconsistently labeled with respect to the input. $\#paths_B \colon \Sigma^* \to \mathbb{N}$ is the number of paths and $\#cuts_B \colon \Sigma^* \to \mathbb{N}$ the number of cuts some input generates.

Based on that we define ZAP for BPs.

Definition 2. *Let $\mathcal{B}$ be a BP-based complexity class. Then ZAP-$\mathcal{B}$ is the set of all functions counting paths and cuts of some input in a BP B which is in the bounds of $\mathcal{B}$, i.e. all functions of the form $\#acc_B := \#paths_B - \#cuts_B$.*

By this definition we get i.e. **BP** as well as **BWBP**, **PBP**, and **PBWBP**. Obviously such functions are of the desired form $\Sigma^* \rightarrow \mathbb{Z} \setminus \{0\}$ because a path exists iff no cut exists. Our goal is to find relations between ZAP circuits and ZAP BPs.

We find upper and lower bounds for $\mathbf{NC^1}$ in BPs, which are similar to the # case.

Theorem 3. ZAP-**BWBP** $\subseteq$ ZAP-**NC^1** $\subseteq$ ZAP-**PBP**

The first inclusion holds because we can use the idea for counting paths by matrix multiplication to count cuts. For the second inclusion we have a procedure so transform an circuit into a BP. The idea is that AND gates correspond to serial computation in BPs and OR gates to parallel computation. We then use the observation that this construction results in planar graphs. This is convenient, since paths and cuts are dual in taking the dual graph which makes each face to an vertex and draws edges between adjacent faces. Paths and cuts switch places in the dual graph. This shows us that cuts are really the counterpart to negative proof trees. This gives us the following:

Corollary 1. ZAP-**PBP** *is closed under inversion*[3] *and* ZAP-**PBWBP** *stays planar under inversion.*

If we apply the construction of theorem 3 on arbitrary BPs, then we see, that we can make a BP planar by admitting it quasi-polynomial size.

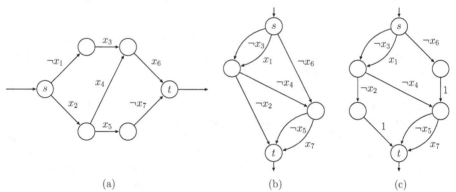

(a)   (b)   (c)

Fig. 1. Inversion in BPs. (a) is the original BP. (b) is the dual graph of (a), hence its inversion. (c) is the layered version of (b).

Figure 1 visualizes inversion in planar BPs. It remains a very interesting question how boundedness and planarity are related exactly and how to catch ZAP-$\mathbf{NC^1}$ in terms of BPs.

---

[3] Inversion means multiplication with -1. That some circuit (or BP) is the inversion of some other circuit (BP) is a stronger requirement than that it is the negation.

4 Proofs

In this section we will provide the proofs for the theorems stated in the results section.

Proof (Theorem 1: ZAP and arithmetic circuits). We show the result by induction on the gates of the circuit. For input gates, we have one positive proof tree if the input is 1 and one negative proof tree if it is 0. Negation gates obviously realize an inversion of the sign. Now consider AND gates. Let g be some AND gate and assume that the inputs $x_1, \ldots x_k$ already transport the right value. If the inputs are all true, g will output one and have a positive number of proof trees. In this case all input values have to be multiplied and this is what happens in this case. If one of the outputs is false, g must be assigned a negative number of proof trees by adding all negative inputs. The construction assures this. The OR case is shown analogously.   □

The next Lemma allows us to split ZAP-$\mathcal{C}$ functions into a positive and a negative part. For a function f, we define $f^+ = \max\{f, 0\}$ and $f^- = \min\{f, 0\}$.

Lemma 1. *In the case of* $\mathcal{C} \in \{\mathbf{NC^i}, \mathbf{AC^i} | i \geq 0\}$, *it holds that if* $f \in$ ZAP-$\mathcal{C}$, *then* f^+ *and* $-f^-$ *are in* $\#\mathcal{C}$.

Proof. Let C be a family of circuits so that $\#acc_C = f$. We can assume that C is in negation normal form, since any negations can be pushed up to input level which does not change the arithmetic interpretation of the circuit. Calculating the number of positive proofs in a monotone circuit is exactly $\#$ counting as we know it, i.e. $f^+ \in \#\mathcal{C}$. The (positive) number of negative proofs is $-f^-$. In this case we use C and attach a negation gate after the output gate and push it to input level to get an equivalent negation normal form whose arithmetic interpretation counts negative proofs. So we get $-f^- \in \#\mathcal{C}$.   □

The next result is the following lemma which states that circuits can be transformed in such a way that the realized ZAP function only takes absolute value 1. We use the regular definition for the sign function.

Definition 3. *The sign function* sgn: $\mathbb{Z} \to \{-1, 0, 1\}$ *maps negative values to* -1, *positive ones to* 1 *and* 0 *to itself.*

Now we prove that ZAP-$\mathcal{C}$ functions are closed under application of the sign function. This is the only way we know to reduce the absolute value. The lemma originates in [Fla12] and a related construction can be found in [Lan93].

Lemma 2. *If* f *is in* ZAP-$\mathcal{C}$, *then* sgn $\circ f$ *is also.*

Proof. Assume an arithmetic circuit C which computes f for some fixed input length. We inductively transform it into a circuit which computes sgn $\circ f$.
Input nodes by definition only take values in $\{-1, 1\}$. NEGATION gates leave values in this set. Now consider an OR gate v with inputs $x_1, \ldots, x_k$ and assume

inductively all inputs to only take values in $\{-1, 1\}$. For each input we insert a NEGATION gate neg_i and an AND gate and_i. We make the following connections for all i (The construction is also pictured in following figure):

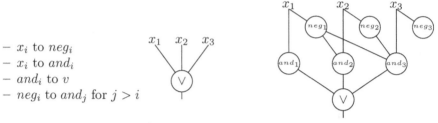

- x_i to neg_i
- x_i to and_i
- and_i to v
- neg_i to and_j for $j > i$

Construction of sgn in the case of OR with fan-in 3.

Now v only outputs values in $\{-1, 1\}$: Assume all x_i take value -1. Then all and_i output -1 and so does v. If there is at least one input which takes value 1, then let x_l be the one of those with smallest index l. In the construction only and_l will take the value 1 and hence v will output 1.

The construction for the case when v is an AND gate is easily adapted.

To stay in some complexity class, we need to note that this construction enlarges the circuit by a constant factor in depth as well as in size. □

An interesting interpretation of the previous lemma is that counting proof trees tells us nothing about efficiency of a circuit as one's intuition may suggest. In particular we made the circuit larger and the result is that we only get minimal numbers of proof trees: -1 and 1.

Next, we prove theorem 2, which gave us the relation to GAP-$\mathcal{C}$.

Proof (Theorem 2: ZAP-$\mathcal{C}$ functions as difference of #$\mathcal{C}$ functions). For a function $f \in$ ZAP-$\mathcal{C}$ we choose $g = f^+$ and $h = -f^-$. By Lemma 1, we know that $g, h \in$ #$\mathcal{C}$ and $g(x) = 0 \Leftrightarrow h(x) \neq 0$.

For the opposite direction, we are given g and h in #$\mathcal{C}$ so that $g(x) = 0 \Leftrightarrow h(x) \neq 0$. We construct a circuit whose arithmetic interpretation computes $g - h$. Let C_g and C_h be the corresponding monotonic Boolean circuits, whose arithmetization results in g respectively h. We construct a new Boolean

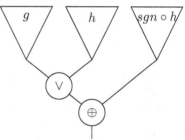

Fig. 2. Construction for the ZAP circuit computing $g - h$

circuit C in the way described in figure 2. The construction first adds the values of g and h using an OR gate. We show, that the construction using sgn and XOR ensures that the output's sign is inverted if $h(x) > 0$. Let ϕ be the sub-circuit $C_g \vee C_h$ (here and in the following we abuse notation by pretending we are dealing with formulas) and ψ be $sgn \circ h$. So the arithmetic interpretation of $\phi \oplus \psi$ gives us the end result. For the result we rewrite: $\phi \oplus \psi = (\phi \wedge \neg\psi) \vee (\neg\phi \wedge \psi)$. The first case is $g(x) > 0$. That means $(\phi \wedge \neg\psi)(x) = g(x)$ and $(\neg\phi \wedge \psi)(x) < 0$, hence the result is $g(x)$. The second case is $h(x) > 0$, so we get $(\phi \wedge \neg\psi)(x) = -1$ and

$(\neg\phi \wedge \psi)(x) = -h(x)$. By applying the arithmetic semantic of the OR gate, we get $-h(x)$ as the desired result. Hence in arithmetic interpretation the circuit computes $g - h$. □

Proof (Theorem 3). We will only prove the first part of the theorem. (ZAP-**BWBP** $\subseteq$ ZAP-**NC^1**). Let B be a BWBP. To prove the result, we have to show that $\#paths_B$ and $\#cuts_B$ are in $\#\mathbf{NC^1}$. By application of theorem 2, we have the desired result. $\#paths_B$ however is trivially in $\#\mathbf{NC^1}$ because that is how counting is defined in BPs and $\#\mathrm{BWBP} \subseteq \#\mathbf{NC^1}$.

We are left proving that counting cuts is possible in $\#\mathbf{NC^1}$. To do this we modify the construction for counting paths in $\#\mathbf{NC^1}$ which we will explain briefly. In the construction one defines matrices for each layer A_i. We can assume that the BP is not only layered but rasterized, i.e. layered also horizontally. If the BWBP is bounded by k then each node is addressable by the layer i and the position in the layer j for $1 \le j \le k$ as (i, j). Here we let $s = (1, 1)$ and $t = (p(n), 1)$. $p(n)$ is the polynomial bound for the number of layers. Now each matrix codes the edges between the layers. If we have a vector v_i which tells us in layer i how many paths are there from s to each node, then $v_i A_{i+1} = v_{i+1}$. It is $v_1 = (1, 0, \ldots, 0)$. Consider the matrix $A_i = (a_{pq})_{1 \le p, q \le k}$. We then have that a_{pq} is the sum of all labels for edges from $(i-1, p)$ to (i, q). The first element of the vector $v_1 A_2 \ldots A_{p(n)}$ is then the number of paths between s and t. Calculating this is possible in $\#\mathbf{NC^1}$, since multiplying two constant-sized matrices is possible in constant depth and if we multiply $p(n)$ many such matrices, we need a circuit if depth $\mathcal{O}(\log n)$, i.e. $\#cuts_B$ is calculable in $\#\mathbf{NC^1}$.

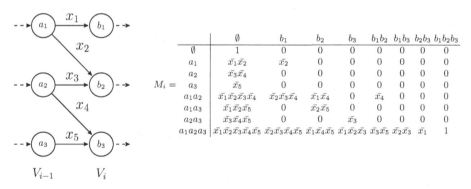

$M_i =$		$\emptyset$	b_1	b_2	b_3	b_1b_2	b_1b_3	b_2b_3	$b_1b_2b_3$
	$\emptyset$	1	0	0	0	0	0	0	0
	a_1	$\bar{x}_1\bar{x}_2$	$\bar{x}_2$	0	0	0	0	0	0
	a_2	$\bar{x}_3\bar{x}_4$	0	0	0	0	0	0	0
	a_3	$\bar{x}_5$	0	0	0	0	0	0	0
	a_1a_2	$\bar{x}_1\bar{x}_2\bar{x}_3\bar{x}_4$	$\bar{x}_2\bar{x}_3\bar{x}_4$	$\bar{x}_1\bar{x}_4$	0	$\bar{x}_4$	0	0	0
	a_1a_3	$\bar{x}_1\bar{x}_2\bar{x}_5$	0	$\bar{x}_2\bar{x}_5$	0	0	0	0	0
	a_2a_3	$\bar{x}_3\bar{x}_4\bar{x}_5$	0	0	$\bar{x}_3$	0	0	0	0
	$a_1a_2a_3$	$\bar{x}_1\bar{x}_2\bar{x}_3\bar{x}_4\bar{x}_5$	$\bar{x}_2\bar{x}_3\bar{x}_4\bar{x}_5$	$\bar{x}_1\bar{x}_4\bar{x}_5$	$\bar{x}_1\bar{x}_2\bar{x}_3$	$\bar{x}_3\bar{x}_5$	$\bar{x}_2\bar{x}_3$	$\bar{x}_1$	1

Fig. 3. Example for the construction of the cut-counting matrix M_i for layer i

Back to counting cuts: We do the same as in the path calculation, so that the first element of the vector $v_1 A_2 \ldots A_{p(n)}$ holds the desired value. Here we also calculate from layer to layer but this time we have matrices of size $2^k \times 2^k$. The reason is that we must regard the power-set of nodes in a layer. Recall that a cut is a partition of the vertices which separates s and t and so that the only edges between the two sets are inconsistent ones from V_s to V_t. An entry in a vector v_i corresponds to a set of nodes X in the layer i. The value is the number of cuts separating s from layers beyond i, so that $X \subseteq V_s$ and $\overline{X} \cap V_s = \emptyset$. By applying

$A_i = (a_{pq})_{1 \le p,q \le 2^k}$ we get this vector for the next layer. Say a_{pq} corresponds to sets X_1 and X_2. Then a_{pq} codes the possibility of extending a cut containing X_1 from the layer i to a cut containing X_2. If there are edges from $\overline{X}_1$ to X_2 then $a_{pq} = 0$ and else a_{pq} is the product of all negated labels of edges going from X_1 to $\overline{X}_2$. If there are no such edges at all then $a_{pq} = 1$. Figure 3 shows the construction.

We sketch the proof for the correctness of this construction based on a induction upon the layers of the BP. Let v_1 be the initial vector for the first layer which only includes s and let $v_i = v_{i-1}A_i$. Each component in such a vector stands for a subset X of nodes in a layer and we want to show that it holds the value how many cuts there are, so that exactly the nodes X in this layer belong to V_s. We assume inductively that this calculation is correct for layer i, i.e. $v_1 A_1 \ldots A_i$ has the right values for each subset. Then $v_1 A_1 \ldots A_i A_{i+1}$ is correct for layer $i + 1$. Each subset Y in layer $i + 1$ can become part of V_s if there is an extendable subset X from the previous layer. That means there are no edges going from V_t to V_s. The number of cuts with $Y \subseteq V_s$ is the sum of cuts of all subsets of the previous layer having no forbidden edges. Whether a forbidden edge occurs is dependent on the input; the matrix holds exactly the required constraints to the input so that a cut with some subset in layer i can be extended using exactly some subset in layer $i + 1$. One can verify that this is exactly what happens of v_i is multiplied with A_{i+1}.   □

5   Discussion and Future Work

In this work we introduced a meaningful counterpart to (positive) proofs trees in circuits, which are negative proof trees. Counting positive and negative proofs seems to be a natural thing to do. Based on some complexity class $\mathcal{C}$, ZAP-$\mathcal{C}$ is the class of functions counting positive and negative proofs. We were also able to adapt our notion of negative proofs to branching programs. We were able to prove that ZAP-**BWBP** $\subseteq$ ZAP-**NC**^{1} $\subseteq$ ZAP-**PBP**.

We think the notion of negative proofs is a neglected aspect of counting. We will outline possible applications.

The first one is the question whether #**BWBP** equals #**NC**^{1}. By looking at ZAP-**NC**^{1}, we know that f is in ZAP-**NC**^{1} iff f^+ and $-f^-$ are in #**NC**^{1}. It seems rather unlikely that this also holds for BWBPs. If one proved that cuts in BWBPs cannot be counted in #**BWBP** then #**BWBP** would be separated from #**NC**^{1}.

Another possible application is the well-known fact that **NL** is closed under complement. If we look at **NL**/*poly* which is equal to **BP** then we have the following implication: If ZAP-**BP** is closed under inversion then **NL**/*poly* is closed under complement. Note that we could formulate such an implication also using #**BP**, but this seems less promising because of the absence of information concerning rejection of some input. As a first step one could try to re-prove the complement closure of **NL**/*poly* using our approach. Also **NL**/*poly* = **UL**/*poly* [RA97] could be a candidate to be re-proved. New proofs to those theorems would naturally give us new insights.

References

[All04]   Allender, E.: Arithmetic circuits and counting complexity classes. In: Krajek, J. (ed.) In Complexity of Computations and Proofs. Quaderni di Matematica (2004)

[Bar89]   Barrington, D.A.M.: Bounded-width polynomial-size branching programs recognize exactly those languages in NC^1. J. Comput. Syst. Sci. 38(1), 150–164 (1989)

[Bei93]   Beigel, R.: The polynomial method in circuit complexity. In: Structure in Complexity Theory Conference, pp. 82–95. IEEE Computer Society (1993)

[BLMS98]   Barrington, D.A.M., Lu, C.-J., Miltersen, P.B., Skyum, S.: Searching constant width mazes captures the AC^0 hierarchy. In: Meinel, C., Morvan, M., Krob, D. (eds.) STACS 1998. LNCS, vol. 1373, pp. 73–83. Springer, Heidelberg (1998)

[CMTV98]   Caussinus, H., McKenzie, P., Thérien, D., Vollmer, H.: Nondeterministic NC^1 computation. J. Comput. Syst. Sci. 57(2), 200–212 (1998)

[Dor13]   Dorzweiler, O.: Zap-Klassen für Schaltkreise und Branching Programs. Masterarbeit, Universität Tübingen (2013)

[FFK94]   Fenner, S.A., Fortnow, L., Kurtz, S.A.: Gap-definable counting classes. J. Comput. Syst. Sci. 48(1), 116–148 (1994)

[Fla12]   Flamm, T.: Zap-C Schaltkreise. Diplomarbeit, Universität Tübingen (2012)

[Han08]   Hansen, K.A.: Constant width planar branching programs characterize ACC^0 in quasipolynomial size. In: IEEE Conference on Computational Complexity, pp. 92–99. IEEE Computer Society (2008)

[Jun85]   Jung, H.: Depth efficient transformations of arithmetic into boolean circuits. In: Budach, L. (ed.) FCT 1985. LNCS, vol. 199, pp. 167–174. Springer, Heidelberg (1985)

[Lan93]   Lange, K.-J.: Unambiguity of circuits. Theor. Comput. Sci. 107(1), 77–94 (1993)

[Pap94]   Papadimitriou, C.H.: Computational complexity. Addison-Wesley (1994)

[RA97]   Reinhardt, K., Allender, E.: Making nondeterminism unambiguous. In: FOCS, pp. 244–253. IEEE Computer Society (1997)

[Raz91]   Razborov, A.A.: Lower bounds for deterministic and nondeterministic branching programs. In: Budach, L. (ed.) FCT 1991. LNCS, vol. 529, pp. 47–60. Springer, Heidelberg (1991)

[Ven92]   Venkateswaran, H.: Circuit definitions of nondeterministic complexity classes. SIAM J. Comput. 21(4), 655–670 (1992)

[Vol99]   Vollmer, H.: Introduction to circuit complexity - a uniform approach. Texts in theoretical computer science. Springer (1999)

[VT89]   Venkateswaran, H., Tompa, M.: A new pebble game that characterizes parallel complexity classes. SIAM J. Comput. 18(3), 533–549 (1989)

Regularity and Size of Set Automata

Martin Kutrib, Andreas Malcher, and Matthias Wendlandt

Institut für Informatik, Universität Giessen
Arndtstr. 2, 35392 Giessen, Germany
{kutrib,malcher,matthias.wendlandt}@informatik.uni-giessen.de

Abstract. The descriptional complexity of deterministic and nondeterministic set automata is investigated. Set automata are basically finite automata equipped with a set as an additional storage medium. The basic operations on the set are the insertion of elements, the removing of elements, and the test whether or not an element is in the set. We show that regularity is decidable for deterministic set automata and describe a conversion procedure into deterministic finite automata which leads to a double exponential upper bound. This bound is proved to be tight in the order of magnitude by presenting also a double exponential lower bound. In contrast to these recursive bounds we obtain non-recursive trade-offs when nondeterministic set automata are considered.

1 Introduction

Descriptional complexity is a field of theoretical computer science in which one main interest is to investigate how the size of description of a formal language varies under the description by different formalisms. A fundamental result is the exponential trade-off between nondeterministic and deterministic finite automata [7]. Additional exponential or double exponential trade-offs are known, for example, between unambiguous and deterministic finite automata, between alternating and deterministic finite automata, and between deterministic pushdown automata (DPDA) and deterministic finite automata. Beside these recursive trade-offs, which are bounded by recursive functions, it is known that also non-recursive trade-offs which are not bounded by any recursive function exist. Such trade-offs were first shown in [7] between context-free grammars generating regular languages and finite automata. Another non-recursive trade-off is known to exist between nondeterministic PDA and deterministic PDA [3]. Nowadays, many non-recursive trade-offs have been established. For a survey we refer to [2,4], where also references for the above-mentioned recursive trade-offs can be found.

In this paper, we investigate the descriptional complexity of nondeterministic and deterministic *set automata*. The latter model, abbreviated as DSA, has been introduced in [6] and extends deterministic finite automata by adding the storage medium of a set which allows to store words of arbitrary length. As operations on the set it is possible to add elements, to remove elements, and to test whether some element is in the set. To prepare a set operation the DSA can write on

H. Jürgensen et al. (Eds.): DCFS 2014, LNCS 8614, pp. 282–293, 2014.

a one-way write-only tape. For the set operation the contents of that tape are taken and added to the set, removed from the set, or tested. In order to keep the mode of operation simple, there is no extra erase operation that empties the tape. Instead, after a set operation, the writing tape is reset to the empty tape and a new set operation may be prepared. The main results obtained on DSA in [6] are incomparability with deterministic context-free languages and deterministic queue languages, closure under complementation, non-closure under union and intersection, and the decidability of emptiness. Thus, DSA might be seen as an interesting model that is able to accept non-context-free languages, but still has a decidable emptiness problem. In this paper, we extend the decidability of emptiness for DSA to regularity and show that it is decidable for an arbitrary DSA whether or not it accepts a regular language. If the language accepted is regular the decision procedure delivers an effective construction of the equivalent deterministic finite automaton whose size turns out to be at most double exponential in the size of the given DSA. We will also present a double exponential lower bound which gives the tightness of the construction in the order of magnitude. We then consider *nondeterministic* set automata (NSA) and yield as main difference to the recursive bounds in the deterministic case that there are non-recursive trade-offs between NSA and DSA, between NSA and DPDA, and vice-versa between PDA and DSA. This means, for example, for the first case that the increase in size when changing from an NSA description to a DSA description might not be bounded by any recursive function. Another result is that for NSA the questions of universality, equivalence, inclusion, and, in particular, regularity are undecidable which is in contrast to the deterministic case.

2   Preliminaries and Definitions

We write Σ^* for the set of all words over the finite alphabet Σ. The empty word is denoted by λ, and we set $\Sigma^+ = \Sigma^* \setminus \{\lambda\}$. The reversal of a word w is denoted by w^R, and for the length of w we write $|w|$. We use $\subseteq$ for inclusions and $\subset$ for strict inclusions.

A set automaton is a system consisting of a finite state control, a data structure *set* where words of arbitrary length can be stored, and a one-way writing tape where words for the set operations are assembled. At each time step, it is possible to either extend the writing tape inscription at its end, to insert or remove the word written on the tape to or from the set, or to test whether the tape inscription belongs to the set. Each time a set operation {in, out, or test} is done, the content of the writing tape is erased and its head is reset to the left end.

Formally, a *nondeterministic set automaton*, abbreviated as NSA, is a system $M = \langle S, \Sigma, \Gamma, \lhd, \delta, s_0, F \rangle$, where S is the finite set of *internal states*, Σ is the finite set of *input symbols*, Γ is the finite set of *tape symbols*, $\lhd \notin \Sigma$ is the *right endmarker*, $s_0 \in S$ is the *initial state*, $F \subseteq S$ is the set of *accepting states*, and δ is the partial transition function mapping $S \times (\Sigma \cup \{\lambda, \lhd\})$ to the finite subsets of $(S \times (\Gamma^* \cup \{\text{in}, \text{out}\})) \cup (S \times \{\text{test}\} \times S)$, where **in** is the instruction to add the content of the tape to the set, **out** is the instruction to remove the content of the tape from the set, and **test** is the instruction to test whether or not the content of the tape is in the set.

A *configuration* of a NSA $M = \langle S, \Sigma, \Gamma, \triangleleft, \delta, s_0, F \rangle$ is a quadruple $(s, v, z, \mathbb{S})$, where $s \in S$ is the current state, $v \in \{\Sigma^* \triangleleft\} \cup \{\lambda\}$ is the unread part of the input, $z \in \Gamma^*$ is the content of the tape, and $\mathbb{S} \subseteq \Gamma^*$ is the finite set of stored words. The *initial configuration* for an input string w is set to $(s_0, w\triangleleft, \lambda, \emptyset)$. During the course of its computation, M runs through a sequence of configurations. One step from a configuration to its successor configuration is denoted by $\vdash$. Let $s, s', s'' \in S$, $x \in \Sigma \cup \{\lambda, \triangleleft\}$, $v \in \{\Sigma^* \triangleleft\} \cup \{\lambda\}$, $z, z' \in \Gamma^*$, and $\mathbb{S} \subseteq \Gamma^*$. We set

1. $(s, xv, z, \mathbb{S}) \vdash (s', v, zz', \mathbb{S})$, if $(s', z') \in \delta(s, x)$,
2. $(s, xv, z, \mathbb{S}) \vdash (s', v, \lambda, \mathbb{S} \cup \{z\})$, if $(s', \text{in}) \in \delta(s, x)$,
3. $(s, xv, z, \mathbb{S}) \vdash (s', v, \lambda, \mathbb{S} \setminus \{z\})$, if $(s', \text{out}) \in \delta(s, x)$,
4. $(s, xv, z, \mathbb{S}) \vdash (s', v, \lambda, \mathbb{S})$, if $(s', \text{test}, s'') \in \delta(s, x)$ and $z \in \mathbb{S}$,
5. $(s, xv, z, \mathbb{S}) \vdash (s'', v, \lambda, \mathbb{S})$, if $(s', \text{test}, s'') \in \delta(s, x)$ and $z \notin \mathbb{S}$.

We denote the reflexive and transitive closure of $\vdash$ by $\vdash^*$. It should be noted that an instruction to remove some z from $\mathbb{S}$ does not test whether $z \in \mathbb{S}$; it only ensures that $z \notin \mathbb{S}$ after the operation. The language accepted by the NSA M is the set $L(M)$ of words for which there exists a computation beginning in the initial configuration and ending in a configuration in which the whole input is read and an accepting state is entered. Formally:

$$L(M) = \{\, w \in \Sigma^* \mid (s_0, w\triangleleft, \lambda, \emptyset) \vdash^* (s_f, \lambda, z, \mathbb{S}) \text{ with } s_f \in F, z \in \Gamma^*, \mathbb{S} \subseteq \Gamma^* \,\}.$$

A set automaton is said to be deterministic (DSA) if there is at most one choice of action for any possible configuration. In particular, there must never be a choice of using an input symbol or of using λ input. Formally, $\delta(s, x)$ of a deterministic set automaton contains at most one element, for all x in $\Sigma \cup \{\lambda\}$ and s in S and, if the transition function is defined for some pair (s, λ) with $s \in S$, then it is not defined for any pair (s, a) with $a \in \Sigma \cup \{\triangleleft\}$. The family of all languages accepted by a device of type X is denoted by $\mathscr{L}(X)$.

Example 1. The language

$$L_1 = \{\, w_1 \$ w_2 \$ \cdots w_m \$ w \mid m \geq 0, w, w_1, w_2, \ldots, w_m \in \{a, b\}^*,$$
$$\text{and } \exists 1 \leq i \leq m : w = w_i \,\}$$

is accepted by a DSA. The idea is to iteratively read each sequence of a's and b's up to the letter $\$$ and to copy it to the tape. When the input head arrives at the $\$$, it stores the word x written on the tape in its set. When the input head arrives at the right endmarker, it tests whether the content on the tape is in the set. If this is the case, then the input is accepted and otherwise rejected. The separating symbols $\$$ are used to control the **in**-operations of the DSA. Consider language L_2 which is in a way a variant of L_1 without separating symbols.

$$L_2 = \{\, x \$ w \mid x, w \in \{a, b\}^* \text{ and } w \text{ is a factor of } x \,\}$$

We will now describe how L_2 can be accepted by an NSA. First, we read the input while nothing is written on the tape. At some time step it is guessed that the factor w starts and any input symbol read is written on the tape. At some other time step it is guessed that the factor w has ended, an in-operation is performed, and for the following input nothing is written on the tape. After symbol \$, the input is again written on the tape. When the input head arrives at the right endmarker, the NSA tests whether the content on the tape is in the set. If this is the case, then the input is accepted and otherwise rejected.   □

3   Deciding Regularity for Deterministic Set Automata

In this section, we turn to show that regularity is decidable for deterministic set automata. In addition, the test procedure reveals an upper bound for the descriptional complexity of the DSA to DFA conversion. The upper bound is double exponential where the exponent is quadratic. We also show a double exponential lower bound.

Normal form. For the regularity test we will assume that set automata are in a certain normal form. We say DSA M is in *action normal form*, if the initial state of M is only visited once at the beginning of the computation and each other state indicates uniquely which action the automaton M did in the last computation step. The states are marked with a corresponding subscript test+ (for a successful test), test- (for an unsuccessful test), in, or out. Non-marked states are interpreted as states where the last action was a write operation on the tape.

Let $M = \langle S, \Sigma, \Gamma, \lhd, \delta, s_0, F \rangle$ be a DSA in action normal form. Thus, the state set is partitioned as $S = \{s_0\} \cup S_{\text{in}} \cup S_{\text{out}} \cup S_{\text{test-}} \cup S_{\text{test+}} \cup S_{\text{write}}$. We note that the tape is empty at the beginning of the computation as well as after each operation on the set. Now we build sets of the form L_{s_i, s_j} with $s_i \in \{s_0\} \cup S_{\text{in}} \cup S_{\text{out}} \cup S_{\text{test-}} \cup S_{\text{test+}}$ and $s_j \in S_{\text{in}} \cup S_{\text{out}} \cup S_{\text{test-}} \cup S_{\text{test+}}$ that describe all words that can be written on the tape when the computation starts in state s_i with empty tape and ends in state s_j, and in between only states from S_{write} are entered. Formally we define

$$L_{s_i, s_j} = \{\, w \in \Gamma^* \mid \text{there is } u \in \Sigma^* \text{ such that } (s_i, u, \lambda, \mathbb{S}) \vdash (s_{i+1}, u_1, w_1, \mathbb{S})$$
$$\vdash^* (s_{i+(n-1)}, u_{n-1}, w_{n-1}, \mathbb{S}) \vdash (s_{i+n}, u_n, w_n, \mathbb{S}) \vdash (s_j, \lambda, \lambda, \mathbb{S}'), \text{ where}$$
$$w = w_n, w_1, w_2, \ldots, w_n \in \Gamma^*, \text{ and } s_{i+1}, s_{i+2}, \ldots, s_{i+n} \in S_{\text{write}} \,\}.$$

All these sets L_{s_i, s_j} are regular, since an equivalent finite automaton M_{s_i, s_j} can be built from M as follows. Automaton M_{s_i, s_j} has S as state set, s_i as initial state, and s_j as only accepting state. We consider all transitions in M from some state $s \in \{s_i\} \cup S_{\text{write}}$ to some state $s' \in \{s_j\} \cup S_{\text{write}}$ writing some word $z \in \Gamma^*$ on the tape. For every such transition we add to M_{s_i, s_j} a transition from s to s' on input z. We say that a DSA M is in *infinite action normal form* if M is in action normal form and all sets L_{s_i, s_j} are infinite. It is shown in [6] that any DSA can effectively be converted into an equivalent DSA in infinite action normal form.

Meta automaton. Next, the DSA $M = \langle S, \Sigma, \Gamma, \vartriangleleft, \delta, s_0, F \rangle$ in infinite action normal form with $S = \{s_0\} \cup S_{\text{in}} \cup S_{\text{out}} \cup S_{\text{test-}} \cup S_{\text{test+}} \cup S_{\text{write}}$ is transformed into a meta automaton M' whose states are the initial state and the in-, out-, test--, and test+-states of M, that is, the write-states are deleted. An edge connecting some state s_i with some s_j is included in M' whenever L_{s_i,s_j} is defined. The edge is labeled with the infinite regular languages L_{s_i,s_j}. It is worth mentioning that we do not care about the actual input here, but for the argumentation it is understood that there always is a suitable input. The set of accepting states of M' is defined as follows. First, all states from S' that are accepting in M are accepting in M' as well. Second, for every edge connecting state s_i with s_j in M', state s_i is defined to be accepting in M' if some accepting state is passed through in a path from s_i to s_j in M containing only write-states in between. Similarly, a state s_i of M' is defined to be accepting, if there is some path in M starting in s_i that never reaches any other state of M', passes through an accepting state of M, and reaches a state of M in which the computation of M ends. So, the meta automaton M' can effectively be constructed from M.

Regularity test. Let $M = \langle S, \Sigma, \Gamma, \vartriangleleft, \delta, s_0, F \rangle$ be a DSA in infinite action normal form and M' its meta automaton. Basically, to test the regularity of $L(M)$ all possible paths in the state graph of M' up to length $(t + 1) \cdot e \cdot |S| + |S|$ are examined, where t denotes the number of edges in M' that connect to some test+-state and e denotes the number of edges in M' that connect to some in-state. So, we consider the computation tree built from the state graph of M' and perform a depth-first traversal to evaluate the paths. We start at the root which is associated with the initial state of M'. Whenever the current state of a path is s and $\ell \geq 1$ states can be reached from s, that is, there are edges between s and states $s_1, s_2, \ldots, s_\ell$, the corresponding node of the tree has the successor nodes $s_1, s_2, \ldots, s_\ell$.

In a first phase the nodes of the tree are labeled by sets of elements of the form (x, R), where $x \geq 1$ and R is a regular language. Such an element of a label represents the information that x strings from the language R are in the set $\mathbb{S}$. The root is labeled by the empty set. When the traversal extends the current path by visiting a node associated with state s_j reached from a node associated with state s_i, the label of s_j is determined as follows.

If s_j is an in-state the label of s_i is copied to s_j. If this label contains already an element of the form (x, L_{s_i,s_j}) the first component is increased by one. Otherwise, the pair $(1, L_{s_i,s_j})$ is added to the label.

If s_j is a test+-state reached from s_i for the first time, we have to verify that the test can be positive. So, the intersection $L_{s_i,s_j} \cap L_m$ is built for each of the $m \geq 0$ pairs (x_m, L_m) in the label of s_i. If all these intersections are empty, the examination of the current path is terminated. Otherwise, the information in the label of s_i can be made more precise when it is copied to s_j. To this end, first the node associated with s_j currently reached in the tree is removed. Second, let $I_1, I_2, \ldots, I_n$, $n \geq 1$, be the non-empty intersections. Then n new nodes associated with s_j are inserted as successors of s_i. The label of s_i is copied to each of the new nodes. In addition, the label of node k, $1 \leq k \leq n$, is updated by adding the tuple $(1, I_k)$ and decreasing the first component of (x_k, L_k) by one. If $x_k = 0$, the tuple (x_k, L_k) is removed from the label.

If s_j is a test+-state reached again from s_i, the same word as before can be used for the successful test. So, only the label of s_i is copied to s_j.

If s_j is an out- or test--state the label of s_i is copied to s_j.

Now we turn to conditions for terminating the examination of a branch of the depth-first traversal. The examination of a path is terminated if there are no outgoing edges from the current state, all intersections constructed are empty after entering a test+-state, or the length of the path exceeds $(t + 1) \cdot e \cdot |S| + |S|$. The first condition is obvious. The second criterion applies if there never can be words in the set for which the test is positive and, thus, no computation can use this path. The third condition is discussed later.

Once the examination of the whole computation tree is finished, in a second phase, any of the finitely many paths from the root to some leaf in the tree is tested. Language $L(M)$ is regular if and only if all tests are positive.

If the path to be tested does not contain any accepting state, its test is positive. Otherwise, we consider the initial part of the path up to the last occurring accepting state. From left to right along the path we consider edges connecting some state s_i to a test+-state s_j. Let $\{(x_k, L_k) \mid m \geq 0, 1 \leq k \leq m\}$ be the label of the node associated with s_i. Then we build the intersection $L_{s_i, s_j} \cap \bigcup_{1 \leq k \leq m} L_k$. If the intersection is infinite the test for the current path is negative. Otherwise the test continues by considering the next edge connecting to some test+-state along the path in the same way. If there is no further test+-state, the test for the whole path is positive.

Theorem 2. *It is decidable whether or not a given deterministic set automaton accepts a regular language.*

Proof. Let s be a node in the tree labeled $\{(x_k, L_k) \mid m \geq 0, 1 \leq k \leq m\}$. For easier writing we denote the union of the regular languages $\bigcup_{1 \leq k \leq m} L_k$ in the label by P_s.

In order to show that the regularity test works correctly, first assume that the test for a path is negative. Then there is an edge in the path connecting some state s_i to a test+-state s_j so that the intersection $P_{s_i} \cap L_{s_i, s_j}$ is infinite. If s_i is not an in-state, we consider the predecessor s_i' of s_i. Since $P_{s_i} \cap L_{s_i, s_j}$ is infinite, $P_{s_i'} \cap L_{s_i, s_j}$ must be infinite as well. If s_i' is neither an in-state the argumentation is iterated. In this way the path is searched from s_j backwards for an in-state, say s_r, so that $P_{s_r} \cap L_{s_i, s_j}$ is infinite. Let s_q be the predecessor of s_r. If $L_{s_q, s_r} \cap L_{s_i, s_j}$ is infinite the search stops. Otherwise, we conclude that $P_{s_q} \cap L_{s_i, s_j}$ must be infinite since $P_{s_r} \cap L_{s_i, s_j}$ is, and the search continues. Since the initial state has an empty label and $P_{s_i} \cap L_{s_i, s_j}$ is infinite, there must exist an in-state that can insert infinitely many different strings in the set. More precisely, there exist some in-state s_r' with predecessor s_q' such that $L_{s_q', s_r'} \cap L_{s_i, s_j}$ is infinite. Thus, the search is always successful.

Next we expand the path to paths in M by inserting the write-states again. Moreover, we fix some input u_0 that drives M from the initial state along the path to state s_q, some input v_0 that drives M from s_r to s_i, and some input w_0 that drives M from s_j to an accepting state. Assume that $L(M)$ is regular. Then

$$L' = L(M) \cap u_0 \Sigma^* v_0 \Sigma^* w_0$$
$$= \{ u_0 z_1 v_0 z_2 w_0 \mid pr_{s_q,s_r}(z_1) = pr_{s_i,s_j}(z_2) \in L_{s_q,s_r} \cap L_{s_i,s_j} \}.$$

is regular as well. Here $pr_{s_1,s_2}(z)$ denotes the string that M writes on the tape while it passes from state s_1 to s_2 reading z so that in between only write-states are used. If z does not drive M in this way from s_1 to s_2, $pr_{s_1,s_2}(z)$ is undefined. In order to disprove the regularity of L' we will insert a separating symbol in the words of L' by transduction. To this end, a finite state transducer N is constructed as follows. At the beginning of its computation, N reads and outputs the input prefix u_0. Then it starts to simulate the state transitions of M beginning in state s_q and ending in state s_r, where in between only write-states of M are allowed. In this way, exactly the inputs z_1 with $pr_{s_q,s_r}(z_1) \in L_{s_q,s_r}$ are read. Since in between s_q and s_r there are only write-states but s_r is an in-state, during this part of the simulation s_r is entered exactly once, that is, at the end. In this phase, transducer N outputs every symbol read. In addition it outputs a new symbol \$ once when it enters state s_r. Subsequently, N continues to read the input whereby every symbol read is output. Since the family of regular languages is closed under finite-state transductions,

$$L'' = N(L') = \{ u_0 z_1 \$ v_0 z_2 w_0 \mid pr_{s_q,s_r}(z_1) = pr_{s_i,s_j}(z_2) \in L_{s_q,s_r} \cap L_{s_i,s_j} \}$$

is regular as well. A simple application of the pumping lemma on words with $|pr(z_1)| > |u_0 v_0 w_0|$ yields a contradiction. So, $L(M)$ is not regular, and we have shown that the regularity test is right when it says that M is not regular.

For the second part of the proof assume that the tests for all paths are positive. So, for all edges in all paths up to the last occurrence of an accepting state we know that if the edge connects some state s_i to a test+-state s_j, then $P_{s_i} \cap L_{s_i,s_j}$ is finite. Consider all these intersections in the whole tree, and let c be the length of the longest word occurring in them. Then any test in any path can be computed, only by the knowledge of the tape inscription if it does not exceed c, and the knowledge which of the finitely many words up to length c are currently in the set. All this information can be tracked in a finite number of states. Therefore, a deterministic finite automaton simulating M can effectively be constructed. So, $L(M)$ is regular, and we have shown that the regularity test is right when it says that M is regular, provided that we do not have overlooked possible computation paths by the way the computation tree was labeled.

On the one hand, every path in M' can be expanded to computation paths of M. On the other hand, we did not consider paths of M whose lengths are exceeding $(t+1) \cdot e \cdot |S| + |S|$, and we always labeled out-nodes as their predecessors. Let us first discuss the latter point. Copying the label of the predecessor to an out-node represents a computation of M where the string on the tape does not belong to the set. Such a computation always exists, since at every time step there are only finitely many strings in the set, but every edge connecting two states in M' is labeled by an infinite language. Therefore, it is always possible to find some word in the language that is not in the set. Now assume a computation of M so that a string is removed from the set when entering an out-state, say s_j. We continue with this path and compare it with the path we considered. The write-, in-, and out-operations possibly following s_j are not affected by the removal and are the same as in the path we considered. When a test+-state appears in the path, it clearly also appears in the path we considered. Moreover, again, since at every time step there are only finitely many strings in the set, but every edge connecting two states in M' is labeled by an infinite language, it is always possible to find some word in the language that is not in the set. So, also the negative branch of the test exists in the path we considered. We conclude that all computations of M including a positive removal are already represented by some path in M'.

We turn to the former point, that is, we did not consider paths whose length exceeds $(t + 1) \cdot e \cdot |S| + |S|$, where t denotes the number of edges in M' that connect to some test+-state and e denotes the number of edges in M' that connect to some in-state. Now assume an accepting computation path of M whose length exceeds $(t + 1) \cdot e \cdot |S| + |S|$. We recall that for testing the path from left to right, only the intersections $P_{s_i} \cap L_{s_i,s_j}$ for some edge connecting some state s_i to a test+-state s_j are used. out-, and test--operations do not change the labels of the nodes. An edge connecting to a test+-operation changes the labels only when it is used for the first time. The changes caused by an in-operation may either increase the first component of some label element or else may include a new label element. Suppose that an edge connecting to an in-state has been used more than t times. Then the first component of the label element this edge introduces can never be decreased to zero again by test+-operations. Therefore, the status of the test is not affected by any further use of this edge. Next we analyze the path from left to right and keep track in some vector $(\{0, 1, \ldots, t\} \cup \{t^>\})^e$ how many times each edge connecting to an in-state occurs. We start with zero vector and increase a component by one if the corresponding edge occurs. If the component is t or $t^>$ it is set to $t^>$.

Since the path is longer than $|S|$ it contains a loop. If there is any loop that does not change the vector, we can remove the loop from the path, since the status of the test is not affected by the loop. Moreover, removing the loop does not affect the existence of the rest of the path for the reasons given in the first point above. So, we have found a shorter accepting path whose test result is the same as for the original path. If the shorter path is still longer than $(t + 1) \cdot e \cdot |S| + |S|$ we repeat the reasoning. On the other hand, assume that all loops increase at least one component of the vector. Then, after at most $(t+1)\cdot e$ many loops, each of length at most $|S|$, the vector cannot change anymore. So, the total length of the path is at most $(t + 1) \cdot e \cdot |S| + |S|$. We conclude that if there is a path whose test is negative then there is a path of length at most $(t + 1) \cdot e \cdot |S| + |S|$ whose test is negative. This concludes the proof that the regularity test is correct, and the theorem follows. $\square$

Finally, we turn to the descriptional complexity of the conversion from a DSA accepting a regular language to a DFA.

Theorem 3. *Let M by an n-state DSA with set of tape symbols Γ that accepts a regular language. Then an equivalent DFA with at most $2^{|\Gamma|^{O(n^2)}}$ states can effectively be constructed.*

Proof. Let $M = \langle S, \Sigma, \Gamma, \lhd, \delta, s_0, F \rangle$ be a DSA. The conversion into action normal form requires at most to double the number of states. Basically, for the conversion into infinite action normal form the meta automaton as above is constructed, where each edge is labeled with a regular language. These regular languages are all accepted by DFA having no more than $2|S|$ states. Consequently, if an edge is labeled by a finite language, the length of its longest word is at most $O(|S|)$. Then the edges labeled with finite languages are simulated in such a way that the automaton tracks all finitely many words up to that length in its state, whereby nothing is written on the tape (see [6] for further details). In particular, this implies that all edges of the meta automaton used above for the regularity test are labeled with regular languages that are accepted by DFA with at most $O(|S|)$ states, despite the fact that we assumed the DSA already to be in infinite action normal form.

Since M accepts a regular language the regularity test is successful. Therefore, we know that whenever a positive test is performed, the intersection of the language of words tested with the languages of words possibly in the set is finite. All involved regular languages are accepted by DFA with at most $O(|S|)$ states. So, all the intersections of the language labeling the edge (say L_{s_i,s_j}) with the languages in the elements of the node labels (say the languages whose union is P_{s_i}) can be accepted by DFA with at most $O(|S|^2)$ states. Since the intersections are finite the length of their longest words is at most $O(|S|^2)$. Whenever a longer word is written on the tape we do not have to care about it, because it never will be tested. So, a DFA M' simulating M can be constructed in such a way that it tracks all possible words at most of this length in its states. Since this clearly includes the words to be tracked for the conversion into infinite action normal form, both simulations can be superimposed.

The number of states of M' is calculated as follows. There are $|\Gamma|^{O(|S|^2)}$ many words to be tracked. Every word may or may not be present, which gives $2^{|\Gamma|^{O(|S|^2)}}$ possibilities. In addition, the state transitions of M have to be simulated. But $|S| \cdot 2^{|\Gamma|^{O(|S|^2)}}$ is already of order $2^{|\Gamma|^{O(|S|^2)}}$. $\qquad\square$

To show a lower bound, we will use a "regular" variant of the language L_1 discussed in Example 1. For every $n \geq 1$, let

$$L_n = \{\, \$^* w_1 \$^+ w_2 \$^+ \cdots w_m \$^+ w \mid m \geq 0, w, w_1, w_2, \dots, w_m \in \{a, b\}^n,$$
$$\text{and } \exists 1 \leq i \leq m : w = w_i \,\}$$

Theorem 4. *For $n \geq 1$, language L_n is accepted by an $(n+2)$-state DSA, but any equivalent DFA needs at least 2^{2^n} states.*

We would like to note that we immediately obtain the following corollary when we consider for the language family L_n larger alphabets than $\{a, b\}$.

Corollary 5. *For every $n \geq 1$, there are regular languages L_n which are accepted by $(n + 2)$-state DSA with tape alphabet Γ such that any equivalent DFA needs at least $2^{|\Gamma|^n}$ states.*

4   Nondeterminism and Non-recursive Trade-Offs

This section is devoted to nondeterministic set automata and to the non-recursive trade-offs between NSA and DSA and between NSA and deterministic PDA. We start with recalling the necessary notions and definitions. Following [4] we say that a *descriptional system $\mathcal{S}$* is a set of finite descriptors such that each $D \in \mathcal{S}$ describes a formal language $L(D)$, and the underlying alphabet alph(D) over which D represents a language can be read off from D. The *family of languages represented* (or *described*) by $\mathcal{S}$ is $\mathscr{L}(\mathcal{S}) = \{ L(D) \mid D \in \mathcal{S} \}$. For every language L, the set $\mathcal{S}(L) = \{ D \in \mathcal{S} \mid L(D) = L \}$ is the set of its descriptors in $\mathcal{S}$. A *complexity measure* for a descriptional system $\mathcal{S}$ is a total computable mapping $c : \mathcal{S} \to \mathbb{N}$.

Here we only use complexity measures that (with respect to the underlying alphabets) are related to length by a computable function. If there is a total computable function $g : \mathbb{N} \times \mathbb{N} \to \mathbb{N}$ such that length$(D) \leq g(c(D), |\text{alph}(D)|)$, for all $D \in \mathcal{S}$, then c is said to be an *s-measure*. If, in addition, for any alphabet A, the set of descriptors in $\mathcal{S}$ describing languages over A is recursively enumerable in order of increasing size, then c is said to be an *sn-measure*. Clearly, length and trans are sn-measures for set automata.

Let $\mathcal{S}_1$ and $\mathcal{S}_2$ be descriptional systems with complexity measures c_1 and c_2, respectively. A total function $f : \mathbb{N} \to \mathbb{N}$ is an *upper bound* for the increase in complexity when changing from a descriptor in $\mathcal{S}_1$ to an equivalent descriptor in $\mathcal{S}_2$, if for all $D_1 \in \mathcal{S}_1$ with $L(D_1) \in \mathscr{L}(\mathcal{S}_2)$, there exists a $D_2 \in \mathcal{S}_2(L(D_1))$ such that $c_2(D_2) \leq f(c_1(D_1))$.

If there is no recursive, that is, computable function serving as upper bound, the *trade-off is said to be non-recursive*. For establishing non-recursive trade-offs the following generalization of a result by Hartmanis [3] is useful.

Theorem 6 ([4]). *Let $\mathcal{S}_1$ and $\mathcal{S}_2$ be two descriptional systems for recursive languages such that any descriptor D in $\mathcal{S}_1$ and $\mathcal{S}_2$ can effectively be converted into a Turing machine that decides $L(D)$, and let c_1 be a measure for $\mathcal{S}_1$ and c_2 be an sn-measure for $\mathcal{S}_2$. If there exists a descriptional system $\mathcal{S}_3$ and a property P that is not semi-decidable for descriptors from $\mathcal{S}_3$, such that, given an arbitrary $D_3 \in \mathcal{S}_3$, (i) there exists an effective procedure to construct a descriptor D_1 in $\mathcal{S}_1$, and (ii) D_1 has an equivalent descriptor in $\mathcal{S}_2$ if and only if D_3 does not have property P, then the trade-off between $\mathcal{S}_1$ and $\mathcal{S}_2$ is non-recursive.*

In the following we show all non-recursive trade-offs by reduction of the finiteness problem for deterministic one-tape one-head Turing machines (DTM). Let

$M = \langle Q, \Sigma, T, \delta, q_0, B, F \rangle$ be a DTM, where T is the set of tape symbols including the set of input symbols Σ and the blank symbol B, Q is the finite set of states and $F \subseteq Q$ is the set of final states. The initial state is q_0 and δ is the transition function. Without loss of generality, we assume that Turing machines can halt only after an odd number of moves, accept by halting, make at least three moves, and cannot print blanks. At any instant during a computation, M can be completely described by an *instantaneous description* (ID) which is a string $tqt' \in T^*QT^*$ with the following meaning: M is in the state q, the non-blank tape content is the string tt', and the head scans the first symbol of t'. The initial ID of M on input $x \in \Sigma^*$ is $w_0 = q_0 x$. An ID is accepting whenever it belongs to T^*FT^*. The set VALC(M) of valid (accepting) computations of M consists of all finite strings $w_0 \# w_1^R \# w_2 \# w_3^R \# \cdots \# w_{2n} \# w_{2n+1}^R$ such that ID w_i leads to its successor ID w_{i+1} according to δ, and w_{2n+1} is an accepting ID. Similarly, the set VALC$'$(M) consists of all finite strings $w_0 \# w_1 \# w_2 \# \cdots \# w_{2n+1}$ such that w_i leads to its successor w_{i+1} according to δ, and w_{2n+1} is an accepting ID. The set of *invalid computations* INVALC(M) resp. INVALC$'$(M) is the complement of VALC(M) resp. VALC$'$(M) with respect to the alphabet $T \cup Q \cup \{\#\}$.

Lemma 7. *Let M be a Turing machine. Then, the following holds:*

(1) INVALC(M) belongs to $\mathscr{L}$(PDA) and a PDA can be effectively constructed.
(2) INVALC(M) belongs to $\mathscr{L}$(DSA) if and only if $L(M)$ is a finite set.
(3) INVALC$'$(M) belongs to $\mathscr{L}$(NSA) and an NSA can be effectively constructed.
(4) INVALC$'$(M) belongs to $\mathscr{L}$(DSA) if and only if $L(M)$ is a finite set.
(5) INVALC$'$(M) belongs to $\mathscr{L}$(DPDA) if and only if $L(M)$ is a finite set.

Proof. Statement (1) is well known and shown, for example, in [5]. It is also shown in [5] that INVALC(M) is a regular set, if $L(M)$ is a finite set. Thus, INVALC(M) belongs to $\mathscr{L}$(DSA) in this case. On the other hand, we show that INVALC(M) is not in $\mathscr{L}$(DSA) if $L(M)$ is an infinite set. Assume that INVALC(M) $\in \mathscr{L}$(DSA). Since $\mathscr{L}$(DSA) is closed under complementation [6], the set VALC(M) would belong to $\mathscr{L}$(DSA) as well. With standard arguments (see, for example, [5]) it can be shown that then emptiness is not semi-decidable for DSA which is a contradiction to the result given in [6] that emptiness is decidable for DSA. Altogether, we obtain statement (2). With similar considerations it is possible to prove statements (4) and (5). It remains for us to show statement (3). An input belongs to INVALC$'$(M) if it has a wrong format, if it does not start with an initial ID, if it does not end with an accepting ID, or if ID w_{i+1} is not the successor of ID w_i for some $0 \le i \le 2n$. An NSA M' for INVALC$'$(M) guesses at first which of the above possibilities M' is going to test. The first three possibilities concern the structure of the input only and can be tested by M' without any set operation. For the latter possibility, we additionally guess the position $i \ge 0$ and have to ensure that w_{i+1} is not the successor ID of w_i. Having guessed the position i, we read the following input w_i and write the successor ID on the tape. This can be achieved by M', since the changes between two consecutive IDs are only local. When reading the separating symbol $\#$, the calculated successor ID is added to the set. The following input w_{i+1} is written on the tape. When reading the separating symbol # again, we perform a `test`-operation. If the test is negative, we know that w_{i+1} is not the successor ID of w_i and the input is accepted. If the test is positive, then the input is rejected. By this effective construction M' accepts $\text{INVALC}'(M)$.   □

Theorem 8. *The following trade-offs are non-recursive. (1) From* NSA *to* DSA. *(2) From* NSA *to* DPDA. *(3) From* PDA *to* DSA.

Proof. In order to apply Theorem 6, we use as property P the infiniteness of DTM which is known to be not semi-decidable. Next, given an arbitrary DTM M, that is, a descriptor $D_3 \in \mathcal{S}_3$, we must construct an NSA, that is, a descriptor D_1 in $\mathcal{S}_1$, that has an equivalent DSA, that is, a descriptor in $\mathcal{S}_2$, if and only if M accepts a finite language, that is, D_3 does not have property P. Thus, we obtain the non-recursive trade-offs (1) and (2) by applying statements (3)–(5) of Lemma 7. Similarly, the non-recursive trade-off (3) is obtained by applying statements (1) and (2) of Lemma 7.   □

The fact that NSA can accept the set of invalid computations of a Turing machine implies that the following questions for NSA are not semi-decidable. The proof of the results is similar to the corresponding results for pushdown automata which are shown in [5].

Theorem 9. *For* NSA *the questions of universality, equivalence with regular sets, equivalence, inclusion, and regularity are not semi-decidable. Furthermore, it is not semi-decidable whether the language accepted by some* NSA *belongs to* $\mathscr{L}(\text{DSA})$.

References

1. Geffert, V., Mereghetti, C., Palano, B.: More concise representation of regular languages by automata and regular expressions. Inf. Comput. 208, 385–394 (2010)
2. Goldstine, J., Kappes, M., Kintala, C.M.R., Leung, H., Malcher, A., Wotschke, D.: Descriptional complexity of machines with limited resources. J. UCS 8, 193–234 (2002)
3. Hartmanis, J.: On the succinctness of different representations of languages. SIAM J. Comput. 9, 114–120 (1980)
4. Holzer, M., Kutrib, M.: Descriptional complexity – An introductory survey. In: Martín-Vide, C. (ed.) Scientific Applications of Language Methods, pp. 1–58. Imperial College Press (2010)
5. Hopcroft, J.E., Ullman, J.D.: Introduction to Automata Theory, Languages, and Computation. Addison-Wesley, Reading (1979)
6. Kutrib, M., Malcher, A., Wendlandt, M.: Deterministic set automata. In: Developments in Language Theory (DLT 2014). LNCS. Springer (2014)
7. Meyer, A.R., Fischer, M.J.: Economy of description by automata, grammars, and formal systems. In: Symposium on Switching and Automata Theory (SWAT 1971), pp. 188–191. IEEE (1971)

Operational State Complexity
under Parikh Equivalence

Giovanna J. Lavado[1],*, Giovanni Pighizzini[1],*, and Shinnosuke Seki[2,3],**

[1] Dipartimento di Informatica, Università degli Studi di Milano, Italy
{lavado,pighizzini}@di.unimi.it
[2] Helsinki Institute for Information Technology (HIIT),
P.O. Box 15600, 00076, Aalto, Finland
[3] Department of Information and Computer Science, Aalto University,
P.O. Box 15400, FI-00076, Aalto, Finland
shinnosuke.seki@aalto.fi

Abstract. We investigate, under Parikh equivalence, the state complexity of some language operations which preserve regularity. For union, concatenation, Kleene star, complement, intersection, shuffle, and reversal, we obtain a polynomial state complexity over any fixed alphabet, in contrast to the intrinsic exponential state complexity of some of these operations in the classical version. For projection we prove a superpolynomial state complexity, which is lower than the exponential one of the corresponding classical operation. We also prove that for each two deterministic automata A and B it is possible to obtain a deterministic automaton with a polynomial number of states whose accepted language has as Parikh image the intersection of the Parikh images of the languages accepted by A and B. Finally, we prove that for each finite set there exists a small context-free grammar defining a language with the same Parikh image.

Keywords: state complexity, regular languages, Parikh equivalence, context-free grammars, semilinear sets.

1 Introduction

The investigation of the state complexity of regular languages and their operations is extensively reported in the literature (e.g., [13,17,18]). In a previous work [11], we proposed to extend that investigation by considering the classical notion of Parikh equivalence [12], which have been extensively studied in the literature (e.g., [1,7]) even for the connections with semilinear sets [9] and with other fields such as Presburger Arithmetics [6], Petri Nets [4], logical formulas [16], and formal verification [15]. We remind the reader that two words over

* Their works are partially supported by MIUR under the project PRIN "Automi e Linguaggi Formali: Aspetti Matematici e Applicativi" H41J12000190001.
** His work is supported in part by Academy of Finland, Postdoctoral Researcher Grant No. 13266670/T30606.

H. Jürgensen et al. (Eds.): DCFS 2014, LNCS 8614, pp. 294–305, 2014.

a same alphabet Σ are Parikh equivalent if and only if they are equal up to a permutation of their symbols or, equivalently, for each letter $a \in \Sigma$, the number of occurrences of a in the two words is the same (the vector consisting of these numbers is also called *Parikh image*). This notion extends in a natural way to languages (two languages L_1 and L_2 are Parikh equivalent when for each word in L_1 there is a Parikh equivalent word in L_2 and vice versa) and to formal systems which are used to specify languages as, for instance, grammars and automata. A well-known result by Parikh states that context-free and regular languages are indistinguishable under Parikh equivalence [12]. More precisely, the Parikh image of a context-free language is a semilinear set and from each semilinear set a Parikh equivalent NFA can be immediately obtained.

In particular, in [11] we treated the conversion of one-way nondeterministic finite automata (NFAs) into Parikh equivalent one-way deterministic finite automata (DFAs). We proved that the state cost of this conversion is smaller than the exponential cost of the classical conversion. In fact, we showed that from each n-state NFA we can build a Parikh equivalent DFA with $e^{O(\sqrt{n \cdot \ln n})}$ states. Furthermore, this cost is tight. Quite surprisingly, this cost is due to the unary words in the language, i.e., to the words consisting only of occurrences of a same symbol. In fact, if the given NFA accepts only words containing at least two different letters then the cost reduces to a polynomial.

Motivated by the interest in regular languages, here we continue the same line of research by considering basic operations on regular languages and on DFAs. We reformulate under Parikh equivalence some classical questions on the state complexity of operations as, for instance, the following: given two arbitrary DFAs A and B of n_1 and n_2 states, respectively, how many states are sufficient and necessary in the worst case (as a function of n_1 and n_2) for a DFA to accept the concatenation of the languages accepted by A and B? For this question an exponential cost is known [18]. Using our above mentioned bound on the conversion of NFAs into Parikh equivalent DFAs, this exponential bound can be reduced, under Parikh equivalence, to a superpolynomial upper bound. In this paper we further reduce it to a polynomial, namely we show that there exists a DFA with a number of states polynomial in n_1 and n_2 accepting a language that is Parikh equivalent to the concatenation of the languages accepted by A and B. We obtain a similar result for the Kleene star operation while, for the union, the cost is polynomial even in the classical case. We also present results for other operations as intersection, complement, reversal, shuffle and projection.

Concerning intersection and complement, we observe that these operations do not commute with Parikh image, e.g., the Parikh image of the complement of a language L does not necessarily coincide with the complement of the Parikh image of L. However, semilinear sets are closed under intersection and complement [5]. Hence, we can formulate state complexity questions about intersections and complements of Parikh images of languages accepted by given DFAs. We solve the question for the intersection by proving, in a constructive way, that for each two DFAs there exists a DFA of polynomial size accepting a language whose Parikh image is the intersection of the Parikh images of the languages accepted by the two given DFAs, while the analogous question for the complement will be the subject of future investigations.

We take into consideration a further question: under Parikh equivalence, can we use context-free grammars (CFG) to represent, in a more succinct way, regular languages? In the unary case (hence without recurring to Parikh equivalence) a positive answer to this question has been given in [3]. We extend the techniques used in that paper to the case of a general alphabet, obtaining a positive answer for finite languages.

2   Preliminaries

We assume the readers to be familiar with the basic notions and properties from automata and formal language theory. We recall just a few notions, referring the reader to standard textbooks (e.g., [8,14]) for further details.

Throughout the paper, let us fix an alphabet $\Sigma = \{a_1, a_2, \ldots, a_m\}$ of m letters. As usual, let us denote by Σ^* the set of all words over Σ including the *empty word* ε and, given a word $w \in \Sigma^*$, by $|w|$ its *length* and, for $a \in \Sigma$, by $|w|_a$ the *number of occurrences* of a in w.

A word is said to be *unary* if it consists of $k \geq 0$ occurrences of a same symbol, otherwise it is said to be *nonunary*. A language L is *unary* if $L \subseteq \{a\}^*$ for some letter a. In a similar way, automata and CFGs are *unary* when their input and terminal alphabets, respectively, consist of just one symbol.

Given a language $L \subseteq \Sigma^*$, the *unary parts* of L are the languages $L_1 = L \cap \{a_1\}^*$, $L_2 = L \cap \{a_2\}^*, \ldots$, $L_m = L \cap \{a_m\}^*$, while the *nonunary part* is the language $L_0 = L \setminus (L_1 \cup L_2 \cup \ldots \cup L_m)$, i.e., the set which consists of all nonunary words belonging to the language L. Clearly, $L = \bigcup_{i=0}^m L_i$.

Given an automaton A, we can readily build automata accepting the unary and nonunary parts of $L(A)$, respectively. A proof can be found in [11].

Lemma 1. *For each n-state NFA A over an m-letter alphabet, there exist $m+1$ NFAs $A_0, A_1, \ldots, A_m$ such that A_0 has $n(m+1)+1$ states and accepts the nonunary part of $L(A)$ while, for $i = 1, \ldots, m$, A_i is a unary NFA of n states which accepts the unary part $L(A) \cap \{a_i\}^*$. Furthermore, if A is deterministic then so are $A_0, A_1, \ldots, A_m$.*

Let $\mathbb{N}$ be the set of nonnegative integers. Then $\mathbb{N}^m$ denotes the corresponding sets of m-dimensional nonnegative integer vectors including the *null vector* $\mathbf{0} = (0, 0, \ldots, 0)$. For $1 \leq i \leq m$, we denote the i-th component of a vector $\mathbf{v}$ by $\mathbf{v}[i]$.

A *linear set* in $\mathbb{N}^m$ is a set of the form

$$\{\mathbf{v}_0 + n_1 \mathbf{v}_1 + n_2 \mathbf{v}_2 + \cdots + n_k \mathbf{v}_k \mid n_1, n_2, \ldots, n_k \in \mathbb{N}\}, \tag{1}$$

where $k \geq 0$ and $\mathbf{v}_0, \mathbf{v}_1, \mathbf{v}_2, \ldots, \mathbf{v}_k \in \mathbb{N}^m$. The vector $\mathbf{v}_0$ is called the *offset* (a.k.a. *constant*), while the vectors $\mathbf{v}_1, \ldots, \mathbf{v}_k$ are called *generators* (a.k.a. *periods*). A *semilinear set* in $\mathbb{N}^m$ is a finite union of linear sets in $\mathbb{N}^m$.

The *Parikh map* $\psi : \Sigma^* \to \mathbb{N}^m$ associates with a word $w \in \Sigma^*$ the vector $\psi(w) = (|w|_{a_1}, |w|_{a_2}, \ldots, |w|_{a_m})$, which counts the occurrences of each letter of

Σ in w. The vector $\psi(w)$ is also called *Parikh image* of w. One can naturally generalize this map for a language $L \subseteq \Sigma^*$ as $\psi(L) = \{\psi(w) \mid w \in L\}$. The set $\psi(L)$ is called the *Parikh image* of L. Two languages $L, L' \subseteq \Sigma^*$ are said to be *Parikh equivalent* to each other if and only if $\psi(L) = \psi(L')$. Parikh equivalence can be defined not only between languages but among languages, grammars, and finite automata by referring, in the last two cases, to the defined languages.

In the paper we will make use of the following conversion from NFAs to Parikh equivalent DFAs:

Theorem 2 ([11]). *For each n-state* NFA *over an m-letter alphabet Σ, there exists a Parikh equivalent* DFA *with $e^{O(\sqrt{n \cdot \ln n})}$ states. Furthermore, this cost is tight. If the language under consideration does not contain any unary word, then the cost reduces to $O(n^{3m^3 + 6m^2} m^{m^3/2 + m^2 + 2m + 5})$.*

3   Regular Operations under Parikh Equivalence

In this section, we investigate the state complexity of regular operations under Parikh equivalence. Regular operations include concatenation, Kleene star, reversal, shuffle, projection, union, intersection, and complementation. The problems we will work on in this section are in the following general form:

Problem 3. For DFAs A and B of n_1 and n_2 states, respectively, solve the following problems:

1. For a unary operation f, how small can we make a DFA M that is Parikh equivalent to $f(L(A))$?
2. For a binary operation g, how small can we make a DFA M that is Parikh equivalent to $g(L(A), L(B))$?

3.1   Union, Intersection, and Complement

The state complexity of union and intersection is in the low order $n_1 n_2$ even in the conventional sense over both unary and nonunary alphabets. Moreover, it is known to be tight already over a unary alphabet [17]. Similar considerations hold for the complement. The next result hence follows.

Proposition 4. *Given two* DFAs *A and B of n_1 and n_2 states, respectively, there exist two* DFAs *of $n_1 n_2$ states that accept languages (Parikh) equivalent to $L(A) \cup L(B)$ and to $L(A) \cap L(B)$, and a* DFA *of n_1 states that accepts a language (Parikh) equivalent to the complement of $L(A)$. These bounds are tight.*

3.2   Concatenation, Star, and Shuffle

Unlike union or intersection, both concatenation and star are known to cost an exponential number of states on DFAs. In fact, the number of states which is necessary and sufficient in the worst case for a DFA to accept the concatenation of an n_1-state DFA language and an n_2-state DFA language over a binary alphabet is $(2n_1 - 1)2^{n_2-1}$ [18]; over a unary alphabet, the cost decreases to $n_1 n_2$ [17]. As for star of an n-state DFA language, the tight bound is $2^{n-1} + 2^{n-2}$ over a binary alphabet, whereas it is $(n-1)^2 + 1$ over a unary alphabet [18].

We will show that under Parikh equivalence, the state complexity of concatenation and star decreases to polynomial in n_1 and n_2 over a nonunary alphabet (over a unary alphabet, the cost is already polynomial as seen above). The Parikh equivalent conversion of NFAs to DFAs in Theorem 2 makes a great contribution to this purpose. In order to avoid a superpolynomial blowup in the number of states caused by this conversion when being applied to the unary part of an NFA, we give *ad hoc* constructions below.

Theorem 5. *Given an n_1-state DFA A and n_2-state DFA B over an m-letter alphabet with $m \geq 2$, there is a DFA of $O\left(n_1 n_2(n_1 + n_2)^{3m^3 + 6m^2} m^{m^3/2 + m^2 + 2m + 6}\right)$ states that is Parikh equivalent to $L(A)L(B)$.*

Proof. Let $L = L(A)L(B)$. For each $a_i \in \Sigma$, let us denote by L_i the language $L \cap \{a_i^*\}$ and by L_0 the language $L \setminus (\bigcup_{i=1}^m L_i)$, namely the nonunary part of L.

Let $A_1, \ldots, A_m$ be the unary DFAs of n_1 states obtained by applying Lemma 1 to the DFA A. Let $B_1, \ldots, B_m$ be the unary DFAs of n_2 states thus obtained from B. Let us denote by M the $(n_1 + n_2)$-state NFA obtained by applying the standard construction for the product to A and B. Then we proceed as follows:

- From M we derive an NFA accepting L_0 and we convert it according to Theorem 2 into a polynomial size Parikh equivalent DFA M_0 that consists of $O((n_1 + n_2)^{3m^3 + 6m^2} m^{m^3/2 + m^2 + 2m + 5})$ states.
- For $i = 1, \ldots, m$, from A_i and B_i we obtain a unary DFA M_i accepting the language L_i. According to results in [13] the number of states of M_i can be bounded by $n_1 n_2$.
- We build a DFA M' with at most $1 + m n_1 n_2$ states accepting all the unary words in L, namely the set $\bigcup_{i=1}^m L_i$. This consists of an initial state q_0 and one copy of each automaton M_i obtained in the previous step. A transition from q_0 on a_i to an appropriate state of M_i simulates the transition from the initial state of M_i.
- Finally, applying the standard construction for the union to M_0 and M' results in a DFA that accepts a language Parikh equivalent to $L = L(A)L(B)$. It consists of $O(n_1 n_2(n_1 + n_2)^{3m^3 + 6m^2} m^{m^3/2 + m^2 + 2m + 6})$ states.   □

With this result in mind, we can address the state complexity of shuffle under Parikh equivalence. Let A and B be DFAs of n_1 and n_2 states, respectively. In the conventional sense, shuffle involves the exponential cost $2^{n_1 n_2} - 1$ and this bound is tight [2]. In contrast, we can construct a DFA of polynomial number of states in n_1 and n_2 that accepts a language Parikh equivalent to the shuffle of $L(A)$ and $L(B)$, and in fact, the DFA we engineered in Theorem 5 for concatenation is such a DFA. This is because the Parikh image of the shuffle of two languages is equal to that of their concatenation.

Corollary 6. *Given an n_1-state DFA A and n_2-state DFA B over an m-letter alphabet with $m \geq 2$, there is a DFA of $O\left(n_1 n_2 (n_1 + n_2)^{3m^3 + 6m^2} m^{m^3/2 + m^2 + 2m + 6}\right)$ states that is Parikh equivalent to the shuffle of $L(A)$ and $L(B)$.*

Let us develop the above conversion of polynomial overhead for star.

Theorem 7. *Given an n-state DFA A over an m-letter alphabet with $m \geq 2$, there is a DFA of $O(2^{3m^3 + 6m^2 + 1} n^{3m^3 + 6m^2 + 2} m^{7m^3/2 + 7m^2 + 2m + 6})$ states that is Parikh equivalent to $L(A)^*$.*

Proof. We employ a similar technique to the one for concatenation.

From the given DFA A, Lemma 1 extracts DFAs $A_1, \ldots, A_m$ (of n states) for the unary parts. From them, we can construct DFAs $M_1, \ldots, M_m$ of $(n-1)^2 + 1$ states accepting $L(A_1)^*, \ldots, L(A_m)^*$, respectively [18]. We further combine them into a DFA M_{unary} of $m((n-1)^2 + 1) + 1$ states that accepts $L(A_1)^* \cup \cdots \cup L(A_m)^*$.

From A, we also construct an n-state NFA recognizing $L(A)^*$, extract its nonunary part by Lemma 1 (with $n(m+1) + 1$ states), and convert it into a Parikh equivalent DFA M_0 using Theorem 2. The resulting DFA consists of $O(2^{3m^3 + 6m^2} n^{3m^3 + 6m^2} m^{7m^3/2 + 7m^2 + 2m + 5})$ states. Applying the standard construction for the union to M_0 and M_{unary} results in a DFA that is Parikh equivalent to $L(A)^*$. The DFA consists of $O(2^{3m^3 + 6m^2 + 1} n^{3m^3 + 6m^2 + 2} m^{7m^3/2 + 7m^2 + 2m + 6})$ states[1]. □

3.3 Reversal

Reversal is also expensive for DFAs. In fact, the tight bound 2^n is known for reversal [18]. Under Parikh equivalence, however, nothing need be said since Parikh image is invariant under this operation.

3.4 Projection

Given a word $w \in \Sigma^*$, the *projection* of w over an alphabet $\Sigma' \subseteq \Sigma$, is the word $P_{\Sigma'}(w)$ obtained by removing from w all the symbols which are not in Σ'. We can extend this notion to languages in a standard way. It is easy to see that projection preserves regularity. However, transforming a DFA A of n states into a DFA for the projection can require an exponential number of states in n (a detailed investigation on this operation is presented in [10]). Even in this case, the bound can be reduced if we want to obtain a Parikh equivalent DFA: from A we can obtain an NFA of n states for the projection, and then we can transform it into a Parikh equivalent DFA of $e^{O(\sqrt{n \cdot \ln n})}$ states. By using a projection over a unary alphabet we can show that this bound cannot be reduced.

---

[1] Note that the number of states of M_{unary}, $m((n-1)^2 + 1) + 1$, is at most $2mn^2$.

3.5 Intersection and Complement, Revisited

We consider one more time the intersection and the complement. In fact, the noncommutativity of those operations with the Parikh mapping brings us a second problem of interest. The noncommutativity in the case of intersection is illustrated in the inequality $\psi(a^+b^+ \cap b^+a^+) \neq \psi(a^+b^+) \cap \psi(b^+a^+)$; the left-hand side is the empty set, while the right-hand side is the linear set $\mathbb{N} \times \mathbb{N}$. In the case of complement the reader may consider the language $(ab)^*$.

Note that each of the other operations examined so far is either commutative with the Parikh mapping (i.e., union and projection) or not defined naturally over the set of nonnegative integer vectors (i.e., concatenation, star, shuffle, and reversal). The problem of interest asks: given two DFAs A and B of n_1 and n_2 states, respectively, how small can we make a DFA whose Parikh image is equal to $\psi(L(A)) \cap \psi(L(B))$? We can formulate a similar problem in the case of the complement. The fact that the Parikh image of a language accepted by an NFA is semilinear and the closure property of semilinear sets under intersection and complement [5] makes these problems meaningful. Here, we solve the problem for intersection, leaving the one for complement for future investigations.

Over a unary alphabet, the problem is degenerated into the problem addressed in Proposition 4 because over such an alphabet, intersection commutes with the Parikh mapping. Therefore, in the following, we examine the problem over a nonunary alphabet, and solve it by showing that polynomial number of states in n_1 and n_2 are sufficient. The proof consists of revisiting the Ginsburg and Spanier's proof [5] of the closure property of semilinear sets under intersection with a careful analysis of the size of the resulting semilinear set.

Let us present some notation and results necessary for the proof. Given $C, P \subseteq \mathbb{N}^k$ for some $k \geq 1$, let $L(C; P)$ be the set of all $\boldsymbol{w} \in \mathbb{N}^k$ which can be represented in the form $\boldsymbol{w} = \boldsymbol{w}_0 + \boldsymbol{w}_1 + \cdots + \boldsymbol{w}_\ell$ with $\boldsymbol{w}_0 \in C$ and $\boldsymbol{w}_1, \ldots, \boldsymbol{w}_\ell \in P$ for some $\ell \geq 0$.

Lemma 8 ([11]). *There exists a polynomial p such that for each n-state NFA A over Σ, the Parikh image of $L(A)$ can be written as $\psi(L(A)) = Y \cup \bigcup_{i \in I} Z_i$, where*

- *$Y \subseteq \mathbb{N}^m$ is a finite set of vectors whose components are bounded by $p(n)$;*
- *I is a set of at most $p(n)$ indices;*
- *for each $i \in I$, $Z_i \subseteq \mathbb{N}^m$ is a linear set of the form:*

$$Z_i = \{\boldsymbol{v}_{i,0} + n_1 \boldsymbol{v}_{i,1} + \cdots + n_{k_i} \boldsymbol{v}_{i,k_i} \mid n_1, \ldots, n_{k_i} \in \mathbb{N}\},$$

with $0 \leq k_i \leq m$, the components of the offset $\boldsymbol{v}_{i,0}$ are bounded by $p(n)$, and the generators $\boldsymbol{v}_{i,1}, \ldots, \boldsymbol{v}_{i,k_i}$ are linearly independent vectors from $\{0, 1, \ldots, n\}^m$.

Furthermore, if all the words in $L(A)$ are nonunary, then for each $i \in I$, we can choose a nonunary vector $\boldsymbol{x}_i$ that is component-wise smaller than or equal to $\boldsymbol{v}_{i,0}$ such that all those chosen vectors are pairwise distinct.

Let A be a $k \times \ell$ matrix with entries in $\mathbb{Z}$ and $k \leq \ell$, and let $\boldsymbol{b} \in \mathbb{Z}^k$. Consider a system of linear Diophantine equations

$$A\boldsymbol{x} = \boldsymbol{b}. \tag{2}$$

Let $S_{\min}(A, \boldsymbol{b})$ be the set of minimal nonnegative integer solutions to (2), where the minimality is with respect to the component-wise comparison. It is well-known that $S_{\min}(A, \boldsymbol{b})$ is finite. Hence, we let $S_{\min}(A, \boldsymbol{b}) = \{s_1, s_2, \ldots, s_r\}$ for some $r \geq 1$ and $s_1, \ldots, s_r \in \mathbb{N}^\ell$. Define $\|S_{\min}(A, \boldsymbol{b})\| = \max_{1 \leq i \leq r} \|s_i\|_\infty$, where $\|s_i\|_\infty$ refers to the maximum norm. Huynh bounded $\|S_{\min}(A, \boldsymbol{b})\|$ as follows.

Lemma 9 ([9]). *Let α be the rank of A and M be the maximum of the absolute values of the $\alpha \times \alpha$ subdeterminants of the extended matrix $(A; \boldsymbol{b})$. Then $\|S_{\min}(A, \boldsymbol{b})\| \leq (\ell + 1)M$. Thus, $r \leq ((\ell + 1)M + 1)^\ell$.*

Theorem 10. *Given an n_1-state DFA A and n_2-state B over an m-letter alphabet $\Sigma = \{a_1, \ldots, a_m\}$, there is a DFA of $O(n^{(2m-1)(3m^3+6m^2)+2}p(n)^{2(3m^3+6m^2)+m})$ states whose Parikh image is equal to $\psi(L(A)) \cap \psi(L(B))$, where n is defined as $\max\{n_1, n_2\}(m+1) + 1$.*

Proof. As usual, we begin with converting the given DFAs A and B into $2(m+1)$ DFAs $A_0, A_1, \ldots, A_m, B_0, B_1, \ldots, B_m$ according to Lemma 1. The nonunary DFAs A_0 and B_0 contain $n_1(m+1) + 1$ and $n_2(m+1) + 1$ states, respectively, while the other unary ones A_i and B_i contain only n_1 and n_2 states for $1 \leq i \leq m$, respectively. It is clear from the definition of these automata that

$$\psi(L(A)) \cap \psi(L(B)) = \big(\psi(L(A_0)) \cap \psi(L(B_0))\big) \cup \bigcup_{1 \leq i \leq m} \big(\psi(L(A_i)) \cap \psi(L(B_i))\big).$$

As observed before the proof, for any $1 \leq i \leq m$, we can construct a DFA M_i of $n_1 n_2$ states such that $\psi(L(M_i)) = \psi(L(A_i)) \cap \psi(L(B_i))$. We combine them into one DFA M_{unary} with $mn_1 n_2 + 1$ states that accepts $\bigcup_{i=1}^m L(M_i)$.

What we have to consider now is the intersection between the Parikh images of the nonunary DFAs A_0 and B_0. In order to simplify the notation below, let $n = \max\{n_1, n_2\}(m+1) + 1$. Then, A_0 and B_0 consist of at most n states each. Lemma 8 implies that the Parikh image $\psi(L(A_0))$ can be represented as:

$$\psi(L(A_0)) = Y_A \cup \bigcup_{i \in I} Z_{A,i},$$

where $Y_A \subseteq \{0, 1, \ldots, p(n)\}^m$, I is a set of at most $p(n)$ indices, and for each $i \in I$, $Z_{A,i} \subseteq \mathbb{N}^m$ is a linear set whose offset is in $\{0, 1, \ldots, p(n)\}^m$ and whose generators are linearly independent vectors in $\{0, 1, \ldots, n\}^m$. The Parikh image of $L(B_0)$ also admits an analogous representation with Y_B, J, and $Z_{B,j}$ $(j \in J)$. The intersection $\psi(L(A_0)) \cap \psi(L(B_0))$ can be expressed as:

$$\big(Y_A \cap \psi(L(B_0))\big) \cup \left(\bigcup_{i \in I} Z_{A,i} \cap Y_B\right) \cup \left(\bigcup_{i \in I} Z_{A,i} \cap \bigcup_{j \in J} Z_{B,j}\right).$$

Since both Y_A and Y_B are finite, the first two terms are finite, and can be computed even by hand. Let $P_Y = \big(Y_A \cap \psi(L(B_0))\big) \cup \big(\bigcup_{i \in I} Z_{A,i} \cap Y_B\big)$, that is, their union. Note that $P_Y \subseteq Y_A \cup Y_B \subseteq \{0, 1, \ldots, p(n)\}^m$. We can construct a DFA M_Y of $O(p(n)^m)$ states that accepts the bounded language $\{a_1^{i_1} a_2^{i_2} \cdots a_m^{i_m} \mid (i_1, \ldots, i_m) \in P_Y\}$. It is clear that $\psi(L(M_Y)) = P_Y$.

Now we shift our attention to the third term $\big(\bigcup_{i \in I} Z_{A,i} \cap \bigcup_{j \in J} Z_{B,j}\big)$. What we actually do is to construct a DFA $B_{i,j}$ whose Parikh image is equal to $Z_{A,i} \cap Z_{B,j}$, for each $i \in I$ and $j \in J$. According to Lemma 8, for some $p, q \leq m$, we can let

$$Z_{A,i} = \{u_0 + n_1 u_1 + \cdots + n_p u_p \mid n_1, \ldots, n_p \in \mathbb{N}\},$$
$$Z_{B,j} = \{v_0 + m_1 v_1 + \cdots + m_q v_q \mid m_1, \ldots, m_q \in \mathbb{N}\},$$

where $u_0, v_0 \in \{0, 1, \ldots, p(n)\}^m$ and $u_1, \ldots, u_p, v_1, \ldots, v_q \in \{0, 1, \ldots, n\}^m$.
Let

$$X_C = \left\{ (n_1, \ldots, n_p, m_1, \ldots, m_q) \,\middle|\, u_0 + \sum_{i=1}^p n_i u_i = v_0 + \sum_{j=1}^q m_j v_j \right\}$$

$$X_P = \left\{ (n_1, \ldots, n_p, m_1, \ldots, m_q) \,\middle|\, \sum_{i=1}^p n_i u_i = \sum_{j=1}^q m_j v_j \right\}$$

Let $\tau : \mathbb{N}^{p+q} \to \mathbb{N}^m$ be a function defined as $\tau((n_1, \ldots, n_p, m_1, \ldots, m_q)) = \sum_{i=1}^p n_i u_i$, and we generalize this function over a set $X \subseteq \mathbb{N}^{p+q}$ as $\tau(X) = \{\tau(x) \mid x \in X\}$. Then

$$Z_{A,i} \cap Z_{B,j} = \{u_0 + w \mid w \in \tau(X_C)\}. \tag{3}$$

Note that if X_C is semilinear, then so is $\tau(X_C)$ [5]. Ginsburg and Spanier proved that X_C is semilinear [5]. More strongly, they gave a representation of X_C as $L(C; P)$, where C and P are the set of minimal elements of X_C and $X_P \setminus \{0^{p+q}\}$, respectively. Lemma 9 gives the following bounds:

$$\|C\| \leq (p + q + 1)m!n^{m-1}p(n),$$
$$\|P\| \leq (p + q + 1)m!n^m.$$

They bound the cardinality of the sets C and P as follows:

$$|C| \leq ((p + q + 1)m!n^{m-1}p(n) + 1)^m,$$
$$|P| \leq ((p + q + 1)m!n^m + 1)^m.$$

Using $\tau(X_C) = \tau(L(C; P)) = L(\tau(C); \tau(P))$ (see the proof of Lemma 6.3 in [5]), we rewrite (3) as

$$Z_{A,i} \cap Z_{B,j} = u_0 + L(\tau(C); \tau(P)).$$

Note that $|\tau(C)| \leq |C|$, $|\tau(P)| \leq |P|$, and we have

$$\|\tau(C)\| \leq nm\|C\| \leq m(p + q + 1)m!n^m p(n)$$
$$\|\tau(P)\| \leq nm\|P\| \leq m(p + q + 1)m!n^{m+1}.$$

We construct an NFA whose Parikh image is equal to this semilinear set. Specifically, it is to accept the language $\{f(\boldsymbol{u}_0)f(\boldsymbol{v})f(\boldsymbol{w}) \mid \boldsymbol{v} \in \tau(C), \boldsymbol{w} \in L(\mathbf{0}; \tau(P))\}$, where $\mathbf{0}$ is the zero vector and $f : \mathbb{N}^m \to \Sigma^*$ is defined as: $f((x_1, x_2, \ldots, x_m)) = a_1^{x_1} a_2^{x_2} \cdots a_m^{x_m}$. It first recognize $f(\boldsymbol{u}_0)$ using $mp(n) + 1$ states (recall that any coordinate of $\boldsymbol{u}_0$ is at most $p(n)$). Then it recognizes $f(\boldsymbol{v})$ for some $\boldsymbol{v}$ nondeterministically chosen from $\tau(C)$, using $m||\tau(C)|| \cdot |\tau(C)|$ states. Finally it recognizes $f(\boldsymbol{w})$ for some $\boldsymbol{w} \in L(\mathbf{0}; \tau(P))$ using $m||\tau(P)|| \cdot |\tau(P)|$ states. In total, it consists of $O(m^2(2m+1)^2(m!)^2n^{2m-1}p(n)^2)$ states. Now that we have at most $p(n)^2$ NFAs $M_{0,0}, \ldots, M_{|I|-1,|J|-1}$ of such size. The standard construction combines them into an NFA with $O(m^2(2m+1)^2(m!)^2n^{2m-1}p(n)^4)$ states and it is nonunary. The nonunarity allows Theorem 2 to convert this NFA into a Parikh equivalent DFA M_0 with $O(n^{(2m-1)(3m^3+6m^2)}p(n)^{2(3m^3+6m^2)})$ states.

Now that the standard construction for union combines this, M_Y, and M_{unary} into the DFA with $O(n^{(2m-1)(3m^3+6m^2)+2}p(n)^{2(3m^3+6m^2)+m})$ states, and its Parikh image is equal to $\psi(L(A)) \cap \psi(L(B))$. $\square$

4   From Finite Languages to Small Grammars

Each finite set of words defined over an alphabet Σ of m symbols can be easily represented by a grammar in Chomsky normal form (CNFG) with at most $N+m$ symbols, where N is the total length of the words. This upper bound cannot be significantly reduced. In fact, in [3, Lemma 3.1], the authors provide, for each integer $k \geq 1$, a language L_k consisting of just one word of length $2^k + k - 1$ over the alphabet $\{0, 1\}$ such that any CNFG generating L_k requires more than $c2^k/k$, for a constant c.

However, as we show in this section, the bound can be reduced if we are allowed to replace the original finite language by a set which is Parikh equivalent to it. The proof is obtained by adapting some techniques introduced for unary sets in [3].

A grammar G is in *binary normal form* (BNFG) if every production is in one of the following forms: $A \to a$, $A \to \varepsilon$, $A \to B$, or $A \to BC$, where A, B, C are variables and $a \in \Sigma$. It is known that if $G = (V, \Sigma, P, S)$ is a BNFG, then there exists a CNFG $G' = (V', \Sigma, P', S)$ such that $L(G') = L(G) \setminus \{\varepsilon\}$ and $V' \subseteq V$.

Lemma 11. *Let $n \geq 1$ and $T \subseteq \{0, 1, 2, \ldots, n-1\}^m$. Then there exists a BNFG G of $O(n^{m/3})$ variables such that $\psi(L(G)) = T$. This bound is asymptotically tight.*

Proof. This proof is a modification of the proof for Lemma 2.1 in [3]. Let $r = \lceil n^{1/3} \rceil$. Any integer less than n can be expressed in base r using at most 3 digits as $ir^2 + jr + k$ with $0 \leq i, j, k < r$.

For each $1 \leq \ell \leq m$, we first define $A_\ell \to a_\ell$. We introduce r^m variables $G_{k_1, k_2, \ldots, k_m}$ for $0 \leq k_1, k_2, \ldots, k_m < r$, and define the productions

$$G_{k_1,k_2,\ldots,k_m} \to A_1 G_{k_1-1,k_2,\ldots,k_m}$$
$$G_{0,k_2,\ldots,k_m} \to A_2 G_{0,k_2-1,\ldots,k_m}$$
$$\vdots$$
$$G_{0,0,\ldots,0,k_m} \to A_m G_{0,0,\ldots,0,k_m-1}$$
$$G_{0,0,\ldots,0} \to \varepsilon.$$

Then we have $L(G_{k_1,k_2,\ldots,k_m}) = \{a_1^{k_1} a_2^{k_2} \cdots a_m^{k_m}\}$.

For $1 \le \ell \le m$, let $B_\ell \to A_\ell G_{0,\ldots,0,r-1,0,\ldots,0}$, where the sole nonzero index is the ℓ-th one. Then $L(B_\ell) = \{a_\ell^r\}$. According to the same principle as above, we define the following productions:

$$F_{j_1,j_2,\ldots,j_m} \to B_1 F_{j_1-1,j_2,\ldots,j_m}$$
$$\vdots$$
$$F_{0,0,\ldots,0,j_m} \to B_m F_{0,0,\ldots,0,j_m-1}$$
$$F_{0,0,\ldots,0} \to \varepsilon.$$

Then we have $L(F_{j_1,j_2,\ldots,j_m}) = \{a_1^{j_1 r} a_2^{j_2 r} \cdots a_m^{j_m r}\}$.

Furthermore, we define $C_\ell \to B_\ell F_{0,\ldots,0,r-1,0,\ldots,0}$ for $1 \le \ell \le m$ and

$$E_{i_1,i_2,\ldots,i_m} \to C_1 E_{i_1-1,i_2,\ldots,i_m}$$
$$\vdots$$
$$E_{0,0,\ldots,0,i_m} \to C_m E_{0,0,\ldots,0,i_m-1}$$
$$E_{0,0,\ldots,0} \to \varepsilon.$$

We have $L(C_\ell) = \{a_\ell^{r^2}\}$ and $L(E_{i_1,i_2,\ldots,i_m}) = \{a_1^{i_1 r^2} a_2^{i_2 r^2} \cdots a_m^{i_m r^2}\}$. Up to now, we have introduced $O(r^m)$ variables. Finally, we define the remaining productions:

$$S \to E_{i_1,i_2,\ldots,i_m} S_{i_1,i_2,\ldots,i_m} \text{for each } 0 \le i_1,\ldots,i_m < r$$
$$S_{i_1,\ldots,i_m} \to F_{j_1,\ldots,j_m} G_{k_1,\ldots,k_m} \text{for all } 0 \le j_1,\ldots,j_m,k_1,\ldots,k_m < r$$
$$\text{such that } (i_1 r^2 + j_1 r + k_1,\ldots,i_m r^2 + j_m r + k_m) \in T.$$

The resulting grammar is a BNFG, and the total number of variables is $4r^m + 3m = O(n^{m/3})$.

We conclude the proof by showing the asymptotic tightness of the bound. Over ℓ variables, there are $\ell^3 + \ell^2 + \ell(m+1)$ possible productions, so there are $2^{\ell^3+\ell^2+\ell(m+1)}$ BNFGs. On the other hand, there are 2^{n^m} (finite) subsets of the set of all m-dimensional vectors any of whose component is less than n. In order for each subset to admit a BNFG, $2^{\ell^3+\ell^2+\ell(m+1)} \ge 2^{n^m}$ must hold. Hence, $\ell = \Omega(n^{m/3})$. $\qquad\square$

From Lemma 11 we obtain the following result:

Theorem 12. *Let L be a language containing only words on which each letter occurs less than n times. Then there exists a CNFG with $O(n^{m/3})$ variables which is Parikh equivalent to L. This bound is asymptotically tight.*

We conclude this section by observing that to represent a finite language L as in Theorem 12 by a Parikh equivalent DFA A (or even NFA) we need, in the worst case, more than $(n-1)^m$ states. In fact, since L is finite, all the states on each accepting path of A cannot belong to any loop. Thus, if $(n-1)^m \in \psi(L)$ then A should contain a path of at least $(n-1)^m + 1$ states.

References

1. Aceto, L., Ésik, Z., Ingólfsdóttir, A.: A fully equational proof of Parikh's theorem. RAIRO-Theor. Inf. Appl. 36(2), 129–153 (2002)
2. Câmpeanu, C., Salomaa, K., Yu, S.: Tight lower bound for the state complexity of shuffle of regular languages. Journal of Automata, Languages and Combinatorics 7(3), 303–310 (2002)
3. Domaratzki, M., Pighizzini, G., Shallit, J.: Simulating finite automata with context-free grammars. Inform. Process. Lett. 84(6), 339–344 (2002)
4. Esparza, J.: Petri nets, commutative context-free grammars, and basic parallel processes. Fund. Inform. 31(1), 13–25 (1997)
5. Ginsburg, S., Spanier, E.H.: Bounded ALGOL-like languages. T. Am. Math. Soc. 113, 333–368 (1964)
6. Ginsburg, S., Spanier, E.H.: Semigroups, Presburger formulas, and languages. Pac. J. Math. 16(2), 285–296 (1966)
7. Goldstine, J.: A simplified proof of Parikh's theorem. Discrete Math 19(3), 235–239 (1977)
8. Hopcroft, J.E., Ullman, J.D.: Introduction to Automata Theory, Languages and Computation. Addison-Wesley (1979)
9. Huynh, D.T.: The complexity of semilinear sets. In: de Bakker, J.W., van Leeuwen, J. (eds.) ICALP 1980. LNCS, vol. 85, pp. 324–337. Springer, Heidelberg (1980)
10. Jirásková, G., Masopust, T.: On a structural property in the state complexity of projected regular languages. Theor. Comput. Sci. 449, 93–105 (2012)
11. Lavado, G.J., Pighizzini, G., Seki, S.: Converting nondeterministic automata and context-free grammars into Parikh equivalent one-way and two-way deterministic automata. Inform. Comput. 228, 1–15 (2013)
12. Parikh, R.J.: On context-free grammars. J. ACM 13(4), 570–581 (1966)
13. Pighizzini, G., Shallit, J.: Unary language operations, state complexity and jacobsthal's function. Int. J. Found. Comput. S. 13(1), 145–159 (2002)
14. Shallit, J.O.: A Second Course in Formal Languages and Automata Theory. Cambridge University Press, Cambridge (2008)
15. To, A.W.: Model Checking Infinite-State Systems: Generic and Specific Approaches. Ph.D. thesis, School of Informatics, University of Edinburgh (August 2010)
16. Verma, K., Seidl, H., Schwentick, T.: On the complexity of equational Horn clauses. In: Nieuwenhuis, R. (ed.) CADE 2005. LNCS (LNAI), vol. 3632, pp. 337–352. Springer, Heidelberg (2005)
17. Yu, S.: State complexity of regular languages. Journal of Automata, Languages and Combinatorics 6, 221–234 (2000)
18. Yu, S., Zhuang, Q., Salomaa, K.: The state complexity of some basic operations on regular languages. Theor 125, 315–328 (1994)

Complexity of Checking Whether Two Automata Are Synchronized by the Same Language

Marina Maslennikova*

Ural Federal University, Ekaterinburg, Russia
maslennikova.marina@gmail.com

Abstract. A deterministic finite automaton $\mathscr{A}$ is said to be *synchronizing* if it has a *reset* word, i.e. a word that brings all states of the automaton $\mathscr{A}$ to a particular one. We prove that it is a **PSPACE**-complete problem to check whether the language of reset words for a given automaton coincides with the language of reset words for some particular automaton.

Keywords: ideal language, synchronizing automaton, reset word, reset complexity, **PSPACE**-completeness.

Introduction

Let $\mathscr{A} = \langle Q, \Sigma, \delta \rangle$ be a *deterministic finite automaton* (DFA), where Q is the *state set*, Σ stands for the *input alphabet*, and $\delta : Q \times \Sigma \to Q$ is the totally defined *transition function* defining the action of the letters in Σ on Q. The function δ is extended uniquely to a function $Q \times \Sigma^* \to Q$, where Σ^* stands for the free monoid over Σ. The latter function is still denoted by δ. In the theory of formal languages the definition of a DFA usually includes the *initial state* $q_0 \in Q$ and the set $F \subseteq Q$ of *terminal states*. We will use this definition when dealing with automata as devices for recognizing languages. A language $L \subseteq \Sigma^*$ is *recognized* (or *accepted*) by an automaton $\mathscr{A} = \langle Q, \Sigma, \delta, q_0, F \rangle$ if $L = \{w \in \Sigma^* \mid \delta(q_0, w) \in F\}$. We denote by $L[\mathscr{A}]$ the language accepted by the automaton $\mathscr{A}$.

A DFA $\mathscr{A} = \langle Q, \Sigma, \delta \rangle$ is called *synchronizing* if there exists a word $w \in \Sigma^*$ whose action leaves the automaton in one particular state no matter at which state in Q it is applied: $\delta(q, w) = \delta(q', w)$ for all $q, q' \in Q$. Any word w with this property is said to be *reset* for the DFA $\mathscr{A}$. For the last 50 years synchronizing automata received a great deal of attention. In 1964 Černý conjectured that every synchronizing automaton with n states possesses a reset word of length at most $(n-1)^2$. Despite intensive efforts of researchers this conjecture still remains open. For a brief introduction to the theory of synchronizing automata we refer the reader to the recent surveys [11, 13].

---

* The author acknowledges support from the Presidential Programm for young researchers, grant MK-3160.2014.1.

H. Jürgensen et al. (Eds.): DCFS 2014, LNCS 8614, pp. 306–317, 2014.

In the present paper we focus on some complexity aspects of the theory of synchronizing automata. We denote by $\mathrm{Syn}(\mathscr{A})$ the language of reset words for a given automaton $\mathscr{A}$. It is well known that $\mathrm{Syn}(\mathscr{A})$ is regular [13]. Furthermore, it is an *ideal* in Σ^*, i.e. $\mathrm{Syn}(\mathscr{A}) = \Sigma^* \mathrm{Syn}(\mathscr{A})\Sigma^*$. On the other hand, every regular ideal language L serves as the language of reset words for some automaton. For instance, the minimal automaton recognizing L is synchronized exactly by L [7]. Thus synchronizing automata can be considered as a special representation of an ideal language. Effectiveness of such a representation was addressed in [7]. The *reset complexity* $rc(L)$ of an ideal language L is the minimal possible number of states in a synchronizing automaton $\mathscr{A}$ such that $\mathrm{Syn}(\mathscr{A}) = L$. Every such automaton $\mathscr{A}$ is called a *minimal synchronizing automaton* (for brevity, MSA). Let $sc(L)$ be the number of states in the minimal automaton recognizing L. For every ideal language L we have $rc(L) \leq sc(L)$ [7]. Moreover, there are languages L_n for every $n \geq 3$ such that $rc(L_n) = n$ and $sc(L_n) = 2^n - n$ [7]. Thus the representation of an ideal language by means of a synchronizing automaton can be exponentially more succinct than the "traditional" representation via the minimal automaton. However, no reasonable algorithm is known for computing an MSA of a given language. One of the obstacles is that an MSA is not uniquely defined. For instance, there is a language with at least two different MSAs [7].

Let L be an ideal regular language over Σ with $rc(L) = n$. The latter equality means that there exists some n-state DFA $\mathscr{B}$ such that $\mathrm{Syn}(\mathscr{B}) = L$, and $\mathscr{B}$ is an MSA for L. Now it is quite natural to ask the following question: how hard is it to verify the condition $\mathrm{Syn}(\mathscr{B}) = L$? It is well known that the equality of the languages accepted by two given DFAs can be checked in polynomial of the size of automata time. However, the problem of checking the equality of the languages of reset words of two synchronizing DFAs turns out to be hard. Moreover, it is hard to check whether one particular ideal language serves as the language of reset words for a given synchronizing automaton. We state formally the SYN-EQUALITY problem:

–*Input:* synchronizing automata $\mathscr{A}$ and $\mathscr{B}$.

–*Question:* is $\mathrm{Syn}(\mathscr{A}) = \mathrm{Syn}(\mathscr{B})$?

We prove that SYN-EQUALITY is a **PSPACE**-complete problem. Actually, we prove a stronger result, that it is a **PSPACE**-complete problem to check whether the language $\mathrm{Syn}(\mathscr{A})$ for a given automaton $\mathscr{A}$ coincides with the language $\mathrm{Syn}(\mathscr{B})$ for some particular automaton $\mathscr{B}$. Also it is interesting to understand how hard it is to verify a strict inclusion $\mathrm{Syn}(\mathscr{A}) \subsetneq \mathrm{Syn}(\mathscr{B})$. We prove that it is not easier than to check the precise equality of the languages $\mathrm{Syn}(\mathscr{A})$ and $\mathrm{Syn}(\mathscr{B})$. So the problem of constructing an MSA for a given ideal language is unlikely to be an easy task. Also we obtain that the problem of checking the inequality $rc(L) \leq \ell$, for a given positive integer number ℓ, is **PSPACE**-complete. Here an ideal language L is presented by a DFA, for which L serves as the language of reset words. Actually, we prove that the problem of checking the equalities $rc(L) = 1$ or $rc(L)$ is trivial, however it is a **PSPACE**-complete problem to verify whether $rc(L) = 3$.

The paper is organized as follows. In Section 1 we introduce some definitions and state formally the considered problems. In Section 2 we prove main results about **PSPACE**-completeness of the problem SYN-EQUALITY and **PSPACE**-completeness of the problem of checking whether the reset complexity of a given ideal language is not greater than ℓ.

1   Preliminaries

A standard tool for finding the language of synchronizing words of a given DFA $\mathscr{A} = \langle Q, \delta, \Sigma \rangle$ is the *power automaton* $\mathcal{P}(\mathscr{A})$. Its state set is the set $\mathcal{Q}$ of all nonempty subsets of Q, and the transition function is defined as a natural extension of δ on the set $\mathcal{Q} \times \Sigma$ (the resulting function is also denoted by δ), namely, $\delta(S, a) = \{\delta(s, a) \mid s \in S\}$ for $S \subseteq Q$ and $a \in \Sigma$. If we take the set Q as the initial state and singletons as final states in $\mathcal{P}(\mathscr{A})$, then we obtain an automaton recognizing $\mathrm{Syn}(\mathscr{A})$. It is easy to see that if all the singletons are merged into a unique sink state s (i.e. s is fixed by all letters in Σ), the resulting automaton still recognizes $\mathrm{Syn}(\mathscr{A})$. Throughout the paper the term *power automaton* and the notation $\mathcal{P}(\mathscr{A})$ will refer to this modified version.

One may notice now that the problem SYN-EQUALITY can be solved by the following naive algorithm. Indeed, we construct the power automata $\mathcal{P}(\mathscr{A})$ and $\mathcal{P}(\mathscr{B})$ for DFAs $\mathscr{A}$ and $\mathscr{B}$. Now it remains to verify that automata $\mathcal{P}(\mathscr{B})$ and $\mathcal{P}(\mathscr{A})$ accept the same language. However, the automaton $\mathcal{P}(\mathscr{A})$ has $2^n - n$ states, where n is the number of states in the DFA $\mathscr{A}$. So we cannot afford to construct directly the corresponding power automata. Now we state formally the SYN-INCLUSION problem. It will be shown that SYN-INCLUSION is in **PSPACE**.

SYN-INCLUSION
 −*Input:* synchronizing automata $\mathscr{A}$ and $\mathscr{B}$.
 −*Question:* is $\mathrm{Syn}(\mathscr{A}) \subseteq \mathrm{Syn}(\mathscr{B})$?

Since SYN-INCLUSION belongs to the class **PSPACE**, we obtain that SYN-EQUALITY is in **PSPACE** as well. Further we prove that the SYN-EQUALITY problem is complete for the class **PSPACE**. Now it is interesting to consider the SYN-STRICT-INCLUSION problem:

 −*Input:* synchronizing automata $\mathscr{A}$ and $\mathscr{B}$.
 −*Question:* is $\mathrm{Syn}(\mathscr{A}) \subsetneq \mathrm{Syn}(\mathscr{B})$?

It will be shown that SYN-STRICT-INCLUSION is a **PSPACE**-complete problem.

Recall that the word $u \in \Sigma^*$ is a *prefix* (*suffix* or *factor*, respectively) of the word w if $w = us$ ($w = tu$ or $w = tus$, respectively) for some $t, s \in \Sigma^*$. A reset word w for a DFA $\mathscr{A}$ is called *minimal* if none of its proper prefixes nor suffixes is reset. We will denote by $w[i]$ the ith letter of w and by $|w|$ the length of the word w. In what follows the word $w[i]w[i + 1]...w[j]$, for $i < j$, will be denoted by $w[i..j]$.

2   PSPACE-Completeness

Theorem 1. *SYN-INCLUSION is in **PSPACE**.*

Proof. Savitch's theorem states that **PSPACE=NPSPACE** [12]. Therefore, it is enough to prove that SYN-INCLUSION belongs to **NPSPACE**, i.e. it suffices to solve the problem by a non-deterministic algorithm within polynomial space. Let $\mathscr{A} = \langle Q_1, \Sigma, \delta_1 \rangle$ and $\mathscr{B} = \langle Q_2, \Sigma, \delta_2 \rangle$ be synchronizing automata over Σ. We have to prove that the language $\mathrm{Syn}(\mathscr{B})$ contains the language $\mathrm{Syn}(\mathscr{A})$, or equivalently the following equality takes place: $\mathrm{Syn}(\mathscr{A}) \cap \mathrm{Syn}(\mathscr{B})^c = \emptyset$, where $\mathrm{Syn}(\mathscr{B})^c$ is the complement language of $\mathrm{Syn}(\mathscr{B})$. An obstacle is that we cannot afford to construct the automaton recognizing the language $\mathrm{Syn}(\mathscr{A}) \cap \mathrm{Syn}(\mathscr{B})^c$ directly. Instead we provide an algorithm that guesses a word w which is reset for $\mathscr{A}$ and is not reset for $\mathscr{B}$. Let us notice that w may turn out to be exponentially long in $|Q_1| + |Q_2|$. Hence even if our algorithm correctly guesses w, it would not have enough space to store its guess. Thus the algorithm should guess w letter by letter.

Let $Q_1 = \{q_1, \ldots, q_n\}$ and $Q_2 = \{p_1, \ldots, p_m\}$. The algorithm guesses the first letter $w[1]$ of w, applies $w[1]$ at every state in Q_1 and Q_2 and stores two sets of images, namely, $\delta_1(Q_1, w[1])$ and $\delta_2(Q_2, w[1])$. These sets clearly require only $O(n+m)$ space. Further the algorithm guesses the second letter $w[2]$ and updates the sets of images re-using the space, and so on. Note that $\delta_1(Q_1, w[1..k]) = \delta_1(\delta_1(Q_1, w[1..k-1]), w[k])$, where $k \geq 2$. So we do not need to store the whole word w in order to build the sets $\delta_1(Q_1, w)$ and $\delta_2(Q_2, w)$. At the end of the guessing steps the algorithm gets two sets $\delta_1(Q_1, w)$ and $\delta_2(Q, w)$. It remains to check the following conditions:
– the first set is a singleton;
– the second set is not singleton.

The latter checking does not require any additional space. Thus the problem SYN-INCLUSION is in **PSPACE**. $\square$

Since $\mathrm{Syn}(\mathscr{A}) = \mathrm{Syn}(\mathscr{B})$ if and only if $\mathrm{Syn}(\mathscr{A}) \subseteq \mathrm{Syn}(\mathscr{B})$ and $\mathrm{Syn}(\mathscr{B}) \subseteq \mathrm{Syn}(\mathscr{A})$, we obtain the following corollary.

Corollary 1. *SYN-EQUALITY is in **PSPACE**.*

To prove that SYN-EQUALITY is a **PSPACE**-complete problem we reduce the following well-known **PSPACE**-complete problem to the complement of SYN-EQUALITY. This problem deals with checking emptiness of the intersection of languages accepted by DFAs from a given collection [5].
FINITE AUTOMATA INTERSECTION
   −*Input:* given n DFAs $M_i = \langle Q_i, \Sigma, \delta_i, q_i, F_i \rangle$, for $i = 1, \ldots, n$.
   −*Question:* is $\bigcap_i L[M_i] \neq \emptyset$?
Given an instance of FINITE AUTOMATA INTERSECTION, we can suppose without loss of generality that each initial state q_i has no incoming edges and $q_i \notin F_i$. Indeed, excluding the case for which the empty word ε is in $L[M_i]$ we can always build a DFA $M_i' = \langle Q_i', \Sigma, \delta_i', q_i', F_i \rangle$, which recognizes the same

language of M_i, such that the initial state q_i' has no incoming edges. This can be easily achieved by adding a new initial state q_i' to the state set Q_i and defining the transition function δ_i' by the rule: $\delta_i'(q_i', a) = \delta_i(q_i, a)$ for all $a \in \Sigma$ and $\delta_i'(q, a) = \delta_i(q, a)$ for all $a \in \Sigma$, $q \in Q_i$. Furthermore, we may assume that the sets Q_i, for $i = 1, \ldots, n$, are pairwise disjoint.

To build an instance of SYN-EQUALITY from DFAs M_i, $i = 1, \ldots, n$, we construct a DFA $\mathscr{A} = \langle Q, \Delta, \varphi, \rangle$ with $Q = \bigcup_{i=1}^{n} Q_i \cup \{s, h\}$, where s and h are new states not belonging to any Q_i. We add three new letters to the alphabet Σ and get in this way $\Delta = \Sigma \cup \{x, y, z\}$. The transition function φ of the DFA $\mathscr{A}$ is defined by the following rules:

$$\begin{aligned}
\varphi(q, a) &= \delta_i(q, a) && \text{for all } i = 1, \ldots, n, \text{for all } q \in Q_i \text{ and } a \in \Sigma; \\
\varphi(q, x) &= q_i && \text{for all } i = 1, \ldots, n, \text{for all } q \in Q_i; \\
\varphi(q, z) &= s && \text{for all } i = 1, \ldots, n, \text{for all } q \in F_i; \\
\varphi(q, z) &= h && \text{for all } i = 1, \ldots, n, \text{for all } q \in (Q_i \setminus F_i); \\
\varphi(q, y) &= s && \text{for all } i = 1, \ldots, n, \text{for all } q \in Q_i; \\
\varphi(h, a) &= s && \text{for all } a \in \Delta; \\
\varphi(s, a) &= s && \text{for all } a \in \Delta.
\end{aligned}$$

The resulting automaton $\mathscr{A}$ is shown schematically in Fig. 1. The action of letters from Σ on the states $p \in Q_i$ is not shown. Denote by G_i the set $Q_i \setminus (F_i \cup \{q_i\})$. All the states from the set G_i are shown as the node labeled by G_i. All the states from the set F_i are shown as the node labeled by F_i.

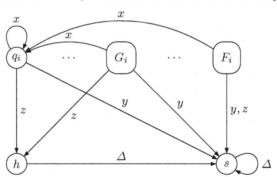

Fig. 1. Automaton $\mathscr{A}$

It can be easily seen that by the definition of the transition function φ we get $\varphi(Q, w) \cap Q_i \neq \emptyset$ if and only if $w \in (\Sigma \cup \{x\})^*$. From this observation and the definition of φ we obtain the following lemma.

Lemma 1. *For any $w \in \Delta^*$ we have $\varphi(Q, w) \cap Q_i \neq \emptyset$ for all $i = 1, \ldots, n$ if and only if there is some $j \in \{1, \ldots, n\}$ such that $\varphi(Q, w) \cap Q_j \neq \emptyset$.*

Consider the languages L_1 and L_2:

$$L_1 = (\Sigma \cup \{x\})^* y \Delta^*;$$
$$L_2 = (\Sigma \cup \{x\})^* z \Delta^+.$$

Consider the language $I = L_1 \cup L_2$.

Lemma 2. $\bigcap_{i=1}^{n} L[M_i] = \emptyset$ *if and only if* $\mathrm{Syn}(\mathscr{A}) = I$.

Proof. Let $\bigcap_{i=1}^{n} L[M_i] = \emptyset$. We take a word $w \in \mathrm{Syn}(\mathscr{A})$. Since s is a sink state in $\mathscr{A}$, we have $\varphi(Q, w) = \{s\}$. If $w \in (\Sigma \cup \{x\})^+$, then $\varphi(Q, w) \cap Q_i \neq \emptyset$, which is a contradiction. Therefore, w contains some factor belonging to $\{y, z\}^+$. Thus we can factorize w as $w = uav$, where u is a maximal prefix of w belonging to $(\Sigma \cup \{x\})^*$, $a \in \{y, z\}$ and $v \in \Delta^*$.

Case 1: $a = y$.

Since y maps all the states of $\mathscr{A}$ to a sink state s, we have that w is a reset word and $w \in L_1$.

Case 2: $a = z$.

2.1. Let $u \in \Sigma^*$, i.e. u does not contain a factor belonging to $\{x\}^+$. By lemma 1, $\varphi(Q, u) \cap Q_i \neq \emptyset$ for all $i = 1, \ldots, n$. Note that $u \notin \bigcap_i L[M_i]$, that is $u \notin L[M_j]$ for some j. It means that $\varphi(q_j, u) \in Q_j \setminus F_j$. Hence $h \in \varphi(Q, uz)$. More precisely, $\varphi(Q, uz) = \{h, s\}$. It remains to apply a letter from Δ in order to map h to s. So we have $w \in L_2$.

2.2. Let u contain a factor belonging to $\{x\}^+$. Obviously, we may factorize u as $u = u'xt$, where t is the maximal suffix of u belonging to Σ^*. By lemma 1, $\varphi(Q, u') \cap Q_i \neq \emptyset$ for all $i = 1, \ldots, n$. Thus $q_i \in \varphi(Q, u'x)$ for all $i = 1, \ldots, n$. By the argument from the previous case, $\varphi(q_j, u'xt) \in Q_j \setminus F_j$ for some index j. Thus $\varphi(Q, u'xtz) = \{h, s\}$ and $w \in L_2$.

So we obtain that if $\bigcap_{i=1}^{n} L[M_i] = \emptyset$, then $\mathrm{Syn}(\mathscr{A}) \subseteq I$. The opposite inclusion $I \subseteq \mathrm{Syn}(\mathscr{A})$ follows easily from the construction of $\mathscr{A}$. Assume now that the equality $\mathrm{Syn}(\mathscr{A}) = I$ takes place. Arguing by contradiction, assume that $\bigcap_{i=1}^{n} L[M_i] \neq \emptyset$, thus there is some $w' \in \bigcap_{i=1}^{n} L[M_i]$. By the definition of φ we get that the word $w = xw'z$ is a reset word for $\mathscr{A}$. However, $w \notin I$. So we came to a contradiction. $\qquad \square$

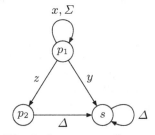

Fig. 2. Automaton $\mathscr{B}$

Now we build a 3-state automaton $\mathscr{B} = \langle P, \Delta, \tau \rangle$ (see Fig. 2). Its state set is $P = \{p_1, p_2, s\}$, where s is the unique sink state. Further we verify that I serves as the language of reset words for $\mathscr{B}$. Moreover, $\mathscr{B}$ is an MSA for I.

Lemma 3. $\mathrm{Syn}(\mathscr{B}) = I$.

Proof. It is clear that $I \subseteq \mathrm{Syn}(\mathscr{B})$. Let $w \in \mathrm{Syn}(\mathscr{B})$. Obviously, $w \notin (\Sigma \cup \{x\})^*$. Thus we may factorize w as $w = uav$, where $(\Sigma \cup \{x\})^*$, $a \in \{y, z\}$ and $v \in \Delta^*$. If $a = y$ then $w \in L_1$. If $a = z$ then $\tau(Q, uz) = \{p_2, s\}$. Since w is a reset word for $\mathscr{B}$ we obtain that $w \in L_2$. So we have the inclusion $\mathrm{Syn}(\mathscr{B}) \subseteq I$.   □

For each instance of FINITE AUTOMATA INTERSECTION one may construct the corresponding automaton $\mathscr{A}$ and the DFA $\mathscr{B}$. It is easy to check that I does not serve as the language of reset words for a synchronizing automaton of size at most two over the same alphabet Δ. So $\mathscr{B}$ is an MSA for I and $rc(\mathrm{Syn}(\mathscr{B})) = 3$. Furthermore, $\mathscr{B}$ is a finitely generated synchronizing automaton, that is its language of reset words $\mathrm{Syn}(\mathscr{B})$ can be represented as $\mathrm{Syn}(\mathscr{B}) = \Delta^* U \Delta^*$ for some finite set of words U. Namely, $U = y \cup z\Delta$. Finitely generated synchronizing automata and its languages of reset words were studied in [4,9,10]. In particular, it was shown in [9] that recognizing finitely generated synchronizing automata is a **PSPACE**-complete problem. Finally, by lemmas 2 and 3, we have the following claim.

Lemma 4. $\bigcap_{i=1}^n L[M_i] = \emptyset$ *if and only if* $\mathrm{Syn}(\mathscr{A}) = \mathrm{Syn}(\mathscr{B})$.

Now we are in position to prove the main result.

Theorem 2. *SYN-EQUALITY is **PSPACE**-complete.*

Proof. By Corollary 1, SYN-EQUALITY is in **PSPACE**. Since the construction of automata $\mathscr{A}$ and $\mathscr{B}$ can be performed in polynomial time from the automata M_i $(i = 1, \ldots, n)$, by Lemmas 2, 3 and 4, we can reduce FINITE AUTOMATA INTERSECTION to co-SYN-EQUALITY.   □

Theorem 3. *SYN-STRICT-INCLUSION is **PSPACE**-complete.*

Proof. By theorem 1, SYN-STRICT-INCLUSION is in **PSPACE**.   □
    PSPACE-hardness of SYN-STRICT-INCLUSION follows easily from the proof of Lemma 2. The proof of the Lemma implies that $\bigcap_{i=1}^n L[M_i] \neq \emptyset$ if and only if $I = \mathrm{Syn}(\mathscr{B}) \subsetneq \mathrm{Syn}(\mathscr{A})$.

Let us note that we build synchronizing automata $\mathscr{A}$ and $\mathscr{B}$ over at least 5-letter alphabet to obtain an instance of SYN-EQUALITY. What about alphabets of size less than five? It can be easily seen that, for automata over a unary alphabet, SYN-EQUALITY can be solved in polynomial time. Indeed, if $\mathscr{A} = \langle Q_1, \{a\}, \delta_1 \rangle$ is a synchronizing DFA, then $\mathrm{Syn}(\mathscr{A}) = a^* a^k$, where k is the length of the shortest reset word for $\mathscr{A}$. Furthermore, it is easy to check that $k < |Q_1|$. Analogously, the language of reset words for a DFA $\mathscr{B} = \langle Q_2, \{a\}, \delta_2 \rangle$ is $\mathrm{Syn}(\mathscr{B}) = a^* a^m$, where $m < |Q_2|$. Finally, positive integer numbers k and m

can be found in polynomial time. So it is interesting to consider automata over alphabets of size at least two.

We have reduced the problem FINITE AUTOMATA INTERSECTION to the problem SYN-EQUALITY. By construction of DFAs $\mathscr{A}$ and $\mathscr{B}$, we have $\Delta = \{y, z, a, b, x\}$. We build DFAs $\mathscr{C} = \langle C, \{\mu, \lambda\}, \varphi_2 \rangle$ and $\mathscr{D} = \langle D, \{\mu, \lambda\}, \tau_2 \rangle$ with unique sink states ζ_1 and ζ_2 respectively. It will be shown that $\mathrm{Syn}(\mathscr{A}) = \mathrm{Syn}(\mathscr{B})$ if and only if $\mathrm{Syn}(\mathscr{C}) = \mathrm{Syn}(\mathscr{D})$. A standard technique is applied here and also was used in [1, 6, 7]. Namely, we define morphisms $h : \{\lambda, \mu\}^*\lambda \to \Delta^*$ and $\overline{h} : \Delta^* \to \{\lambda, \mu\}^*\lambda$ preserving the property of being a reset word for the corresponding automaton. Let $d_1 = y$, $d_2 = z$, $d_3 = a$, $d_4 = b$ and $d_5 = x$. We put $h(\mu^k\lambda) = d_{k+1}$ for $k = 0, \ldots, 4$ and $h(\mu^k\lambda) = d_5 = x$ for $k \geq 5$. Every word from the set $\{\lambda, \mu\}^*\lambda$ can be uniquely factorized by words $\mu^k\lambda$, $k = 0, 1, 2, \ldots$, thus the mapping h is totally defined. We also consider the morphism $\overline{h} : \Delta^* \to \{\lambda, \mu\}^*\lambda$ defined by the rule $\overline{h}(d_k) = \mu^{k-1}\lambda$. Note that for every word $u \in \Delta^*$ we have $h(\overline{h}(u)) = u$.

Now we take the constructed above DFA $\mathscr{B} = \langle P, \Delta, \tau \rangle$ with the state set $P = \{p_1, p_2, s\}$. We build the DFA $\mathscr{D} = \langle D, \{\mu, \lambda\}, \tau_2 \rangle$ with a unique sink state ζ_2.

We associate each state p_i of the automaton $\mathscr{B}$ with the 5-element set of states $P_i = \{p_{i,1}, \ldots, p_{i,5}\}$ of the automaton $\mathscr{D}$. Namely, the states $p_{i,2}, p_{i,3}, p_{i,4}, p_{i,5}$ are copies of the state p_i associated with $p_{i,1}$. The action of the letter μ is defined in the following way: $\tau_2(p_{i,k}, \mu) = p_{i,k+1}$ for $k \leq 4$, and $\tau_2(p_{i,5}, \mu) = p_{i,5}$. We put $D = P_1 \cup P_2 \cup \{\zeta_2\}$, where ζ_2 is a unique sink state. The action of the letter λ is defined by the rules:

– if $\tau(p_i, d_k) = s$, then $\tau_2(p_{i,k}, \lambda) = \zeta_2$;
– if $\tau(p_i, d_k) = p_j$, then $\tau_2(p_{i,k}, \lambda) = p_{j,1}$.

The latter rule means that if there is the transition from p_i to p_j labeled by the letter d_k, then there is the transition from $p_{i,1}$ to $p_{j,1}$ labeled by the word $\mu^{k-1}\lambda$.

P_i is called the *i-th colomn* of the set D. For each $k = 1, \ldots, 5$, one may take the set $R_k = \{p_{1,k}, p_{2,k}\}$. The set R_k is called the *k-th row* of the set D.

The DFA $\mathscr{C}$ is constructed in analogous way. Finally, note that the resulting automata $\mathscr{C}$ and $\mathscr{D}$ have $O(5|Q_1|)$ and $O(5|Q_2|)$ states respectively, where $|Q_1|$ and $|Q_2|$ are the cardinalities of the state sets of $\mathscr{A}$ and $\mathscr{B}$ respectively. Figure 3 illustrates the automaton $\mathscr{D}$. The action of the letter μ is shown in solid lines, the action of the letter λ is shown in dotted lines.

Lemma 5. $\mathrm{Syn}(\mathscr{A}) = \mathrm{Syn}(\mathscr{B})$ *if and only if* $\mathrm{Syn}(\mathscr{C}) = \mathrm{Syn}(\mathscr{D})$.

Proof. It is convenient to organize the constructed DFA $\mathscr{D}$ as a table. The k-th row contains copies of all states corresponding to the k-th letter from Δ. The i-th column contains the state $p_{i,1}$ corresponding to the state p_i and its copies $p_{i,2}, \ldots, p_{i,5}$. Each state from the i-th column maps under the action of μ to a state from the same column. The k-th row R_k maps under the action of μ to the $k + 1$-st row R_{k+1} ($k \leq 4$). The 5-th row is fixed by μ, that is $\tau_2(R_5, \mu) = R_5$. The state set D maps under the action of λ to a subset of R_1. The DFA $\mathscr{C}$ possesses such properties as well.

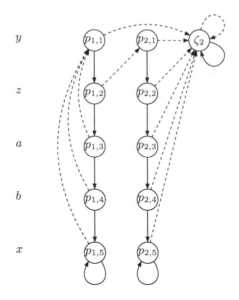

Fig. 3. Automaton $\mathscr{D}$

It can be easily checked that the word $u \in \Delta^*$ is reset for the DFA $\mathscr{B}$ if and only if $\overline{h}(u)$ is reset for $\mathscr{D}$. An analogous property takes place for DFAs $\mathscr{A}$ and $\mathscr{C}$.

Assume that $\mathrm{Syn}(\mathscr{A}) \neq \mathrm{Syn}(\mathscr{B})$. From the proof of lemma 2 it follows that the word $w = xw'z$ with $w' \in \bigcap_i L[M_i]$ is reset for $\mathscr{A}$ and it is not reset for $\mathscr{B}$. Thus $\overline{h}(w) \in \mathrm{Syn}(\mathscr{C})$ and $\overline{h}(w) \notin \mathrm{Syn}(\mathscr{D})$. So $\mathrm{Syn}(\mathscr{C}) \neq \mathrm{Syn}(\mathscr{D})$.

Assume now that $\mathrm{Syn}(\mathscr{A}) = \mathrm{Syn}(\mathscr{B})$. We show that every minimal reset word of $\mathscr{C}$ is reset for $\mathscr{D}$ and every minimal reset word of $\mathscr{D}$ is reset for $\mathscr{A}$. Let u be a minimal reset word of $\mathscr{C}$. Any word $u \in \{\mu\}^*$ is not in $\mathrm{Syn}(\mathscr{C})$, since μ brings each column to its subset. Thus we have $u \in \{\lambda, \mu\}^* \setminus \{\mu\}^*$. The automaton $\mathscr{C}$ possesses a unique sink state ζ_1. Hence $\mathscr{C}$ is synchronized to ζ_1. Furthermore, all the transitions leading to ζ_1 are labeled by λ, and ζ_1 is fixed by μ and λ. Thus if u does not end with λ then it is not a minimal reset word. We have $u \in \{\lambda, \mu\}^*\lambda$. Consider the word $w = h(u)$. Since $\overline{h}(w) = u$, we have that w is a reset word for $\mathscr{A}$ and $\mathscr{B}$. Hence $u \in \mathrm{Syn}(\mathscr{D})$. So we obtain that $\mathrm{Syn}(\mathscr{C}) \subseteq \mathrm{Syn}(\mathscr{D})$. The opposite inclusion is verified analogously. □

Lemma 5 gives the desired result on **PSPACE**-completeness of the problem SYN-EQUALITY restricted to a binary alphabet case. Analogously it is shown that SYN-STRICT-INCLUSION is also a **PSPACE**-complete problem for automata over a binary alphabet.

Theorem 4. *SYN-EQUALITY restricted to a binary alphabet case is **PSPACE**-complete.*

Proposition 1. *Let ℓ be a positive integer number, L an ideal language, and $\mathcal{A}$ a synchronizing DFA for which L serves as the language of reset words. The problem of checking the inequality $rc(L) \leq \ell$ is in* **PSPACE**.

Proof. If ℓ is greater or equal to the size of $\mathcal{A}$, then the answer is "yes" and there is nothing to prove. Let ℓ be less than the size of $\mathcal{A}$. One may non-deterministically guess a DFA $\mathcal{B}$ with at most ℓ states and check the equality $\mathrm{Syn}(\mathcal{B}) = \mathrm{Syn}(\mathcal{A})$ within polynomial space.   □

Lemma 6. *Let L be an ideal language and $\mathcal{A}$ some automaton with $\mathrm{Syn}(\mathcal{A}) = L$. The equalities $rc(L) = 1$ and $rc(L) = 2$ can be checked in polynomial of the size of $\mathcal{A}$ time.*

Proof. Let $\mathcal{A} = \langle Q, \Sigma, \delta \rangle$. Denote by k the size of the alphabet Σ. It is easy to see that $rc(L) = 1$ if and only if $L = \Sigma^*$, so it is required that $n = 1$.

Let us notice that some 2-state automaton $\mathcal{B} = \langle P, \Sigma, \delta \rangle$ is synchronizing if and only if some letter brings the automaton to a singleton and each letter $a \in \Sigma$ either maps the state set P to a singleton or acts as a permutation on P. So we find the set $\Gamma = \{a \mid a \in \mathrm{Syn}(\mathcal{A})\}$ in time $O(kn)$ and obtain the DFA $\mathcal{A}' = \langle Q, \Sigma \setminus \Gamma, \delta \rangle$ from $\mathcal{A}$ removing the transitions labeled by letters from Γ. It remains to check that $\mathcal{A}'$ is not synchronizing. The latter checking can be done in time $O(kn^2)$ [3]. We have that $rc(L) = 2$ if and only if $rc(L) \neq 1$ and $\mathcal{A}'$ is not synchronizing.   □

We have constructed for each instance of FINITE AUTOMATA INTERSECTION the corresponding automaton $\mathcal{A}$ over the alphabet $\Delta = \Sigma \cup \{x, y, z\}$ in order to prove that SYN-EQUALITY is a **PSPACE**-complete problem. We will use that automaton again to prove the following theorem.

Theorem 5. *Let $J = \mathrm{Syn}(\mathcal{A})$. $\bigcap_{i=1}^{n} L[M_i] \neq \emptyset$ if and only if $rc(J) > 3$.*

Proof. Let $\bigcap_{i=1}^{n} L[M_i] = \emptyset$. In this case we have $\mathrm{Syn}(\mathcal{A}) = I$. As it was mentioned above $\mathcal{B}$ is an MSA for I, so we have the equality $rc(\mathrm{Syn}(\mathcal{A})) = 3$. Let us assume now that $\bigcap_{i=1}^{n} L[M_i] \neq \emptyset$. Since y is the unique letter which resets $\mathcal{A}$ and the automaton $\mathcal{A}'$ (which is obtained from $\mathcal{A}$ by removing all transitions labeled by y) is still synchronizing, we get that $rc(J) \geq 3$. Now we verify that $rc(J) \neq 3$, that is no 3-state synchronizing automaton is synchronized exactly by J. It is easy to see that if $\bigcap_{i=1}^{n} L[M_i] \neq \emptyset$, then $J = I \cup I_3 \cup I_4$ where

$$I_3 = \{wz \mid w \in \bigcap_{i=1}^{n} L[M_i] \text{ and } \delta_i(Q_i, w) \subseteq F_i, \text{ for all } i \text{ from 1 to } n\};$$

$$I_4 = \{uxwz \mid u \in (\Sigma \cup \{x\})^*, w \in \bigcap_{i=1}^{n} L[M_i]\}.$$

Arguing by contradiction, assume that $\mathrm{Syn}(\mathcal{B}) = J$ for some 3-state automaton $\mathcal{B}$ over Δ. Denote the state set of $\mathcal{B}$ by $P = \{0, 1, 2\}$ and the transition function by τ. Since y is a reset letter for $\mathscr{B}$, we have that y brings P to a singleton, say $\{2\}$. Letter z is not reset for $\mathscr{A}$ and $z^2 \in J$, hence z maps P to a 2-element subset. It is easy to check that there are just three possible different ways of defining the action of z on the state set P (see Fig.4).

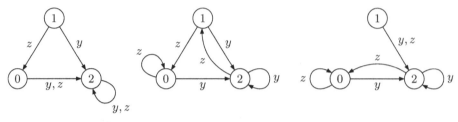

Fig. 4. Possible ways of defining the action of y and z in $\mathscr{B}$

Let us assume as above that $\Sigma = \{a, b\}$. It remains to define the action of x, a and b on the state set P. One may see that the words za, zb and zx are reset for $\mathscr{A}$. So letters x, a and b should map the set $\tau(P, z)$ to singletons. However, any word from $(\Sigma \cup \{x\})^*$ is not reset for $\mathscr{A}$. In particular, $xx, aa, bb \notin \mathrm{Syn}(\mathscr{A})$. Consider, for instance, the first automaton in Fig.4. We have that $\tau(P, z) = \{0, 2\}$, thus the action of x, a and b is defined in such a way that $|\tau(\{0,2\}, x)| = |\tau(\{0,2\}, a)| = |\tau(\{0,2\}, b)| = 1$. So there are six possible ways of defining the action of x on the states of $\mathscr{B}$. Since $xx \notin \mathrm{Syn}(\mathscr{A})$, we have that the following two ways of defining the transitions under the action of x are impossible:

$$012$$
$$x \; 020$$
$$x \; 202.$$

Indeed, in both cases the word xx brings the set $\{0, 1, 2\}$ to a singleton. So, actually, there are just four possible ways of defining the action of x on the states of $\mathscr{B}$. The same arguments can be provided for letters a and b. Thus the definition of the action of x, a, and b is chosen in one of the following ways:

$$012$$
$$x_1 \; 010$$
$$x_2 \; 212$$
$$x_3 \; 101$$
$$x_4 \; 121.$$

For instance, one may say that x acts on P as x_1, a acts as x_2 and b acts as x_3. It is sufficient to consider only those cases where all the letters x, a and b act on P differently. There remains four ways of choosing a triple $\{x_i, x_j, x_k\}$ defining the action of letters x, a and b. It can be easily checked that $J \neq \mathrm{Syn}(\mathscr{B})$ in each case. Analogous arguments are provided for the remaining two automata in Fig.4. It means that the reset complexity of the language of reset words of the DFA $\mathscr{A}$ is at least 4, that is $rc(J) \geq 4$. □

Theorem 6. *Let L be an ideal language and $\mathscr{A}$ a synchronizing DFA with at least 5 letters such that* $\mathrm{Syn}(\mathscr{A}) = L$. *The problem of checking the inequality* $rc(L) \leq 3$ *is* **PSPACE**-*complete.*

This result follows immediately from the Theorem 5.

Acknowledgment. The author is grateful to participants of the seminar "Theoretical Computer Science" held in the Ural Federal University for valuable comments.

References

1. Ananichev, D.S., Gusev, V.V., Volkov, M.V.: Slowly Synchronizing Automata and Digraphs. In: Hliněný, P., Kučera, A. (eds.) MFCS 2010. LNCS, vol. 6281, pp. 55–65. Springer, Heidelberg (2010)
2. Cerný, J.: Poznámka k homogénnym eksperimentom s konečnými automatami. Mat.-Fyz. Cas. Slovensk. Akad. Vied 14, 208–216 (1964)
3. Eppstein, D.: Reset sequences for monotonic automata. SIAM J. Comput. 19, 500–510 (1990)
4. Gusev, V.V., Maslennikova, M.I., Pribavkina, E.V.: Finitely generated ideal languages and synchronizing automata. In: Karhumäki, J., Lepistö, A., Zamboni, L. (eds.) WORDS 2013. LNCS, vol. 8079, pp. 143–153. Springer, Heidelberg (2013)
5. Kozen, D.: Lower bounds for natural proof systems. In: Proc. of the 18th FOCS, pp. 254–266 (1977)
6. Martygin, P.: Computational Complexity of Certain Problems Related to Carefuuly Synchronizing Words for Partial Automata and Directing Words for Nondeterministic Automata. Theory Comput. Sci. 54, 293–304 (2014)
7. Maslennikova, M.I.: Reset Complexity of Ideal Languages. In: Bieliková, M. (ed.) Int. Conf. SOFSEM 2012, Proc. V. II, arXiv: 1404.2816, pp. 33–44. Institute of Computer Science Academy of Sciences of the Czech Republic (2014)
8. Pin, J.-E.: On two combinatorial problems arising from automata theory. Ann. Discrete Math. 17, 535–548 (1983)
9. Pribavkina, E.V., Rodaro, E.: Recognizing synchronizing automata with finitely many minimal synchronizing words is PSPACE-complete. In: Löwe, B., Normann, D., Soskov, I., Soskova, A. (eds.) CiE 2011. LNCS, vol. 6735, pp. 230–238. Springer, Heidelberg (2011)
10. Pribavkina, E., Rodaro, E.: Synchronizing automata with finitely many minimal synchronizing words. Inf. and Comput. 209(3), 568–579 (2011)
11. Sandberg, S.: Homing and synchronizing sequences. In: Broy, M., Jonsson, B., Katoen, J.-P., Leucker, M., Pretschner, A. (eds.) Model-Based Testing of Reactive Systems. LNCS, vol. 3472, pp. 5–33. Springer, Heidelberg (2005)
12. Savitch, W.J.: Relationships between nondeterministic and deterministic tape classes. J.CSS 4, 177–192 (1970)
13. Volkov, M.V.: Synchronizing automata and the Černý conjecture. In: Martín-Vide, C., Otto, F., Fernau, H. (eds.) LATA 2008. LNCS, vol. 5196, pp. 11–27. Springer, Heidelberg (2008)

On the Descriptional Complexity
of Deterministic Ordered Restarting Automata

Friedrich Otto

Fachbereich Elektrotechnik/Informatik
Universität Kassel, 34109 Kassel, Germany
otto@theory.informatik.uni-kassel.de

Abstract. We show that the deterministic ordered restarting automaton is polynomially related in size to the weight-reducing Hennie machine. Accordingly, it allows very compact representations of (some) regular languages. In addition, we investigate the descriptional complexity of the operations of reversal, complementation, intersection, and union for regular languages that are given through stateless deterministic ordered restarting automata.

Keywords: restarting automaton, ordered rewriting, descriptional complexity, language operations.

1 Introduction

The restarting automaton was introduced in [4] as a formal device to model the linguistic technique of *analysis by reduction*. Since then many variants and extensions of the basic model have been introduced and studied (for an overview, see, e.g., [7]), and several classical families of formal languages have been characterized by certain types of restarting automata.

Here we study the descriptional complexity of the *deterministic ordered restarting automaton* (or *det-ORWW-automaton*) that was introduced in [6] in the setting of picture languages. A det-ORWW-automaton has a finite-state control, a tape with end markers that initially contains the input, and a window of size three. Based on its state and the content of its window, the automaton can perform one of three types of operations: it may perform a *move-right step* that shifts the window one position to the right and changes the state, or it may perform a combined *rewrite/restart step* that replaces the symbol in the middle of the window by a symbol that is smaller with respect to a predefined ordering on the working alphabet, that moves the window back to the left end of the tape, and that resets the state to the initial state, or it may perform an *accept step* that causes the automaton to halt and accept. It has been shown in [6] that the nondeterministic variant of the ordered restarting automaton accepts some languages that are not even context-free. Here, however, we only consider the deterministic variant, which is known to accept exactly the regular languages, and study its descriptional complexity.

H. Jürgensen et al. (Eds.): DCFS 2014, LNCS 8614, pp. 318–329, 2014.

First we show that each det-ORWW-automaton can be simulated by an automaton of the same type that has only a single state, which means that for these automata, states are actually not needed. Accordingly, we restrict our attention to the stateless det-ORWW-automaton (stl-det-ORWW-automaton). For such an automaton, we take the size of its working alphabet as the measure for its descriptional complexity, and we show that these automata are polynomially related in size to the weight-reducing Hennie machines studied by Průša in [8]. This implies that there is a double exponential trade-off when changing from a deterministic finite-state acceptor (DFA) to an equivalent stl-det-ORWW-automaton.

If M_1 and M_2 are stl-det-ORWW-automata that accept languages L_1 and L_2, respectively, and if $\circ$ is a binary operation on languages, as e.g. union or intersection, then there exists a stl-det-ORWW-automaton M for the language $L_1 \circ L_2$. We provide upper bounds for the size of M in terms of the sizes of M_1 and M_2 for these operations and some others. For the operations of union and intersection, the bounds obtained are comparable to those for DFAs, while for the operation of reversal, we get a better bound as for DFAs. Finally, we provide upper bounds for the sizes of stl-det-ORWW-automata for languages of the form $L_1 \circ L_2$, where L_1 and L_2 are given through DFAs. Here we obtain better bounds for the operations of reversal, union, intersection, and product as for DFAs.

This paper is structured as follows. In Section 2, we introduce the det-ORWW-automaton, we present a detailed example showing that this automaton can describe certain regular languages in a much more succinct way than DFAs or even than deterministic RR(1)-automata (see, e.g., [3]), and we show that each det-ORWW-automaton can be converted into an equivalent automaton of the same type that is stateless. In Section 3, we relate the stl-det-ORWW-automaton to the weight-reducing Hennie machine, and in the next section we study the descriptional complexity of the operations of reversal, union, and intersection for languages given through stl-det-ORWW-automata. Finally, in Section 5, we study the corresponding problem for languages that are given through DFAs. The paper closes with Section 6 which summarizes our results in short and states a number of open problems for future work.

2   Deterministic Ordered Restarting Automaton

A *deterministic ordered restarting automaton* (det-ORWW-automaton) is a one-tape machine that is described by an 8-tuple $M = (Q, \Sigma, \Gamma, \triangleright, \triangleleft, q_0, \delta, >)$, where Q is a finite set of states, Σ is a finite input alphabet, Γ is a finite tape alphabet such that $\Sigma \subseteq \Gamma$, the symbols $\triangleright, \triangleleft \notin \Gamma$ serve as markers for the left and right border of the work space, respectively, $q_0 \in Q$ is the initial state,

$$\delta : Q \times (((\Gamma \cup \{\triangleright\}) \cdot \Gamma \cdot (\Gamma \cup \{\triangleleft\})) \cup \{\triangleright\triangleleft\}) \to (Q \times \{\mathsf{MVR}\}) \cup \Gamma \cup \{\mathsf{Accept}\}$$

is the *transition function*, and $>$ is a *partial ordering* on Γ. The transition function describes three different types of transition steps:

(1) A *move-right step* has the form $\delta(q, a_1 a_2 a_3) = (q', \mathsf{MVR})$, where $q, q' \in Q$, $a_1 \in \Gamma \cup \{\triangleright\}$ and $a_2, a_3 \in \Gamma$. It causes M to shift the window one position to

the right and to enter state q'. Observe that no move-right step is possible, if the window contains the symbol $\lhd$.

(2) A *rewrite/restart step* has the form $\delta(q, a_1a_2a_3) = b$, where $q \in Q$, $a_1 \in \Gamma \cup \{\rhd\}$, $a_2, b \in \Gamma$, and $a_3 \in \Gamma \cup \{\lhd\}$ such that $a_2 > b$ holds. It causes M to replace the symbol a_2 in the middle of its window by the symbol b and to restart.

(3) An *accept step* has the form $\delta(q, a_1a_2a_3) = \mathsf{Accept}$, where $q \in Q$, $a_1 \in \Gamma \cup \{\rhd\}$, $a_2 \in \Gamma$, and $a_3 \in \Gamma \cup \{\lhd\}$. It causes M to halt and accept. In addition, we allow an accept step of the form $\delta(q_0, \rhd\lhd) = \mathsf{Accept}$.

If $\delta(q, u) = \emptyset$ for some pair (q, u), then M necessarily halts, when it is in state q seeing u in its window, and we say that M *rejects* in this situation. Further, the letters in $\Gamma \setminus \Sigma$ are called *auxiliary symbols*.

The det-ORWW-automaton M is called *stateless* if $Q = \{q_0\}$. For a stateless det-ORWW-automaton, we drop the components Q and q_0 from its description, and we abbreviate it as stl-det-ORWW-automaton.

A *configuration* of a det-ORWW-automaton M is a word $\alpha q \beta$, where $q \in Q$ and $|\beta| \geq 3$, and either $\alpha = \lambda$ (the empty word) and $\beta \in \{\rhd\} \cdot \Gamma^+ \cdot \{\lhd\}$ or $\alpha \in \{\rhd\} \cdot \Gamma^*$ and $\beta \in \Gamma \cdot \Gamma^+ \cdot \{\lhd\}$; here $q \in Q$ represents the current state, $\alpha\beta$ is the current content of the tape, and it is understood that the window contains the first three symbols of β. In addition, we admit the configuration $q_0 \rhd \lhd$. A *restarting configuration* has the form $q_0 \rhd w \lhd$; if $w \in \Sigma^*$, then $q_0 \rhd w \lhd$ is also called an *initial configuration*. A configuration that is reached by an accept step is an *accepting configuration*, and a configuration of the form $\alpha q \beta$ such that $\delta(q, \beta_1) = \emptyset$, where β_1 is the current content of the window, is a *rejecting configuration*. A *halting configuration* is either an accepting or a rejecting configuration.

Any computation of a det-ORWW-automaton M consists of certain phases. A phase, called a *cycle*, starts in a restarting configuration, the head is moved along the tape by MVR steps until a rewrite/restart step is performed and thus, a new restarting configuration is reached. If no further rewrite operation is performed, any computation necessarily finishes in a halting configuration – such a phase is called a *tail*. By $\vdash_M^c$ we denote the execution of a complete cycle, and $\vdash_M^{c*}$ is the reflexive transitive closure of this relation. It can be seen as the *rewrite relation* that is realized by M on the set of restarting configurations.

An input $w \in \Sigma^*$ is accepted by M, if the computation of M which starts with the initial configuration $q_0 \rhd w \lhd$ ends with an accept step. The language consisting of all words that are accepted by M is denoted by $L(M)$.

As each cycle ends with a rewrite operation, which replaces a symbol a by a symbol b that is strictly smaller than a with respect to the given ordering $>$, we see that each computation of M on an input of length n consists of at most $(|\Gamma| - 1) \cdot n$ many cycles. Thus, M can be simulated by a deterministic single-tape Turing machine in time $O(n^2)$. The following example illustrates the way in which a stl-det-ORWW-automaton works.

Example 1. Let $n \geq 2$ be a fixed integer, and let $M = (\Sigma, \Gamma, \rhd, \lhd, \delta, >)$ be defined by taking $\Sigma = \{a, b\}$ and $\Gamma = \Sigma \cup \{a_i, b_i, x_i \mid 1 \leq i \leq n - 1\}$, by choosing the ordering $>$ such that $a > a_i > x_j$ and $b > b_i > x_j$ hold for all $1 \le i, j \le n - 1$, and by defining the transition function δ in such a way that M proceeds as follows: On input $w = w_1 w_2 \ldots w_m$, $w_1, \ldots, w_m \in \Sigma$, M numbers the first $n - 1$ letters of w from left to right, by replacing $w_i = a$ (b) by a_i (b_i) for $i = 1, \ldots, n - 1$. If $w_n \ne a$, then the computation fails, but if $w_n = a$, then M continues by replacing the last $n - 1$ letters of w from right to left using the letters x_1 to x_{n-1}. If the n-th last letter is b or some b_i, then M accepts, otherwise the computation fails again.

Then $L(M) = \{\, w \in \{a, b\}^m \mid m > n, w_n = a, \text{ and } w_{m+1-n} = b \,\}$. It is shown in [3] that this language is accepted by a det-mon-nf-RR(1)-automaton with $O(n)$ states, while every det-RR(1)-automaton for it has at least $O(2^n)$ states. Observe that M has just a single state and an alphabet of size $3n - 1$.

While nondeterministic ORWW-automata are quite expressive, it is known that the deterministic variants are fairly weak.

Theorem 2. [6] REG $= \mathcal{L}(\text{det-ORWW})$.

Further, stateless det-ORWW-automata can simulate those with states.

Theorem 3. *For each det-ORWW-automaton $M = (Q, \Sigma, \Gamma, \rhd, \lhd, q_0, \delta, >)$, there exists a stl-det-ORWW-automaton $M' = (\Sigma, \Delta, \rhd, \lhd, \delta', >')$ such that $L(M') = L(M)$ and $|\Delta| = |Q| \cdot |\Gamma|^2 + 2 \cdot |\Gamma|$.*

Proof. From M we can construct M' as follows:

- $\Delta = \Gamma \cup \{\, [a, q, b] \mid a \in \Gamma, q \in Q, b \in \Gamma \,\} \cup \{\, [\rhd, q_0, a] \mid a \in \Gamma \,\}$,
- for all $a, b \in \Gamma$, $c \in \Gamma \cup \{\rhd\}$, and $q \in Q$, if $a > b$, then $a >' [c, q, a] >' b$,
- the transition function δ' of M' is defined as follows, where $a, b, A, B, C \in \Gamma$, $c, d \in \Gamma \cup \{\lhd\}$, and $p, q, r \in Q$:

(1) $\delta'(\rhd u \lhd)$ $= \mathsf{Accept}$ for all $u \in (\Sigma \cup \{\lambda\}) \cap L(M)$,
(2) $\delta'(\rhd ab)$ $= [\rhd, q_0, a]$,
(3) $\delta'(\rhd [\rhd, q_0, a] b)$ $= A$,   if $\delta(q_0, \rhd ab) = A$,
(4) $\delta'(\rhd [\rhd, q_0, a] b)$ $= \mathsf{MVR}$,   if $\delta(q_0, \rhd ab) = (p, \mathsf{MVR})$,
(5) $\delta'([\rhd, q_0, a] bc)$ $= [a, p, b]$, if $\delta(q_0, \rhd ab) = (p, \mathsf{MVR})$,
(6) $\delta'([\rhd, q_0, a][a, p, b] c)$ $= B$,   if $\delta(p, abc) = B$,
(7) $\delta'([\rhd, q_0, a][a, p, b] c)$ $= \mathsf{MVR}$,   if $\delta(p, abc) = (q, \mathsf{MVR})$,
(8) $\delta'([a, p, b] cd)$ $= [b, q, c]$, if $\delta(p, abc) = (q, \mathsf{MVR})$,
(9) $\delta'(\rhd [\rhd, q_0, a][a, p, b])$ $= \mathsf{MVR}$,
(10) $\delta'([a, p, b][b, q, A][A, r, B])$ $= \mathsf{MVR}$,
(11) $\delta'([a, p, b][b, q, A] B)$ $= C$,   if $\delta(q, bAB) = C$,
(12) $\delta'([a, p, b][b, q, A] B)$ $= \mathsf{MVR}$,   if $\delta(q, bAB) = (r, \mathsf{MVR})$,
(13) $\delta'([b, q, A] Bc)$ $= [A, r, B]$, if $\delta(q, bAB) = (r, \mathsf{MVR})$,
(14) $\delta'([a, p, b][b, q, A] c)$ $= \mathsf{Accept}$, if $\delta(q, bAc) = \mathsf{Accept}$.

It remains to show that $L(M') = L(M)$ holds. Given an input $u \in \Sigma \cup \{\lambda\}$, M' will accept immediately if and when $u \in L(M)$ holds. So, let us consider an input $w = a_1 a_2 \ldots a_n$, where $n \geq 2$ and $a_1, \ldots, a_n \in \Sigma$. M will scan the input from left to right until it detects the first letter, say a_i, that is to be rewritten, that is, M executes the following cycle:

$$q_0 \triangleright a_1 a_2 \ldots a_n \triangleleft \vdash_M \triangleright q_1 a_1 a_2 a_3 \ldots a_n \triangleleft$$
$$\vdash_M^{i-2} \triangleright a_1 \ldots a_{i-2} q_{i-1} a_{i-1} a_i a_{i+1} \ldots a_n \triangleleft$$
$$\vdash_M q_0 \triangleright a_1 a_2 \ldots a_{i-1} b a_{i+1} \ldots a_n \triangleleft .$$

This cycle is simulated by M' as follows:

$$\triangleright a_1 a_2 \ldots a_n \triangleleft \vdash_{M'}^c \triangleright [\triangleright, q_0, a_1] a_2 \ldots a_n \triangleleft$$
$$\vdash_{M'}^c \triangleright [\triangleright, q_0, a_1][a_1, q_1, a_2] a_3 \ldots a_n \triangleleft$$
$$\vdash_{M'}^c \triangleright [\triangleright, q_0, a_1] \ldots [a_{i-2}, q_{i-2}, a_{i-1}] a_i a_{i+1} \ldots a_n \triangleleft$$
$$\vdash_{M'}^c \triangleright [\triangleright, q_0, a_1] \ldots [a_{i-2}, q_{i-2}, a_{i-1}][a_{i-1}, q_{i-1}, a_i] a_{i+1} \ldots a_n \triangleleft$$
$$\vdash_{M'}^c \triangleright [\triangleright, q_0, a_1] \ldots [a_{i-2}, q_{i-2}, a_{i-1}] b a_{i+1} \ldots a_n \triangleleft,$$

that is, preceding from left to right, M' replaces the symbol a_j ($1 \leq j \leq i$) by the triple $[a_{j-1}, q_{j-1}, a_j]$, where q_{j-1} is the state in which M reaches the window contents $a_{j-1} a_j a_{j+1}$. This continues until M reaches the symbol a_{i+1}. At that point, M' realizes that it must simulate a rewrite step of M, and accordingly it replaces the triple $[a_{i-1}, q_{i-1}, a_i]$ by the symbol b. By induction on the number of cycles that M executes it can now be shown that $L(M') = L(M)$ holds. Just observe that after the rewrite step above, M must scan the tape from left to right until its window contains the newly written symbol b, as M is deterministic.   □

Because of Theorem 3 we restrict our attention to stl-det-ORWW-automata for the rest of this paper.

Proposition 4. *For each DFA $A = (Q, \Sigma, q_0, F, \varphi)$, there is a stl-det-ORWW-automaton $M = (\Sigma, \Gamma, \triangleright, \triangleleft, \delta, >)$ such that $L(M) = L(A)$ and $|\Gamma| = |Q| + |\Sigma|$.*

Proof. We take $\Gamma = \Sigma \cup Q$, define $a > q$ for all $a \in \Sigma$ and all $q \in Q$, and define the transition function δ as follows, where $a, b \in \Sigma$ and $p, q, q' \in Q$:

$$
\begin{aligned}
&(1)\ \delta(\triangleright \triangleleft) && = \text{Accept}, && \text{if } \lambda \in L(A),\\
&(2)\ \delta(\triangleright a \triangleleft) && = \text{Accept}, && \text{if } a \in L(A),\\
&(3)\ \delta(\triangleright a b) && = q, && \text{if } \varphi(q_0, a) = q,\\
&(4)\ \delta(\triangleright q b) && = \text{MVR},\\
&(5)\ \delta(q b a) && = p, && \text{if } \varphi(q, b) = p,\\
&(6)\ \delta(\triangleright q p) && = \text{MVR},\\
&(7)\ \delta(p q q') && = \text{MVR},\\
&(8)\ \delta(p q a) && = \text{MVR},\\
&(9)\ \delta(q b \triangleleft) && = \text{Accept}, && \text{if } \varphi(q, b) \in F.
\end{aligned}
$$

Thus, given $w = a_1 \ldots a_n$ as input, where $n \geq 2$ and $a_1, \ldots, a_n \in \Sigma$, M rewrites w from left to right into the word $q_1 \ldots q_{n-1} a_n$, where $q_i = \varphi(q_0, a_1 \ldots a_i)$, $1 \leq i \leq n-1$, and the word $q_1 q_2 \ldots q_{n-1} a_n$ is then accepted in a tail computation if $\varphi(q_{n-1}, a_n) \in F$, that is, M accepts on input w iff $\varphi(q_0, a_1 \ldots a_n) = \varphi(\varphi(q_0, a_1 \ldots a_{n-1}), a_n) = \varphi(q_{n-1}, a_n) \in F$, that is, iff A accepts on input w. Hence, we see that $L(M) = L(A)$ holds.   □

3   Descriptional Complexity

A single-tape Turing machine is *bounded* if it works on a tape that is delimited by a left end marker $\triangleright$ and a right end marker $\triangleleft$. A bounded deterministic Turing machine T is called a *Hennie machine*, if there exists a constant k such that the number of transitions that T performs on any tape field is bounded from above by the number k. Hennie has shown in [2] that Hennie machines can only accept regular languages. Observe, however, that it is undecidable in general whether a given Turing machine is a Hennie machine (see, e.g., [8]). A Hennie machine T with tape alphabet Γ is called *weight-reducing* if there is a weight function $\mu : \Gamma \to \mathbb{N}$ such that, in each step, T replaces the currently read symbol $a \in \Gamma \setminus \{\triangleright, \triangleleft\}$ by a symbol $b \in \Gamma \setminus \{\triangleright, \triangleleft\}$ of strictly less weight. Obviously, it is easily decidable whether a given Hennie machine is weight-reducing. Further, for each Hennie machine T, there exists a weight-reducing Hennie machine T' that accepts the same language [8].

We are interested in the descriptional complexity of stl-det-ORWW-automata. Here we can exploit Průša's results on the descriptional complexity of weight-reducing Hennie machines, where we use the number of states as a measure for the size of a DFA.

Theorem 5. [8]

(a) *For each weight-reducing Hennie machine T with n states and an alphabet of size m, there is a DFA A of size $2^{2^{O(m \log n)}}$ such that $L(A) = L(T)$ holds.*
(b) *For each $n \geq 1$, there exists a regular language $B_n \subseteq \{0, 1, \$\}^*$ that is accepted by a weight-reducing Hennie machine with $O(1)$ states and with $O(n)$ working symbols, but each DFA for accepting B_n has at least 2^{2^n} many states.*

Thus, there is a double exponential trade-off for converting a weight-reducing Hennie machine into an equivalent DFA. We claim that the same holds for the conversion of a stl-det-ORWW-automaton into a DFA. For proving this result we present conversions from weight-reducing Hennie machines to stl-det-ORWW-automata and back.

Proposition 6. *For each stl-det-ORWW-automaton M with n letters, there exists a weight-reducing Hennie machine T with $O(n^2)$ many states on some alphabet of size $O(n^2)$ such that $L(T) = L(M)$ holds.*

Proof. In the proof of Theorem 2 given in [6], a det-ORWW-automaton $M = (Q, \Sigma, \Gamma, \triangleright, \triangleleft, q_0, \delta, >)$ is simulated by a weight-reducing Hennie machine T, which must occasionally store a letter from Γ in its finite-state control, and which stores triples of the form $(a, q, b) \in \Gamma \times Q \times \Gamma$ on its tape. In addition, each time T visits a tape field, which may occur up to $4 \cdot |\Gamma|$ many times, it must replace the current letter by another letter with less weight, which increases the size of the alphabet by another factor $|\Gamma|$. It follows that T has $O(|Q| \cdot |\Gamma|)$ many states and that it uses $O(|Q| \cdot |\Gamma|^3)$ many letters. Thus, if M is stateless, and if $|\Gamma| = n$, then T has only $O(n)$ many states, and it uses $O(n^3)$ many letters. To obtain the desired result, we change T as follows. Instead of replacing the content of a cell by a triple (a, q_0, b), the Hennie machine now stores the contents of the two neighbouring cells of the currently visited cell in its finite-state control. Thus, it uses an alphabet of size $O(n^2)$ only, but now it has $O(n^2)$ many states.   □

For the opposite conversion we have the following result.

Proposition 7. *For each weight-reducing Hennie machine T with n states and an alphabet of size m, there exists a stl-det-ORWW-automaton M on some alphabet of size $O(m \cdot n)$ such that $L(M) = L(T)$ holds.*

Proof. The stl-det-ORWW-automaton M stores the current state of T together with the current letter into the tape field that T is currently scanning. Because of its window of size 3, M can detect this information when sweeping across the tape from left to right, and it can then store the new state into the appropriate field. In the next cycle, it deletes the state information from the 'old' position and simulates the corresponding rewrite step. This is done in exactly the same way as a deterministic Sgraffito automaton is simulated by a det-2D-3W-ORWW-automaton (see [6]). Hence, M needs an alphabet of size $O(m \cdot n)$.   □

Together with Theorem 5 these propositions give the following results.

Corollary 8.

(a) *For each stl-det-ORWW-automaton M with an alphabet of size n, there exists a DFA A of size $2^{2^{O(n^2 \log n)}}$ such that $L(A) = L(M)$ holds.*

(b) *For each $n \geq 1$, there exists a regular language $B_n \subseteq \{0, 1, \$\}^*$ such that B_n is accepted by a stl-det-ORWW-automaton over an alphabet of size $O(n)$, but each DFA for accepting B_n has at least 2^{2^n} many states.*

Thus, there is a double exponential trade-off for converting a stl-det-ORWW-automaton into a DFA. Observe, however, that the gap between the lower bound of 2^{2^n} and the upper bound of $2^{2^{O(n^2 \log n)}}$ is still huge.

4   The Descriptional Complexity of Language Operations

Many results have been obtained on the descriptional complexity of language operations (see [1] for a recent survey): given a DFA of size n for accepting a language L, what is the minimal size of a DFA for the language op(L), where op is an operation like reversal, complement, or Kleene star, and correspondingly for binary operations like union, intersection, or product? For example, it is known that the bound for reversal is 2^n [5], for union and intersection, it is $m \cdot n$, and for product, it is $(m - 1) \cdot 2^n + 2^{n-1}$ [9] , where the two languages used are accepted by DFAs of size m and n, respectively. We now ask these questions for stl-det-ORWW-automata, where we use the number of letters as the complexity measure for our automata. Here we have the following upper bound results.

Theorem 9. *If a language L is accepted by a stl-det-ORWW-automaton with n letters, then its reversal L^R is accepted by a stl-det-ORWW-automaton over an alphabet of size $n^2 + 2n$.*

Proof. Let $M = (\Sigma, \Gamma, \triangleright, \triangleleft, \delta, >)$ be a stl-det-ORWW-automaton such that $L(M) = L$, and let $n = |\Gamma|$. From M we obtain a stl-det-ORWW-automaton M^R for L^R by taking $M^R = (\Sigma, \Delta, \triangleright, \triangleleft, \delta', >')$, where

- $\Delta = \Gamma \cup (\Gamma \times (\Gamma \cup \{\triangleright\}))$, which shows that $|\Delta| = n + (n \cdot (n+1)) = n^2 + 2n$,
- the ordering $>'$ on Δ is defined as follows:

$$\forall a, b \in \Gamma : \text{ if } a > b, \text{ then } a >' [a, c] >' b \text{ for all } c \in \Gamma \cup \{\triangleright\},$$

- and the transition function δ' is defined as follows, where $a, b, c, c' \in \Gamma$ and $d \in \Gamma \cup \{\triangleright\}$:

    (1) $\delta'(\triangleright a \triangleleft)$ $= \delta(\triangleright a \triangleleft)$ for all $a \in \Gamma \cup \{\lambda\}$,
    (2) $\delta'(\triangleright ab)$ $= \mathsf{MVR}$,
    (3) $\delta'(abc)$ $= \mathsf{MVR}$,
    (4) $\delta'(ab\triangleleft)$ $= [b, \triangleright]$,   if $\delta(\triangleright ba) = \mathsf{MVR}$,
    (5) $\delta'(ab\triangleleft)$ $= c$,   if $\delta(\triangleright ba) = c$,
    (6) $\delta'(ab\triangleleft)$ $= \mathsf{Accept}$, if $\delta(\triangleright ba) = \mathsf{Accept}$,
    (7) $\delta'(ab[c, d])$ $= [b, c]$,   if $\delta(dcb) = \mathsf{MVR}$,
    (8) $\delta'(ab[c, d])$ $= \mathsf{MVR}$,   if $\delta(dcb) = c'$,
    (9) $\delta'(b[c, d]A)$ $= c'$,   if $\delta(dcb) = c'$, and $A \in (\Delta \setminus \Gamma) \cup \{\triangleleft\}$,
    (10) $\delta'(ab[c, d])$ $= \mathsf{Accept}$, if $\delta(dcb) = \mathsf{Accept}$,
    (11) $\delta'(\triangleright b[c, d])$ $= [b, c]$,   if $\delta(dcb) = \mathsf{MVR}$,
    (12) $\delta'(\triangleright b[c, d])$ $= \mathsf{MVR}$,   if $\delta(dcb) = c'$,
    (13) $\delta'(b[c, d]A)$ $= c'$,   if $\delta(dcb) = c'$, and $A \in (\Delta \setminus \Gamma) \cup \{\triangleleft\}$,
    (14) $\delta'(\triangleright b[c, d])$ $= \mathsf{Accept}$, if $\delta(dcb) = \mathsf{Accept}$,
    (15) $\delta'(\triangleright [a, b][b, c]) = c'$,   if $\delta(ba\triangleleft) = c'$,
    (16) $\delta'(\triangleright [a, b][b, c]) = \mathsf{Accept}$, if $\delta(ba\triangleleft) = \mathsf{Accept}$.

While M scans a given input w from left to right, M^R scans the corresponding input w^R from left to right, which would correspond to M scanning its input from right to left. M^R uses the letters of the form $[a, b]$ to mark the position in w^R to which M has proceeded. Thus, during a computation the tape contents of M^R will be of the form $\triangleright ua[b, c]v \triangleleft$, where $u \in \Gamma^*$, $a, b \in \Gamma$, $c \in \Gamma \cup \{\triangleright\}$, and $v \in (\Delta \setminus \Gamma)^*$. This factorization tells us that M would move right across the prefix corresponding to $(bv)^R$. Now if $\delta(cba) = b'$, then M would replace b by b' in the next cycle, and correspondingly, M^R will rewrite $\triangleright ua[b, c]v \triangleleft$ into $\triangleright uab'v \triangleleft$. It can now be shown that $L(M^R) = L^R$ holds.   □

The next result follows easily by interchanging accept steps with undefined steps in the stl-det-ORWW-automaton considered.

Theorem 10. *If a language L is accepted by a stl-det-ORWW-automaton with n letters, then its complement L^c is accepted by a stl-det-ORWW-automaton over the same alphabet.*

Finally, we look at the operation of intersection.

Theorem 11. *If the languages L_1 and L_2 are accepted by stl-det-ORWW-auto-mata with n letters, then their intersection $L_1 \cap L_2$ is accepted by a stl-det-ORWW-automaton over an alphabet of size $n^2 + 2n$.*

Proof. Let $M_1 = (\Sigma, \Gamma_1, \triangleright, \triangleleft, \delta_1, >_1)$ be a stl-det-ORWW-automaton such that $L(M_1) = L_1$, and let $M_2 = (\Sigma, \Gamma_2, \triangleright, \triangleleft, \delta_2, >_2)$ be a stl-det-ORWW-automaton such that $L(M_2) = L_2$, where $|\Gamma_1| = |\Gamma_2| = n$. From M_1 and M_2 we obtain a stl-det-ORWW-automaton M for $L = L_1 \cap L_2$ on the alphabet

$$\Delta = \Sigma \cup \{ [a, b] \mid a \in \Gamma_1, b \in \Gamma_2 \} \cup \{ \bar{b} \mid b \in \Gamma_2 \}$$

of size $|\Sigma| + n^2 + n \leq n^2 + 2n$ that works as follows:

1. From right to left, M rewrites each input letter a into the pair $[a, a]$.
2. Then M simulates M_1 using the first component of each letter. When M_1 accepts, then M replaces each pair of the form $[a, b]$ by the letter $\bar{b}$, again proceeding from right to left.
3. Finally, M simulates M_2 interpreting each letter of the form $\bar{b}$ just as M_2 would interpret the letter b. If and when M_2 accepts, then so does M.

Obviously, $L(M) = L_1 \cap L_2$ follows.   □

From the last two results, we immediately obtain the following consequence.

Corollary 12. *If the languages L_1 and L_2 are accepted by stl-det-ORWW-auto-mata with n letters, then their union $L_1 \cup L_2$ is accepted by a stl-det-ORWW-automaton over an alphabet of size $n^2 + 2n$.*

Thus, for realizing the Boolean operations, stl-det-ORWW-automata are essentially just as efficient as DFAs, while for the operation of reversal, they are much more efficient. So far, no results are known on how to realize the operations of product and Kleene star efficiently by stl-det-ORWW-automata.

5   Simulating DFAs by stl-det-ORWW-Automata

If $\circ$ is a binary operation on languages, and if, for $i = 1, 2$, A_i is a DFA of size n_i, what is the size of a stl-det-ORWW-automaton for $L(A_1) \circ L(A_2)$? Here we present upper bounds for the operations of intersection, union, and product.

Theorem 13. *If the languages L_1 and L_2 over an alphabet of size k are accepted by DFAs of size m and n, respectively, then their intersection $L_1 \cap L_2$ is accepted by a stl-det-ORWW-automaton over an alphabet of size $k \cdot (m + 1) + n$.*

Proof. Let $A_1 = (Q_1, \Sigma, q_0, F_1, \delta_1)$ be a DFA such that $L(A_1) = L_1$, where $|Q_1| = m$ and $|\Sigma| = k$, and let $A_2 = (Q_2, \Sigma, p_0, F_2, \delta_2)$ be a DFA such that $L(A_2) = L_2$, where $|Q_2| = n$. From A_1 and A_2 we construct a stl-det-ORWW-automaton $M = (\Sigma, \Gamma, \triangleright, \triangleleft, \delta, >)$ for $L = L_1 \cap L_2$ by taking $\Gamma = \Sigma \cup \{ [q, a] \mid q \in Q_1 \text{ and } a \in \Sigma \} \cup Q_2$, that is, $|\Gamma| = k + k \cdot m + n = k \cdot (m + 1) + n$, by choosing the ordering $>$ through $a > [q, b] > p$ for all $a, b \in \Sigma$, $q \in Q_1$, and $p \in Q_2$, and by defining the transition function δ such that M works as follows:

1. First M simulates two steps of A_1 by turning the first two input letters ab into the pairs $[q, a][q', b]$, where $q = \delta_1(q_0, a)$ and $q' = \delta_1(q, b)$.
2. Then on the prefix $[q, a][q', b]$ the first step of A_2 is simulated by rewriting $[q, a]$ into p, where $p = \delta_2(p_0, a)$.
3. Now M alternates between simulating a step of A_1 and a step of A_2.
4. If the last two symbols are of the form $p[q, c]$ for some $p \in Q_2$ and $q \in F_1$, and if $\delta_2(p, c) = p' \in Q_2$, then $[q, c]$ is rewritten into p'.
5. M accepts if the tape contents ends with $p' \lhd$ for some $p' \in F_2$.

It can be shown that $L(M) = L(A_1) \cap L(A_2)$ holds.   □

Observe that by first building the product automaton of A_1 and A_2 and by then turning this into an equivalent stl-det-ORWW-automaton, we would obtain a stl-det-ORWW-automaton for the language $L = L(A_1) \cap L(A_2)$ over an alphabet of size $k + m \cdot n$. As typically m and n are large, while k is small, Theorem 13 gives a much better bound. The construction above can easily be extended to the intersection of more than two DFAs.

Corollary 14. *For all $t \geq 3$, if A_i is a DFA with n_i states over an alphabet of size k for all $1 \leq i \leq t$, then there exists a stl-det-ORWW-automaton M on an alphabet of size $k \cdot (1 + n_1 + \cdots + n_{t-1}) + n_t$ such that $L(M) = \bigcap_{i=1}^{t} L(A_i)$.*

Obviously, this construction can easily be adopted to the operation of union.

Corollary 15. *For all $t \geq 3$, if A_i is a DFA with n_i states over an alphabet of size k for all $1 \leq i \leq t$, then there exists a stl-det-ORWW-automaton M on an alphabet of size $k \cdot (1 + n_1 + \cdots + n_{t-1}) + n_t$ such that $L(M) = \bigcup_{i=1}^{t} L(A_i)$.*

By turning the left-to-right simulation of a DFA as described in the proof of Proposition 4 into a right-to-left simulation, we obtain the following result.

Proposition 16. *If a language L over an alphabet of size k is accepted by a DFA of size n, then the language L^R is accepted by a stl-det-ORWW-automaton over an alphabet of size $k + n$.*

Next we turn to the operation of product.

Theorem 17. *If the languages L_1 and L_2 over an alphabet of size k are accepted by DFAs of size m and n, respectively, then their product $L_1 \cdot L_2$ is accepted by a stl-det-ORWW-automaton over an alphabet of size $k \cdot (m + 1) + 2^n$.*

Proof. Let $A_i = (Q_i, \Sigma, q_0^{(i)}, F_i, \delta_i)$ be a DFA for L_i, $i = 1, 2$, where $|Q_1| = m$, $|Q_2| = n$, and $|\Sigma| = k$. From A_1 and A_2 we obtain a stl-det-ORWW-automaton M for $L = L_1 \cdot L_2$ on the alphabet $\Delta = \Sigma \cup \{ [q, b] \mid q \in Q_1, b \in \Sigma \} \cup 2^{Q_2}$ of size $k + k \cdot m + 2^n = k \cdot (m + 1) + 2^n$ that works as follows:

1. First M simulates A_1 on the given input just as in the proof of Proposition 4. However, instead of replacing a letter $b \in \Sigma$ read by the corresponding state $q \in Q_1$, we now replace b by the combined letter $[q, b]$. When A_1 reaches the right sentinel $\lhd$, then the given input $w = a_1 a_2 \ldots a_r$ ($r \geq 1$, $a_1, \ldots, a_r \in \Sigma$) has been rewritten into the word $W = [q_1, a_1] \ldots [q_r, a_r]$, where $q_i = \delta_1(q_0^{(1)}, a_1 \ldots a_i)$, $1 \leq i \leq r$.

2. By interpreting each letter of the from $[q, b]$ simply as the letter b, M now simulates A_2 in reverse preceding from right to left, starting from the set of final states F_2. During this part of the computation the tape content will be of the form $u \cdot v$, where $u \in \{ [q, b] \mid q \in Q_1, b \in \Sigma \}^*$ and $v \in (2^{Q_2})^*$.

3. When M reaches a window content of the form $[q', a][q, b]P$, where $q', q \in Q_1$, $a, b \in \Sigma$, and $P \subseteq Q_2$, such that $q \in F_1$ and $q_0^{(2)} \in P$, then we have found a factorization $w = w_1 w_2$, where w_1 ends with the letters ab displayed, such that $w_1 \in L_1$ and $w_2 \in L_2$. Thus, in this situation M halts and accepts.

If $\lambda \in L_2$, then M also accepts in step 1 in case that the last letter a_r of the input in rewitten into a letter of the form $[q, a_r]$ such that $q \in F_1$, and if $\lambda \in L_1$, then M accepts, if a rewrite in step 2 replaces the first letter by a set $P \subseteq Q_2$ such that $q_0^{(2)} \in P$. It now follows that $L(M) = L_1 \cap L_2$ holds.   □

As pointed out by one of the reviewers, the above bound can be replaced by a better one, which, however, is less precise. It is known that $L_1 \cdot L_2$ is accepted by a nondeterministic finite-state acceptor (NFA) A with $O(m + n)$ states. By Theorem 11 of [8], A can be simulated by a weight-reducing Hennie machine T with a polynomial number of transitions in k and $m + n$. From T we obtain a stl-det-ORWW-automaton for $L_1 \cdot L_2$ with a polynomial number of letters in k, m, and n by Proposition 7. However, the bound given in Theorem 17 is still preferable in the case that n is small in comparison to k and m.

A still better result is obtained for the following variant of the product operation.

Theorem 18. *If the languages L_1 and L_2 over an alphabet of size k are accepted by DFAs of size m and n, respectively, then the product $L_1 \cdot L_2^R$ is accepted by a stl-det-ORWW-automaton over an alphabet of size $k \cdot (m + 1) + n$.*

Proof. Let $A_i = (Q_i, \Sigma, q_0^{(i)}, F_i, \delta_i)$, $i = 1, 2$, be a DFA for L_i, where $|Q_1| = m$, $|Q_2| = n$, and $|\Sigma| = k$. From A_1 and A_2 we can construct a stl-det-ORWW-automaton $M = (\Sigma, \Gamma, \triangleright, \triangleleft, \delta, >)$ on the alphabet $\Gamma = \Sigma \cup \{ [q, a] \mid q \in Q_1, a \in \Sigma \} \cup Q_2$ of size $|\Gamma| = k + k \cdot m + n$ that works as follows:

1. An input of length at most one is accepted immediately, if it belongs to $L_1 \cdot L_2^R$.
2. An input $w = a_1 \ldots a_r$, where $r \geq 2$ and $a_1, \ldots, a_r \in \Sigma$, is first processed from left to right be replacing the letter a_i by the pair $[q_i, a_i]$, where $q_i = \delta_1(q_0^{(1)}, a_1 \ldots a_i)$. Hence, $q_i \in F_1$ iff the prefix $a_1 \ldots a_i$ belongs to L_1.
3. If $q_n \in F_1$ and $\lambda \in L_2$, then $w = w \cdot \lambda \in L_1 \cdot L_2^R$, and accordingly, M accepts.
4. Otherwise, M now processes the word $[q_1, a_1] \ldots [q_r, a_r]$ from right to left. It replaces the pair $[q_i, a_i]$ by the state symbol $p \in Q_2$ if $\delta_2(q_0^{(2)}, a_r \ldots a_i) = p$. In particular, $p \in F_2$, if the suffix $a_i \ldots a_r$ belongs to the language L_2^R. Accordingly, M accepts, if the prefix $a_1 \ldots a_{i-1}$ belongs to L_1, and if the corresponding suffix $a_i \ldots a_r$ belongs to L_2^R.

It follows that $L(M) = L_1 \cdot L_2^R$ holds.   □

6 Concluding Remarks

We have seen that stl-det-ORWW-automata can provide much more succinct representations for regular languages than DFAs, as the corresponding trade-off is double exponential. Also for the operations of reversal, union, and intersection, the representation by stl-det-ORWW-automata is much better suited than the representation by DFAs. However, many open problems remain:

- Are the given upper bounds sharp, that is, can they be obtained by example languages?
- How can two stl-det-ORWW-automata for L_1 and L_2 be combined into a 'small' stl-det-ORWW-automaton for the product $L_1 \cdot L_2$?
- How can a stl-det-ORWW-automaton for L be transformed into a 'small' stl-det-ORWW-automaton for L^*?

Finally, because of its window of size three, a stl-det-ORWW-automaton with an alphabet of size n can have $O(n^3)$ many transitions, while a DFA of size n over an alphabet of size k has at most $k \cdot n$ many transitions. Thus, when measuring the sizes of these types of automata in terms of their number of transitions, then the comparison is less favourable for the stl-det-ORWW-automaton, although the double exponential trade-off of Corollary 8 also holds for this measure.

References

1. Brzozowski, J.: In search of the most complex regular languages. In: Moreira, N., Reis, R. (eds.) CIAA 2012. LNCS, vol. 7381, pp. 5–24. Springer, Heidelberg (2012)
2. Hennie, F.C.: One-tape, off-line Turing machine computations. Inform. Contr. 8, 553–578 (1965)
3. Hundeshagen, N., Otto, F.: Characterizing the regular languages by nonforgetting restarting automata. In: Mauri, G., Leporati, A. (eds.) DLT 2011. LNCS, vol. 6795, pp. 288–299. Springer, Heidelberg (2011)
4. Jančar, P., Mráz, F., Plátek, M., Vogel, J.: Restarting automata. In: Reichel, H. (ed.) FCT 1995. LNCS, vol. 965, pp. 283–292. Springer, Heidelberg (1995)
5. Mirkin, B.G.: On dual automata. Cybernetics 2, 6–9 (1966)
6. Mráz, F., Otto, F.: Ordered restarting automata for picture languages. In: Geffert, V., Preneel, B., Rovan, B., Štuller, J., Tjoa, A.M. (eds.) SOFSEM 2014. LNCS, vol. 8327, pp. 431–442. Springer, Heidelberg (2014)
7. Otto, F.: Restarting automata. In: Ésik, Z., Martín-Vide, C., Mitrana, V. (eds.) Recent Advances in Formal Languages and Applications. SCI, vol. 25, pp. 269–303. Springer, Berlin (2006)
8. Průša, D.: Weight-reducing Hennie machines and their descriptional complexity. In: Dediu, A.-H., Martín-Vide, C., Sierra-Rodríguez, J.-L., Truthe, B. (eds.) LATA 2014. LNCS, vol. 8370, pp. 553–564. Springer, Heidelberg (2014)
9. Yu, S., Zhuang, Q., Salomaa, K.: The state complexities of some basic operations on regular languages. Theoret. Comput. Sci. 125, 315–328 (1994)

State Complexity of Unary Language Operations for NFAs with Limited Nondeterminism

Alexandros Palioudakis, Kai Salomaa, and Selim G. Akl

School of Computing, Queen's University, Kingston, Ontario K7L 3N6, Canada
{alex,ksalomaa,akl}@cs.queensu.ca

Abstract. We study the state complexity of language operations for unary NFAs with limited nondeterminism. We consider the operations of concatenation, Kleene star, and complement. We give upper bounds for the state complexity of these language operations and lower bounds that are fairly close to the upper bounds. Our constructions rely on the fact that minimal unary NFAs with limited nondeterminism can be found in Chrobak normal form.

Keywords: finite automata, limited nondeterminism, state complexity, language operations, unary regular languages.

1 Introduction

Finite automata is a basic model of computation and it has been studied for more than half a century. State complexity is one important aspect in the study of finite automata [20,21]. The state complexity of language operations was first considered by Maslov [19], later, Yu, Zhuang, and Salomaa systematically studied the deterministic state complexity of language operations [28]. Pighizzini and Shallit considered the state complexity of unary language operations [26]. Nondeterministic state complexity of basic language operations was investigated by Holzer and Kutrib [9] and the nondeterministic state complexity of unary language operations was studied by the same authors in [10]. Good recent surveys on the descriptional complexity of finite automata can be found in [11,12,16] and operational state complexity is discussed in more detail in [5].

Motivated by the exponential trade off between deterministic finite automata (DFA) and nondeterministic finite automata (NFA) there has been much work on limited nondeterminism. The degree of ambiguity counts the number of accepting computations [17,18]. Leung showed in [18] an exponential trade off between unambiguous NFAs and general NFAs as well as an exponential trade off between DFAs and unambiguous NFAs. Okhotin [22] studied the state complexity of unambiguous unary NFAs. The *branching measure* [8] is the product of the nondeterministic choices an NFA makes along a best accepting computational path. Goldstine, Kintala, and Wotschke showed that there are regular languages where NFAs with finite branching require roughly the same number of states with a DFA and exponential more than an NFA. The *tree width measure*, which

H. Jürgensen et al. (Eds.): DCFS 2014, LNCS 8614, pp. 330–341, 2014.

is called *leaf size* in [13,14], counts the total number of computation paths corresponding to a given input. We showed in [23] that there is a polynomial trade off between finite tree width NFAs and DFAs. In [24], we introduced the *trace measure* which is a worst variant of branching and we showed a relation between NFAs with finite tree width and NFAs with finite trace. In [25], we considered limited nondeterminism for unary NFAs.

In [23] we studied the state complexity of union and intersection for languages recognized by finite tree width NFAs over general alphabets. The lower bound constructions were given using unary regular languages and hence the same results apply also in our case. In [23] we also gave an upper bound for the state complexity of concatenation of finite tree width NFAs over general alphabets. Finding a matching lower bound for the concatenation of binary languages (or languages over a general alphabet) defined by NFAs with finite tree width has turned out to be a very hard problem, and the same applies to complementation. It can be noted that, similarly, the operational state complexity of unambiguous NFAs over general alphabets is a hard problem: for complementation the only known upper bound is 2^n while the known lower bound is $n^{2-o(1)}$ [22] and also the state complexity of concatenation of unambiguous NFAs remains wide open.

In this paper, as a special case, we study the state complexity of concatenation, Kleene star, and complement for unary NFAs with limited nondeterminism. The measures of nondeterminism which are considered in this paper are tree width, ambiguity, and trace. The structure of the paper is as follows. In Section 3 some auxiliary results are given. In Section 4, we give an upper bound on the number of states required for the concatenation of two languages for NFAs with limited nondeterminism, the required number of states depend on the state complexity of the two given languages and as well as the degree of nondeterminism that we allow. In Section 5 we study the state complexity of Kleene star for NFAs with limited nondeterminism, the difference here is that the bounds do not depend on the allowed degree of nondeterminism. Finally, in Section 6 we study the state complexity of complement of a language for NFAs with limited nondeterminism. We show that all the above bounds cannot be essentially improved, by giving lower bound constructions that are fairly close, but not exactly matching, to the upper bounds. Naturally some improvements in the estimations remain for future work.

2   Preliminaries

We assume that the reader is familiar with the basic definitions concerning finite automata [27,29] and descriptional complexity [7,12]. Here we just fix some notation needed in the following.

The set of strings, or words, over a finite alphabet Σ is Σ^*, the length of $w \in \Sigma^*$ is $|w|$ and ε is the empty string. In this paper, we focus on alphabets containing only one letter, we call these alphabets unary and we assume that $\Sigma = \{a\}$. The set of positive integers is denoted by $\mathbb{N}$. The cardinality of a finite set S is $\#S$.

A nondeterministic finite automaton (NFA) over a unary alphabet is a 4-tuple $A = (Q, \delta, q_0, F)$, where Q is a finite set of states, $\delta : Q \to 2^Q$ is the transition function, q_0 is the initial state and $F \subseteq Q$ is the set of accepting states. We define the transition function δ in the form of a relation $\mathrm{rel}(\delta) = \{(q, q') \mid q' \in \delta(q)\}$. The function δ is extended in the usual way as a function $Q \times \mathbb{N} \to 2^Q$, where $\delta(q, 0) = \{q\}$, $\delta(q, 1) = \delta(q)$, and $\delta(q, i) = \{p \in Q \mid \exists q', q' \in \delta(q, i-1) \text{ and } p \in \delta(q')\}$ for $2 \leq i$. The language recognized by A, denoted by $L(A)$, consists of all the strings $a^m \in \{a\}^*$ such that $\delta(q_0, m) \cap F \neq \emptyset$. An NFA A is called deterministic finite automaton (DFA) if for every state q of A the transition function goes to at most one state, i.e. $\#\delta(q) \leq 1$. Note that a unary DFA always consists of a sequence of states connected by transitions, called the tail, followed by a loop (where the loop may be empty). Unless otherwise mentioned, we assume that any state q of an NFA A is reachable from a start state and some computation originating from q reaches a final state. The *size of A* is the number of states of A, i.e. $\mathrm{size}(A) = \#Q$.

A *computation* of a unary NFA $A = (Q, \delta, q_0, F)$ with underlying word a^m, for a no negative integer m, from a state s_1 to a state s_2 is a sequence of transitions $(q_0, \ldots, q_m)$, where for all $0 \leq i < m$ we have $q_{i+1} \in \delta(q_i)$ and $s_1 = q_0$ and $s_2 = q_m$. For the word x, $\mathrm{comp}_A(x)$ denotes the set of all computation of A with underlying word x, starting from the initial state of A. We call a computation of A accepting if it starts from the initial state and it finishes at a final state.

By a transition of A we mean a triple (q, a, p) where $p \in \delta(q)$ The *trace of a transition* (q, a, p) of an NFA A, denoted by $\tau_A((q, a, p))$, is the number $\#\delta(q)$ and the *trace of a computation* C, denoted by $\tau_A(C)$, is the product of the traces of each transition in C. The *trace of a word* x is the maximum trace among all computations by reading the word x, the trace of a word x is given by the type $\tau_A(x) = \max\{\tau_A(C) \mid C \in \mathrm{comp}_A(y), y \text{ is a prefix of } x\}$. The trace of an NFA A, denoted by $\tau(A)$, is the maximum trace of A on any string, assuming this quantity is bounded.

The computation tree of an NFA A on string w is defined in the natural way and denoted as $T_{A,w}$. The tree width of A on w, denoted by $\mathrm{tw}_A(w)$, is the number of leaves of $T_{A,w}$ and the tree width of A, denoted by $\mathrm{tw}(A)$, (if it is finite) is the maximum tree width of A on any string w. The formal definitions associated with computation trees and tree width of an NFA can be found in [23,24].[1] The ambiguity of A on w, denoted by $\mathrm{amb}_A(w)$, is the number of accepting leaves of $T_{A,w}$ and the ambiguity of A, denoted by $\mathrm{amb}(A)$, (if it is finite) is the maximum ambiguity of A on any string w. Ambiguity is a well studied measure of nondeterminism, more details on ambiguity in NFAs can be found in [7].

The minimal size of a DFA (respectively, an NFA) recognizing a regular language is called the state complexity (respectively, the nondeterministic state complexity) of L and denoted $\mathrm{sc}(L)$ (respectively, $\mathrm{nsc}(L)$). Note that we allow DFAs to be incomplete and, consequently, the deterministic state complexity of

---

[1] Note that the tree width of an NFA is unrelated to the notion of tree width as used in graph theory.

L may differ by one from a definition using complete DFAs. Now, we want to consider questions that involve the state complexity of classes of NFAs of limited nondeterminism. To formalize such questions we have to define the following notation, where $sNFA$ is the set of all NFAs, α is a measure of nondeterminism, and c a constant.

$$nsc_{\alpha \leq c}(L) = \min_{A \in sNFA} \{size(A) \mid L = L(A) \text{ and } \alpha(A) \leq c\}$$

Now the numbers $nsc_{tw \leq k}(L)$ and $nsc_{\tau \leq k}(L)$ have a meaning. The number $nsc_{tw \leq k}(L)$ is the size of a smallest NFA A such that $L = L(A)$ and $tw(A) \leq k$. The number $nsc_{\tau \leq k}(L)$ is the smallest number of states required from an automaton B such that $\tau(B) \leq k$. From [24] we have that for every NFA A $tw(A) \leq \tau(A)$, and hence $nsc_{tw \leq k}(L) \leq nsc_{\tau \leq k}(L)$, for every positive integer k.

A useful form that unary NFA have is called Chrobak normal form [4]. A unary NFA A is in Chrobak normal form if initially the states of A form a 'tail' and later, at the end of the tail, are followed nondeterministically by disjoint deterministic cycles. Note, that the only state with nondeterministic choices is the last state of the tail. Formally, the NFA $M = (Q, \delta, q_0, F)$ is in Chrobak normal form if it has the following properties:

1. $Q = \{q_0, \ldots, q_{t-1}\} \cup C_1 \cup \cdots \cup C_k$, where $C_i = \{p_{i,0}, p_{i,1}, \ldots, p_{i,y_i-1}\}$ for $i \in \{1, \ldots, k\}$,
2. $rel(\delta) = \{(q_i, q_{i+1}) \mid 0 \leq i \leq t-2\} \cup \{(q_{t-1}, p_{i,0}) \mid 1 \leq i \leq k\} \cup \{(p_{i,j}, p_{i,j+1}) \mid 1 \leq i \leq k, 1 \leq j \leq y_i - 2\} \cup \{(p_{i,y_i-1}, p_{i,0}) \mid 1 \leq i \leq k\}$.

Finally, for some integers $m_1, \ldots, m_k$ we denote by $gcd(m_1, \ldots, m_k)$ to be their greatest common divisor and $lcm(m_1, \ldots, m_k)$ to be their least common multiple. Two numbers m, n are called co-prime when $gcd(m, n) = 1$.

3   Unary Finite Tree Width NFAs

We present in this section some basic results on unary finite tree width NFAs that we will use in the later sections of this paper. Our state complexity bounds rely crucially on the fact that a minimal finite tree width NFA can always be chosen to be in Chrobak normal form.

Proposition 3.1 ([25]). *Let A be a unary n-state NFA with tree width k. Then there exists an equivalent Chrobak normal form NFA B with at most n states and tree width k.*

Lemma 3.1 ([2,28]). *Let m, n be two positive integers such that $gcd(m, n) = 1$.*

(i) *The largest integer that cannot be presented as $c_1 \cdot m + c_2 \cdot n$ for any integers $c_1, c_2 > 0$ is $m \cdot n$.*
(ii) *The largest integer that cannot be presented as $c_1 \cdot m + c_2 \cdot n$ for any integers $c_1 > 0$ and $c_2 \geq 0$ is $m \cdot n - n$.*
(iii) *The largest integer that cannot be presented as $c_1 \cdot m + c_2 \cdot n$ for any integers $c_1, c_2 \geq 0$ is $m \cdot n - m - n$.*

Note that the item (iii) of the previous lemma is the solution of the Frobenius problem (coin problem) for $N = 2$. For more on that problem the reader is referred to [1].

4   State Complexity of Concatenation

The main question of this section is to find the state complexity of the concatenation of two unary NFAs with limited nondeterminism. The state complexity of concatenation for unary DFAs has been studied in [28], it has been shown that an upper bound for the concatenation of an m state language with an n state language is $m \cdot n$, this bounds can be reached when $gcd(m, n) = 1$. We see in the following results that the case of limited nondeterminism is similar to DFAs. Before we study the general case of concatenation, we give a technical result that gives an upper bound for the size of a DFA equivalent to an NFA that consists of two cycles and additionally transitions from the first cycle to the second cycle.

Let C_1 and C_2 be two cycles, possibly containing final states and both C_1 and C_2 have a state that is specified as the first state. Then by merge(C_1, C_2) we mean the NFA where the state set consists of the union of C_1 and C_2 and includes the transitions of the two cycles and additionally transitions from the final states of C_1 to the second state of C_2. The initial state of merge(C_1, C_2) is the first state of C_1 and the final states are all states that were specified to be final in C_1 and C_2. An example of the NFA $merge(C_1, C_2)$ for two cycles C_1 and C_2 appears in Figure 1.

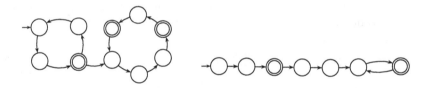

Fig. 1. On the left an NFA as a result of merging two cycles and on the right its equivalent DFA

Lemma 4.1. *Let C_1 and C_2 be two disjoint cycles, of sizes c_1 and c_2 respectively. Then, there is a DFA recognizing $L(merge(C_1, C_2))$ such that:*

i. *The size of its tail is at most $lcm(c_1, c_2)$,*
ii. *The size of its loop is at most c_1 if c_2 divides c_1, and at most $gcd(c_1, c_2)$, otherwise.*

The following technical lemma will be used in the proof of Theorem 4.1. The lemma states the simple observation that in a Chrobak normal form NFA A we can reduce the number of cycles as long as, for each cycle C of A, the new Chrobak normal form NFA has a cycle whose length is a multiple of the length of C.

Lemma 4.2. *Let A be a Chrobak normal form NFA with size of tail t and k cycles of lengths $c_1, \ldots, c_k$, respectively. Let $d_1, \ldots, d_m \in \mathbb{N}$ be integers such that*

$$(\forall i)(1 \leq i \leq k \implies (\exists \ell)(1 \leq \ell \leq m \implies (c_i \text{ divides } d_\ell)))$$

Then $L(A)$ has a Chrobak normal form NFA with size of tail t and m cycles of lengths $d_1, \ldots, d_m$, respectively.

Now we are ready to prove the main results of this section.

Theorem 4.1. *Let L_1 and L_2 be unary regular languages. Let k_1 and k_2 be positive integers, $k_1, k_2 \geq 2$, such that $nsc_{\text{tw} \leq (k_1-1)}(L_1) > nsc_{\text{tw} \leq k_1}(L_1)$ and $nsc_{\text{tw} \leq (k_2-1)}(L_2) > nsc_{\text{tw} \leq k_2}(L_2)$, then we have $nsc_{\text{tw} \leq \min\{k_1,k_2\}}(L_1 \cdot L_2) \leq (nsc_{\text{tw} \leq k_1}(L_1) - \frac{(k_1-1) \cdot (k_1+2)}{2}) \cdot (nsc_{\text{tw} \leq k_2}(L_2) - \frac{(k_2-1) \cdot (k_2+2)}{2}).$*

Proof outline. Let us have the two state-minimal NFAs A_1 and A_2 recognizing the languages L_1 and L_2 respectively, with tree width of k_1 and k_2 respectively. From Proposition 3.1 we can assume that the NFAs A_1 and A_2 are in Chrobak normal form and from the assumptions of the theorem we have that A_1 has exactly k_1 cycles and A_2 has exactly k_2 cycles.

From the NFA A_1 and the NFA A_2 we can construct a new NFA C which recognizes the language $L_1 \cdot L_2$. We construct the NFA C by adding a transition from every final state of A_1 to the second state of the tail of the NFA A_2. Since C is a unary NFA we can change the order of the cycles of A_1 and the tail of A_2 by also changing the final states for the resulting NFA C' to recognize $L_1 \cdot L_2$. The NFA C' will consist of a tail by attaching the tails of A_1 and A_2 together, followed by the merge operation of two cycles (the first one will be a cycle from the NFA A_1 and the second one from the NFA A_2). We can further transform the resulting NFA with the help of Lemma 4.1 and Lemma 4.2. The resulting NFA C'' will be in Chrobak normal form with cycles same size as the cycles of A_1 and a tail of length at most $t_1 + t_2 - 1 + \max\{i \cdot j \mid i \in c(A_1) \text{ and } j \in c(A_2)\}$, where t_i is the tail of A_i and $c(A_i)$ is the the set of the sizes of cycles of A_i, for $i = 1, 2$.

Note that the size of any cycle $c_{i,j} \in c(A_i)$ is the number $size(A_i) - t_i - \sum_{l=1/l \neq j}^{k_i} c_{i,l}$, for $i = 1, 2$. Hence the total size of the NFA C'' is at most $t_1 + t_2 - 1 + (\max_{1 \leq i \leq k_1}\{size(A_1) - t_1 - \sum_{l=1/l \neq i}^{k_1} c_{1,l}\} \cdot \max_{1 \leq j \leq k_2}\{size(A_2) - t_2 - \sum_{l=1/l \neq j}^{k_2} c_{2,l}\}) + size(A_1)$, which is at most $(\max_{1 \leq i \leq k_1}\{size(A_1) - \sum_{l=1/l \neq i}^{k_1} c_{1,l}\} \cdot \max_{1 \leq j \leq k_2}\{size(A_2) - \sum_{l=1/l \neq j}^{k_2} c_{2,l}\})$.

We also note that all the cycles of A_1 cannot divide any other cycle of A_1, the same applies for A_2. In other words all cycles have size greater than 1 and are different from each other, which means that the sum of k such cycles is at least $\sum_{i=2}^{k+1} i$. Then, we have that the size of C'' is at most $(size(A_1) - \sum_{i=2}^{k_1} i) \cdot (size(A_2) - \sum_{i=2}^{k_2} i)$. $\qquad \square$

Theorem 4.2. *Let L_1 and L_2 be unary regular languages. Let k be a positive integer, $k \geq 2$, such that $nsc_{\text{tw} \leq (k-1)}(L_2) > nsc_{\text{tw} \leq k}(L_2)$, then for every positive integer k' we have $nsc_{\text{tw} \leq k'}(L_1 \cdot L_2) \leq sc(L_1) \cdot (nsc_{\text{tw} \leq k}(L_2) - \frac{(k-1) \cdot (k+2)}{2}).$*

Note that a similar bound of Theorem 4.2 also holds the number $nsc_{\text{tw} \leq k'}(L_2 \cdot L_1)$. The concatenation of unary languages is commutative, then $L_1 \cdot L_2 = L_2 \cdot L_1$, which implies that $nsc_{\text{tw} \leq k'}(L_1 \cdot L_2) = nsc_{\text{tw} \leq k'}(L_2 \cdot L_1)$.

It is known from [15,25] that every unary NFA A with limited ambiguity (or trace) there is an NFA B recognizing the language $L(A)$ which has at most $size(A)$ states and the same ambiguity (or trace) with the NFA A. With similar proofs of Theorem 4.1 and Theorem 4.2, these theorems also hold with tree width replaced by ambiguity or trace.

The following two results show that the upper bound of Theorem 4.1 cannot be essentially improved.

Theorem 4.3. *For any $n_0 \in \mathbb{N}$, there exists $n_1, n_2 \geq n_0$ and regular languages L_i, with $nsc_{\text{tw} \leq k_i}(L_i) = n_i$ and $nsc_{\text{tw} \leq (k_i - 1)}(L_i) > nsc_{\text{tw} \leq k_i}(L_i)$, $i = 1, 2$, such that $nsc_{\text{tw} \leq \min\{k_1, k_2\}}(L_1 \cdot L_2) \geq (n_1 - 2 \cdot 3 \cdot (k_1 - 1) \cdot (k_1 + 10) \cdot \ln(k_1 + 4)) \cdot (n_2 - 2 \cdot 3 \cdot (k_2 - 1) \cdot (k_2 + 10) \cdot \ln(k_2 + 4))$.*

Corollary 4.1. *Let $k_1, k_2, n_0 \in \mathbb{N}$. There exists $n_1, n_2 > n_0$ and unary regular languages L_i, $i = 1, 2$, such that $nsc_{\text{tw} \leq k_i}(L_i) = n_i$, any minimal NFA for L_i has tree width at least k_i and $nsc_{tw \leq \min(k_1, k_2)}(L_1 \cdot L_2) = (nsc_{tw \leq k_1}(L_1) - \Omega(k_1^2 \cdot \log k_1)) \cdot (nsc_{tw \leq k_2}(L_2) - \Omega(k_2^2 \cdot \log k_2))$*

Note that similar statements of Theorem 4.3 also hold for finite ambiguity and finite trace.

Theorem 4.3 gives a lower bound for state complexity of the concatenation of tree width k_1 and k_2 unary NFAs that is fairly close to the upper bound of Theorem 4.1. Next we observe that by choosing languages where the minimal NFA is, in fact, a DFA we get a lower bound result that completely omits the minus terms in the lower bound of Theorem 4.3. The constructions used for the proof of Theorem 4.4 is essentially the same as the lower bound construction for unary DFAs. Thus, if only count the total number of states, the absolute worst case of concatenation of finite tree width NFAs is given by the deterministic case.

Theorem 4.4. *For any $n_1, n_2 \in \mathbb{N}$, there exist unary regular languages L_i with $sc(L_i) = n_i$, $i = 1, 2$, such that for any $k \in \mathbb{N}$*

$$nsc_{tw \leq k}(L_1 \cdot L_2) = n_1 \cdot n_2.$$

The tree width measure of the statement of Theorem 4.4 can be replaced with the measures of ambiguity or trace.

5   Kleene Star

Let L be a unary language such that $sc(L) = n$. From [28] we have that the deterministic state complexity of Kleene star of L is at most $(n-1)^2 + 1$ and there are instances where we cannot improve this bound. We will show here that we can not do better than the deterministic case when we have limited tree width (or trace, or ambiguity). Let us start first with the upper bound.

In [4] Chrobak showed that for every unary NFA A with n states, there is a unary NFA A' in Chrobak normal form with n states participating in cycles and $O(n^2)$ states in its tail. For our purposes, we need a more accurate estimation on the size of the tail, which is due to Gawrychowski in [6].

Lemma 5.1 ([6]). *For each unary NFA A with n states, there is a unary NFA A' in Chrobak normal form with at most n states participating in cycles and with a tail with at most $n \cdot (n-1)$ states.*

Theorem 5.1. *Let k be a positive integer and let L be a unary regular language. Then, we have that $nsc_{tw \leq k}(L^*) \leq (nsc_{tw \leq k}(L))^2$.*

Proof. Let us have a minimal NFA N for L with tree width k. For $k = 1$, we have that $nsc_{tw \leq k}(L^*) \leq (nsc_{tw \leq k}(L) - 1)^2 + 1$ from [28]. Next consider the case $k \geq 2$. From Proposition 3.1 we can assume that the NFA N is in Chrobak normal form. Now, let us create the NFA N' which is the same as the NFA N plus we add transitions from all the final states in cycles to the second state of its tail. We make the initial state a final state, if it is not already. The NFA N' recognizes the language L^*.

We transform the NFA N' in an equivalent NFA M in Chrobak normal form. By Lemma 5.1 the NFA N' has at most $n + n \cdot (n-1)$ states. $\qquad \square$

Corollary 5.1. *Let k be a positive integer and let L be a unary regular language. Then, we have that $nsc_{\tau \leq k}(L^*) \leq (nsc_{tw \leq k}(L))^2$ and $nsc_{amb \leq k}(L^*) \leq (nsc_{tw \leq k}(L))^2$.*

Note that if a general unary NFA A has n states, $L(A)^*$ can be recognized by an NFA with $n + 1$ states [10]. In the above proof we relied on Proposition 3.1 just in order to save the one additional state.

In the next theorem, we show that the worst case of the state complexity of Kleene star of NFAs with limited tree width coincides with the deterministic state complexity of Kleene star. The proof of Theorem 5.2 is similar to the lower bound proof of the Kleene star of DFAs.

Theorem 5.2. *For every $n \in \mathbb{N}$ there exists a unary language L with $sc(L) = n$ such that for every $k \in \mathbb{N}$, $nsc_{tw \leq k}(L^*) = (n-1)^2 + 1$.*

Corollary 5.2. *For every $n \in \mathbb{N}$ there exists a unary language L with $sc(L) = n$ such that for every $k \in \mathbb{N}$, $nsc_{\tau \leq k}(L^*) = nsc_{amb \leq k}(L^*) = (n-1)^2 + 1$.*

Note that the worst case languages L used in the proof of Theorem 5.2 have the property that any finite tree width NFA for L has at least the same size as the minimal DFA. In fact, when A is a Chrobak normal form NFA, the largest integer not in $L(A)^*$ can be approximated by solutions to a Frobenius problem defined in terms of cycle sizes of A, and by relying on estimates for the size of solutions for the Frobenius problem [3] it is easy to see that the worst-case bound of Theorem 5.2 cannot be matched by any Chrobak normal form NFA with tree width greater than one.

6   Boolean Operations

The state complexity of union and intersection of NFAs with limited nondeterminism was studied in [23]. There the authors considered finite tree width NFAs over general alphabets, however, the lower bound constructions were given over a unary alphabet and hence the same state complexity results apply directly to unary regular languages.

Proposition 6.1 ([23]). *For unary regular languages L_i and $k_i \geq 1$, $i = 1, 2$, we have that* $\mathrm{nsc}_{\mathrm{tw} \leq k_1 + k_2}(L_1 \cup L_2) \leq \mathrm{nsc}_{\mathrm{tw} \leq k_1}(L_1) + \mathrm{nsc}_{\mathrm{tw} \leq k_2}(L_2) + 1$ *and* $\mathrm{nsc}_{\mathrm{tw} \leq k_1 \cdot k_2}(L_1 \cap L_2) \leq \mathrm{nsc}_{\mathrm{tw} \leq k_1}(L_1) \cdot \mathrm{nsc}_{\mathrm{tw} \leq k_2}(L_2)$.

Moreover, for every $k_i, n_i \in \mathbb{N}$, $i = 1, 2$, there exists unary regular languages L_i and L_i' such that $\mathrm{nsc}_{\mathrm{tw} \leq k_i}(L_i) \geq n_i$, $\mathrm{nsc}_{\mathrm{tw} \leq k_i}(L_i') \geq n_i$, $\mathrm{nsc}_{\mathrm{tw} \leq k_1 + k_2}(L_1 \cup L_2) \geq \mathrm{nsc}_{\mathrm{tw} \leq k_1}(L_1) + \mathrm{nsc}_{\mathrm{tw} \leq k_2}(L_2) - 1$, *and* $\mathrm{nsc}_{\mathrm{tw} \leq k_1 \cdot k_2}(L_1 \cap L_2) \geq (\mathrm{nsc}_{\mathrm{tw} \leq k_1}(L_1) - 1) \cdot (\mathrm{nsc}_{\mathrm{tw} \leq k_2}(L_2) - 1) + 1$.

The state of Proposition 6.1 could be replaced for NFAs with finite ambiguity or trace instead of NFAs with finite tree width. In this section we study the state complexity of complement of unary NFAs with limited nondeterminism.

Theorem 6.1. *Let k_1 be a positive integer and let L be a unary regular language. Then, for every positive integer $k_2 \geq 2$ such that $nsc_{\mathrm{tw} \leq k_2}(L) < nsc_{\mathrm{tw} \leq k_2 - 1}(L)$, by denoting $nsc_{\mathrm{tw} \leq k_2}(L) = n$, we have*

$$nsc_{\mathrm{tw} \leq k_1}(L^c) \leq \begin{cases} \prod_{j=0}^{\frac{k-1}{2}}(\lceil \frac{n-1}{k_2}\rceil - j) \cdot \prod_{j=1}^{\frac{k-1}{2}}(\lceil \frac{n-1}{k_2}\rceil + j) & \text{if } k_2 \text{ is odd,} \\ \prod_{j=1}^{\frac{k}{2}}(\lceil \frac{n-1}{k_2}\rceil - j) \cdot \prod_{j=1}^{\frac{k}{2}}(\lceil \frac{n-1}{k_2}\rceil + j) & \text{if } k_2 \text{ is even.} \end{cases}$$

Proof. Let us have a state minimal NFA A with tree width k_2 which recognizes L. From Proposition 3.1, we can assume that A is in Chrobak normal form, with at most k_2 cycles. We know that the NFA A has exactly k_2 cycles since $nsc_{\mathrm{tw} \leq k_2}(L) < nsc_{\mathrm{tw} \leq k_2 - 1}(L)$. We determinize A, the tail will remain the same and now we have a loop which is of the size at most of the product of sizes of the cycles of A. We know that in the worst case the size of the tail is 1 and the sum of the cycles of A is $size(A) - 1$. By analyzing the function $f(x_1, \ldots, x_{k_2}) = x_1 \cdot \ldots \cdot x_{k_2}$ where $2 \leq x_i \leq size(A) - 1$, $1 \leq i \leq k_2$, and $\sum_{i=1}^{k_2} x_i = size(A) - 1$, we have that the maximum value for f is in the point where $x_i = \frac{size(A) - 1}{k_2}$ for all $1 \leq i \leq k_2$.

Note here that the k_2 cycles can not divide each other otherwise we could just omit some of them, which implies that there are no two cycles with the same size. The sizes of the cycles are positive integers and then we are searching for k_2 distinct positive integers that maximize the above function $f(x_1, \ldots, x_{k_2})$ where $\sum_{i=1}^{k_2} x_i = nsc_{\mathrm{tw} \leq k_2}(L) - 1$. The solution is given by the k_2 distinct positive integers that are closer to the number $\frac{nsc_{\mathrm{tw} \leq k_2}(L) - 1}{k_2}$. These numbers are $\prod_{j=0}^{\frac{k-1}{2}}(\lceil \frac{nsc_{\mathrm{tw} \leq k_2}(L) - 1}{k_2}\rceil - j) \cdot \prod_{j=1}^{\frac{k-1}{2}}(\lceil \frac{nsc_{\mathrm{tw} \leq k_2}(L) - 1}{k_2}\rceil + j)$ when k_2 is odd and $\prod_{j=1}^{\frac{k-1}{2}}(\lceil \frac{nsc_{\mathrm{tw} \leq k_2}(L) - 1}{k_2}\rceil - j) \cdot \prod_{j=1}^{\frac{k-1}{2}}(\lceil \frac{nsc_{\mathrm{tw} \leq k_2}(L) - 1}{k_2}\rceil + j)$ when k_2 is even. □

The proof of Theorem 6.1 uses the fact that minimal unary NFAs with finite tree width can be found in Chrobak normal form. The same applies for unary NFAs with finite ambiguity or trace and, hence, the statement of Theorem 6.1 also applies by replacing the tree width with ambiguity or trace.

In the following result we try to approach the upper bound given in Theorem 6.1.

Theorem 6.2. *For every positive integer k there exists a constant n_0 such that for infinitely many values $n \geq n_0$ the following holds.*

There exists a unary NFA A with $\mathrm{tw}(A) = k$ and $\mathrm{size}(A) = n$ such that any NFA recognizing $L(A)^c$ has size at least $(\frac{n-k \log k-1}{k})^k$.

Proof. We use a construction inspired by Theorem 7 of [10], with the difference that now the tree width of the given NFA has to be fixed. Let $p_1, \ldots, p_k$ be the first k prime numbers. We choose $m_1, \ldots, m_k$ such that for any $i \neq j$, $1 \leq i, j \leq k$,

$$|p_i^{m_i} - p_j^{m_j}| \leq k \log k. \tag{1}$$

This is possible because the kth prime has size roughly $k \log k$ [2]. Now as in Theorem 7 of [10] choose

$$L = ((a^{p_1^{m_1}})^*)^c + \ldots + ((a^{p_k^{m_k}})^*)^c.$$

Now L has an NFA B with $n = 1 + \sum_{i=1}^{k} p_i^{m_i}$ states. Since the p_i's are distinct primes,

$$L^c = (a^{\prod_{i=1}^{k} p_i^{m_i}})^*.$$

By (1), we have $a^{\prod_{i=1}^{k} p_i^{m_i}} \geq (\frac{n-k \log k-1}{k})^k$.   □

There is a gap between the upper bound that we get from Theorem 6.1 and Theorem 6.2. On the upper bound of Theorem 6.1 we used the fact that all the cycles of the minimum Chrobak normal form NFA are distinct numbers, however we know that a stronger statement is true. We know that the sizes of the cycles do not divide each other. We can even assume that the size of the cycles are co-primes since this is the case when the least common multiple is maximized. Similarly, if we want to improve the lower bound of Theorem 6.2 we need to find a set of co-primes that are as close to each other as possible.

7   Conclusion

We have studied the state complexity of Boolean language operations for unary NFAs with limited nondeterminism. We have considered the nondeterministic measures tree width, ambiguity, and trace. We know that for these measures minimal NFAs with limited nondetermism can be always found in Chrobak normal form, which is a basic condition for the results of this paper.

Fairly close upper and lower bounds have been given for the state complexity of these language operations. The upper and lower bounds are roughly within the order of a logarithmic factor and further improvement on these bounds seems difficult as hard number theoretical problems would need to be solved.

The main open question for the state complexity of unary language operations for NFAs with limited nondeterminism is to determine reasonable upper and lower bounds when the unary languages are defined using NFAs with finite branching. Unlike the case of unary NFAs with finite tree width (ambiguity, trace), it is not known whether NFAs with finite branching can be found in any simple form.

References

1. Aliev, I.M., Gruber, P.M.: An optimal lower bound for the Frobenius problem. Journal of Number Theory 123(1), 71–79 (2007)
2. Bach, E., Shallit, J.: Algorithmic Number Theory: Efficient Algorithms, vol. 1. MIT Press (1996)
3. Beck, M., Diaz, R., Robins, S.: The Frobenius problem, rational polytopes, and fourier–dedekind sums. Journal of Number Theory 96(1), 1–21 (2002)
4. Chrobak, M.: Finite automata and unary languages. Theor. Comput. Sci. 47(3), 149–158 (1986)
5. Gao, Y., Moreira, N., Reis, R., Sheng, Y.: A review of state complexity of individual operations. Technical report, Universidade do Porto, Technical Report Series DCC-2011-08, Version 1.1 (2012), http://www.dcc.fc.up.pt/Pubs
6. Gawrychowski, P.: Chrobak normal form revisited, with applications. In: Bouchou-Markhoff, B., Caron, P., Champarnaud, J.-M., Maurel, D. (eds.) CIAA 2011. LNCS, vol. 6807, pp. 142–153. Springer, Heidelberg (2011)
7. Goldstine, J., Kappes, M., Kintala, C.M.R., Leung, H., Malcher, A., Wotschke, D.: Descriptional complexity of machines with limited resources. J. UCS 8(2), 193–234 (2002)
8. Goldstine, J., Kintala, C.M.R., Wotschke, D.: On measuring nondeterminism in regular languages. Inf. Comput. 86(2), 179–194 (1990)
9. Holzer, M., Kutrib, M.: Nondeterministic descriptional complexity of regular languages. Int. J. Found. Comput. Sci. 14(6), 1087–1102 (2003)
10. Holzer, M., Kutrib, M.: Unary language operations and their nondeterministic state complexity. In: Ito, M., Toyama, M. (eds.) DLT 2002. LNCS, vol. 2450, pp. 162–172. Springer, Heidelberg (2003)
11. Holzer, M., Kutrib, M.: Nondeterministic finite automata - recent results on the descriptional and computational complexity. Int. J. Found. Comput. Sci. 20(4), 563–580 (2009)
12. Holzer, M., Kutrib, M.: Descriptional and computational complexity of finite automata - a survey. Inf. Comput. 209(3), 456–470 (2011)
13. Hromkovič, J., Karhumäki, J., Klauck, H., Schnitger, G., Seibert, S.: Measures of nondeterminism in finite automata. In: Welzl, E., Montanari, U., Rolim, J.D.P. (eds.) ICALP 2000. LNCS, vol. 1853, pp. 199–210. Springer, Heidelberg (2000)
14. Hromkovic, J., Seibert, S., Karhumäki, J., Klauck, H., Schnitger, G.: Communication complexity method for measuring nondeterminism in finite automata. Inf. Comput. 172(2), 202–217 (2002)
15. Jiang, T., McDowell, E., Ravikumar, B.: The structure and complexity of minimal nfa's over a unary alphabet. In: Biswas, S., Nori, K.V. (eds.) FSTTCS 1991. LNCS, vol. 560, pp. 152–171. Springer, Heidelberg (1991)

16. Kutrib, M., Pighizzini, G.: Recent trends in descriptional complexity of formal languages. Bulletin of EATCS 3(111) (2013)
17. Leung, H.: Separating exponentially ambiguous finite automata from polynomially ambiguous finite automata. SIAM J. Comput. 27(4), 1073–1082 (1998)
18. Leung, H.: Descriptional complexity of nfa of different ambiguity. Int. J. Found. Comput. Sci. 16(5), 975–984 (2005)
19. Maslov, A.N.: Estimates of the number of states of finite automata. Doklady Akademii Nauk SSSR 194, 1266–1268 (1970)
20. Meyer, A.R., Fischer, M.J.: Economy of description by automata, grammars, and formal systems. In: SWAT (FOCS), pp. 188–191. IEEE Computer Society (1971)
21. Moore, F.R.: On the bounds for state-set size in the proofs of equivalence between deterministic, nondeterministic, and two-way finite automata. IEEE Transactions on Computers C-20(10), 1211–1214 (1971)
22. Okhotin, A.: Unambiguous finite automata over a unary alphabet. Inf. Comput. 212, 15–36 (2012)
23. Palioudakis, A., Salomaa, K., Akl, S.G.: State complexity and limited nondeterminism. In: Kutrib, M., Moreira, N., Reis, R. (eds.) DCFS 2012. LNCS, vol. 7386, pp. 252–265. Springer, Heidelberg (2012)
24. Palioudakis, A., Salomaa, K., Akl, S.G.: Comparisons between measures of nondeterminism on finite automata. In: Jurgensen, H., Reis, R. (eds.) DCFS 2013. LNCS, vol. 8031, pp. 217–228. Springer, Heidelberg (2013)
25. Palioudakis, A., Salomaa, K., Akl, S.G.: Unary nfas with limited nondeterminism. In: Geffert, V., Preneel, B., Rovan, B., Štuller, J., Tjoa, A.M. (eds.) SOFSEM 2014. LNCS, vol. 8327, pp. 443–454. Springer, Heidelberg (2014)
26. Pighizzini, G., Shallit, J.: Unary language operations, state complexity and jacobsthal's function. Int. J. Found. Comput. Sci. 13(1), 145–159 (2002)
27. Shallit, J.O.: A Second Course in Formal Languages and Automata Theory. Cambridge University Press (2008)
28. Yu, S., Zhuang, Q., Salomaa, K.: The state complexities of some basic operations on regular languages. Theoret. Comput. Sci. 125(2), 315–328 (1994)
29. Sheng, Y.: Regular Languages. In: Handbook of Formal Languages, vol. 1, pp. 41–110. Springer (1998)

A Note on Pushdown Automata Systems

Holger Petersen

Reinsburgstr. 75
D-70197 Stuttgart
Germany

Abstract. In (Csuhaj-Varjú et. al. 2000) Parallel Communicating Systems of Pushdown Automata (PCPA) were introduced and shown to be able to simulate nondeterministic one-way multi-head pushdown automata in returning mode, even if communication is restricted to be one-way having a single target component. A simulation of such centralized PCPA by one-way multi-head pushdown automata (Balan 2009) turned out to be incomplete (Otto 2012). Subsequently it was shown that centralized returning PCPA are universal even if the number of components is two (Petersen 2013) and thus are separated from one-way multi-head pushdown automata. Another line of research modified the definition of PCPA such that communication is asynchronous (Otto 2013). While the simulation of one-way multi-head pushdown automata is still possible, now a converse construction shows this model in returning mode to be equivalent to the one-way multi-head pushdown automaton in a very precise sense. It was left open, whether non-centralized returning PCPA of degree two are universal. In the first part of the paper we show this to be the case.

Then we turn our attention to Uniform Distributed Pushdown Automata Systems (UDPAS). These systems of automata work in turn on a single tape. We show that UPDAS accepting with empty stack do not form a hierarchy depending on the number of components and that the membership problem is complete in NP, answering two open problems from (Arroyo et. al. 2012).

1 Introduction

Growing interest in distributed computing has lead to generalizations of classical concepts of Formal Language Theory including context-free grammars and pushdown automata. Several components each consisting of a grammar or automaton communicate and together decide about the acceptance of an input.

Parallel Communicating Systems of Pushdown Automata (PCPA) were introduced in [5] as systems of automata communicating by transferring their pushdown contents following several different protocols. In [5] it was shown that all recursively enumerable languages can be accepted by general PCPA of degree two (number of components) and returning mode PCPA of degree three (the source pushdown is emptied after a transfer).

H. Jürgensen et al. (Eds.): DCFS 2014, LNCS 8614, pp. 342–351, 2014.

In [2] it was claimed that centralized PCPA (having a single target automaton) of degree k working in returning mode can be simulated by nondeterministic one-way k-head pushdown automata. It had been shown previously in [5] that PCPA of degree k can simulate nondeterministic one-way k-head pushdown automata, which would complement the universal power of other variants of non-centralized or non-returning PCPA.

Otto [8] pointed out a flaw in the construction from [2]. Thus the power of centralized PCPA working in returning mode was open. In [10] it was shown that centralized PCPA of degree two working in returning mode are universal, a result that is optimal since PCPA of degree one accept the context-free languages.

In another line of research, Otto [9] modified the definition of communication of PCPA leading to the model of asynchronous PCPA. He could show that centralized asynchronous PCPA working in returning mode can be simulated by nondeterministic one-way k-head pushdown automata. All variants of asynchronous PCPA of degree two were shown to be universal except for non-centralized returning asynchronous PCPA. Here we show these PCPA of degree two to be universal even if the pushdown automata are deterministic. The technique is novel in this area, being based on a simulation of a computational model equipped with a queue storage. In contrast the results of [10] were shown with the help of counter automata and the constructions of [9] use two-pushdown automata. The specific computational model we use is the Post machine introduced in the textbook [7]. While the universality of a queue storage is implied by Post's work and is sometimes considered to be folklore, references [12,7] appear to give the earliest formal definitions of machines with a finite control and a queue as their storage.

While a PCPA transfers information via pushdown contents, in Distributed Pushdown Automata Systems (DPAS) several automata work on the same input string and communication takes place by activating components. More precisely an active component keeps working as long as transitions are possible and then nondeterministically another component becomes active. Acceptance can be defined by final state or empty stack. This model has been introduced in [6]. In [1] the restriction to systems consisting of identical pushdown automata is investigated (Uniform Distributed Pushdown Automata Systems, UDPAS) and the resulting classes of languages are compared with the iterated shuffle of several copies of the underlying context-free language. The authors relate different modes of acceptance showing that acceptance by final state is stronger than by empty stack. Some decision and closure properties are also presented. In the Final Remarks of [1] three open problems are mentioned:

Open Problem 1: What conditions should a context-free language L satisfy such that the shuffle of several copies of L is accepted by a UDPAS with empty stacks?

Open Problem 2: Is there any hierarchy depending on the number of components?

Open Problem 3: What is the complexity of the membership problem for UDPAS accepting with empty stacks?

While we cannot report any progress on Open Problem 1, we will present solutions of the other two problems in Section 4.

2   Preliminaries

We assume the reader to be familiar with basic concepts of Automata Theory and Formal Languages, see e.g. [11].

Several variants of PCPA were defined in [5,2,9]. A PCPA of degree k consists of k nondeterministic pushdown automata defined in the standard way. These automata (called components) work in parallel reading the same input string. The components communicate using special pushdown store symbols. In the asynchronous mode a communication symbol has to be on top of the pushdown store of the target component and a response symbol is required on top of the pushdown store of the source component. Then the contents of the source pushdown store are copied onto the target pushdown store replacing the topmost symbol. The source pushdown store is emptied if the PCPA is working in returning mode. An input is accepted if at least one component reaches a final state when all components have read the entire input string.

Formally, an asynchronous PCPA of degree k [9] is a tuple

$$A = (V, \Delta, A_1, A_2, \ldots, A_k, K, R)$$

where

- V is a finite input alphabet,
- Δ is a finite alphabet of pushdown symbols,
- A_i is a component as defined below for $1 \leq i \leq k$,
- $K = \{K_1, \ldots, K_k\} \subseteq \Delta$ is a set of query symbols.
- $R \in V \setminus K$ is a response symbol (different from all bottom symbols of the component pushdown automata).

Each component $A_i = (Q_i, V, \Delta, f_i, q_i, Z_i, F_i)$ is a pushdown automaton where

- Q_i is a finite set of states,
- f_i is a function from $Q_i \times (V \cup \varepsilon) \times \Delta$ to the finite subsets of $Q_i \times \Delta^*$,
- $q_i \in Q_i$ is the initial state,
- $Z_i \in \Delta$ is the bottom symbol,
- $F_i \subseteq Q_i$ is the set of final states.

If only function f_1 of the first component maps to sets with members containing query symbols, the system is called centralized.

A configuration of a PCPA of degree k is a $3k$-tuple

$$(s_1, x_1, \alpha_1, \ldots, s_k, x_k, \alpha_k)$$

where

- $s_i \in Q_i$ is the state of component A_i,

- $x_i \in V^*$ is the part of the input not yet processed by A_i,
- $\alpha_i \in \Delta^*$ is the word on the pushdown store of A_i with its topmost symbol on the left.

In returning mode the step relation $\vdash_r$ between configurations is defined by:

$$(s_1, x_1, B_1\alpha_1, \ldots, s_k, x_k, B_k\alpha_k) \vdash_r (s_1', x_1', \alpha_1', \ldots, s_k', x_k', \alpha_k'),$$

if one of the following conditions holds:

Communication step: There are $1 \leq i, j \leq k$ such that $B_i = K_{j_i}$ and $B_{j_i} = R$ we have $\alpha_i' = \alpha_{j_i}\alpha_i$, $\alpha_{j_i}' = Z_{j_i}$, and $\alpha_m' = B_m\alpha_m$ for all other indices m. States and input are not modified: $s_i' = s_i$ and $x_i' = x_i$ for $1 \leq i \leq k$.

Internal step: If there is no pair as defined above an internal step is carried out:

- If $x_i = a_i x_i'$ with $a_i \in K \cup \{R\}$ then $\alpha_i' = \alpha_i$, $x_i' = x_i$, and $s_i' = s_i$.
- If $x_i = a_i x_i'$ with $a_i \in (V \setminus K \setminus \{R\}) \cup \{\varepsilon\}$ then $(s_i', \beta) \in f_i(s_i, a_i, B_i)$ with $\alpha_i' = \beta\alpha_i$.

The PCPA accepts exactly those words w that admit a sequence of steps from the initial configuration

$$(q_1, w, Z_1, \ldots, q_k, w, Z_k)$$

to a final configuration

$$(s_1, \varepsilon, \alpha_1, \ldots, s_k, \varepsilon, \alpha_k)$$

with $s_i \in F_i$ for some $1 \leq i \leq k$.

A Post machine M [7, p. 24] can be described by a program[1] with a single variable x having as its value a string over a finite alphabet $\Sigma \cup \{\#\}$, where Σ is the input alphabet and $\#$ is an auxiliary symbol. The program consists of instructions of the following types:

HALT statements: ACCEPT and REJECT with the obvious meaning.
TEST statements: conditional statements of the form

```
if x = ε then goto i₀ else
    case  head(x) of
            σ₁: then x := tail(x); goto i₁;
            ⋮
            σₙ: then x := tail(x); goto iₙ;
```

where $n = |\Sigma| + 1$, $\sigma_1, \ldots, \sigma_n \in \Sigma \cup \{\#\}$, and $i_0, \ldots, i_n$ are instructions of M.
ASSIGNMENT statements: $x := x\sigma_k$ for $\sigma_k \in \Sigma \cup \{\#\}$.

---

[1] In [7] the program takes the form of a directed graph, which is obviously equivalent.

Execution of the program starts with x holding the input at the first instruction. The input is accepted if the ACCEPT instruction is reached.

A *Distributed Pushdown Automata System* (DPAS) of degree k consists of k pushdown automata (components) working in turn on a common input string. At each point in time one of the pushdown automata is active and may perform a transition. If the active automaton has no transition (blocks), another pushdown automaton becomes active. Formally a DPAS A of degree k is a tuple

$$A = (V, \Delta, A_1, A_2, \ldots, A_k)$$

where the components are defined as for PCPA. An instantaneous description (ID) of the DPAS A is a tuple

$$(x, s_1, \delta_1, \ldots, s_k, \delta_k)$$

where $x \in V^*$ is the part of the (common) input string, that has not yet been read. Two instantaneous descriptions

$$(ax, s_1, \delta_1, \ldots, s_i, \delta_i, \ldots, s_k, \delta_k) \vdash_i (x, s_1, \delta_1, \ldots, s_i', \delta_i', \ldots, s_k, \delta_k)$$

are in successor relation $\vdash_i$ if A_i with its current stack contents can process a in one step, $(s_i', \beta) \in f_i(s_i, a, \alpha)$ for $a \in V \cup \{\varepsilon\}$, $\delta_i = \alpha y$, and $\delta_i' = \beta y$. We say that instantaneous description C_1 derives C_2 in t-mode ($C_1 \vdash^t C_2$) if there is an i with $C_1 \vdash_i^* C_2$ and there is no instantaneous description C_3 with $C_2 \vdash_i C_3$. A word is accepted by a DPAS with final states (empty stacks) if after a sequence of instantaneous descriptions in relation $\vdash^t$ all components are in final states (have empty stacks). Further details can be found in [1]. If all components $A_1, A_2, \ldots, A_k$ are equal, we call the DPAS *uniform* (UDPAS).

For words w, x over an alphabet Σ we define their *shuffle* in the following way:

$$w \sqcup \varepsilon = \varepsilon \sqcup w = \{w\}$$
$$aw \sqcup bx = a(w \sqcup bx) \cup b(aw \sqcup x)$$

For languages L_1, L_2 we define

$$L_1 \sqcup L_2 = \bigcup_{w_1 \in L_1, w_2 \in L_2} w_1 \sqcup w_2$$

and

$$\sqcup^p(L_1) = \begin{cases} L_1 & p = 1 \\ L_1 \sqcup (\sqcup^{p-1}(L_1)) & p \geq 2 \end{cases}$$

as the *iterated shuffle*.

3 Universality of Non-centralized Deterministic PCPA of Degree Two in Returning Mode

Theorem 1. *Every recursively enumerable language can be accepted by a non-centralized PCPA of degree two working in returning mode.*

Proof. We make use of the fact that every recursively enumerable language can be accepted by a Post machine [7, Theorem 1-3] and that recursively enumerable languages are closed under reversal.

Let L be a recursively enumerable language and M a Post machine accepting L. We will describe a system A simulating M and accepting L^R. The main task of the simulation is carried out by component 1. It first reads the input and puts it onto its pushdown store. Then it starts a cycle simulating a single step of M consisting of the following tasks:

1. It checks if M has reached a HALT statement and accepts resp. rejects accordingly.
2. It pops the topmost symbol of the pushdown (if the store is non-empty).
3. It puts the response symbol R on top of its pushdown store.
4. By having the response symbol on its pushdown store, component 1 of A stops until the contents of its pushdown store have been transferred to component 2.
5. After having resumed its operation, component 1 of A puts a string (possibly empty) on its (now empty) pushdown store, depending on the simulated state of M and the information from step 2.
6. Component 1 puts the communication symbol for a communication from component 2 on top of the pushdown store.
7. After having resumed its operation, component 1 starts the next cycle.

Component 2 first reads the input string and then repeatedly executes the following steps:

1. It puts the communication symbol from component 1 on top of the pushdown store.
2. It puts the response symbol on top of the pushdown store.

No state of component 2 is accepting, such that acceptance depends on component 1 only. By construction component 1 simulates M on the reversal of its input.

□

Remark 1. In the proof of Theorem 1 the Post machine being simulated is a deterministic model and the determinism carries over to the simulating PCPA. There is however a problem with component 1 reading the entire input in a deterministic way, since the last symbol cannot be detected. A way to overcome this difficulty is to introduce an end-marker (which is part of the input alphabet) and relax the requirement to accept all recursively enumerable languages with the extra symbol (and possibly more) attached. Another solution is to modify the definition of PCPA such that end of input can be detected.

4 Results for Uniform Distributed Pushdown Automata Systems

In this section we address two of the three open problems mentioned in the final remarks of [1]. The answers show that the classes of languages accepted by UDPAS have a complex structure (they do not form a hierarchy) and the computational complexity of the non-uniform word-problem is the same as for the shuffle of two context-free languages, namely complete in NP.

Theorem 2. *There is no hierarchy of languages accepted by UDPAS depending on the number of components.*

Proof. Let $M \subseteq a^*$ be a finite, non-empty language over the single letter alphabet $\{a\}$ with $M \neq \{\epsilon\}$. Clearly, M is a context-free language and can thus be accepted by a UDPAS with one component. Define $m = \max(\{|x| \mid x \in M\}$. Suppose that M is accepted by UPDAS $\mathcal{A}$ with $n = |m| + 1$ components. By Lemma 1 of [1] there is a language L such that $L^n = M$. If $u \in L$ is any non-empty word, then u^n is accepted by A. But $|u^n| > m$ and therefore A cannot accept M. We conclude that $L \subseteq \{\epsilon\}$ and $M = L^n = \{\epsilon\}^n = \{\epsilon\}$, contradicting the choice of M. ☐

Theorem 3. *The non-uniform word-problem for languages accepted by UDPAS is complete in NP.*

Proof. The problem is in NP, since for n copies of a given pushdown automaton A a nondeterministic Turing-machine can guess a distribution of all symbols of an input word among the copies of A and check membership in $L(A)$ for each of the interleaved subwords.

For NP-hardness we reduce the NP-complete membership-problem for the shuffle of two context-free languages [4] to the problem at hand. Let A and B be two pushdown-automata. Without loss of generality, A and B are over a common input alphabet Σ and pushdown alphabet Δ. Let $\#, \$ \notin \Sigma$ be two new symbols. We define A' and B' as automata having the finite controls of A and B with self-loops on $\#$ added to every state. In A' we duplicate every state and its transitions originally in A (thus omitting the self-loops on $\#$), while in B' we duplicate every state and add a transition on $\#$ to the original state. We finally add ε-transitions from every state in A or B to its copy. The automata obtained will be called A'' and B''. Notice that $L(A') = L(A'') = L(A) \sqcup \{\#\}^*$ and $L(B') = L(B'') = L(B) \sqcup \{\#\}^*$, since the additional transitions in A'' and B'' do not influence the accepted languages.

Formally let

$$A = (Q_A, \Sigma, \Delta, f_A, q_A, Z_A, F_A)$$

and

$$B = (Q_B, \Sigma, \Delta, f_B, q_B, Z_B, F_B)$$

be the initial pushdown-automata. Then

$$A' = (Q_A, \Sigma \cup \{\#\}, \Delta, f_A \cup \{(q, \#, d, \{(q, d)\}) \mid q \in Q_A, d \in \Delta\}, q_A, Z_A, F_A)$$

and

$$B' = (Q_B, \Sigma \cup \{\#\}, \Delta, f_B \cup \{(q, \#, d, \{(q, d)\}) \mid q \in Q_B, d \in \Delta\}, q_B, Z_B, F_B).$$

Further

$$A'' = (Q_A \cup \{\hat{q} \mid q \in Q_A\}, \Sigma \cup \{\#\}, \Delta,$$
$$f_A \cup \{(q, \#, d, \{(q, d)\}) \mid q \in Q_A, d \in \Delta\} \cup$$
$$\{(\hat{q}, \sigma, d, S) \mid (q, \sigma, d, S) \in f_A\} \cup \{(q, \varepsilon, d, \{\hat{q}\}) \mid q \in Q_A, d \in \Delta\},$$
$$q_A, Z_A, F_A)$$

and

$$B'' = (Q_B \cup \{\hat{q} \mid q \in Q_B\}, \Sigma \cup \{\#\}, \Delta,$$
$$f_B \cup \{(q, \#, d, \{(q, d)\}) \mid q \in Q_B, d \in \Delta\} \cup$$
$$\{(\hat{q}, \#, d, \{q\}) \mid q \in Q_B\} \cup \{(q, \varepsilon, d, \{\hat{q}\}) \mid q \in Q_B, d \in \Delta\},$$
$$q_B, Z_B, F_B).$$

We now form a new pushdown-automaton C consisting of the union of the finite controls of A'' and B'' plus a new state q_C, which is the initial state of C. On $\#$ there is a transition from q_C to the initial state of A'', on $\$$ there is a transition from q_C to B''. All new transitions (not in A or B) do not affect the pushdown-store.

For a given word $w = w_1 w_2 \cdots w_n$ with $w_i \in \Sigma$ for which the membership-problem of the shuffle of the languages accepted by A and B has to be decided, we form the word $w' = \#\$\#w_1\#w_2\# \cdots \#w_n$ and ask whether a system of two copies of C accepts w'.

Suppose w is a member of the shuffle of $L(A)$ and $L(B)$. We fix a distribution of the symbols of w among A and B. On the prefix $\#\$$ of w' the initial states of A'' and B'' are reached in the copies of C with the initial state of B'' being active. Let us call these two copies C_A and C_B depending on the initial state. Notice that all states reachable in C_A (C_B) will be from A'' (B''). Now the system is repeatedly about to read the symbols $\#w_i$. If the symbol w_i is part of the input of A and C_A is active in a state from A', then using the self-loop C reads $\#$ and then w_i. If w_i is part of the input of B and C_A is active, then C_A jumps into the corresponding state without the self-loop and blocks since there is no transition on $\#$. Then C_B becomes active and can skip $\#$ either by the self-loop or by a transition from the copy of a state to the state from B. The computation of C_B continues on w_i. If w_i is part of the input of B and C is in a state from B'', then C_B reads $\#$ either by a self-loop or by a transition to a state from B and processes w_i. If w_i is part of the input of A and C_B is active, the $\#$ is skipped by a self-loop and then C_B blocks using an ε-transition to the copy of the current state. This strategy shows, that for every word w in the shuffle of $L(A)$ and $L(B)$ the modified input $w' = \#\$\#w_1\#w_2\# \cdots \#w_n$ can be accepted by C.

If conversely an input $w' = \#\$\#w_1\#w_2\# \cdots \#w_n$ is accepted by C, we can identify two words from $L(A)$ and $L(B)$ forming w' by recording the sequence of states from C_A and C_B. □

5   Conclusion

We have shown that non-centralized returning asynchronous PCPA of degree two are universal.

The following table summarizes results characterizing the power of PCPA working in different modes. Notice that in the cases where two components are universal a reduction to one component is impossible, since systems with one component accept the context-free languages.

PCPA with k components	Power of variant	Reference
synchronous non-returning	universal for $k \geq 2$	[5]
synchronous centralized non-returning	universal for $k \geq 3$	[3]
	universal for $k \geq 2$	[10]
synchronous returning	universal for $k \geq 3$	[5]
	universal for $k \geq 2$	[10]
synchronous centralized returning	universal for $k \geq 2$	[10]
asynchronous non-returning	universal for $k \geq 2$	[9]
asynchronous centralized non-returning	universal for $k \geq 2$	[9]
asynchronous returning	universal for $k \geq 3$	[9]
	universal for $k \geq 2$	Theorem 1
asynchronous centralized returning	k-head PDA	[9]

Uniform Distributed Pushdown Automata Systems have a membership problem that is complete in NP. The technique from the proof of Theorem 3 of letting an automaton block by nondeterministically moving to a copy of a state having only a subset of the original transitions seems to be promising for solving Open Problem 1 of [1] asking for conditions that a context-free language L should satisfy such that $\sqcup\!\sqcup^p(L)$ is accepted by a UPDAS with empty stacks.

Acknowledgements. Thanks are due to Friedrich Otto for remarks on a first draft of this paper and to the anonymous referees for suggesting several improvements.

References

1. Arroyo, F., Castellanos, J., Mitrana, V.: Uniform distributed pushdown automata systems. In: Kutrib, M., Moreira, N., Reis, R. (eds.) DCFS 2012. LNCS, vol. 7386, pp. 64–75. Springer, Heidelberg (2012)
2. Balan, M.S.: Serializing the parallelism in parallel communicating pushdown automata systems. In: Dassow, J., Pighizzini, G., Truthe, B. (eds.) 11th International Workshop on Descriptional Complexity of Formal Systems, DCFS 2009, pp. 59–68 (2009), http://dx.doi.org/10.4204/EPTCS.3.5
3. Balan, M.S., Krithivasan, K., Mutyam, M.: Some variants in communication of parallel communicating pushdown automata. Journal of Automata, Languages and Combinatorics 8(3), 401–416 (2003)

4. Berglund, M., Björklund, H., Högberg, J.: Recognizing shuffled languages. In: Dediu, A.-H., Inenaga, S., Martín-Vide, C. (eds.) LATA 2011. LNCS, vol. 6638, pp. 142–154. Springer, Heidelberg (2011)

5. Csuhaj-Varjú, E., Martín-Vide, C., Mitrana, V., Vaszil, G.: Parallel communicating pushdown automata systems. Int. J. Found. Comput. Sci. 11(4), 633–650 (2000)

6. Csuhaj-Varjú, E., Mitrana, V., Vaszil, G.: Distributed pushdown automata systems: Computational power. In: Ésik, Z., Fülöp, Z. (eds.) DLT 2003. LNCS, vol. 2710, pp. 218–229. Springer, Heidelberg (2003)

7. Manna, Z.: Mathematical Theory of Computation. McGraw-Hill, New York (1974)

8. Otto, F.: Centralized PC systems of pushdown automata versus multi-head pushdown automata. In: Kutrib, M., Moreira, N., Reis, R. (eds.) DCFS 2012. LNCS, vol. 7386, pp. 244–251. Springer, Heidelberg (2012)

9. Otto, F.: Asynchronous PC systems of pushdown automata. In: Dediu, A.-H., Martín-Vide, C., Truthe, B. (eds.) LATA 2013. LNCS, vol. 7810, pp. 456–467. Springer, Heidelberg (2013)

10. Petersen, H.: The power of centralized PC systems of pushdown automata. In: Jurgensen, H., Reis, R. (eds.) DCFS 2013. LNCS, vol. 8031, pp. 241–252. Springer, Heidelberg (2013)

11. Sipser, M.: Introduction to the Theory of Computation, 2nd edn., Thomson (2006)

12. Vollmar, R.: Über einen Automaten mit Pufferspeicherung (On an automaton with buffer-tape). Computing 5, 57–70 (1970), (in German)

Non-recursive Trade-offs between Two-Dimensional Automata and Grammars[*]

Daniel Průša

Czech Technical University, Faculty of Electrical Engineering,
Karlovo náměstí 13, 121 35 Prague 2, Czech Republic
prusapa1@cmp.felk.cvut.cz

Abstract. We study succinctness of descriptional systems for picture languages. Basic models of two-dimensional finite automata and generalizations of context-free grammars are considered. It is shown that non-recursive trade-offs between the systems are very common. The results are based on the ability of the systems to simulate Turing machines.

Keywords: Picture languages, four-way automata, two-dimensional context-free grammars, descriptional complexity.

1 Introduction

Many concepts and techniques from the theory of formal languages and automata have been generalized to two-dimensional (2D) languages, where the basic entity, the *string*, has been replaced by a 2D rectangular array of symbols, called a *picture*. The *four-way finite automaton* of Blum and Hewitt [1] was introduced already in 1967. It has a finite-state control and a head that moves over an input picture by performing movements right, left, up, and down. The early models of picture grammars include matrix and Kolam grammars of Siromoney [14,15].

Several other systems have been proposed, usually having different recognition power [4,3,9,2]. So far, no comparison has been done with respect to their descriptional complexity. The purpose of this paper is to fill this gap. We show that there are many non-recursive trade-offs between the descriptional systems for picture languages.

Examples of such trade-offs are well known in the case of descriptional systems for string languages. The first non-recursive trade-off was presented by Meyer and Fisher [8]. They showed that the gain in economy of description can be arbitrary when the size of finite automata and general context-free grammars generating regular languages is compared. Since this results, other non-recursive trade-offs were reported and unified proof schemes were established [7].

The paper is structured as follows. In Section 2 we give the basic notions and notation on picture languages. In Section 3 we exploit some known properties of four-way automata and show non-recursive trade-off between the non-deterministic

---

[*] The author was supported by the Grant Agency of the Czech Republic under the project P103/10/0783.

H. Jürgensen et al. (Eds.): DCFS 2014, LNCS 8614, pp. 352–363, 2014.

and deterministic four-way automaton (4NFA and 4DFA, respectively). It is also explained how the result can be extended to other automata. Descriptional complexity of a 2D context-free grammar variant from [12] is studied in Section 4. A new technique of a Turing machine simulation using the grammar is presented there. The paper closes with a short summary and some open problems in Section 5.

2   Preliminaries

Here we use the common notation and terms on pictures and picture languages (see, e.g., [3]). If Σ is a finite alphabet, then $\Sigma^{*,*}$ is used to denote the set of rectangular pictures over Σ, that is, if $P \in \Sigma^{*,*}$, then P is a two-dimensional array (matrix) of symbols from Σ. If P is of size $m \times n$, this is denoted by $P \in \Sigma^{m,n}$. We also write $\text{rows}(P) = m$ and $\text{cols}(P) = n$. If P is a square picture $n \times n$, we shortly say P is of size n. The empty picture Λ is defined as the only picture of size 0×0. Moreover, $\Sigma^{+,+} = \Sigma^{*,*} \setminus \{\Lambda\}$.

We use $[a_{ij}]_{m \times n}$ as a notation for a general matrix with m rows and n columns where the element in the i-th row and j-th column is denoted as a_{ij}.

Two (partial) binary operations are introduced to concatenate pictures. Let $A = [a_{ij}]_{k \times \ell} \in \Sigma^{k,\ell}$ and $B = [b_{ij}]_{m \times n} \in \Sigma^{m,n}$. The *column concatenation* $A \oplus B$ is defined iff $k = m$, and the *row concatenation* $A \ominus B$ is defined iff $\ell = n$. These products are specified by the following schemes:

$$A \oplus B = \begin{bmatrix} a_{11} \cdots a_{1\ell} & b_{11} \cdots b_{1n} \\ \vdots \ddots \vdots & \vdots \ddots \vdots \\ a_{k1} \cdots a_{k\ell} & b_{m1} \cdots b_{mn} \end{bmatrix} \quad \text{and} \quad A \ominus B = \begin{bmatrix} a_{11} \cdots a_{1\ell} \\ \vdots \ddots \vdots \\ a_{k1} \cdots a_{k\ell} \\ b_{11} \cdots b_{1n} \\ \vdots \ddots \vdots \\ b_{m1} \cdots b_{mn} \end{bmatrix}.$$

We generalize $\ominus$ and $\oplus$ to a *grid concatenation* which is applied to a matrix of pictures $[P_{ij}]_{m \times n}$ where each $P_{ij} \in \Sigma^{*,*}$. The operation $\bigoplus[P_{ij}]_{m \times n}$ is defined iff

$$\text{rows}(P_{i1}) = \text{rows}(P_{i2}) = \ldots = \text{rows}(P_{in}) \quad \text{for all } i = 1, \ldots, m$$
$$\text{cols}(P_{1j}) = \text{cols}(P_{2j}) = \ldots = \text{cols}(P_{mj}) \quad \text{for all } j = 1, \ldots, n$$

Then, $\bigoplus[P_{ij}]_{m \times n} = P_1 \ominus P_2 \ominus \ldots \ominus P_m$, where $P_k = P_{k1} \oplus P_{k2} \oplus \ldots \oplus P_{kn}$ for $k = 1, \ldots, m$. An example is given in Figure 1. In order to enable all considered finite automata to detect the border of P, they always work over the *boundary picture* $\widehat{P}$ over $\Sigma \cup \{\#\}$ of size $(m + 2) \times (n + 2)$, defined by the scheme in Figure 2. We assume that $\# \notin \Sigma$ for any input alphabet Σ.

3   Two-Dimensional Finite Automata

For a 4NFA $\mathcal{A}$, we define the automaton *size measure* $c(\mathcal{A})$ as the number of states (state complexity) of $\mathcal{A}$.

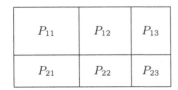

Fig. 1. The product of $\bigoplus[P_{ij}]_{2\times3}$

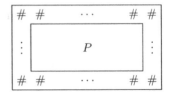

Fig. 2. The boundary picture $\widehat{P}$

Blum and Hewitt [1] proved that 4NFA is more powerful than 4DFA using the language of odd length square pictures over $\{0,1\}$ containing 0 in the center. Here we slightly modify the language to include even length square pictures as well and we follow their proof to estimate the number of states needed by a 4DFA to accept a finite subset of the language.

Let L_{cent} be a picture language over $\Sigma = \{0,1\}$ consisting of all square pictures P that contain 0 at (the central) position ($\lceil\frac{m+1}{2}\rceil, \lceil\frac{m+1}{2}\rceil$) where $m = \text{rows}(P)$. This language is accepted by a 4NFA by checking that the input is a square picture, followed by guessing the trajectory depicted in Figure 3 and verifying that the central field stores 0.

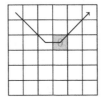

Fig. 3. Two pictures in L_{cent}. 4NFA recognizes the pictures by guessing the shown trajectories (they end in the top-right corner) and checking the highlighted central field.

Lemma 1. For $m \in \mathbb{N}$, let $\mathcal{A}$ be a 4DFA such that $L(\mathcal{A}) \cap \Sigma^{m,m} = L_{\text{cent}} \cap \Sigma^{m,m}$. Then, $\mathcal{A}$ has $\Omega(m/\log m)$ states.

Proof. Let Q be the set of states of $\mathcal{A}$ and let $P \in L_{\text{cent}}$ be a picture over $\{0,1\}$ of size $m \times m$. Define $k = \lceil\frac{m-1}{2}\rceil$ and take a block (sub-picture) B of size $k \times k$ in P which does not contain the top left corner of P. Consider the behavior of $\mathcal{A}$ when working over B. It can enter it at one of $4k - 4$ positions of the perimeter being in one of $|Q|$ states. After performing some steps, it either leaves the block, again at some position and in one of the states, or it accepts or rejects (rejecting includes cycling inside the block). Thus the block defines a mapping from $\{1, \ldots, 4k - 4\} \times Q$ to $(\{1, \ldots, 4k - 4\} \times Q) \cup \{\text{acc, rej}\}$.

Observe it is not possible to have two different blocks B_1, B_2 with the same mapping. Consider that (i, j) is a position where the blocks differ. Construct the picture P_1 of size $m \times m$ which includes B_1, having its field at (i, j) placed in the center of P_1. The fields of P_1 outside B_1 are filled by 0. Similarly, construct P_2 by extending B_2. If the mappings for B_1 and B_2 are identical, pictures P_1 and P_2 are either both accepted by $\mathcal{A}$ or both rejected, which is a contradiction.

There are $(|Q|(4k - 4) + 2)^{|Q|(4k-4)}$ mappings defined by blocks of size $k \times k$. On the other hand, a two-letter alphabet forms 2^{k^2} pictures of size $k \times k$. This results in the inequality $(|Q|(4k - 4)) \log_2 (|Q|(4k - 4) + 2) \geq k^2$ which implies a lower bound of the form

$$|Q| = \Omega \left(\frac{k}{\log_2 k} \right) = \Omega \left(\frac{m}{\log m} \right).$$

$\square$

As has been observed by Kari and Moore [6], a 4DFA can operate as a 2-counter machine [10], consisting of a finite state control unit and two integer registers, called counters. It is possible to represent the value of the first and second register by the horizontal and vertical position of the automaton head, respectively. Moving the head by one tape field increments/decrements a register. To check whether a register is of zero value means to test whether the head is in the first row/column. Naturally, the simulated registers are not unbounded, the maximal value they can store is limited by the size of the input picture. The content of the tape can be arbitrary, it is not involved in the simulation.

Minsky showed in [10] that the 2-counter machine is equivalent to a Turing machine. Namely, a Turing machine can be simulated by two stacks, a stack can be simulated by two counters and four counters can be simulated by two counters. The input to the Turing machine is assumed to be encoded in unary.

Proposition 2. *Let Σ be an alphabet. There is an infinite sequence of 4DFAs $\{\mathcal{A}_k\}_{k=1}^{\infty}$, a recursive function $f : \mathbb{N} \to \mathbb{N}$ and a non-recursive function $g : \mathbb{N} \to \mathbb{N}$, growing faster than any recursive function, such that $c(\mathcal{A}_k) \leq f(k)$ and $\mathcal{A}_k$ accepts a finite picture language which includes all pictures over Σ of size $g(k) \times g(k)$.*

Proof. We construct $\mathcal{A}_k$ accepting some pictures whose size is larger than $S(k)$, where $S : \mathbb{N} \to \mathbb{N}$ is a busy beaver function [13], growing faster than any recursive function. Let $\mathcal{T}_1, \ldots, \mathcal{T}_s$ be an enumeration of all one-dimensional (1D) deterministic binary Turing machines with k states. Each $\mathcal{T}_i$ (started over a blank tape) is simulated by a 2-counter machine $\mathcal{C}_i$. We assume $\mathcal{C}_i$ increases the length of $\mathcal{T}_i$'s configuration encoding each time it simulates a step of $\mathcal{T}_i$. This ensures that $\mathcal{C}_i$ never goes into a cycle. Design $\mathcal{A}_k$ as it follows.

- $\mathcal{A}_k$ checks whether the input P is a square picture, if not it rejects it.
- $\mathcal{A}_k$ follows computations of $\mathcal{C}_1, \ldots, \mathcal{C}_s$ to simulate machines $\mathcal{T}_1, \ldots, \mathcal{T}_s$ one by one. If there is a simulation which ends at the moment when the head of $\mathcal{A}_k$ scans the last row or column of P, then $\mathcal{A}_k$ accepts P. If the simulation exceeds the area of P or finishes without reaching the last row or column of P, the simulation of the next $\mathcal{T}_i$ is launched.

– If all simulations of machines $\mathcal{T}_i$ pass without accepting P, $\mathcal{A}_k$ rejects.

$\mathcal{A}_k$ must memorize in states instructions for each $\mathcal{C}_i$ as well as the index of the currently simulated $\mathcal{T}_i$. However, this is no problem, since the number of states of automata $\mathcal{A}_k$ can grow as fast as any recursive function.

We define $g(k)$ to be the size of the largest (square) picture accepted by $\mathcal{A}_k$. It is obvious that $g(k) \geq S(k)$. $\square$

Theorem 3. *The trade-off between 4NFAs and 4DFAs is non-recursive.*

Proof. Let $\mathcal{M}$ be a 4NFA accepting L_{cent}. For each k, combine $\mathcal{A}_k$ from Proposition 2 and $\mathcal{M}$ to obtain a 4NFA $\mathcal{M}_k$ accepting the finite language $L_{\text{cent}} \cap L(\mathcal{A}_k)$. By Lemma 1 applied for $m = g(k)$, a 4DFA needs $\Omega(g(k)/\log g(k))$ states to recognize $L(\mathcal{M}_k)$. $\square$

To obtain more non-recursive trade-offs, we can exploit other results related to the mutual power of two-dimensional automata. For example, the one-marker four-way automaton is more powerful than 4DFA [1], the four-way alternating automaton is more powerful than 4NFA [5] and it is incomparable with online-tessellation automaton [4] which defines the same family of recognizable picture languages (REC) as tiling systems [3]. All the results are proved using some states counting arguments similar to that one presented in the proof of Lemma 1, hence non-recursive trade-offs are induced between the mentioned pairs of models.

4   Two-Dimensional Context-Free Grammars

In this section we give details on two-dimensional context-free grammars from [12] which are closely related to other variants mentioned in the introduction. We compare succinctness of the grammars with that of four-way automata. Non-recursive trade-offs are demonstrated in both directions.

The result by Meyer and Fisher regarding the non-recursive trade-off between (1D) context-free grammars and finite automata can be directly applied to 2D context-free grammars and four-way automata, since both models collapse to their 1D counterparts when working over one-row inputs. However, there are still two cases to study. The trade-off in the one-dimensional setting is recursive if unary [11] or finite [8] languages are considered. We show that such restrictions do not play any role in the case of 2D systems.

A *two-dimensional context-free grammar* is a tuple $\mathcal{G} = (V_N, V_T, \mathcal{P}, S_0)$, where V_N is a finite set of nonterminals, V_T is a finite set of terminals, $S_0 \in V_N$ is the initial nonterminal and $\mathcal{P}$ is a finite set of productions. Productions are of the form $N \to W$ where N is a nonterminal in V_N and W is a non-empty matrix whose elements are terminals and nonterminals, i.e., $W \in (V_N \cup V_T)^{+,+}$. $\mathcal{P}$ can optionally contain the production $S_0 \to \Lambda$. If so, S_0 can not be a part of the right-hand side of any production.

For each $N \in V_N$, let $L(\mathcal{G}, N)$ denote the set of pictures *generated* by $\mathcal{G}$ from N. Two recurrent rules are applied to establish the set. A picture $P \in \{V_T\}^{*,*}$ belongs to $L(\mathcal{G}, N)$ iff

1. $N \to P$ is a production, or
2. There is a production $N \to [A_{ij}]_{m \times n}$ and a picture $P = \bigoplus [P_{ij}]_{m \times n}$ where

$$P_{ij} = A_{ij} \qquad \text{if } A_{ij} \in V_T,$$
$$P_{ij} \in L(\mathcal{G}, A_{ij}) \quad \text{if } A_{ij} \in V_N.$$

The picture language generated by $\mathcal{G}$ is defined as $L(\mathcal{G}) = L(\mathcal{G}, S_0)$.

The definition covers Kolam grammars [15,9] which are less powerful and have only productions $N \to [A_{ij}]_{p \times q}$ where $p = 1$ or $q = 1$ and both $p, q \leq 2$. On the other hand, regional tile grammars [2] are more powerful. They combine pictures into bigger ones using a more general concatenation, the alignment of subpictures into a grid is not required.

Example 4. Let $\mathcal{G} = (V_N, V_T, \mathcal{P}, R)$ be a 2D context-free grammar where $V_N = \{H, V, A, R\}$, $V_T = \{a\}$ and $\mathcal{P}$ consists of the following productions

$$H \to a, \qquad H \to a\,H, \qquad V \to a, \qquad V \to \frac{a}{V},$$

$$A \to V, \qquad A \to V\,A, \qquad R \to a, \qquad R \to \frac{R\ V}{H\ a}.$$

Then, $L(\mathcal{G}, H)$ consists of all one-row pictures of a's, $L(\mathcal{G}, V)$ consists of all one-column pictures, $L(\mathcal{G}, A) = \{a\}^{+,+}$ and $L(\mathcal{G}, R) = L(\mathcal{G})$ contains all non-empty square pictures.

Note that 2D context-free grammars do not have any equivalent to Chomsky normal form. Limiting the size of production's right-hand sides affects the power [12]. A size measure taking into account the number of elements in productions is preferable here over a simple counting of productions. For a production $\pi = N \to [A_{ij}]_{m \times n}$, define its size as $c(\pi) = mn$. Beside that, define $c(S_0 \to A) = 1$. Let $\mathcal{P} = \{\pi_1, \ldots, \pi_k\}$ be the set of productions of $\mathcal{G}$. Define $c(\mathcal{G}) = \sum_{i=1}^{k} c(\pi_i)$.

We will utilize a slight modification of the set of productions specified in the following lemma.

Lemma 5. *Every 2D context-free grammar* $\mathcal{G} = (V_N, V_T, \mathcal{P}, S_0)$ *has an equivalent grammar* $\mathcal{G}' = (V_N', V_T, \mathcal{P}', S_0)$ *such that* $c(\mathcal{G}') \leq 3c(\mathcal{G})$ *and each production* $S_0 \to [A_{ij}]_{p \times q}$ *in* $\mathcal{P}'$ *fulfills*

- *If* $p = q = 1$ *then* A_{11} *is a terminal.*
- *If* $p > 1$ *or* $q > 1$ *then each* A_{ij} *is a nonterminal.*

Proof. First, for each $a \in V_T$ figuring at a right-hand side of some production, introduce new nonterminal N_a, add production $N_a \to a$ and replace all occurrences of a in production's right-hand sides containing two or more elements. Second, replace $A \in V_N$ in $S_0 \to A$ by right-hand sides of other productions reachable from A. Each of the two steps increases size of $\mathcal{G}$ by a maximum of $c(\mathcal{G})$. $\qquad \square$

1	1	0	1	0	1
0	·	·	·	·	0
0	·	·	·	·	0
0	·	·	·	·	0
1	·	·	·	·	1
1	1	0	1	0	1

Fig. 4. A picture in L_{perim}, the content of inner fields is hidden, they store symbol 0

Define a language L_{perim} over $\Sigma = \{0,1\}$ where $P \in \Sigma^{*,*}$ is in L_{perim} if and only if the first row equals the last row, the first column equals the last column and all inner fields contain 0, as depicted in Figure 4.

L_{perim} can be easily recognized by a 4DFA. The following lemma indicates that it cannot be generated by a 2D context-free grammar.

Lemma 6. *For $m \in \mathbb{N}$, let $\mathcal{G} = (V_N, \{0,1\}, \mathcal{P}, S_0)$ be a 2D context-free grammar such that $L(\mathcal{G}) \cap \Sigma^{m,m} = L_{\text{perim}} \cap \Sigma^{m,m}$. Then, $c(\mathcal{G})$ is $\Omega(m/\log m)$.*

Proof. W.l.o.g., $\mathcal{G}$ is in the form specified in Lemma 5 and $m \geq 2$. Let L be the set of all square pictures in L_{perim} of size m. It holds $|L| = 2^{2m-3}$. There is a production proving for at least $\lceil \frac{|L|}{|\mathcal{P}|} \rceil$ pictures from L that they belong to $L(\mathcal{G})$. Let $S_0 \to [A_{ij}]_{k \times \ell}$ be such a production. W.l.o.g., $k \geq \ell$ (implying $k \geq 2$).

Consider there are pictures $P, Q \in L(\mathcal{G})$, $P \neq Q$, $P = \bigoplus[P_{ij}]_{k \times \ell}$, $Q = \bigoplus[Q_{ij}]_{k \times \ell}$ and each P_{ij} and Q_{ij} is in $L(\mathcal{G}, A_{ij})$. Observe that if all the pairs P_{1j}, Q_{1j} have the same size, then the first row of P and Q have to be necessarily identical. If not, $\mathcal{G}$ would generate a picture which is not in L_{perim}. Indeed, it would be possible to create P' from P by replacing subpictures P_{1j} by Q_{1j}. However, the first row in P' does not match its last row.

Sizes of pictures P_{1j}, $j = 1, \ldots, \ell$ are determined by one vertical and $\ell - 1$ horizontal coordinates in P. The number of such possibilities is certainly bounded by m^k. The number of pictures in L_{perim} having the same first and last row is 2^{m-2}. We derive the inequality

$$\frac{2^{2m-3}}{|\mathcal{P}|} \leq m^k 2^{m-2} \quad \Rightarrow \quad |\mathcal{P}| \geq \frac{2^{m-1}}{m^k}.$$

If $k \geq m/(2\log_2 m)$, then $c(\mathcal{G}) \geq c(S_0 \to [A_{ij}]_{k \times \ell}) \geq m/(2\log_2 m)$. If $k < m/(2\log_2 m)$, then

$$c(\mathcal{G}) \geq |\mathcal{P}| \geq \frac{2^{m-1}}{m^{\frac{m}{2\log_2 m}}} = \frac{2^{m-1}}{2^{\frac{m}{2\log_2 m}\log_2 m}} = 2^{\frac{m}{2}-1} = \Omega\left(\frac{m}{\log m}\right),$$

which completes the proof. □

Corollary 7. *Transforming 4DFAs to 2D context-free grammars yields a non-recursive trade-off.*

Proposition 8. *Let $\mathcal{T}$ be a 1D deterministic Turing machine and w be an input to $\mathcal{T}$. There is a 2D context-free grammar $\mathcal{G}$ over $\{a\}$ such that $L(\mathcal{G})$ is empty if $\mathcal{T}$ does not halt on w, else if $\mathcal{T}$ halts in $t(w)$ steps, $L(\mathcal{G})$ is a one-element set containing a square picture of size at least $t(w)$. Moreover, size of $\mathcal{G}$ is recursive in size of $\mathcal{T}$ and $|w|$.*

Proof. Let the set of states of $\mathcal{T}$ be $Q = \{q_1, \ldots, q_s\}$ and let the tape alphabet be $\Gamma = \{a_1, \ldots, a_t\}$, containing the blank symbol $\#$. We assume $\mathcal{T}$ is single-tape and the tape is infinite rightwards only. Moreover, w.l.o.g., each instruction of $\mathcal{T}$ moves the head.

Let $\mathcal{C}_0, \mathcal{C}_1, \ldots$ be configurations of $\mathcal{T}$ when computing over w. We will construct a 2D context-free grammar $\mathcal{G} = (\{a\}, V_N, \mathcal{P}, S_0)$ such that

- V_N contains nonterminal C and $L(\mathcal{G}, C)$ consists of square pictures whose sizes encode configurations $\mathcal{C}_i$.
- $L(\mathcal{G}, S_0)$ contains at most one picture – that one which encodes the final configuration.

For convenience, we treat integers also as binary strings and vice versa. Define $\text{code}(q_i) = 0^{i-1}10^{|Q|-i}$ and $\text{code}(a_i) = 0^{i-1}10^{|\Gamma|-i}$. For some $\mathcal{C}_i$, let $b_1 b_2 \ldots$ be the content of the tape and let the control unit of $\mathcal{T}$ be in state q. We encode each $\mathcal{C}_i$ by a binary string (integer)

$$\text{code}(\mathcal{C}_i) = 1B_1 B_2 \ldots B_{|w|+i+2}$$

where

$$B_j = \begin{cases} 1\,\text{code}(b_j)\,\text{code}(q) & \text{if } \mathcal{T} \text{ scans the } j\text{-th tape field}, \\ 1\,\text{code}(b_j)0^{|Q|} & \text{otherwise}. \end{cases}$$

Note that the suffix $B_{|w|+i+2}$ always equals $\text{code}(\#)0^{|Q|}$, even if $|w| = 0$.

Let us compare codes of two subsequent configurations $\mathcal{C}_k, \mathcal{C}_{k+1}$. Let the head of $\mathcal{T}$ scan the m_1-th and m_2-th tape field in $\mathcal{C}_k$ and $\mathcal{C}_{k+1}$, respectively. Since $\mathcal{T}$ always moves the head, $m_1 \neq m_2$. Denote $m = \min\{m_1, m_2\}$. We can write

$$\mathcal{C}_k = 1B_1 B_2 \ldots B_{|w|+k+1}B_{|w|+k+2},$$
$$\mathcal{C}_{k+1} = 1B_1 \ldots B_{m-1}B'_m B'_{m+1}B_{m+2} \ldots B_{|w|+k+2}B_{|w|+k+3}.$$

When comparing $\mathcal{C}_{k+1}$ to $\mathcal{C}_k$, we see it differs in B'_m, B'_{m+1} and it is prolonged by the suffix $B_{|w|+k+3} = 1\,\text{code}(\#)0^{|Q|}$. Denote $x_1 = B_m B_{m+1}$, $x_2 = B'_m B'_{m+1}$ (codes of local changes performed by the applied instruction of $\mathcal{T}$, note that x_2 is determined by x_1), $\ell = |\Gamma| + |Q| + 1$ (the length of each block B_i), $y = |w| + k + 1 - m$ (the number of blocks following B_{m+1} in $\mathcal{C}_k$), $c_\# = \text{code}(\#)0^{|Q|}$ (the constant suffix of each $\mathcal{C}_i$). Then, $\mathcal{C}_{k+1}$ is expressed as

$$\mathcal{C}_{k+1} = (\mathcal{C}_k + (x_2 - x_1)2^{y\ell})2^\ell + c_\#.$$

We define productions that generate a representation of the difference $\mathcal{C}_{k+1} - \mathcal{C}_k$ and combine it together with the square picture of size $\mathcal{C}_k$ to produce the square picture of size C_{k+1}. We include all nonterminals and productions from Example 4 to $\mathcal{G}$. Next, the following productions are defined.

$$(1) \quad C \rightarrow \left.\begin{matrix} \overbrace{a \cdots a} \\ \vdots \ddots \vdots \\ a \cdots a \end{matrix}\right\} C_0 \qquad (2) \quad C \rightarrow \begin{matrix} C & J \\ A & R \end{matrix} \qquad (3) \quad J \rightarrow \begin{matrix} \overbrace{F \cdots F}^{2^\ell - 1} & A & \overbrace{V \cdots V}^{c_\#} \\ A \cdots A & X & V \cdots V \end{matrix}$$

Production (1) ensures that C generates the square picture of size C_0. Recall languages generated by A, V and R (see Example 4). Assume that X generates all pictures of size $x_1' 2^{y'\ell} \times (x_2' 2^\ell - x_1')2^{y'\ell}$ where pairs x_1', x_2' encode possible local changes determined by instructions of $\mathcal{T}$ and $y' \in \mathbb{N}$. Moreover, assume that F generates all square pictures of size $1B_1 \ldots B_{p-1}0^\ell 0^\ell B_{p+2}\ldots B_q$, where $q \geq 2$, $p < q$ and each $B_i = 1\,\mathrm{code}(c_i)0^{|Q|}$ for some $c_i \in \Gamma$. This means, $L(\mathcal{G}, F)$ includes pictures whose sizes encode all configurations C_i with two neighboring blocks erased (filled by zeros) where one of the erased blocks encodes (in C_i) the head's position and the state of the control unit.

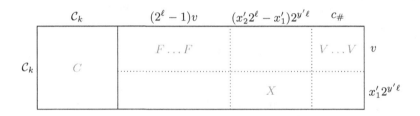

Fig. 5. Substitution of pictures to CJ with sizes of components given along the border

Let us examine which pictures generated by J can be substituted to the right-hand side of production (2) provided that the square picture $C_k \times C_k$ is substituted to C. The situation is illustrated in Figure 5. The picture substituted for J is decomposed by production (3) there. The number of rows enforces the equality $C_k = v + x_1' 2^{y'\ell}$. Observe that it determines parameters x_1', y' and v uniquely. Comparing to C_k, value v, which is the size of a square picture in $L(\mathcal{G}, F)$, has two blocks erased and misses the representation of the state of $\mathcal{T}$ in C_k. This needs to be supplied by adding $x_1' 2^{y'\ell}$. Since all blocks start by bit 1, there is only one possibility at which position to add x_1', it holds $y' = y$ and v has the same number of bits as C_k. Further, C_k determines if the head position is encoded in the first or second block of x_1' ($\mathcal{T}$ is deterministic), hence x_1' must be equal to x_1. Value of parameter x_2' is determined by x_1', thus $x_2' = x_2$. Moreover, it is also obvious that the non-erased blocks of v have to match the corresponding blocks in C_k. Summing the counts of columns in Figure 5 gives

$$C_k + (2^\ell - 1)(C_k - x_1 2^{y\ell}) + (x_2 2^\ell - x_1)2^{y\ell} + c_\# = C_{k+1}.$$

The second row of the right-hand side of production (2) takes care for completing a square picture.

We have to list productions generating pictures in $L(\mathcal{G}, X)$ and $L(\mathcal{G}, F)$. This is mainly a technical issue. Let us start by productions related to X.

$$
(4)\quad U \to a \qquad (5)\quad U \to \left.\begin{matrix} U & \cdots & U \\ \vdots & \ddots & \vdots \\ U & \cdots & U \end{matrix}\right\} 2^\ell \overset{2^\ell}{} \qquad (6)\quad X \to \left.\begin{matrix} U & \cdots & U \\ \vdots & \ddots & \vdots \\ U & \cdots & U \end{matrix}\right\} x_1 \overset{x_2 2^\ell - x_1}{}
$$

Production (6) is added for each possible pair x_1, x_2. Pictures in $L(\mathcal{G}, F)$ are generated in three phases: nonterminal D generates square pictures of sizes $1B_1 \ldots B_{p-1}$, E generates pictures representing $1B_1 \ldots B_{p-1}0^\ell 0^\ell$ and F generates pictures of sizes $1B_1 \ldots B_{p-1}0^\ell 0^\ell B_{p+2} \ldots B_q$. Productions related to D are as follows.

$$(7)\quad D \to a$$

For each $i = 1, \ldots, |\Gamma|$, add productions

$$
(8)\quad D \to D_{i,\ell} \qquad (9)\quad D_{i,i+1} \to \begin{matrix} D_{i,i} & D_{i,i} & V \\ H & H & a \end{matrix} \qquad (10)\quad D_{i,1} \to \begin{matrix} D & D & V \\ H & H & a \end{matrix}
$$

$$
(11)\quad D_{i,j} \to \begin{matrix} D_{i,j-1} & D_{i,j-1} \\ D_{i,j-1} & D_{i,j-1} \end{matrix} \qquad \text{for each } j \in \{2, \ldots, \ell\} \smallsetminus \{i+1\}
$$

Nonterminals H, V come from Example 4. $L(\mathcal{G}, D_{i,\ell})$ consists of square pictures of size $1B_1 \ldots B_{p-2}\,\mathrm{code}(a_i)0^{|Q|}$. Production (9) or (10) can be interpreted as appending bit 1 to size of a picture in $L(\mathcal{G}, D_{i,i})$ or $L(\mathcal{G}, D)$, respectively. Analogously, production (11) appends bit 0 to size of a picture in $L(\mathcal{G}, D_{i,j-1})$. Productions related to E and F follow similar patterns.

$$(12)\quad E \to E_\ell$$

$$
(13)\quad E_i \to \begin{matrix} E_{i-1} & E_{i-1} \\ E_{i-1} & E_{i-1} \end{matrix} \qquad \text{for each } i = 2, 3, \ldots, \ell
$$

$$
(14)\quad E_1 \to \begin{matrix} D & D \\ D & D \end{matrix} \qquad (15)\quad F \to E
$$

For each $i = 1, \ldots, |\Gamma|$, add productions

$$
(16)\quad F \to F_{i,\ell} \qquad (17)\quad F_{i,i+1} \to \begin{matrix} F_{i,i} & F_{i,i} & V \\ H & H & a \end{matrix} \qquad (18)\quad F_{i,1} \to \begin{matrix} F & F & V \\ H & H & a \end{matrix}
$$

$$
(19)\quad F_{i,j} \to \begin{matrix} F_{i,j-1} & F_{i,j-1} \\ F_{i,j-1} & F_{i,j-1} \end{matrix} \qquad \text{for each } j \in \{2, \ldots, \ell\} \smallsetminus \{i+1\}
$$

It remains to add productions generating the final configuration provided that $\mathcal{T}$ halts. We make clones of productions (2), (3) and (5). The production will be applicable only to the configuration preceding the final configuration.

$$(20)\ S_0 \rightarrow \begin{matrix} C\ J_0 \\ A\ R \end{matrix} \quad (21)\ J_0 \rightarrow \begin{matrix} \overbrace{F\ \cdots\ F}^{2^\ell-1}\ \overbrace{A\ V\ \cdots\ V}^{c_\#} \\ A\ \cdots\ A\ X_0\ V\ \cdots\ V \end{matrix} \quad (22)\ X_0 \rightarrow \left.\begin{matrix} \overbrace{U\ \cdots\ U}^{x_2 2^\ell-x_1} \\ \vdots\ \ddots\ \vdots \\ U\ \cdots\ U \end{matrix}\right\} x_1$$

Pattern (22) applies to those pairs x_1, x_2, where x_2 encodes a final state of $\mathcal{T}$.

$\square$

Theorem 9. *Transforming 2D context-free grammars generating finite unary picture languages to* 4NFAs *yields a non-recursive trade-off.*

Proof. Results in [5] imply that $L = \{P \in \{a\}^{*,*} \mid \mathrm{cols}(P) = \mathrm{rows}^2(P)\}$ is not accepted by any 4NFA. Moreover, if a 4NFA $\mathcal{A}$ accepts a unary picture of size $m \times m^2$ $(m \geq 3)$ and rejects other unary pictures with m rows, then $\mathcal{A}$ has at least m states.

For $k \in \mathbb{N}$, construct 2D context-free grammar $\mathcal{G}_k$ as the union of grammars given by Proposition 8 for every k-state deterministic Turing machine with the tape alphabet $\{\#, 1\}$, started over a blank tape. Let N be the initial nonterminal of $\mathcal{G}_k$. Modify $\mathcal{G}_k$ to generate rectangular pictures of size $m \times m^2$ instead of square pictures of size m. Nonterminals and productions from Example 4 are included in $\mathcal{G}_k$. Add productions

$$S \rightarrow \begin{matrix} a\ a \\ a\ a \end{matrix}, \quad S \rightarrow \begin{matrix} S\ R\ R \\ H\ H\ H \end{matrix}, \quad S_0 \rightarrow N\ S.$$

Change the initial nonterminal from N to S_0. It is easy to verify (by induction on n) that S generates every picture of size $n \times (n^2 - n)$ where $n \geq 2$. Thus, by the third production, $L(\mathcal{G}_k, S_0)$ if formed of translations of square pictures in $L(\mathcal{G}_k, N)$ to pictures with the desired size.

Finite unary picture languages $\{L(\mathcal{G}_i)\}_{i=1}^{\infty}$ are accepted by some 4NFAs, however, their sizes grow faster than any recursive function. $\square$

5   Conclusion

We have showed that the non-recursive trade-offs are a common phenomenon in the world of picture languages.

In the proofs we utilized the ability of four-way automata and 2D context-free grammars to simulate Turing machines. This allowed us to come up with constructive witnessing sequences of systems describing finite picture languages.

Besides the proven non-recursive trade-offs, the simulation of a Turing machine by a 2D context-free grammar has additional consequences. For example, it shows that the emptiness problem is not decidable for the two-dimensional context-free grammars as well as for the tile regional grammars.

There remain some interesting open problems. The presented simulation of a Turing machine by a 2D context-free grammar benefited from productions with right-hand sides of size at least 2×2. Can be the simulation performed using only productions of Kolam grammars? What is the succinctness of Kolam grammars generating finite picture languages, comparing to the succinctness of 4DFAs?

References

1. Blum, M., Hewitt, C.: Automata on a 2-dimensional tape. In: FOCS 1967 Proceedings of the 8th Annual Symposium on Switching and Automata Theory (SWAT 1967), pp. 155–160. IEEE Computer Society, Washington, DC (1967)
2. Cherubini, A., Crespi Reghizzi, S., Pradella, M.: Regional languages and tiling: A unifying approach to picture grammars. In: Ochmański, E., Tyszkiewicz, J. (eds.) MFCS 2008. LNCS, vol. 5162, pp. 253–264. Springer, Heidelberg (2008)
3. Giammarresi, D., Restivo, A.: Two-dimensional languages. In: Rozenberg, G., Salomaa, A. (eds.) Handbook of Formal Languages, vol. 3, pp. 215–267. Springer, New York (1997)
4. Inoue, K., Nakamura, A.: Some properties of two-dimensional on-line tessellation acceptors. Inf. Sci. 13(2), 95–121 (1977)
5. Kari, J., Moore, C.: New results on alternating and non-deterministic two-dimensional finite-state automata. In: Ferreira, A., Reichel, H. (eds.) STACS 2001. LNCS, vol. 2010, pp. 396–406. Springer, Heidelberg (2001)
6. Kari, J., Moore, C.: Rectangles and squares recognized by two-dimensional automata. In: Karhumäki, J., Maurer, H., Păun, G., Rozenberg, G. (eds.) Theory Is Forever (Salomaa Festschrift). LNCS, vol. 3113, pp. 134–144. Springer, Heidelberg (2004)
7. Kutrib, M.: The phenomenon of non-recursive trade-offs. International Journal of Foundations of Computer Science 16(05), 957–973 (2005)
8. Meyer, A.R., Fischer, M.J.: Economy of description by automata, grammars, and formal systems. In: SWAT (FOCS). pp. 188–191. IEEE Computer Society (1971)
9. Matz, O.: Regular expressions and context-free grammars for picture languages. In: Reischuk, R., Morvan, M. (eds.) STACS 1997 Proceedings . LNCS, vol. 1200, pp. 283–294. Springer, Heidelberg (1997)
10. Minsky, M.L.: Computation: Finite and Infinite Machines. Prentice Hall, Inc. (1967)
11. Pighizzini, G., Shallit, J., Wang, M.: Unary context-free grammars and pushdown automata, descriptional complexity and auxiliary space lower bounds. Journal of Computer and System Sciences 65(2), 393–414 (2002)
12. Průša, D.: Two-dimensional Languages. Ph.D. thesis, Faculty of Mathematics and Physics. Charles University, Prague, Czech Republic (2004)
13. Radó, T.: On non-computable functions. Bell System Technical Journal 41(3), 877–884 (1962)
14. Siromoney, G., Siromoney, R., Krithivasan, K.: Abstract families of matrices and picture languages. Computer Graphics and Image Processing 1(3), 284–307 (1972)
15. Siromoney, G., Siromoney, R., Krithivasan, K.: Picture languages with array rewriting rules. Information and Control 22(5), 447–470 (1973)

Author Index